Applications of Neuroscience:

Breakthroughs in Research and Practice

Information Resources Management Association
USA

IGI Global
DISSEMINATOR OF KNOWLEDGE

Published in the United States of America by
IGI Global
Medical Information Science Reference (an imprint of IGI Global)
701 E. Chocolate Avenue
Hershey PA, USA 17033
Tel: 717-533-8845
Fax: 717-533-8661
E-mail: cust@igi-global.com
Web site: http://www.igi-global.com

Library of Congress Cataloging-in-Publication Data

Names: Information Resources Management Association, editor.
Title: Applications of neuroscience : breakthroughs in research and practice / Information Resources Management Association, editor.
Description: Hershey PA : Medical Information Science Reference, [2018] | Includes bibliographical references.
Identifiers: LCCN 2017045700| ISBN 9781522554783 (hardcover) | ISBN 9781522554790 (ebook)
Subjects: | MESH: Neurosciences--methods | Decision Making | Medical Informatics Applications
Classification: LCC QP355.2 | NLM WL 100 | DDC 612.8--dc23 LC record available at https://lccn.loc.gov/2017045700

British Cataloguing in Publication Data
A Cataloguing in Publication record for this book is available from the British Library.

All work contributed to this book is new, previously-unpublished material. The views expressed in this book are those of the authors, but not necessarily of the publisher.

For electronic access to this publication, please contact: eresources@igi-global.com.

List of Contributors

Adu-Brobbey, Veronica / *University of Education – Winneba, Ghana* 317
Araujo, Bruno H.S. / *Universidade Federal de São Paulo, Brazil* 70
Babiloni, Fabio / *Sapienza University, Italy & BrainSigns srl, Italy* 391,427
Boesch, Becky / *Portland State University, USA* 100
Breuer, Christoph / *German Sport University Cologne, Germany* 337
Bulley, Cynthia A. / *Central University, Ghana* 317
Çakar, Tuna / *Acıbadem University, Turkey* 371
Cartocci, Giulia / *Sapienza University, Italy & BrainSigns srl, Italy* 391
Chandan, Harish C. / *Argosy University, USA* 445
Cherubino, Patrizia / *IULM University, Italy & BrainSigns srl, Italy* 391,427
Di Flumeri, Gianluca / *Sapienza University, Italy & BrainSigns srl, Italy* 391
Dimitriadis, Nikolaos / *The University of Sheffield International Faculty, City College, Greece* ... 350
Duodu, Esther O. / *ABL Intel Consulting Ltd., Ghana* 317
Falsarone, Alessia / *PineBridge Investments, USA* 477
Feitosa-Santana, Claudia / *Federal University of ABC, Brazil & Albert Einstein Israelite Hospital, Brazil & School of The Art Institute of Chicago, USA* 40
Garvey, Gregory P. / *Quinnipiac University, USA* 176
Gez, Kaan / *Anatolian University, Turkey* 371
Gorgiev, Anka / *The University of Sheffield International Faculty, City College, Greece* 350
Graziani, Ilenia / *BrainSigns srl, Italy & Sapienza University, Italy* 427
Huang, Qiang / *Beijing Institute of Technology, China* 118
Iancu, Ioana / *Babeş-Bolyai University, Romania* 296
Kennedy, Dennis T. / *La Salle University, USA* 242
Lahmiri, Salim / *ESCA, Morocco* 286,466
Leote, Rosangella / *Universidade Estadual Paulista (UNESP), Brazil* 86
Lovell, Kathryn L. / *Michigan State University, USA* 226
Maglione, Anton Giulio / *Sapienza University, Italy & BrainSigns srl, Italy* 391,427
Mancini, Marco / *BrainSigns srl, Italy* 391
Modica, Enrica / *Sapienza University, Italy & BrainSigns srl, Italy* 391
Nasseh, Ibrahim Elias / *Universidade de São Paulo (USP), Brazil* 70
Nield, Geoffrey E / *Neuro-Insight, Australia* 413
Pynta, Peter / *Neuro-Insight, Australia* 413
Rappaport, Jack M. / *Brilliance Consulting, USA* 242
Richter, Stephen B. / *West Chester University, USA* 242
Rossi, Dario / *Sapienza University, Italy & BrainSigns srl, Italy* 391

Table of Contents

Preface

The ever-changing landscape surrounding the diverse applications of different scientific areas can make it very challenging to stay on the forefront of innovative research trends. That is why IGI Global is pleased to offer this one-volume comprehensive reference that will empower students, researchers, practitioners, and academicians with a stronger understanding of neuroscience, the scientific field of study, and how it can be applied to numerous industries and disciplines.

This compilation is designed to act as a single reference source on conceptual, methodological, and technical aspects, and will provide insight into emerging topics, including but not limited to cognitive neuroscience, neurophysiological methods, neurobiology, neuromarketing, and neuroeconomics. The chapters within this publication are sure to provide readers with the tools necessary for further research and discovery in their respective industries and/or fields.

Applications of Neuroscience: Breakthroughs in Research and Practice is organized into three sections that provide comprehensive coverage of important topics. The sections are:

1. Art, Creativity, and Social Issues.
2. Computer Science and Information Systems.
3. Marketing and Management.

The following paragraphs provide a summary of what to expect from this invaluable reference source:

Section 1, "Art, Creativity, and Social Issues," opens this extensive reference source by highlighting the latest trends in neuropsychology and psychodynamics. Through perspectives on embodied cognition, neuroplasticity, and consciousness integration, this section demonstrates the importance of the study of human cognition and different therapeutic methods such as creative arts therapy and play therapy to improve behavior and emotional processes. The presented research facilitates a better understanding of how new therapies and techniques are helping us to better understand the mind and perceptual processes.

Section 2, "Computer Science and Information Systems," includes chapters on emerging innovations in neuro-information systems and human-computer interfaces. Including discussions on eye tracking, cognitive load therapy, and robotics, this section presents research on the impact of neuroscience on computer engineering and the development of advanced intelligent systems.

Section 3, "Marketing and Management," presents coverage on novel strategies and research methods to enhance consumers' cognitive and emotional engagement toward advertising and branding in a variety of channels and platforms. The chapters within this section include innovative discussions on the use of focus groups, brain imaging techniques, and social media tactics.

Although the primary organization of the content in this work is based on its three sections, offering a progression of coverage of the important concepts, methodologies, technologies, applications, social issues, and emerging trends, readers can also identify specific content by utilizing the extensive indexing system listed at the end.

Section 1

Art, Creativity, and Social Issues

Chapter 1
Creativity and the Arts

Richard L. Tietze
Marymount Manhattan College, USA

ABSTRACT

The chapter explores creativity from a psychological viewpoint, as spontaneous and reflective activities in several self-defining human efforts, such as telling stories and forming an identity. New discoveries from neuroscience and creative arts therapies combine to understand creativity as a range of normative human functioning, which blends with professional creative discovery in the arts at the higher extremes of this range. The chapter focuses on visual art and music, followed by a normative application of creativity as utilized in creative arts therapies. Finally, the chapter concludes by introducing normative creative tools to incorporate into everyday life.

OVERVIEW/INTRODUCTION

Recent discoveries in modern neuroscience have radically altered, and greatly expanded our understanding of the human brain. Many of these changes also affect our insights into human creativity, and the present chapter will include a perspective informed by neuroscience (Gazzaniga, et al., 2014; Juslin & Sloboda, 2012; Levitin, 2006; Myers & deWall, 2015).

Two major overarching conceptual models from psychology will be used to organize the following chapter discussion. The Biological-Psychological-Social, or BPS, model of psychology integrates information from many disciplines and adds a conceptual framework for incorporating new evidence and driving new hypotheses regarding human meaning. Psychology is described as a "hub" scientific discipline (Myers, & deWall, 2015, p.11), with levels of analysis that reach out to include biology and neuroscience as well as psychological cognition, emotion and behavior, all within a social/cultural context. These complementary outlooks help psychology to gain the broadest perspective on human mind, behavior and self-understanding (Myers & deWall, 2015).

The Life-Span Development Model was explored early in psychology's history by Carl Jung (1957, 1964), and advanced to become a widely-accepted view through Erik Erikson's later psychosocial model of development. He describes the model as "epigenetic," or built on a biological ground plan, and culminating in the creative organization of experience known as identity. Adult growth proceeds to balance the continuity of identity within ongoing changes of life throughout adulthood (Erikson, 1960, 1997).

DOI: 10.4018/978-1-5225-5478-3.ch001

HISTORY OF CREATIVITY STUDY

Early 20th century studies attempted to study creativity using creative products (i.e.; acknowledged works of art) and later attempted to explore the creative process by asking artists to introspect and describe their thoughts as they were creating. This was a typical strategy of that era, but, unfortunately, individual descriptions varied so much that the research produced unreliable and non-comparable descriptions of the thought process. This ultimately led to a dead end for half a century until modern neuroscience opened new ways to conceptualize how brain and mind connected human cognitive processes (McAdams, 2001; Myers & deWall, 2015). Cognitive neuroscience established a dimension of observation and data from about 1985, with neuroimaging of normal functioning human brains bringing the biological perspective strongly into the field (Gazzaniga, et al., 2014; Levitin, 2006).

Guilford (1950) and Bruner (1986) sought a modern approach to empirical/conceptual study, demonstrating the many ways to understand and express creativity. Bruner also distinguished narrative thinking, which was later realized to be a foundation of human literature as stories, and a way to explain our understanding of how life experience is recorded in the brain, basically by linking external events in sequence with emotional responses to these events. His work also encouraged the paradigm now known as Life-Span Development. (Bruner, 1986; McAdams, 2001). The field of Narrative Psychology was explored in relation to personality by McAdams (2001), and also applied to counseling processes (Nystul, 2011). The storytelling model has become a tool for identity development and/or correcting dysfunction, using the client's own language and experiences to describe meaning, whether for an individual, a couple or a family (Nystul, 2011).

SCIENTIFIC STUDY OF CREATIVITY

Since the mid- 20th century, the scientific study of creativity has been organized historically into "three waves" (Sawyer, 2012, in Beghetto & Breslow, p.417):

1. In the 1950s and 60s, the focus was on individual creativity and personality;
2. The second wave, in the 1970s and 80s, focused on the cognitive aspects of creative behavior. Both of these tended to focus on narrow approaches, using the individual as the unit of study.
3. The "third wave," beginning in the 1980s and 90s, broadened the perspective to examine creativity as a phenomenon of groups within a sociocultural context, including an interdisciplinary approach and many possible forms dependent on context.

Using the metaphor of "waves," rather than linear historical stages, has the implication of rhythms within a process, which is Sawyer's intention. Creativity is a complex function that changes over time as a cultural process, and all forms of investigation, from all three historical "waves," are valid tools for understanding (Sawyer, 2012, in Beghetto & Breslow, 2013). We are reminded again of the major shift in thinking about the brain activity required to explore the complexity and creativity of brain function, from the linear perspective of the neuron, to the simultaneous, parallel processing perspective of modern neuroscience (Gazzaniga, et al., 2014; Myers & deWall, 2015).

HUMANISTIC PERSPECTIVES

Abraham Maslow brought creativity within the bounds of normal human functioning. In "The Farther Reaches of Human Nature" (1971), he created a new conceptual language for psychology that includes normative biology as a foundation in terms of distinguishing survival needs (biological needs for sustaining life), and social needs (security and esteem), from growth (psychological needs for self-expression and personal growth), as well as different cognitive modes corresponding to each need system, all organized into his famous hierarchy (Maslow, 1971).

The climax of self-actualization is the peak experience, "a naturalistic idiom, an earned merging of subject and object, and like the promise of a rainbow, gives the idea of transcendence an empirical ground." (Maslow, 1971, p.xvi –xvii). He also argues for a humanistic-oriented biology:

Science comes from human beings and human passions and interests, and must include human needs and experience, rather than attempt to be morally and value neutral. (Maslow, 1971, p.21)

This assertion underlies his construction of survival needs and growth needs, of which self-expression and spontaneity are important representatives. As a founder of what has become known as "Humanistic Psychology," Maslow acknowledged support from psychologists such as Carl Rogers, whose studies of human relationships had led to similar conclusions about the growth needs of people (Maslow, 1962, 1971). Both approached creativeness from the process of improvising and inspiration rather than from the view of the completed product (Maslow, 1971). He also viewed a more complete view of creativity when we include the creativity of women. He cited Jung's model of anima-animus, and described the antagonism between the sexes as "largely a projection of the unconscious struggle within the person ... to make peace between the sexes, make peace within the person" (Maslow, 1971). Other modern thinkers have also explored the masculine/feminine dimensions within the self, recently in the fields of using music as a healing tool, with a meaningful contrast of approaches (Jung, 1957; 1964; Gioia, 2006).

Maslow's central creative concept of "peak experience" has been developed further. Mihaly Csikszentmihalyi calls this creativity-enhancing tool "Flow" (1990; 1996):

Between the anxiety of being overwhelmed and stressed, and the apathy of being underwhelmed and bored, lies a zone in which people experience "Flow," a completely involved, focused state of consciousness, with diminished awareness of self and time, resulting from optimal engagement of one's skills... These experiences boost our sense of self-esteem, competence and well-being. (Myers & deWall, 2015, p.A-1)

Flow was viewed as meaningful attempts to direct one's consciousness toward "optimal functioning" — activity choices that maintain a balance between extremes of boredom and anxiety. "Flow" is one of the discoveries that comprise a new understanding of the role of human cognition in emotional well-being and resilient responsiveness to adversity (Csikszentmihalyi, 1990, 1996).

Similarly, the creative stimulus has been linked to play, where the process of thinking becomes the focus, not products or performance (Brown, 1998, 2009). Play expands the range of experience that can be included within any human creative process, including the formation of one's individual identity. Art can be viewed as both an adult form of play and as a tool for reflecting upon experience. Play serves many functions in early childhood development, but has also been viewed recently as an adaptive tool that can be useful throughout the Life-Span; as Brown describes, "... healthy varied play in childhood

[is] necessary for the development of empathy, social altruism and the possession of a repertoire of social behaviors enabling the player to handle stress. (Brown, in Bekoff & Byers, 1998, p.250).

This approach is echoed by Greenacre (1971), who cites the rhythmic pattern of alternating stimulation and comfort in play "which allows gradual integration of complexity at a pace chosen by the individual, with performance pleasure in the activity, rather than a focus on the result." She used the term "creative imagination" to imply a capacity for originality and inventiveness, rather than creative in the sense merely of the ability to synthesize to reach a product (Greenacre, 1971, p. 573).

Dreaming

Jung's idea of "active imagination" is another analogue to normative creativity, as part of life-span development and growth, especially in later adult years. For example, rather than Sigmund Freud's assumption that there is basically one emotional language in dreams that reflects a person's conflicts, especially in the areas of sex and aggression (Freud, 1938), Jung sees the unconscious revealed in dreams as a creative way to reorganize experience, sometimes providing insights to life issues (Jung, 1964). Jung's system of psychological types, or preferred styles of behavior related to conscious choice and personality, also has been developed into a normative system of creative expression by Levesque (2001) that views creativity as a number of processes used by each of us, depending one's preferred modes of experiencing the world.

People have explored the significance of dreams throughout recorded history; ancient civilizations saw them as portents of future events. Freud pioneered the study of dreams in assisting awareness of unconscious and emotional experiences and aspects of the self (Freud, 1938; Natterson, 1980). Dreams have also been found to trigger brain activity that mimics an important stage of creativity, that of "brainstorming" a range of possibilities, akin to psychoanalytic "free association." It also seems to evoke a selective pattern of brain areas, with those focused on executive control, decision-making, focused attention and short-term memory functions shut down during dreaming (Hartmann, 1996)

Hartmann (1996) developed a Continuity Theory of Dreams in this manner: Dreams make connections more broadly than the waking mind…when working on a problem… have a dream. It makes new connections, and it helps you make sense of (the problem, which)… can be very

useful (Hartmann, 1996, p.48-9)." More recently, the phenomenon of unconscious neuroprocessing has added a normative dimension to both dreams and creativity; our brains process so much information simultaneously and in parallel process that most of the information is below the threshold of consciousness (Gazzaniga, et al., 2014; Levitin, 2006). This is a much different and larger view of the unconscious than Freud assumed (Freud, 1938).

Many of these ideas have been incorporated into a modern understanding of dreams, especially through the lens of neuroscience (Kandel, 2012; Wroe, 2011). At the end of this chapter, practical suggestions to enhance one's own creativity will be offered, as an approach to expanding our own diverse choices. The neuroscience-based distinction between "Big – C, little c & Mini –c" creativity (Beghetto & Kaufman, 2007), will be used to describe the similar neurological processes of creativity in all humans, while maintaining the distinction between creative products, as a continuum ranging from normal to creative artist/ scientist.

Creativity

Most recent descriptions of the creative process include elements of preparation, usually requiring domain-relevant skills learning and an incubation phase, which is often described as time to let our unconscious processes work on the problem without consciously focusing on it (Beghetto and Breslow, 2013; Chatterjee, 2013; Myers and deWall, 2015). In addition, writers often describe a different process of thinking involved, with boundaries and categories opened to new possibilities of connection, and a developed ability to view things from different perspectives. Jonathan Lehrer (2012), in a book written for a popular audience but informed by neuroscience research, refers to this process as "Imagination.. (which many thinkers have)… separated from other kinds of cognition" (Lehrer, 2012, p. xvii). "… creativity…requires a description from many perspectives… (as) we're always situated in a context and a culture…we need to merge together the outside world and the inside of the mind (p. xviii)". Lehrer also says, "We do some of our best thinking when we're half asleep (p. 33)," because several forms of cognition, and several levels of attention are open to us, without the clear boundaries of waking life. He includes daydreams in this mix. This theme is echoed by Chatterjee (2013), who encourages us all to build more "downtime" into our lives, to allow more mental activity that is not focused on solutions, but on individually different possibilities.

Psychology has developed a number of conceptual/experiential tools that can be useful to anyone who seeks to increase self-awareness and life choices, both foundations of normative creativity. During the search for cognitive understanding of meaning, many of these tools have been validated over the past few decades by discovery of the underlying neurological processes (Peretz & Zatorre, 2007). Essentially, the idea of creativity as a normative human activity has been a valuable practical strategy adopted by the Creative Arts Therapies. It has expanded their range of methods and populations to include a normative/educational perspective to complement the original healing-based foundation (Malchiodi, 2012). From the use of play as an expansive mode, in addition to a rehearsal strategy, to early concepts such as creative dreaming, self-actualization, and modern tools such as "Flow," science has learned much about the brain through exploration of the arts. Important discoveries, as well as practical applications, will be explored later in this chapter.

Basic Neuroscience

With approximately 1,500 neuroimaging studies being published each year, and a continuous process of new measurement tools being developed, cognitive neuroscience has become one of the most rapidly growing fields of study (Gazzaniga, et al., 2014).

The human brain contains about 100 billion neurons, each able to make about 1,000 connections to other neurons. Approximately 75 percent of these neurons are interneurons, whose chief function is to transmit information within the brain. Most of our brain activity, then, is exchanging information between different parts of the brain, rather than transducing ("translating") signals between the outside world and the brain (Myers & deWall, 2015). Our understanding of brain functioning is far different than previously thought, especially since there are so many opportunities for individual differences to be expressed in creating meaning.

Several mid-20th century psychological thinkers were searching for a way to incorporate biological information into psychological study, but empirical success awaited modern concepts, such as "mind within embodied brain," (Gazzaniga, et al., 2014), a symmetry based on the early "mind-body" links studied by the ancient Greek thinkers; our mind experiences the world within a body linked to a physical and emotional environment.

A related idea is that of the "brain as internalized environment" (Myers & deWall, 2015); our brains "translate" physical environmental stimuli into a neurological "picture, or map" of our world, one that changes in relationship to our experience. A good example of this is our homunculus, or map of different body parts represented in the brain. We have larger areas for our lips and hands, since they are so important to our experience of the world (tools and words), and these areas change over time as we develop skills through use and practice. Our bodies essentially project our own experience onto our brains. These insights are now shared by many fields exploring the nature of the human condition, thanks to neuroscience (Myers & deWall, 2015).

Neuroplasticity describes changes to brain circuitry in response to external changes in the environment and to internal changes within the brain. This is a Life-Span process, producing new nerve connections (synapses and dendrites), and new neurons when we learn or connect past experiences differently, while also pruning away redundant circuits that were available but not used to connect experiences. "The brain renews itself through life to an extent not previously thought possible" (Gazzaniga, et al., 2014, p.65)." Most of the brain's growth (quadrupling between birth and adulthood) derives from forming new synapses and dendritic structures, and the peak of this creation is reached in about the mid-20s, when emotionally laden experiences (usually described as mediated by the limbic system) are coordinated and balanced by the frontal cortex, which coordinates potential action with future consequences and decision-making (Gazzaniga, et al., 2014).

Neuroplasticity is now understood as a normative process, but first evidence was given by the brain's attempts to reorganize circuits (healing to regain functioning) after injury, from traumas such as stroke or loss of sensory mechanisms. A central example is loss of speech function after a stroke to the left side of the brain, an important mediator and processor of speech and language information. Using the concepts of neuroplasticity, clinicians develop exercises that stimulate other areas of the brain attempting to "reroute" the information through different circuits as a way to recover functioning (Levitin, 2006; Mannes, 2009).

One of the major ways that our approach to the study of creativity has changed is reflected in this healing function. Clinicians first discovered ways to enhance the rerouting process so that more functioning is regained, more quickly. Our more detailed knowledge of the brain (especially through neuroimaging, a non-invasive and non-harmful way to measure brain functioning in normal and abnormal brains), has enhanced the ability of clinicians and scientists to become partners in reciprocal treatment and enhanced understanding. This cooperative healing and knowledge is in the best tradition of the scientific method. Examples of these methods, such as Melodic Intonation Therapy, or MIT, help the brain create new pathways for speech functions damaged by trauma, and will be discussed in the Music Therapy section (Thompson & Schlaug, 2015; Mannes, 2009).

A clear example of how much our understanding has grown is in the way we hear music. Within roughly 30/1000 of a second (in the blink of an eye), sound enters the ear, and the signal splits into the basic elements of music (melody, harmony, pitch, timbre and loudness), each to be processed in lower areas of the brain before being sent on to the cortex for integration and meaning. The new signal is also compared with other, similar signals drawn from memory, so that we remember, and recognize a melody that has been heard before, even one transposed into a different musical key. Yet all this happens so quickly that our experience is that we hear music as soon as we turn up the volume (Ball, 2010; Levitin, 2006). As the filmmaker Elena Mannes described, "Music plays the brain, and the brain creates music" (Mannes, 2009).

The new field of Creative Arts Therapies focuses on treatments and procedures that help to heal injury as well as to enhance normal functioning through education and experience. Other discoveries expand our understanding of how normal functioning uses creativity to learn and respond to normative situations. Two important examples are: making sense of complex social interactions that involve language, emotion, past experience and present context of meaning; or recognizing a human face that we may not have seen in years, including changes from growth and/or aging. In these ways, the study of individual differences, which has always been a prime focus of the entire field of psychology, reaches new possibilities and an expansion of opportunities (Gazzaniga, et al., 2014; Myers & deWall, 2015).

The Role of Emotion in Embodied Cognition (Awareness)

Another revolution, also informed by neuroscience, offers a new view of the creative person, as part of a cycle of exploration, and integration, of both the Sciences and the Humanities. The once-revolutionary view of the philosopher Descartes, expressed by the statement, "I think; therefore I am," which centered our human search for meaning and identity within the self and helped to launch the Renaissance, also helped to direct thought and research for several centuries. Now Descartes' view gives way to one informed by the research it helped to generate, including technological tools in the form of brain-imaging methods. The modern view is that memories are naturally encoded with an emotional value in addition to the reconstruction of elements of a cognitive stimulus, and that emotion is as natural a part of cognition as thought. Rather than a hierarchy of levels, both types of cognition function in parallel fashion (Gazzaniga, et al., 2014).

Dr. Antonio Damasio (1999) whose research has been a major contributor to our new understanding is one of the leaders in this field. A review article from the "Arts and Ideas" section of the New York Times subtitled "A Thinker Reunites Mind and Body," summarizes his contribution as, "Feeling, it turns out, is not the enemy of reason, but … an indispensable accomplice." (New York Times, April 19, 2003). The reviewer, Emily Eakin, also borrows a quote to further illustrate the range of connection,

'For students of the humanities, the key neurophysiological insight of our time is that which has been expressed so eloquently by Antonio Damasio,' declared Jonathan Bate, a Shakespearean scholar at the University of Liverpool… 'The division between reason and passion, or cognition and emotion (an opposition that goes all the way back to Aristotle), is, from a neurological point of view, a fallacy'. (Eakin, NYT, 2003)

Thus, what used to be viewed as a linear hierarchy of cognitive functions, with reason as the pinnacle of what it means to be human, is now seen to be a cyclic, parallel system stretching the range of explained experience to include emotional experience as a critical element of human cognition (Damasio, 1999). As an educator, I realize that at the basic level of human physiology, the Arts and Humanities form a range of human experience with the Sciences, and the entire range can and must be explored for tools of understanding and for tools of living. To encourage individuals to reach their unique potential and strive for excellence is especially important today as the complexity of life can often be increasingly overwhelming, clouding individual goals and focus (Tietze, 2006, 2008).

Including the emotional/experiential dimension of identity broadens and deepens our understanding and experience of development, and is a common theme uniting adulthood and the continued need for creative expression. Most of adulthood is characterized by socio-emotional and meaningful internal change, rather than the physical changes that mark both earlier and later periods in the life span. Relationships, intimacy and the meaningful integration of experience are the tools of adulthood, to include a view of life complex enough to comprise a perspective on the past as well as the future, and a range beyond one's individual experience (Levinson, et al., 1978, 1996; Jung, 1957, 1964). The importance of emotion in both memory and experience has been emphasized by much of the recent literature on neuroimaging (Gazzaniga, et al., 2014; also reviewed in Kandel & Squire, 2003). Further, the architecture of the human brain is now seen to function in what is usually described as a "narrative" or story-telling structure, so the idea of "telling one's story" has a biological as well as psychological importance, on both the individual level as well as that of all human literature (Mc Adams, 2001).

Although the empirical evidence is recent, the ideas themselves are not new (Wilson, 1998). The weaving of complex themes into a pattern that gives a new overview and methodology to our understanding of the intricacies of our human condition often await the convergence of ideas, experiences and practical measurement, which all contribute to the cyclic learning process of the Scientific Method.

In 1931, John Dewey was invited to give a series of lectures in memory of William James (both members of the respected tradition of philosopher-psychologists) on the philosophy of art. These were combined into what is now a classic exploration titled *Art as Experience* (1934). Dewey traces the historical development of Western views of art, beginning with the Greeks, as a reflective imitation of life, through assessment of art as using function and form to create meaning, through 20th century views of art as an expansion of experience and perspective, with self-understanding a valuable element of knowledge, and emotion viewed as an important element of cognition. We not only remember experience, but organize it according to value. Excellence would then be a creative, holistic balance integrating all elements. In Dewey's own words: "The difference between the esthetic and the intellectual is … one of the place where emphasis falls in the constant rhythm that marks the interaction of the live creature with his/her surroundings (p.15)." Regarding the life-span nature of human development, he says, "Art celebrates with peculiar intensity the moments in which the past reinforces the present and in which the future is a quickening of what now is." (p.18). Finally, "Experience is the result, the sign, and the reward of that interaction of organism and environment which, when it is carried to the full, is a transformation of interaction into participation and communication (Dewey, 1934, p.22)."

Schulkin (2013), a modern writer who builds upon this earlier work, traced a developmental line from Darwin to Dewey. He writes, "Emotional systems are forms of adaptation, and Darwin did not separate emotions and cognition…music is a core human capability" (Schulkin, 2013, p.1). Both Darwin and Dewey understood that emotions are rich in cognitive functions and appraisal processes…"some of these, like music and faces, are affectively opulent. Human meaning is tied to social contact, and music is a fundamental part of human meaning (Schulkin, 2013, p. 2)." Dewey's major goal was, "in understanding, building, and representing affective content (quoted from Schulkin, 2013, p.20)."

Social Brain and Identity

Adult development is marked by an increased demand to integrate and to cope with the complexities of the social world and our remembered experience of it. Jerome Bruner, an early cognitive psychologist, describes two modes of thought in this regard: Paradigmatic - logical, empirical thought, focusing on the denotated meaning of clarity ("say no more than you mean"); Narrative - the thought of human actors striving to do things over time, which often cannot be repeated precisely, but can "give birth" to new meanings ("mean more than you can say"). Parallel to the dual endeavors of human discovery, Art and Science, both modes exert reciprocal shaping influences upon one another, in a circular, rather than linear, fashion. The Narrative mode has become especially meaningful as the brain's architecture has demonstrated that stories, a blending of external events with internal, often emotional, meaning, have been revealed to be a central mode of organizing human experience (Bruner, 1986).

Erik Erikson, in his pioneering theory of life-span development, also uses this reciprocal, circular mode, describing a "reviewing" process as crucial to a continuation of adult development. Going beyond Freud's "child is father to the man" view (and building upon Freud's earlier work), Erikson describes the (adult) man being father to the child — looking back on earlier experience and memories with the broader view of distance, allows one to see different meanings by also seeing the context of earlier events within a personal history of psychosocial relationships. This is the foundation of his Life Span development theory (Erikson, 1960; 1997).

Erikson saw two major peaks in this process; the first, when one begins integrating experience into a whole, is the adolescent identity stage. The second peak he called the challenge of Generativity, for the middle-aged adult to share meaning, and perhaps wisdom, with the next generation, who are in the process of forming their own identities. This later challenge is as important as the first, or else the adult merely grows older without continuing to participate in the process of life, and stagnates before his/her time.

As the person develops, so does the complexity of the narrative story. The basic emotional tone of the story, a general feeling of optimism or pessimism, develops earliest, as a result of infant attachment experiences. One's sense of visual imagery - a fluid, magical sense of depicting the moment - develops along with the pre-logical thinking process of early childhood. The major advance of adolescence is that of theme development; a sequence of images and events, organized around a goal, which also conveys motivation. At this point, one begins using the raw material of experience to construct an identity (Erikson, 1960, 1997). This becomes a reciprocal psychosocial process throughout adult development, as one is challenged to maintain identity's continuity of self while coping with change over the life span (Erikson, 1997; Tietze, 2006).

The importance of culture in understanding psychological identity has become a critical issue among psychologists. It has shaped ethical guidelines for clinical treatment and grown from a unidimensional model, similar to the "melting pot" view of U.S. immigrant cultures (eliminating old cultural influences in order to more effectively acculturate to the new), into a broader model including many patterns of adopting or identifying with multiple cultural influences (Schwartz, et al., 2010). Today, "an implication of studying more forms of culture is that there are many forms of multiculturalism, and that all people are in fact multicultural" (Cohen, 2009, p. 200). Appreciating the complexity of a multicultural viewpoint has also become an important stimulus toward research and understanding in psychology (Powers & Davidov, 2006; Tietze, 2008).

The creative dimension of art attempts to offer balance and wholeness to the meaning-making process of individual and cultural identity; and the multicultural perspective becomes more relevant, more valid, and also far more complex in the world of global information and communication (Arnett, 2015).

The process of creating art, and responding to it, is also an extension of this mode of functioning into adulthood. As Storr puts it, "Becoming 'what one is' (cf. Nietzsche) is a creative act comparable with creating a work of art." (Storr, 1992,p.153). Jung would certainly agree, and called the creation of one's personality "An act of high courage… the most successful adaptation to the universal conditions of existence coupled with the greatest possible freedom for self-determination." (Jung, 1954, as quoted in Storr, 1992, p. 153).

Music and Identity: Multicultural Growth

Recent contributions to the study of identity applied to music (Macdonald, et al., 2002), use a Post-Modern perspective and view language itself as both structure and process in identity formation. The basic argument is that social situations producing language also produce concepts of self-reference, which then change through the uniquely human endeavor of creating new language to describe experience. Research conducted from the experiential perspective reveals that the social-communicative functions of music are paramount in the "management of interpersonal relationships, mood and self-identity (Macdonald, et al, 2002, p.5)." This creates a more fluid conceptualization of identity, connects directly with story-telling structure, and allows for change in events and sometimes in structure as life-span development continues (MacDonald, et al., 2002; Bruner, 1986). However, identity is also constructed from relationships within one's self, including one's body (Le Doux, 1996) which becomes the internalized environmental frame of reference for emotion, and interpersonal relationships with others, in addition to verbal meaning. Our actively constructed remembering process uses both the implicit memory system of procedures, actions and skills, mediated automatically and mostly unconsciously by the hippocampus, as well as by verbal-based explicit memory, which is mediated through the pre-frontal cortex, and is mostly conscious (Myers & deWall, 2015). The complex interactions involved within interpersonal relationships including emotional responses would seem to engage both memory systems (LeDoux, 1996).

Schulkin (2013) also supports the value of music in human development:

Music emerged as part of a communicative capability, a universal feature...Like language, the roots of music may be in the inherent shared features of our social brain, allowing us to communicate with others. Since its development, music has filled many other important roles for humans...Professor Ian Cross has noted that facilitating the transmission of information across a shared social space is the pervasive social milieu...specifying diverse social contexts in relationships...(and) expands our communicative social contact with one another...Our evolutionary ascent is the scaling of communicative competence... built on core biological propensities...like other features about us, (music) became a worthy end for its own sake. (Schulkin, 2013, pp. 35-6)

The life-span perspective of adult development views the increasing complexity of experience as a psychosocial balance to the biological slowing down of aging, with acculturation as a good example. We continue to learn throughout the life-span, changing and adding information that allows us to process familiar ideas faster than new ones. The globalization of culture resulting from the 20th century's incredible increases in the speed and amount of information available adds another level of complexity to identity development. Delay and blending of young adult identity and growth challenges occur, as a result of multicultural interaction in our rapid-flow information age (Arnett, 2000; 2015).

Accepting different perspectives is an important psychological strategy for a broader understanding of human diversity. Exploring the issue of "self - other," is a valuable one complementing our prosocial propensity, and unifying our ability to create a multicultural sensibility within the human community (Mackey, 1998). Jung saw this in his developmental model of adulthood, consisting of many challenges to integrate a split between different aspects of personality, such as male and female, social persona and private shadow and various individual ways of perceiving and responding to environmental stimuli (Jung, 1964, 1998).

Jung would agree with both the importance and the dysfunction of the dynamic of projecting unacceptable feelings, as well as parts of self, onto "the other." He also saw conflict as a natural part of human existence, in some ways previewing our new understanding of the human brain's complexity through the discoveries of neuroscience, and through the importance of recognizing emotion in information-processing. Life choices, then, become a continuous process of balancing opposing forces.

As a psychologist, I view the transatlantic cultural exchange of music and culture as perhaps some attempt to balance the terribly negative results of slavery. W.E.B. duBois' idea of the "dual consciousness" of Black Americans may be the more complex reality of the modern world, which presents a challenge to our consciousness and identity to integrate the "other" with the self to achieve wholeness. This seems to be a way to describe the psychological meaning of a multicultural perspective (Tietze, 2008, 2011). Using the resources of our personal and universal human experience, Jung saw human capability for growth through the Life-Span, and creativity as a potential solution; so art once again has crucial relevance for the human psyche, and the human condition. Musician-scholar Paul Austerlitz has coined this "Jazz Consciousness" (2005).

As an educator teaching Psychology in a small Liberal Arts-based college, I find these ideas compelling for encouraging students to expand their range of experience as well as their thinking. In my college Liberal Arts course, "Jazz and American Identity," jazz is used as a metaphor for listening to one another, and sharing experience with others, as a way to encourage a hopeful multicultural perspective (Tietze, 2008, 2006).

American jazz offers a successful model for including the music of African-Americans who for centuries of American history were not considered to be fully human, yet created a lasting legacy of creative response to oppression that has now become America's gift to the world artistic community. The social-musical theme of "call and response," brought from West African cultures, uses the metaphor of conversation as a way of transforming musical sharing into a dialogue (Clark, 2001; Shipton, 2007; Tietze, 2006, 2008).

Examples that use jazz metaphors for healing the dissociative split of "self-other," include Murray's (1976) contrast between the "Saturday Night Function" and "Sunday Morning worship" (p. 27), "as important aspects of social life, as well as the distinction between the "blues" of living, and blues music, which is a celebration of life." (Murray, 1976, p. 251). The traditional "New Orleans funeral" ([Baraka] Jones, 1963) also represents this inclusive view, with a dirge played on the way to the cemetery, to reflect loss and followed by an upbeat march for the living who must learn to go on. "When the Saints Go Marching In," a tune borrowed from the spirituals, is a popular example ([Baraka] Jones, 1963, p. 74).

The role of humor in this exploration of positive behavior is critical, and the contributions of African-Americans significant, and often unrecognized (Watkins, 1994). The development of minstrelsy was a way to mockingly imitate stereotypes of racial groups but also a way to learn about these groups through imitation. So, 18th and early 19th Century instances of "taking the cake", for the best, and often funniest, imitation of white formal dancing by slaves among each other, was a form of humorous observation. By the 20th century, with Jim Crow segregation, the learning was diminished, increasing the demeaning stereotypes of people of color into Zip Coon (a slick, but ineffective stock character), and Jim Crow, or Step-n-fetchit (a bumbling, slow character who lives on CPT, or "colored peoples' time"). The repeal of Jim Crow laws and the success of black comedians in clubs and on television would revive this form of comic observation, on life in general, rather than only connections between color and culture (Watkins, 1994).

Rather than a dissociative splitting off, or oversimplifying of emotional experience, life-span views of adult development are marked by an increased demand to integrate and to cope with the complexities of both the social world, and our remembered experience of it. Including the emotional dimension of identity broadens and deepens our understanding and experience of development, and is a common theme uniting adulthood and the continued need for creative expression (Fredrickson, 2003; Fredrickson & Losada, 2005). Much of adulthood is characterized by internal socio-emotional change rather than the physical landmarks of earlier growth. Relationships, intimacy and meaningful integration of experience are the tools of adulthood, to compose a view of life complex enough to include perspective on the past as well as the future, with a potential range beyond one's individual experience (Levinson, et al., 1978, 1996; Jung, 1957, 1964). Austerlitz (2005), in his introduction, used the term "jazz consciousness" to describe the music's expansion to the "overlapping contexts of the United States, the African diaspora and the larger world" (Austerlitz, 2005, p. x). He began with DuBois' "double consciousness," and the ambivalence it represents for African-Americans, American versus European culture, and for individuals developing within a global culture. As he puts it, "Ambivalence is a natural reaction to the contradictions that all sensitive people face, and I believe that it is by honestly confronting and working through internal contradictions that we grow" (Austerlitz, 2005, p.xi). He notes the psychological tradition of William James, one of DuBois' professors at Harvard, and his parallel description of the "divided self" that must undergo a process of 'unification' and be 'twice-born' (James, 1902, as quoted in Austerlitz, 2005, p. xii)," viewing this as frequently necessary for growth. Austerlitz also credited anthropologist Melville Herskovits with contributing the concept of syncretism, or blending elements of two or more cultures, as a way to explain the complex dynamics operating in today's multicultural world, which depend on language and tradition balanced against rapid change. Similarly, every known culture has a form of music that is both social, functioning to bond the group, and individual, to express one's emotional experience (Austerlitz, 2005; Levitin, 2006; Patel, 2009).

The concept of the "myth of the Negro past," central to the oversimplification of racism, may also be applied on the multicultural level; this myth asserted that "blacks living in the diaspora had lost contact with their ancestral traditions," which Herskovits proceeded to refute in detail, noting the existence of many "Africanisms," especially in American music (Austerlitz, 2005), and a centuries-long trans-Atlantic trade/cultural exchange, mainly due to the slave trade. The cultural exchange of music seems to have been the silver lining within the dark cloud (Tietze, 2011). The view of American Culture as a "melting pot," where indigenous and immigrant cultures blended into one, is being replaced by more complex views appreciating cultural diversity, which attempt to explain how people retain memory and the experience of their culture of origin while striving to integrate the experience of establishing meaning and identity within our polyglot American culture (Gates, 1988; Jones [Baraka], 1963). Appreciating the complexity of a multicultural view has also become an important stimulus toward research and understanding in the field of psychology (Powers & Davidov, 2006). In DuBois' early example of the African-American's "dual consciousness" (duBois, 1903, p.16), Gates described the process of creating an American literature as a dialogue between African and American cultures. This produces a "mulatto" art form reflecting the call and response of both cultures, which he symbolizes in the "talking book," the interchange between oral and written cultural transmission (Gates, 1988). Albert Murray has taken a broader cultural perspective, and described "[the blues as] the most fundamental of all existential imperatives: affirmation… (p.6), and "blues musicians being agents of affirmation and continuity in the face of adversity" (Murray, 1976, p. 38). Support has also been given by Eric Porter, who referenced jazz as "a hybrid cultural practice, with African, European, Latin American and North American cultural roots; that has been a vehicle for identity formation and self-actualization for members of disparate cultural communities…" (Porter, 2002, p. xvii).

The "dual consciousness" of Afro-Americans described by DuBois may in fact be an early example of the importance of multiculturalism for surviving in a dangerous world, with more possible identifications offering more resilience and creative response to oppression, and may offer a potential model for healing the dissociative split between self and other, a critical challenge of modern adult development (Tietze, 2008, 2011).

Psychologically, the development of a multicultural perspective has been approached from several viewpoints, but all seem to parallel the cultural syncretism first posited by Hershkovitz. DuBois' "dual consciousness," William James' "twice born," and Jung's description of adult development as individuation through integrating the "other's" perspective, seem to evoke similar dynamics to those of Austerlitz' "jazz consciousness," all reaching for a comprehensive understanding of our human challenge to balance differences with communication, both from a listening and an expressive role, as balanced in the "call and response" of African music brought to America and transformed into Blues and Jazz.

POSITIVE PSYCHOLOGY

To reinforce a positive view of individual diversity and normative creativity, the American Psychologist (JAPA) launched its "Decade of Behavior" theme with a special issue called "Positive Psychology: Happiness, Excellence and Optimal Human Functioning," in January 2000. The team of editors, headed by Martin Seligman and Mihalyi Csikszentmihalyi, both accomplished researchers and writers in this area, introduce this issue of the journal with goals drawn from their own attempts to apply research to living experience and human understanding. Two brief quotes will serve as illustration, "[We have] struggled to reconcile the twin imperatives that a science of human beings should include: to understand what is and what could be (p.7)." Describing the goals of this approach, the authors continue,

Treatment is not just fixing what is broken; it is nurturing what is best…[integrating] science and practice will also reorient psychology back to its two neglected missions – making normal people stronger and more productive and making high human potential actual. (Seligman & Csikszentmihalyi, 2000, pp. 7-8)

A follow-up to this initiative was described by Fredrickson (2003), as follows:

Instead of solving problems of immediate survival, positive emotions solve problems concerning personal growth and development... (and) help to build enduring personal resources...making (people) more optimistic, resilient and socially connected. (Fredrickson, 2003, pgs. 88-90)

As David Myers, another author in the "Positive Psychology" issue puts it, "Being attached to friends and partners with whom we can share intimate thoughts has two effects, believed Francis Bacon (1625): 'It redoubleth joys, and cutteth griefs in half.'"

Perhaps the most common neurological example supporting this integrated view is the work of our so-called "Left-Brain, Right-Brain" dichotomy, with the left hemisphere representing a linear, sequential organization, while the right takes a more holistic, aesthetic view of information as input. As Storr (1992) describes with respect to music and language:

The division of function is not so much between words and music as between logic and emotion. When words are directly linked with emotions, as they are in poetry and song, the right hemisphere is perative. But it is the left hemisphere which deals with the language of conceptual thought. (Storr, 1992, p. 35)

This hemispheric model actually depicts roughly half of the brain's activity when processing a stimulus; different aspects of incoming stimuli are initially referred to different hemispheres or localized areas for processing, but each hemisphere sends signals to many other centers, including across the corpus callosum, a huge "trunk line" of fibers carrying signals between hemispheres, translating the information from one to another. The corpus callosum is the largest axon (white matter) structure in the brain, with 250 million fibers creating a fluid balance of hemispheric function, facilitating sharing and inhibiting duplication of the flow of information, so that each hemisphere has similar information, but is able to process signals in its own style, and complement/complete the analytical/sequential style of the left hemisphere with the holistic/parallel style of the right (Gazzaniga, et al., 2014). This is a modern, and far more complex version of the older "left-brain versus right brain" dichotomy.

So initial responses involving immediate needs may be to part of the information, but reflective or thoughtful responses usually integrate information from many sources – a very practical analogue to Erikson's "re-viewing" or reflecting/meditating on life experience in general, as a developmental tool. Since about 75 percent of our neurons function to exchange information between brain areas (Myers & deWall, 2015), one could argue that our brains were built to handle complexity, and this is a function of accumulating experience, thoughtfully integrated.

Psychodynamics and Art

Freud saw the emotional power of the unconscious, in conflict with our conscious ability to repress emotional trauma. He also saw the value of art:

Psychoanalysis, as Freud conceived it, stresses the value of the social function of art, its communication of mind with mind and psyche with psyche. This involves the transmission of the artist's ideas and psychic states by the use of symbols capable of carrying both conscious and unconscious stimuli which together evoke in the appreciator a combined intellectual and emotional response. (Fraiberg, pp. 44-45)

Freud viewed perception and projection as a continuum; limited willingness or ability to perceive, especially with complex issues such as human individual and cultural differences, causes us to project, usually our unresolved and unconscious emotions, onto those we label as "others," whom we see as unlike us (sometimes less human than we are). After World War I, and during the period between the Great Wars, when he was writing many of his cultural applications of psychoanalysis, Freud didn't see much hope for the future of human development. The balance, as he viewed it, was tipped too far in the direction of projection, with too few willing to do the work involved with perception and making meaning of human differences (Freud, 1923).

Schwartz explores what Freud called "transference," highly emotionally charged relationships with significant others that tend to be recognized in psychoanalysis, but appear unconsciously in everyday life, such as strong emotional reactions to those in authority. Schwartz cites this as the nature of the brain's emotional systems, which seems to reflect support of Freud's process. Freud did hope that brain science would someday be able to examine his findings, and this modern exploration (using neuroscience" seems to be showing promise (Schwartz, 2015)."

A modern challenge might be that to the extent that we are willing to open perception and accept Freud's complexity of ambivalence (i.e.; many emotional responses are possible almost simultaneously in a given situation), we can move from the old model of the "American Melting Pot," toward a view that can appreciate cultural diversity, and attempt to explain how people retain memory and experience of their culture of origin while striving to integrate the creative experience of establishing an identity within a multicultural world (Tietze, 2008).

Fraiberg applied psychoanalysis to art and American literary criticism (1960), and described the contribution of Ernst Kris, one of the leaders of Ego Psychology (a later development of Freud's theories), and especially concerned with the meaningfulness of art:

By focusing on the ego, Kris emphasizes the higher mental nature of creativity. Art is a function...of that which makes us human; it does not come from our isolated selves, but from that part of us which reaches out to the rest of humanity. It expresses not only our relationship with the world as it is but also to the world as we wish it to be, and thus opens the way to limitless human aspirations. (Fraiberg, 1960, p. 107)

Rather than Freud's idea of art deriving from basic sublimation (redirection) of the id's impulses, how does the (reality-based) ego transform these impulses into a work of art in its own right, rather than merely a derivative of unconscious feelings? Improvisation would seem to fit this idea clearly and add imagination to the tools of past experience, reaching toward the creative goal of self-understanding.

Put in a positive way by the cultural critic John Simon, who refers to Nietzsche's labeling of Art as a "merry science," added:

Whatever else it means, the statement suggests that the composer could be surprised by his own work, and discover in it things he was not aware of having put there. This is one of the great insights of modern art: the creator's unconscious is greater than his conscious. When the maker joins the critic and the audience, he is no longer in a privileged position, but must approach the work he created with the same humility, the same openness, as the next person... (Simon, 2005, p.4)

The overarching perspective of Positive Psychology encourages a view of emotions as resources for growth and development, helping people become more resilient and socially connected (Friedrickson, 2003; Fredrickson & Losada, 2005) and also supports a broader view of creativity as a tool of "our everyday lives" (Runco, 2004, p. 640), toward psychological health and adult adaptation, in addition to the more traditional results of innovative problem-solving. As Fraiberg describes, "Creative thinking, then, may provide pleasure over and above that which comes from external creative success." (Fraiberg, 1960, p. 99). So the process is important for the listener and for the adult willing to engage the challenges of adult decision-making. The category of "mini – c" creativity was introduced by Beghetto and Kaufman (2007), and described the "developmental process of creative responses through experience…their work converged with ideas about identity as a process of choosing/creating and responding to stories organized from meaningful experiences relevant to self but not necessarily recognized by others as creative products." (Tietze, 2008, p. 246)

A fundamentally social view of the nature and function of art is expressed in Jazz; individual improvisation is created within a group context. Based on the call-and-response dialogue between listening and the creative expression of one's own voice, improvisation builds upon a foundation provided by the group's rhythmic structure (Berendt, 1992; Wilson, 1998, in O'Meally, ed, 1998; Shipton, 2007). Jazz offered not only a unique art form, but a unique perspective on the process of creativity (Runco, 2004). As scholar Robert O'Meally put it, in his introduction to the Smithsonian catalogue for an art exhibit,

The title 'Seeing Jazz' is a layered play on words that refers to artistic & literary works as visualizations of music – visible equivalents to the sound of jazz – and also to "seeing" in the metaphoric sense of understanding…and as a multi-colored blue cornerstone of what the world knows as modernism in art…How does one begin to define this music that was born in the port towns and big cities of black America and adopted by the world as its own? …Three aspects of jazz have emerged as definitive: complexity of rhythm, the magic of improvisation, and conversational call and response. (O'Meally, quoted in Golson, et al, 1987, p. 2-3)

Several life-span models of development employ a similar perspective (Erikson, 1997; Jung, 1964; Levinson, 1978,1996); individual development operates within a social relationship context that encourages a balance among individual choice, intrinsic motivation, and an adequate response to the social relationship structure. Recent work on psychological creativity, joined with adult development and neuroscience, also offered possibilities (Runco, 2004; Tietze, 2006) for developing a music curriculum (Deutsch, et al., 1999, 2003; Peretz & Zatorre, 2005; Levitin, 2006). One of the most valuable findings is that identity formation and its extension through adulthood can be a creative experience, using the tools of reflective choice, play-oriented dialogue, music listening and verbal discussion (Runco, 2004; Storr, 1992).

Research has also supported this line of exploration. The Annual Review of Psychology summary on creativity indicated clear support for a broader view of creativity as a tool of "our everyday lives" (Runco, 2004, p. 640), promoting psychological health, adult adaptation and flexibility, as well as the more traditionally described benefits of innovation and problem-solving (Runco, 2004). This summary supported a more complex view of creativity as a "whole-brain" process rather than earlier views focused on right-brain interpretations. Associative processes and divergent thinking provided further indication that creativity is encouraged by a play mode. Knowledge and memory, both declarative and procedural, serve to coordinate neural functioning and to provide continuity of experience, a critical function of identity. Knowledge provides information, which can be used as a base but also can limit the creative tactics provided by procedural memory; working together as a system, as most complex neural functions require, provides the combination of originality and appropriateness recognized as creative (Myers & deWall, 2015; Levitin, 2006; Runco, 2004).

We're all capable of making decisions within the socio-emotional context of human relationships, both about our own plans and goals, and about building human relationships as mutual resources to support us through our lives. Earlier thinkers saw the creation of identity as a creative process embedded in interpersonal relationships (Erikson,1960, 1997).

Mini "C", Little "C", and Big "C" Creativity: A Broader Human Range

The categories of "little-c" (everyday uses), and "mini-c" (the construction of personal knowledge and understanding)," introduced by Beghetto & Kaufman (2007) in addition to the standard "Big C," argued for a broader conceptualization of creativity and described a process of creative responses developing through experience. Children and "non-professional" artists were subjects for the research as well as models for us all, and this work converged with ideas about identity as a process of choosing/creating and responding to, stories organized from meaningful experiences relevant to self but not necessarily recognized by others as creative products.

Music is an important human creative endeavor, and neuroscience has shaped our new knowledge of a "normal-creative" field to parallel the acknowledged creativity of human language (Patel, 2009). Music has been found in every existing human culture, and speculation about its function ranges from a universal biological adaptation to socio-culturally learned patterns of coordination and expression. The coordination seems to be between mind and body for individuals and within social groups, unifying activity for social bonding (Levitin, 2006).

Metaphor is a useful linguistic tool that can also be used psychologically, to connect novel experiences between people, and to connect emotional content with remembered experiences for an individual. The descriptive term "translating" one's listening experience into words has been a goal in my teaching, especially when exploring connections between Psychology and the Arts, when it is often designed to encourage students to participate in the creative process of forming individual identity. This process uses reflection upon experience as source material to "translate" and integrate the variety of individual experiences into a group of valued and meaningful qualities, which are then, over time, chosen to be included into one's identity. This process allows one to be open to new experiences with the world, and yet maintain a sense of who each one of us is, as an individual person. One suggested metaphor readily used by students was to shape the listening experience in the form of a narrative story.

The creation of a narrative story is one of the basic ways a human brain learns to organize information, especially emotional experience. Research (McAdams, 2001; Bruner, 1986) indicates that our brains are "tuned in" to organizing information as stories, with the beginning choice of a goal that emerges from a general setting or a context including character, and which the protagonist directs energy to accomplish, overcoming obstacles or solving problems along the journey, until the goal is accomplished, or changed in light of one's learning along the way.

Often the lyrics of a song can offer one possible way of telling the story, and, like a poem, often reveal different types of meaning depending on the listener's experience and attention, to both the words and the underlying emotional "music" of the song-story. Human stories can be organized as an action story, but they can also be stories of reflection, as well. So the musician's idea of telling one's story has biological as well as psychological importance, whoever the musician, whatever the training and/or experience (Mc Adams, 2001). As Storr (1992) explained,

It is not trivializing either literature or music to point out that conventional forms in both arts are based on archetypal patterns of a simple kind which are probably encoded in the brain. Symmetry is one such pattern; stories are another. (Storr, 1992, p. 83)

"Big C" Creative Arts: Visual Art (Painting) and Music

A prime example of this type of creativity is offered by Eric Kandel, neuroscientist and Nobel-prize winner, in his recent work, "The Age of Insight." He explored a creative shift in human portraiture, in late 19th century Vienna, and described the phenomenon of a viewer's active visual observation that contributes an "observer's portion" to interpreting painting. Specifically, it made the creation and observation of human portraits part of a shared process, based on perceptual knowledge that was developed in a shared creative explosion among scientists, artists and thinkers who generate new ideas regarding the human condition (Kandel, 2012).

Kandel viewed this interdisciplinary time as a precursor to the 20th century revolution in neuroscience, taking a similar approach to the unification of the science of mind (cognitive psychology) and brain (neuroscience). His prime example throughout the book was what he termed the "beholder's share," first thought of by Riegl, an art historian and cultural critic, and strongly acclaimed by artist Gustav Klimt. The idea was that the viewer's participation is critical to the completion of a work of art. The concept consists of two dimensions: first, the sensory mechanisms built in by evolution and first discovered by the Gestalt Psychologists, and second, the creative cognitive-emotional strategies of hypothesis testing that have become our brain's most creative adaptation through our species' lifetime. He saw the dialogue between science and art yielding new insights that will enable us to perceive unexpected aspects of art that derive from relationships between biological and psychological phenomena. Kandel saw this Post-Romantic Modernism beginning as a response to restrictions and hypocrisies of everyday life but also to the earlier Enlightenment's emphasis on rationality, and the assumption that we can gain control over our feelings. Modernism was also a way of thinking developed during the aftermath of the Industrial Revolution, and its brutalizing effects coincided with the experience of most of humanity. Truth was not always beautiful, nor easily recognized. Irrational emotion had to be included in order to obtain a complete and realistic picture of the world we build experience upon. Viennese Modernism attempted to include emotional irrationality, include self-examination, and integrate/unify knowledge. This was a time of much interactive thinking, based on salon dialogue as well as art and science research among many different explorers in both the Sciences and the Arts (Kandel, 2012).

He saw similarity between Freud's probing the unconscious through language and artists' attempts to depict it. He attempted to include emotion in the cognitive mix and produced many valuable insights. Freud surmised that most brain activity went on below consciousness, and that human irrationality is primitive but normal, with normal and abnormal functioning operating on a continuum. Freud posited an implicit unconscious, not of conflicts and instincts, but concerned with habits, percept and motor skills, which involve procedural (implicit) memory. We now realize this kind of brain activity is far greater than Freud thought.

Riegl, a Viennese art critic and cultural theorist of the time, argued that in art, as in life, truth is not necessarily beautiful. However, for art to be Modern, it had to depict honestly the unconscious strivings that motivated men and women alike (Kandel, 2012). This search for wholeness and meaning to life experience that can attempt to include the horrors of the 20th Century's wars and mankind's destructiveness, was a critical theme in Modern Art. Western Classical Music as well as American Jazz reached to expand the conception of art to embrace truth, a larger frame of reference than that of beauty, which was the basis of the 19th Century's framework.

Alex Ross and William Appel are cultural scholars who have contributed to this type of exploration toward multicultural sources of 20th century music, and appreciate the importance of improvisation and individual interpretation (Appel, 2002; Ross, 2009). Viewed within this new perspective, Western 20th century music and art have also become part of a multi-cultural exchange process; European painters found new inspiration and energy in African tribal art, not from the view of civilized versus primitive, but rather from a universal archetypal view of cultures using different symbols to embody and express life energy (Appel, 2002; Jung, 1964). In parallel with racist beliefs were genuine explorations from the European classical tradition, the new American jazz tradition, and 20th century African cultural traditions. Musicians were exploring how to reflect the experience of the modern world, to make sense of it and help to develop the coping strategies that art offers by broadening our perceptual boundaries. As Ross described, European composers were searching for ways to unite music with life, as a "metaphor for wholeness." (Ross, 2009, p. 84). Dissonance, as both an element of modern life and modern art, pushed the boundaries of acceptable and meaningful experience. Since jazz was fundamentally focused on improvisation, it moved through this expansion process more quickly, having fewer established boundaries to break through. In barely 50 years, Jazz expanded to include the form known as Free Jazz, emphasizing total spontaneity without any preconceived musical structure. (Ross, 2009; Shipton, 2007; Tietze, 2006, 2008).

Ross describes the challenge facing Western composers,

For much of the 19th century, music had been a theater of the mind; now composers would create a (sic) music of the body. Melodies would follow the patterns of speech; rhythms would match the energy of dance; musical forms would be more concise and clear; sonorities would have the hardness of life as it is really lived...the real break came with the First World War. (Ross, 2009, pp. 76-77)

As Cocteau, spokesman for the Parisian modern composers called, "Les Six," would have it, "We need music on the earth, MUSIC FOR EVERY DAY (sic)...I want someone to make music that I can live in, like a house" (Ross, 2009, p. 100).

Perhaps seeking this modern relevant mode through pitch and melody echoed the challenge jazz musicians were facing in their search for a universal "jazz consciousness" (Austerlitz, 2005) that would also reflect the complexity of human experience. Many 20th century European composers were excited by jazz:

If other composers went further in revolutionizing harmony, none rivaled Stravinsky in the realm of rhythm...'Une musique negre,' Debussy called the "Rite (of Spring)." There is no evidence that Stravinsky knew African music...But his notion of a 'great fusion' in the "Rite" might ultimately be widened to mean something more than a thoroughgoing assimilation of folk motifs and modern music. These rhythms were global in reach, and at the time they were global in their impact. Jazz musicians sat up in their seats when Stravinsky's music started playing; he was speaking something close to their language. (Ross, 2009, p. 92)

Appel shared a famous anecdote; Stravinsky went to hear Charlie Parker play, and "Bird quoted Stravinsky's 'Firebird Suite' in the middle of his jazz piece, 'Koko' – Stravinsky roared with laughter!" (Appel, 2002, p.60).

One form of cultural exchange, known in Jazz as "quoting," is an acknowledgement or tribute to another's music. Much 20th century African popular music began as "copycat" versions of American music, such as jazz or ragtime, blended with African elements, but soon developed into a new cultural model. This syncretism, or blending cultural traditions (Collins, 2005) represented a most creative and cooperative exchange, and offered a positive model for the multicultural and hopefully mutual understanding, both of human similarities and the broad range of differences in cultural experience (Tietze, 2006).

The continuity of this exchange process developed during the 300 years of the slave trade, and initially tended to be perceived as linear, especially with African trends seen as "primitive and past" contributions, while American and Caribbean contributions were seen as "New World" phenomena; a very Colonial-sounding description. The importance of informally motivated exchange is now seen as reciprocal, and was especially an exchange of musical languages and culture supporting a multicultural view of human experience (Collins, 1987, 2005; Tietze, 2011).

So, in another context, musicians on both sides of the Atlantic were exploring new means of expression that modernized archetypes to offer holistic meaning for the 20th Century (Shipton, 2007). Ross further explained the contribution of Leonard Bernstein. In his undergraduate honors thesis, Bernstein attempted to describe the American process of multiculturalism in music (Reader, please forgive the length of the quote, but this represents the clearest and earliest description of this creative multicultural process in American critical literature):

Black music is so intertwined with the wider history of American music that the story of one is to a great extent the story of the other...Everything runs along the color line, as W.E.B. DuBois wrote in The Souls of Black Folk. Still, it's worth asking why the music of 10 percent of the population should have had such an influence...Leonard Bernstein tried to give an answer...Great music in the European tradition... had grown organically from national sources, both in a 'material' sense (folk tunes serving as sources of composition), and in a 'spiritual' sense (folkish music speaking for the ethos of a place). Bernstein's two-tiered conception, which acknowledges in equal measure music's autonomy and its social function, makes a good stab at explaining why black music conquered the more open-minded precincts of white America. First, it made a phenomenal sound. The characteristic devices of African-American musicking – the bending and breaking of diatonic scales, the distortion of instrumental timbre, the layering of rhythms, the blurring of distinction between verbal and non-verbal sound – opened new dimensions in musical space, a realm beyond the written notes. Second, black music compelled attention as a document of spiritual crisis and renewal. It memorialized the wound at the heart of the national experience – the crime of slavery – and it transcended that suffering with acts of individual self-expression and collective affirmation. Thus, black music fulfilled Bernstein's demand for a 'common American musical material.' (Bernstein, 1939, as quoted in Ross, 2009, p.122)

The human brain's powerful and immediate response to music (Jourdain, 1997; Storr, 1992) has been used previously with other musical forms to illustrate this mode of neural functioning (Levitin, 2006). For a description of jazz as a psychosocial musical form, see Tietze (2006).

The research area known as the Psychology of Music finds value in using musical materials to explore human meaning-making memory and attention processes, and has contributed to changing views based on our enhanced understanding of the neurophysiology of the human brain (Deutsch,ed., 1999; Hallam, et al, eds. 2009; Deliege & Davidson, 2014). Several of the authors in Deutsch's edited volume support the view that listening and responding to music is a creative process, and that interconnections among complex human functions such as memory and learning are related to sensory and perceptual processes (Deutsch, 1999). More specifically, "...an emerging broader view of the auditory system includes functions such as perception, attention, learning and memory that heretofore have been largely excluded (Weinberger, in Deutsch, 1999, p. 80)."

Weinberger also quipped, "Learning retunes the brain, so that more cells respond best to behaviorally important sounds," (Weinberger, in Deutsch, 1999, p.93).

Dr. Deutsch also edited an issue of *The Bulletin of Psychology and the Arts,* titled "The Psychology of Music" (Vol.4 #1, 2003), presented a short history of the study of both the science and aesthetics of music, and offered an invitation, "We all share...a deep love of music and a strong desire to contribute towards understanding at least part of its great mystery." (2003, p. 2).

Deutsch described the history of Music Psychology as a progression beginning with the ideal rationalism of the Ancient's view, the "Harmonic Music of the Spheres," and through a gradual process of the last 1,000 years toward an empirical, experiential understanding of music, with musicians always pushing the boundaries of consonance to explore broader patterns of creating sound, while critics and sometimes theorists attempted to hold back the reins with negative judgments of dissonance (and sometimes accusations of musicians not being able to play their instruments properly!) (Deutsch, 2003).

Peretz & Zatorre (2005) reviewed work on the brain's processing of music and "confers to music a privileged role in the exploration of the nature and extent of brain plasticity" (p. 90). They support a "whole-brain" model of coordination among systems, including the cerebellum, cortical integration of memory and emotional responses and the coordination of information from both hemispheres, in processing music (Peretz & Zatorre, 2005). Levitin, (2006) concurred; using the perceptual systems model of "bottom-up" and "top-down" processing, he described a process for music assessment that uses feature detectors to localize elements of music, and a "whole-brain" top-down assessment of meaning by coordinating the elements as they are communicated within the brain, resulting in a dynamic, plastic process integrating features of the music, memory, emotional responses and a creative formation of meaning (Levitin, 2006).

Twentieth century human art has expanded far beyond the ancient aesthetic ideal of beauty, to reflect the whole (including the darker side) of human experience; notable examples are painting (non-representational, or abstract art), and modern music (both jazz, and the continuation of the Western classical tradition). Alex Ross described 20th century European composers who were searching for ways to join music with life as a "metaphor for wholeness" (Ross, 2007, p.84). Dissonance, as both an element of modern life and modern art, pushed the boundaries of acceptable and meaningful experience. Since improvisation is so central to Jazz, with change incorporated into its musical history, this form of music moved through the modernization process more rapidly; fewer than 50 years after its beginnings, Jazz was stretching the boundaries of dissonance (Ross, 2007; Shipton, 2007; Tietze, 2008).

For much of the 19th Century, (Western European) music had been a theater of the mind; now composers would create music of the body. Melodies would follow the patterns of speech; rhythms would match the energy of the dance...sonorities would have the hardness of life as it is really lived...the real break came with the First World War. (Ross, 2007, p. 76-77)

So the new music from the older tradition of Western Europe, and the new music of American Jazz, were exploring similar ways to make music more a part of everyday life, and art more directly connected to life. During the early 20th Century, artists were also responding to the other cultural extreme, the horrors of "total world war" (WWI), which appeared to be splitting human existence into total chaos and destructiveness. Many other cultural explorers, including the psychological thinkers Freud and Jung, changed their understanding of the human psyche because of World War I and its unprecedented destructiveness. Artists before and after the Great War, as it was called, had been generating energy toward exploring the totality of life within art as a chaotic theme requiring a new framework of meaning, including archetypal non-Western, "primitive" art, including that of African cultures. Rather than acting these instincts out through aggressive destructiveness, painters and musicians, especially, were attempting to express a broader conception of Life's meaning (Kandel, 2012; Tietze, 2006; 2011).

This new perspective of art as a form of truth within life, united the cognitive and emotional spheres of information processing in our brains and connected them as we humans form identity and meaning within a social context. Small, another explorer of multicultural traditions, described a parallel to African social influences on American Jazz & Blues,

This brings us back to what we have seen as an important aspect of African musicking: the musician regards himself as responsible, not just for the sounds he makes but for the whole social progress of the event, for its success as a human encounter. The musician, as he improvises, responds not only to the inner necessities of the sound world he is creating but also to the dynamics of the human situation as it develops around him. (Small, 1987, p. 295)

Another aspect of reciprocal and creative dialogue can be developed by viewing listeners as one another's audience, in addition to being audience for the music. To explore with the goal of articulating and understanding what one responds to in listening to music, the listener may ask -what in me is responding to this listening experience, and am I able to connect this with other experiences, especially those I define as meaningful to my life and identity?

Derek Bailey's work on musical improvisation offers a helpful explanation of this social creative process, especially as it seems to help describe a similar process found among student group members and myself, in the classroom dialogue of the college-level Psychology and the Arts course called "Jazz and American Identity":

The relationship between any music which is improvised and its audience is of a very special nature. Improvisation's responsiveness to its environment puts the performance in a position to be directly influenced by the audience... to improvise and not being responsive to one's surroundings is a contradiction if not an impossibility. It can affect the creation of that which is being witnessed. And perhaps because of that possibility the audience for improvisation has a degree of intimacy with the music that is not achieved in any other situation. (Bailey, 1992, p.44)

An even broader view of improvisation is as a learning resource, and a basic problem-solving structure; we are built with strong capabilities for social imitation (Pagel, 2012), along with using experience-based, intuitive trial and error, as strategies of flow and play models. These can all be especially useful as ways to coordinate our brain's resources of ideas, imagination, socio-emotional experience and intuition, as tools for creative problem solving of everyday life (Lehrer, 2012; Pang, 2015; Plucker & Beghetto, 2015; Sowden, et al, 2015).

Improvisation has been linked to creativity in many art forms; some surprising. Milman Parry demonstrated that Homer's classic poems were improvised. The poems were a series of story episodes and oral text, but the poet singing is not reciting a memorized text, but

...he is improvising, along known lines, relying on a huge stock of formulaic phrases, lines, and even whole scenes; but he is improvising ...And every time he sings the poem, he may do it differently. The outline remains the same, but the oral text is flexible. The poem is new every time it is performed. (Parry, quoted in Knox, 1996, p.16)

Improvisation has also been strongly linked to imaginative development in educational contexts (Pang, 2015). In a special issue of the journal, *Psychology of Aesthetics, Creativity and the Arts* (PACA), focused on creativity and education, the editors note that the study of creativity "...has surged over the past decade" (p.113)...with an increase in journals devoted to the subject from 2 to 15 (Plucker and Beghetto, 2015)". Sowden, et al. (2015), examined the positive effects of improvisation, divergent thinking and creativity in primary school arts education (Sowden, et al., 2015). The authors make an important distinction; necessary knowledge and domain-specific skills are not sufficient to explain creativity. Divergent thinking and imagination seem to be important additions to explain novel idea generation, and also correlate with activity in the brain's cortex (inferior frontal gyrus), an area known to be involved in creative activity (Sowden, et al., 2015).

CREATIVE ARTS THERAPY: ART THERAPY

The Creative Arts Therapies (referred to as CAT) began to organize non-verbal and bodily forms of expression into treatment and educational modalities toward the end of the 20th century. An important early theorist and clinician who influenced all action-oriented forms of therapy was Jacob L. Moreno, creator of Psychodrama, a mentor of Fritz Perls (creator of Gestalt Therapy), and an acknowledged contributor to all forms of non-verbal, "here and now" forms of psychological treatment. Moreno broadened psychotherapy beyond words, to include human action in social situations, and the human body as a form of expression and stored memory (Moreno, 1946).

A younger Viennese colleague of Freud, Moreno became a psychiatrist, and was drawn to experimental theater, and observing children's imaginative play. He developed the concept of "surplus reality," to provide a group setting for exploring experiences with imagination, in the "here and now" – rather than tell a story about a past experience, he would have the person act out the scene, using other group members as resources. The goal would be producing "spontaneity," a novel response to a situation that was also adequate to both dimensions of reality, and could be used as a form of social learning in the person's everyday life (Moreno, 1969).

He was the first to describe the concept of "group psychotherapy" in clinical journals (Moreno, 1906), and much of what he created as "role-playing," an important aspect of the creative behavioral process, has been adopted by theater and corporate training specialists during the 20th century (1987). Working with human action as well as words, and with the body's expression in non-verbal behavior, Moreno was an early supporter of "whole person" clinical methods.

The Creative Arts Therapies (CAT) used Moreno's broader approach, and an expanded view of creativity as a normative human capability, which was then applied to a specific art form. A professional pioneer in the field of art therapy, and editor of *The Handbook of Art Therapy* (2012), Dr. Malchiodi wrote in the Preface to the second edition:

Art is a powerful tool in communication...Creative activity has also been used in psychotherapy and counseling not only because it serves another language but also because of its inherent ability to help people of all ages explore emotions and beliefs, reduce stress, resolve problems and conflicts, and enhance their sense of well-being... (Malchiodi, 2012,p. viii)

The field has broadened across methods and treatment/education populations, "in part because of how brain studies are informing the collective understanding of many forms of therapy." (p. ix). Malchiodi goes on to assert, "Art therapy supports the belief that all individuals have the capacity to express themselves creatively, and that the product is less important than the therapeutic process involved" (Malchiodi, 2012, p. 1).

Over the last several decades, a growing body of knowledge from science and medicine has redefined mental health interventions. Neuroscience has rapidly influenced both the scope and practice of psychotherapy and mind-body approaches, (so much that) "...The relationship between neuroscience and art therapy ...influences every area of practice. (Malchiodi, 2012, p. 17)." This professional interaction has helped to expand our understanding of the complexity of brain function. For instance, both hemispheres are necessary for art expression, although the older binary view of the brain is being replaced by a more fluid one; many areas of the brain are engaged simultaneously, including cortical, limbic and sensory areas.

Another connection is made through the Attachment Process in toddlers. Siegel (1999) explained attachment as follows: "Attachment is an inborn system in the brain that evolves in ways that influence and organize motivational, emotional and memory processes with respect to significant caregiving figures" (Siegel, 1999, p. 67). She continued with a neurological model – attachment interactions are important to the process of affect regulation, aiding the infant to develop the capacity to "self-regulate stimulating experiences" (Siegel, in Malchiodi, 2012, p.22).

Evidence has been affirming that neuroplasticity operates throughout the life-span, and Malchiodi described a continuing of creativity and new learning well into the eighth decade of life, in her chapter on creativity and aging (2012). The impact of neuroscience on healthcare will literally "repaint the picture, of how art therapy is used in the treatment of emotional and physical disorders in the future. (Malchiodi, 2012,p. 24)".

Silberstein-Storfer echoes this perspective, crediting her art teacher, Victor D'Amico (a noted educator, famous for creating the moveable "Art Barge," housing educational programs that traveled to New York City public schools), with

...Showing us how we could use our studio skills and aesthetic awareness to become better observers in every area of life.... And that parents are their children's first teachers, (and) they too need to learn how to promote creative growth...art is not an isolated fact of daily living, but was essential to human development and fundamental to all learning (pp 8-9).

To fully benefit from this viewpoint,

You come to these art projects with your own unique package of experiences and skills, which you can incorporate into the use of art materials to express your own thoughts, feelings, and impressions. (Silberstein-Storfer, 1982, p.17)

Music Therapy

The use of Music as a healing tool has a long history, applied and developed by the Creative Arts approach called Music Therapy (Ancelle, 2013; Davis, et al., 2008). Mythically attributed to the ancient Greek Orpheus, giver of music, the ancients divided this art form into two basic elements (Wroe, 2012), which reflect much of the Music Therapy and healing of today. Rhythm was described as the basic quality that joins us all together in a unified social group and allows for modern methods such as retraining a patient's walking gait after neurological damage by having the patient walk while listening to a taped rhythm that matches normal gait (Davis, 2008; Mannes, 2009).

Another form of rhythmic therapy uses a musical instrument such as a guitar to provide cues to normal, steady heartbeat, often while singing along at the same pace, to match the rhythm. The other major use of music has been melody, which encourages expression of our individuality through voice pitch patterns, "…In Orpheus, we find the compromise between two antagonistic traditions, in which both masculine and feminine sensibilities find resolution. This is reminiscent of the conflict between Dionysius and Apollo, between rhythm and harmony in healing music (Gioia, 2006 p.83)."

Gioia continued to describe ancient forms, again noting what's now become a very Jungian idea in modern Psychology:

We can sketch two opposing images of healing music. The traditional, shamanic approach relies on drums and low tones; its emphasis is on rhythm; its tradition is linked with feminine elements… [it] embraces ritual and places a priority on the needs of specific patients. The "modern" Western approach is the mirror image… it relies primarily on the higher frequencies, such as those of flutes and lyres; its emphasis is on melody and harmony…it heals through creating a soothing, relaxing state; its tradition is linked with male-dominated cultural institutions; it eschews ritual while endeavoring to promote health by establishing useful scientific precepts. (Gioia, 2006, p.107)

Cross-fertilization is common both to the arts and the creative process; both approaches to music as healing have had value in the modern world (Beghetto & Breslow, 2012). The opposition between artistic and functional values in music seems to be part of the "beholder's share" (perhaps in the form of negative judgment), rather than being intrinsic to the art form itself. In addition, continued neuroscience discoveries advance and complicate our understanding. For example, the links between cortex (coordinating the many elements of music response) and cerebellum (smoothing of movement) are significant; roughly 60 percent of cortical fibers pass through the cerebellum, making a direct neurological link between music and movement (Gazzaniga, et al., 2014).

Melodic Intonation Therapy is an excellent example of this, especially with speech production problems after neurological trauma. This treatment method uses the brain's plasticity to build alternate circuits to reconfigure speech responses using music al melodies, while simultaneously tapping rhythms with the finger, to organize both hemispheres using both major musical elements (Thompon & Schlaug, 2015; Mannes film demonstration, 2009).

This view also reflected a number of ideas linking music to language (Patel, 2009). Koelsch refers to a "Music-Language Continuum," and explored similarities and differences between both major forms of human social communication (Koelsch, 2012). Another complexity is that, "Music is a universal phenomenon, it is not a universal language, the traditions and beliefs of each culture influence music creation and response (Davis, et al., 2008, p.63)."

Although Gioia was somewhat critical of Music Therapy's apparent need for scientific legitimacy by claiming a linear-historical approach to link ancient use with magic and shamanism, while modern usage is assumed to proceed in parallel with science. Actually, there is little continuous evidence of this link (Gioia, 2006). Most recent uses of Music Therapy are based on neuroscience, and the profession of Music Therapy is working to broaden its perspective, with the most recent definition acknowledging it as a "healthcare profession," as well as a means of, "promoting wellness…and enhancing functioning," while "encouraging research support." (Davis, et al., 2008, p. 8).

Gioia had a valid point; one issue with Music Therapy's claim to a scientific approach of medicine reflecting progress throughout history, is that the ancient Greek ideas of Hippocrates, codified by Galen, dominated Western medicine for more than 1,200 years, without much change or scientific development until the Renaissance, esp. Leonardo da Vinci's work. Clearly, the present goal of Music Therapy, and the other Creative Arts Therapies, to incorporate neuroscientific evidence, will greatly assist future exploration in this field.

The educational preparation for a Music Therapist encourages "multidisciplinary training" (Davis, p. 13), acknowledging the many viewpoints represented within the professional community. The historical perspective given in the Introduction to Music Therapy does differentiate pre-scientific shamanistic approaches from more recent scientific ones, but also includes influences such as Aristotle's "emotional catharsis" as a powerful effect of music, and Plato's description of music as the "medicine of the soul (Davis, et al., 2008, p. 20)."

Music's use as a psychological treatment form greatly increased during the 20th century's wars, not just as a way to help relaxation, but for healing trauma, which now would probably come under the general rubric of Post-Traumatic Stress Disorder (PTSD) (Myers & deHall, 2015). In addition, Karl Meninger incorporated a holistic treatment model which became very influential in the medical-psychiatric field in the 1950s, and provided medical support for alternative treatments (Davis, et al., 2008).The authors describe that a major reason music is such a powerful therapeutic tool, is because people respond to music throughout the lifespan. Their list of music's functions is rather complete and is repeated here: "The body's physical systems, cognition, communication, emotions, and socio/cultural affiliation, (are all affected) (Davis, et al., 2008, p.53)." Again, however, we have also seen the tremendous variability of human response to music, which presents many methodological research problems; these are noted in many research sources, as well as my own team's research in the field (Ancelle, J., Hunter, D. & Tietze, 2014; Hallam, et al, 2009; Sloboda, 2013).

Links between music listening, triggering reward systems and the importance of the emotional experience are also major reasons people listen to music (Levitin, 2006). Two theories help shape our understanding of the meaningful response to music. It is interesting to note that both were developed before the advent of modern neuroimaging, a tribute to the creative genius of both explorers. Berlyne (1971), described the most enjoyable music experience as one of "Optimal Complexity"— too much complexity can be confusing and overwhelming, while too much repetition can be boring and unsatisfying," which helps explain the tremendous variety of individual differences in music preferences (Berlyne, 1971).

The other early theorist, Leonard Meyer, sees "expectations" as a shaping influence; we learn a culturally influenced musical language with patterns we recognize and predict. When a pattern doesn't occur, we have an emotional reaction to the music, often of surprise (Meyer, 1956). Again the overall contour is important; patterns that are too predictable, become boring, while unpredictable patterns become confusing.

Levitin, built upon this earlier work, and also discovered contour, in his early career as a musician and record producer. When he later became a neuroscientist, and explored music's meaning for the listener, he developed a new paradigm for listening research. Levitin realized that music usually has meaning within a context; it is not just notes/pitches, but a contoured tune or song, played through a musical instrument that made it have meaning. When Levitin asked subjects to sing their favorite song through a modern recording device, and then played it back on modern sound equipment, he and many colleagues who later replicated these studies, found the average subject is able to reproduce the song with about 96 percent accuracy! (Levitin, 2006; Mannes, 2009). It's clear that a song that arouses pleasant feelings and/or memories will be sung often, or practiced without necessarily being viewed as rehearsal.

A very recent *Scientific American Mind* article combined neuroscience discoveries with modern popular support for music's many effects – on mind and body, on the individual psyche as well as for the social community. As a "whole-brain" activity that utilizes many brain systems simultaneously, music is uniquely effective for treating many forms of neurological impairment, promotes neuroplasticity, gives pleasure and affords new learning. Emotional effects include redistribution of neurochemicals that can induce positive moods and heightened arousal, which speeds rehabilitation. Music is also social and personal, assisting in relationships and maintaining identity (Thompson, W., & Schlaug, G., 2015). This article is a very readable introduction to our new understanding of our brain functions on music, and is recommended as a supplement to this chapter.

NEUROAESTHETICS: THE FUTURE OF CREATIVITY AND NEUROSCIENCE

As a new field of study prompted by neuroscience, neuroaesthetics attempts to bring the deeply subjective experience of art together with the ever-developing study of the human brain, increasing in detail as objective science maps more and more patterns of neurological response (Chatterjee, 2013).

Chatterjee (2013) builds some theoretical propositions on an increasing number of empirical studies that seem to indicate many of the dynamic reciprocating interactions between brain circuits that had originally been developed as an adaptation for survival. The vast majority of our neurons transmit information between many areas within the brain, anchored by coordinating centers that serve as initial receptors, followed by the complex response coordinators that link many circuits together through parallel processing so as to speed a meaningful response.

As our brain continued to grow over the last half-million years (Pagel, 2012), we humans used social responses more frequently for survival and for social imitation; these became a tremendous source of our adaptation to new environments, especially among expanding social groups. The use of "mirror neurons, and mirror circuits" became a more unique quality of our expanding brain, and led to more complex social responses such as language, reflection and aesthetic experience Pagel, 2012).

Chatterjee (2013) viewed the evidence pointing toward a major shift in our response capabilities over this period; survival circuits relaxed and allow the reflective, aesthetic systems to develop; as the selection pressure requirements relax, there is less need for the most efficient solution and more opportunity for individual variability, leading to more subjective ways of organizing experience. To illustrate, he used an example/analogy of birdsong. Roughly 225 years ago, a Japanese collector began breeding a bird called the munia, for its plumage. Over the years, the bird's song, no longer subjected to the adaptive pressures of finding a mate (since plumage became the mating mechanism of the collectors), became more variable and complex, and the birds were able to acquire other songs more easily. Later experiments found that lesions in the brain areas responsible for song reduced its variability toward a stereotype based on the earlier and original function of song – attracting a mate.

In such a new field, where obviously invasive methods cannot be used on humans, analogy becomes a useful way to explore connections; I don't believe he is trying to state a direct analogy between birdsong and human art, but the analogy allows us to explore, using the creative connection of metaphor, another human tool of expression, to bridge individual differences when searching for unity.

The truth of the reality of human experience as context for appreciating art, rather than only the pleasure of beautiful objects, has broadened the range of expression, and may, in fact, result in pressure toward a new adaptation - a way of understanding our external environment as a world in which we share resources with many other species. To continue the growth of our human aesthetic development, marine biologist Carl Safina points to a need for greater compassion to understand interactions between humans and other species, and the consequences of much waste of resources. The social and unifying function of art may perhaps be described psychologically as "self-other," but perhaps in a larger biological context, be better referred to as "multi-species-cultural". He certainly echoes the voices of many, in both the Sciences as well as the Arts; to be able to share resources with any hope of long-term survival, we must learn new adaptations (Safina, 2015).

Social modeling skills, which are so well-developed in humans, can be complemented by our inner-directed aesthetic experiences, toward a new direction for creativity. Chatterjee's suggestion calls for more "downtime" in our busy, planned time management life styles. He described the process of the creative act comprising four stages: preparatory skills (learning the discipline of using tools well), followed by incubation (his word is "downtime"), until something is reconfigured or illuminated (the "Aha" moment), completed by elaboration or the expression of the discovery (Chatterjee, 2012).

Malcolm Gladwell has been perhaps the most popular recent estimator of the amount of tool-disciplined practice necessary for development of complex skills and creativity (about 10,000 hours) (Gladwell, 2014), the process of developing complex skills is not linear. Of course, the "Aha" moment (earlier known as the "Eureka" moment) has long been a part of our cultural description of creativity. Chatterjee, however, also focuses on the incubation period, including the reflection that has freed us to expand our expression of individual differences, and also explains the function of play into human adulthood. It is not just a way for children to practice adult skills, but to encourage their brains to explore possibilities, without the pressures of performance – analogous to Chatterjee's dynamic of exploring the release of adaptive pressure in human functioning to allow expanded opportunities for individual expression, which is his new way to describe creativity as a normative human activity (Brown, 2009; Chatterjee, 2013).

CONCLUSION: NORMAL "MINI – C" CREATIVITY TOOLS

So, there are several psychological tools any individuals may use to develop their creativity, through focused attention and heightened self-awareness:

1. **Flow:** "Between the anxiety of being overwhelmed and stressed, and the apathy of being underwhelmed and bored, lies a zone in which people experience "Flow," a completely involved, focused state of consciousness, with diminished awareness of self and time, resulting from optimal engagement of one's skills … These experiences boost our sense of self-esteem, competence and well-being." (Myers & deWall, 2015, p.A-1). Allowing oneself more time with imaginative activities brings another life-enhancing resource to our quality of life.
2. **Play:** Expands the range of experience that can be included within any human creative process, including the formation of one's individual identity. Art can be viewed as both an adult form of play and as a tool for reflecting upon experience. As a view of childhood play developed into adult art, this echoed Erikson's conceptualization. Play is not just a way for children to practice adult skills, but to encourage their brains to explore possibilities, without the pressures of performance. For adults, this can serve a similar usefulness by developing a "non-work" mode of thinking and activity.
3. **Reflection:** Erik Erikson, in his pioneering theory of life-span development, also uses this reciprocal, circular mode, describing a "reviewing" process as crucial to a continuation of adult development. Going beyond Freud's "child is father to the man" view (and building upon Freud's earlier work), Erikson describes the man (actually, the adult of either sex) being parent to the child within us; reflecting back on earlier experiences and memories with the broader view of distance, this can allow one to see different meanings by also seeing earlier events in the context of time. This is often helpful during a change in our world; birthdays, Holidays, change of seasons, for example.
4. **Incubation:** Or as Chatterjee describes "unscheduled down time." Chatterjee focuses on the importance of making time for this kind of activity, including the idea of reflection that has freed us to expand the expression of individual differences, and also the related function of play for developing one's imagination.
5. **Dreams:** Jung's idea of "active imagination" is another analogue to normative creativity as part of life-span development and growth, especially in the later years. For example, rather than Freud's assumption that there is basically one emotional language in dreams, reflecting one's conflicts especially in the areas of sex and aggression (Freud, 1938), Jung sees the unconscious revealed in dreams as a creative way to reorganize experience, sometimes providing insights to life, or perhaps an interesting narrative story.
6. **Improvisation**: When our activities are firmly rooted in functionality, there are often alternative solutions or individual methods available. Next time you face a non-essential task, try it a different way, just to "see how it feels."

In conclusion, an example is offered by the president of Stanford University, as he presented the value of creativity in the arts as a component of liberal arts education, especially for the 21st century (Hennessey, J, 2006). In his annual address to the Academic Council, he described the major theme of a new Arts initiative as, "Creativity and the Arts," which had been running successfully to address issues and contributions between these two important areas of human exploration. He described,

Looking to the Arts, not only as a key part of our cultural lives, but also as an integral component of the university's educational mission... Artists have always responded to the issues of the day, and challenged our perceptions… (by) developing ways of thinking that can be non-linear and visual rather than verbal…In addition… the arts give us a venue for dealing with the complexities and ambiguities of human existence, helping to build a bridge between diverse cultures and experiences…but achieving this goal will require facilitating more cross-disciplinary collaboration. (Hennessey, 2006, p. 5)

A cautionary note: in the interest of balance, the new developments in our understanding of creativity may indeed have a downside if viewed as a panacea. A lead article in the New York Times Sunday Review (September 6, 2015) bears the title "We're All Artists Now" (Holson, 2015, pg.1), and began "… unlocking one's creativity has become the latest mantra of personal growth and career success … Like mindfulness and meditation before it, creativity has become a mainstream commodity." (Holson,p. 1). While the author's implied caution seems to be that popular does not necessarily mean easy, she went on to describe creativity's benefits as, "…boosting self-esteem and relieving stress…and the communal aspect is appealing, too." (Holson, p.4). Citing examples such as art teacher Robert Henri of the New York Art Students League, and Julia Cameron, author of the popular book, "The Artist's Way," she offered a brief background for this approach to creativity as normative activity, but balanced it with a caveat: "Creativity is marketed as a model for self-fulfillment and a way to get ahead." (Holson, p.4).

Perhaps the balance provided by this chapter will remind us that creativity is usually a quality that needs time, patience and development, rather like the old maxim, "Creativity is 1 percent inspiration, and 99 percent perspiration."

REFERENCES

Ancelle, J. (2013). *The Importance of Music Therapy*. (Unpublished Honors Thesis). Marymount Manhattan College.

Ancelle, J., Hunter, D., & Tietze, R. (2014). *Differences in Psychophysiological Responses to Self-Selected Music*. Abstract presented at Neurological Society National Conference.

Appel, A. (2002). *Jazz Modernism*. New York: Knopf.

Arnett, J. J. (2000). Emerging adulthood: A theory of development from the late teens through the twenties. *The American Psychologist*, *55*(3), 469–480. doi:10.1037/0003-066X.55.5.469 PMID:10842426

Arnett, J. J. (2015). *Emerging Adulthood* (2nd ed.). New York: Oxford University Press. doi:10.1093/oxfordhb/9780199795574.013.9

Austerlitz, P. (2005). *Jazz consciousness: music, race and humanity*. Wesleyan University Press.

Bailey, D. (1992). *Improvisation: Its Nature and Practice in Music*. New York: Da Capo Press.

Baker, D. (Ed.). (1990). *New Perspectives on Jazz*. Smithsonian Institute Press.

Ball, P. (2010). *The Music Instinct: How Music Works and Why We can't do without it*. New York: Oxford University Press.

Baraka, A. (1963). Blues people: Negro music in white America. New York: William Morrow.

Beghetto & Breslow. (2012). *Book Review of Sawyer. Creativity Research*. PACA.

Beghetto & Kaufmann. (2007). Mini c creativity. *Psychology of Aesthetics, Creativity, and the Arts, 1*(2), 73–79.

Berendt, J. E. (Ed.). The jazz book: from ragtime to fusion and beyond. Brooklyn, NY: Lawrence Hill Books.

Berlyne, D. E. (1971). *Aesthetics and Psychobiology*. New York: Appleton, Century, Crofts.

Bernstein, L. (1982). The Absorption of Race Elements into American Music. In Findings. New York: Simon and Schuster.

Brown, S. (1998). Play as an organizing principle. In M. Bekoff & J. A. Byers (Eds.), *Animal play: evolutionary, comparative and ecological perspectives*. doi:10.1017/CBO9780511608575.013

Brown, S. (2009). *Play: How it shapes the brain, Opens the Imagination, and Invigorates the Soul*. New York: Penguin Books.

Bruner, J. (1986). *Actual Minds, Possible Worlds*. Boston: Harvard University Press.

Chatterjee. (2013). *The Aesthetic Brain: Neuroaesthetics*. Academic Press.

Clark, A. (Ed.). (2001). Riffs & Choruses: A New Jazz Anthology. New York: Continuum.

Cohen, A. (2009). Many Forms of Culture. *The American Psychologist, 64*(3), 194–2004. doi:10.1037/a0015308 PMID:19348520

Collins, J. (1987). Jazz Feedback to Africa. *American Music, 9*(2), 176–193. doi:10.2307/3052161

Collins, J. (2005). A Social History of Ghanian Popular Music since Independence. *Transactions of the Historical Society of Ghana. New Series, 9*, 17–40.

Czikszentmihalyi, M. (1990). *Flow: the psychology of optimal experience*. New York: Harper Collins.

Czikszentmihalyi, M. (1996). *Creativity: Flow and the Psychology of Discovery and Invention*. New York: Harper Collins.

Damasio, A. (1999). *The Feeling of What Happens: Body and Emotion in the Making of Consciousness*. New York: Harcourt.

Davis, W., Gfeller, K., & Thaut, M. (2008). *Introduction to Music Therapy: Theory & Practice* (3rd ed.). American Music Therapy Association.

Deliege, I., & Davidson, J. (Eds.). (2014). *The Musical Mind: Tribute to John Sloboda*. New York: Oxford University Press.

Deutsch, D. (Ed.). (1999). *The psychology of music* (2nd ed.). San Diego, CA: Academic Press.

Deutsch, D. (Ed.). (2003). The Psychology of music. Bulletin of Psychology and the Arts, 4(1).

Dewey, J. (1934). Art as Experience. New York: Perigee Books (Putnam).

DuBois, W. E. B. (1903). The Souls of Black Folk. New York: Fawcett World Library.

Eakin, E. (2003, April 19). I Feel Therefore I Am: A Thinker (Antonio Damasio) Reunites Mind and Body. *New York Times (Art & Ideas section).*

Erikson, E. H. (1960). Childhood and Society (2nd ed.). New York: Norton.

Erikson, E. H. (1997). *The Life Cycle Completed.* New York: Norton.

Floyd, S. A. (1995). *The Power of Black Music (Interpreting its History).* New York: Oxford University Press.

Fraiberg, L. (1960). *Psychoanalysis and American Literary Criticism.* Detroit, MI: Wayne State University Press.

Fredrickson, B., & Losada, M. (2005). Positive Affect and the Complex Dynamics of Human Flourishing. American Psychologist, 60(7), 678-686. doi:10.1037/0003-066X.60.7.678

Fredrickson, B. L. (2003). The Value of Positive Emotions. American Scientist, 91, 330-335.

Freud, S. (1923). *The Ego and the Id.* New York: Norton.

Freud, S. (1938). *On Dreams.* New York: Norton.

Gates, H. L. Jr. (1988). *The signifying monkey: A theory of African-American literary criticism.* New York: Oxford University Press.

Gazzaniga, M., Ivry, R., & Mangun, G. (2014). *Cognitive Neuroscience: The Biology of Mind* (4th ed.). New York: Norton Publishing.

Gioia, D. (2006). *Healing Songs.* Durham, NC: Duke University Press. doi:10.1215/9780822387671

Gladwell, M. (2014). *The Tipping Point.* New York: Little, Brown.

Golson, . (1987). *Seeing Jazz: Visual Music.* Washington, DC: Smithsonian Institution Press.

Greenacre, P. (1971). Play in relation to creative imagination. In *Emotional Growth: psychoanalytic studies of the gifted and a great variety of other individuals* (Vol. 2, pp. 555–574). New York: International Universities Press.

Guilford, J. P. (1950). Creativity. *The American Psychologist, 5*(9), 444–454. doi:10.1037/h0063487 PMID:14771441

Hallam, S., Cross, I., & Thaut, M. (2009). *The Oxford Handbook of Music Psychology.* New York: Oxford University Press.

Hartmann, E. (1996). Outline for a Theory on the Nature and Functions of Dreaming. *Dreaming, 6*(2), 46–69. doi:10.1037/h0094452

Hennessey, M. (2006, April 26). The Role of Creativity and the Arts in a 21st Century Education. *The Stanford (University) Report,* 1-9.

Holson, L. (2015, September 6). We're All Artists Now. *New York Times Review*, pp. 1-4.

Jourdain, R. (1997). *Music, the brain and ecstasy.* New York: William Morrow.

Jung, C. G. (1957). *The undiscovered self* (R. F. C. Hull, Trans.). New York: Mentor Paperback.

Jung, C. G. (1964). Man and his symbols. New York: Doubleday.

Jung, C. G. (1989). *Aspects of the Masculine/Aspects of the Feminine* (R. F. C. Hull, Trans. & Ed.). New York: MJF Books.

Juslin, P., & Sloboda, J. (2001). *Music and Emotion: Theory and Research*. New York: Oxford University Press.

Kandel, E. (2012). *The Age of Insight: The Quest to Understand the Unconscious in Art, Mind and Brain; Vienna 1900 to the Present*. New York: Random House.

Kandel, E., & Squire, L. (2005). "Neuroscience: Breaking Down Barriers to the Scientific Study of Brain and Mind," Science, 11/10/2000. In *Annual Editions: Psychology 2005 -2006*. Dubuque, Iowa: McGraw-Hill.

Koelsch, S. (2012). *Brain & Music*. Sussex, UK: Wiley-Blackwell.

LeDoux, J. (1996). *The emotional brain*. New York: Simon & Schuster.

Lehrer, J. (2012). *Imagine: How Creativity Works*. New York: Houghton-Mifflin.

Levesque, L. (2001). *Breakthrough Creativity*. Palo Alto, CA: Davies-Black Publishers.

Levinson, D., & Levinson, J. (1996). *The seasons of a woman's/man's life*. New York: Knopf. (Original work published 1978)

Levitin, D. (2006). *This is Your Brain on Music: The Science of a Human Obsession*. New York: Dutton/Penguin.

Macdonald, R., Hargreaves, D., & Miell, D. (Eds.). (2002). *Musical identities*. Oxford, UK: Oxford University Press.

Mackey, N. (1998). Other: from noun to verb. In R. O'Meally (Ed.), *The jazz cadence of American culture* (pp. 513–534). New York: Columbia University Press.

Malchiodi, C. A. (Ed.). (2012). *Handbook of Art Therapy* (2nd ed.). New York: Guilford Press.

Mannes, E. (2009). *The Music Instinct. Film*. New York: Public Broadcasting System.

Maslow, A. (1962). *Toward a Psychology of Being*. Princeton, NJ: Van Nostrand Insight. doi:10.1037/10793-000

Maslow, A. (1971). *The Farther Reaches of Human Nature*. New York: Viking Press.

Mc Adams, D. (2001). *The person: An introduction to personality psychology* (3rd ed.). New York: Harcourt Brace.

Meyer, L. (1956). *Emotion and Meaning in Music*. Chicago University Press.

Moreno, J. L. (1946). *Psychodrama* (Vol. 1). New York: Beacon House.

Moreno, J. L. (1946). Psychodrama and Group Psychotherapy. *Sociometry*, *9*(2/3), 249–253. doi:10.2307/2785011

Moreno, J. L. (1969). Viennese Origins of the Encounter Movement, Paving the Way for Existentialism, Group Psychotherapy and Psychodrama. *Group Psychotherapy*, *22*(1-2), 7–16.

Moreno, J. L. J. By Fox (Ed.). (1987). *The Essential Moreno: Writings on Psychodrama, Group Method, and Spontaneity*. New York: Springer Publishers.

Murray, A. (1976). Stomping the blues. New York: Vintage (Random House).

Myers, D., & deWall, N. (2015). *Psychology* (11th ed.). New York: Worth Publishers.

Natterson, J. (Ed.). (1980). *The Dream in Clinical Practice*. New York: Jason Aronson.

Nystul, M. (2011). *Introduction to Counseling: An Art and Science Perspective* (4th ed.). Pearson Education.

O'Meally, R. G. (Ed.). (1998). *The jazz cadence of American culture*. New York: Columbia University Press.

Pagel, M. (2012). *Wired for Culture: Origins of the Human Social Mind*. New York: Norton & Co.

Pang, W. (2015). Promoting Creativity in the Classroom: A Generative View. *Psychology of Aesthetics, Creativity, and the Arts*, *9*(2), 122–127. doi:10.1037/aca0000009

Parry, M. (1996). Penguin Oral Reading of Homer's Odyssey. New York: Penguin Publishing.

Patel, A. (2009). *Music, Language and the Brain*. New York: Oxford University Press.

Peretz, I., & Zatorre, R. (2005). Brain organization for music processing. *Annual Review of Psychology*, *56*(1), 89–114. doi:10.1146/annurev.psych.56.091103.070225 PMID:15709930

Plucker & Beghetto (Ed.). (2015). Introduction. Psychology of Aesthetics, Creativity and the Arts, 9(2), 113-114.

Porter, E. (2002). *What is this thing called jazz*. Berkeley, CA: University of California Press.

Powers, B., & Davidov, B. (2006). The virtue of multiculturalism: Personal transformation, character and openness to the other. *The American Psychologist*, *61*(6), 581–594. doi:10.1037/0003-066X.61.6.581 PMID:16953746

Ross, A. (2007). *The Rest is Noise*. New York: Farrar, Strauss and Giroux.

Runco, M. (2004). Creativity. *Annual Review of Psychology*, *55*(1), 657–687. doi:10.1146/annurev.psych.55.090902.141502 PMID:14744230

Safina, C. (2015). The Ocean and Us: Caught in the Same Net. Honors Day Guest Lecture, Marymount Manhattan College.

Sawyer, R. K. (2012). Explaining Creativity: The Science of Human Innovation (2nd Ed.). New York: Oxford University Press.

Schulkin, J. (2013). *Reflections on the Musical Mind: An Evolutionary Perspective*. Princeton, NJ: Princeton University Press. doi:10.1515/9781400849031

Schwartz, C. (2015, June 28). Tell it About Your Mother. *New York Times Magazine*.

Schwartz, S., Unger, J. B., Zamboanga, B. L., & Szapocznik, J. (2010). Rethinking the Concept of Acculturation. *The American Psychologist*, *65*(4), 237–251. doi:10.1037/a0019330 PMID:20455618

Seligman, M., & Czikszentmihalyi, M. (Eds.). (2000, January). Special Issue on Happiness, Excellence, and Optimal Functioning. *The American Psychologist*, *55*(1). PMID:11392865

Shipton, A. (2007). *A new history of jazz: Revised & Updated Edition*. New York: Continuum.

Siegel, D. (1999). The Developing Mind. In Handbook of Art Therapy. Academic Press.

Silberstein-Storfer, M. (1982). Doing Art Together. New York: Simon & Schuster.

Simon, J. (2005). *Criticism*. New York: Harper and Row.

Small, C. (1987). *Music of the common tongue*. London: Calder/Riverrun.

Sowden, P., Clements, L., Redlich, C., & Lewis, C. (2015). Improvisation Facilitates Divergent Thinking and Creativity: Realizing a Benefit of Primary School Education. *Psychology of Aesthetics, Creativity, and the Arts*, *9*(2), 128–138. doi:10.1037/aca0000018

Storr, A. (1992). *Music and the mind*. New York: Macmillan Free Press.

Thompson, W. F., & Schlaug, G. (2015). The Healing Power of Music. *Scientific American Mind*, (March/April), 2015.

Tietze, R. (2006). Jazz and the American Identity. *Psychology of Aesthetics, Creativity and the Arts*, *S*(1), 33-41.

Tietze, R. (2008). Jazz & American Identity: Case Study of a College Course. *Psychology of Aesthetics, Creativity, and the Arts*, *2*(4), 245–255. doi:10.1037/a0010843

Tietze, R. (2011). Review of the Roots of African Jazz Conference: A Multicultural Challenge. *Journal of Performing Arts, University of Ghana*, *4*(2), 80–100.

Watkins, M. (1994). *On the Real Side. Laughing, Lying and Signifying - The Underground Tradition of African-American Humor that Transformed American Culture, from Slavery to Richard Pryor*. New York: Touchstone/ Simon & Schuster.

Weinberger, N. M. (1999). Music and the Auditory System. In The psychology of music. San Diego, CA: Academic Press. doi:10.1016/B978-012213564-4/50004-4

Wilson, O. (1998). Black music as an art form. In The jazz cadence of American culture. New York: Columbia University Press.

Wroe, A. (2011). *Orpheus: The Song of Life*. New York: Overlook Press.

KEY TERMS AND DEFINITIONS

Archetype: Psychologist Carl Jung's idea of universal human inherited forms of experience, created during the long history of our species.

"Beholder's Share": What the viewer or audience brings to an aesthetic experience; usually a combination of our brain's perceptual structuring and each person's individual experience.

Bio-Psycho-Social (BPS) Model: The modern synthesis of understanding represented by Psychology reaching out to both Biology and the Social Sciences, as a means of broadening and connecting human understanding.

"Call and Response": An African cultural concept linking the social and expressive qualities of music, which usually also includes movement; the analogy is often made to a human conversation.

Creativity: A concept studied throughout the history of Psychology, now understood from the perspective of Neuroscience to be a process used by the human brain to reorganize and recombine ideas in such a way as to create an individually-different experience for each of us.

Cognitive: Modern uses of this term include the importance of emotion as an important tool for organizing our understanding of self within world.

Corpus Callosum: The largest collection of axon (white) fibers in the human brain; its function is to transfer information from one hemisphere to the other, in a complex form of sharing.

Diaspora: Usually refers to a "dispersion", or spreading out of people. Typically used to describe forced migrations, such as African slaves sent to the Americas, or Jews forced during several periods of European history.

Dissonance: Things that clash; in music, notes that don't sound good together. This varies over time, and among people, as a matter of artistic taste.

Divergent Thinking: Viewing a situation or idea from the perspective of possibility, rather than right or wrong. Similar to creative imagination, and a foundation of improvisation and the creative process.

Domain-Relevant Skills: Modern views of creativity view it as being built upon skills developed in a domain relevant to the type of creativity (movement or verbal skills, for example). See Incubation, another element in this process.

Dual-Consciousness (or Double): W.E.B. duBois' idea of Afro-American slaves' identity challenge; to be black (and owned), as well as a contributor to American society and culture.

Ego: First developed as a psychic structure by Freud, the ego functions to balance the internal impulses and demands of the id and superego, as well as a bridge between internal psychic functioning and behavioral responses to the outside world.

Embodied Cognition: A modern Neuroscience version of the ancient "mind-body" problem. Humanness involves imagination of things that may never exist, yet also the grounding of a physical body that challenges our imagination. This also refers to the idea that body experience is what our brain translates into neural information.

Epigenetic: Erikson's theory of personality and life-span development was one of the first attempts to use the modern BPS view of Psychology, by positing this biological ground plan that serves as a foundation for psycho-social development.

Explicit (Verbal) Memory: Memory organized by associations between external objects and verbal ideas; often seen as memory of facts. See also implicit memory.

Flow: Sometimes referred to as "being in the Zone", this cognitive idea builds on Maslow's idea of "peak experience", describing activity that balances between the lethargy of boredom and the disorganization of anxiety, so that the individual experience a coordination between activity and external environment that is pleasurable, and where time perception seems to slow down.

Free Association: Freud's process to help a patient link conscious experience with unconscious emotional processes driving anxious or impulsive behavior, and dreams. Asking the patient to feel free to say whatever enters the conscious mind allows the unconscious and often upsetting emotions to emerge, so they can be discussed and better understood.

Generativity: Erikson's reciprocal concept of adult Identity. To continue development through middle age, we must assist the younger generation to create their own Identities, by providing supportive opportunities for them to experience this challenge.

Gestalt Psychology: A set of principles based on experimental results, that explains the complexity of human perception; Gestalt is German for the holistic overview that symbolizes perception as "greater than the sum of its parts", including not just sensory stimuli, but their organization into patterns.

Hippocampus: Part of the brain's limbic system (processor of emotional responses) that also coordinates memory of places. As emotion began to be understood as an important cognitive response, many brain areas are seen to process both memory and emotional responses.

Homunculus: The "little man" projected onto our brain's sensory and motor cortex in a strip of brain circuits paralleling parts of the body, but organized by function rather than body size. Thus, lips, tongue and hands are represented larger than actual size because of their great use by humans for language and tool use.

Identity: Most representative of Erikson's theory, this personality organizing concept begins in adolescence, when an individual begins to ask "Who am I?", and answer through making choices about ideas, skills and experiences (especially in interpersonal relationships), that continues throughout adulthood. Identity must be cohesive enough to create a reliable sense of self through life's changes, yet flexible enough to incorporate new experience, such as the later adult challenge of career or parenthood.

Implicit (Procedures and Actions) Memory: Different from explicit or verbal memory, this process refers to remembered action sequences, often not fully conscious, such as greeting someone, or riding a bike, for example.

Improvisation: The inclusion of individually-different choices in responding to stimuli, whether through experience or artistic expression.

Incubation: Sometimes referred to as "sleeping on an idea", taking a break from active focus on an idea or problem, to let subconscious processes into the process. Chatterjee refers to this as "downtime."

Individual Differences: A basic focus of Psychology.

Interneuron: A type of neuron or brain cell, which functions to exchange information within the brain, and between different brain areas. They comprise roughly 75% of our 100 billion neurons.

Jazz Consciousness: A realization that all humans are capable of expressing their individuality, and that culture is a varied social form that provides the context for this individual expression.

"Jim Crow": Named after 2 stereotyped minstrelsy characters, the name for US racial segregation laws and policies created in the first half of the 2oth Century, until the US Supreme Court declared these laws unconstitutional in 1954.

"Left Brain-Right Brain": An early and loose model of brain functioning, based on the theory that each half of the brain performed different functions. Now seen to be an oversimplified view.

"Melting Pot": A term created by scholar Daniel Patrick Moynihan to describe the immigrant experience in America, especially during the 19th Century, among European immigrants. The metaphor described America as a place where old cultures "melted away", to be replaced by the new culture of the US.

Multicultural Perspective: A more recent model of acculturation, this perspective describes the many individually-different cultural experiences humans have to changing cultures. The model asserts that we are all multicultural, in the sense that we experience and remember experiences of all cultures we have significant contact with. A corollary is that for the human species to continue surviving and developing, we must acknowledge this broad range of human experience.

Narrative Thinking: A form of organization of experience the human brain uses to combine external events with our emotional responses to them, as a way of creating continuous meaning.

Neuroplasticity: The ability of the human brain to add, change, and prune away neurons and connections between neurons, as human experience changes through learning. A new and radically different understanding of the dynamic quality of the brain.

Parallel Processing: The concept that the brain simultaneously processes many signals, most below the level of consciousness, to maintain our functioning and creativity.

Positive Psychology: The view that emotions, thinking, and decision-making strategies can be geared toward positive ends. Emotions have been studied in the past as temporary upsets in human equilibrium; Positive Psychology balances this view by examining intimate emotions that promote our pro-social adaptation.

Post-Modern Perspective: The late 20th Century philosophical and psychological view that there is no external objectivity or truth, but we construct truth as a way to feel more secure, with a meaningful life.

Pre-Frontal Cortex: The outer covering of the human brain responsible for organizing information from many other parts of the brain, to make meaning, and meaningful decisions in response to accumulating information.

Projection: A major defense mechanism posited by Freud. When an emotion arises that is too powerful to be experienced directly, the ego protects the individual by projecting (throwing) the emotion outward, usually onto another person or people, often in a biased or stereotyped manner (i.e.; "all those kind of people feel or act that way").

Re-Viewing: Erikson's reflective tool for providing context for one's identity, and growth throughout the life-span. Looking back over past experience can often provide context for seeing relevance and meaning.

Self-Actualization: Maslow's idea of the human need for self-expression and creativity. His hierarchy of human needs describes levels; biological, security, social, and finally self-actualizing. The peak experience is a classic example, and later concepts such as "flow" are built on this conception.

"Self-Other": Jung's idea that we all have to integrate opposites in order to grow psychologically through the life-span. In a socio-cultural context, this means we challenge ourselves to see beyond our own individual needs, in order to develop human wisdom, by respecting the rights of the other as well as the self.

Syncretism: The anthropological idea that human functioning strategies and experiences often borrow from one another, as creative ideas. Especially in today's modern world of mass information and communication, there is relatively little "pure" or isolated cultural experience.

Transference: Freud's idea that highly emotional relationships, such as early relationships with parents and family members, are readily projected onto present-day people who raise some of those earlier emotions. This can happen, for example, with adult authority figures, but the same mechanism, projection, can also be used to distort perception by evoking powerful emotion.

Unconscious: An early version was posited by Freud, as the repository of powerful emotional experiences, especially those experienced while young, with relatively little life experience. As we are reminded of those experiences, we experience anxiety as adults, and this may impede our functioning. A modern view of the idea is that much of the information our brain processes is unconscious, because our system operates better that way, and frees consciousness to focus on important stimuli or changes in systems.

"Whole-Brain" Processing: A neurological idea of creative human activity. Most recently with understanding human music and creativity, but also including language, another major creative human function. These complex functions aren't localized, but rely on a series of mechanisms and circuits throughout the brain to be processed. Thus, perception of music uses sensory mechanisms to process qualities of the music (pitch, rhythm, etc.), movement mechanisms in the cerebellum (tap your foot, or not), cortical systems to bring all the qualities together for assessment of meaning, and memory mechanisms to compare the present music experience with others from our past.

This research was previously published in Exploring the Benefits of Creativity in Education, Media, and the Arts edited by Nava R. Silton, pages 337-375, copyright year 2017 by Information Science Reference (an imprint of IGI Global).

Chapter 2
Understanding How the Mind Works:
The Neuroscience of Perception, Behavior, and Creativity

Claudia Feitosa-Santana
Federal University of ABC, Brazil & Albert Einstein Israelite Hospital, Brazil & School of The Art Institute of Chicago, USA

ABSTRACT

The understanding of the inner workings of the mind are relevant to enhance curriculum achievements, therefore optimizing the professional practice in general and of the arts and design in particular. The recent birth of neuroscience as a transdisciplinary field poses a challenge to the curriculum and is yet to be included as an integral part of its core. The lessons taught by #TheDress viral Internet phenomenon are here discussed with the intention of enlightening the urgency of a popularization of neuroscience knowledge, from daily life to the professional practice, as a tool to explain how context and experience influence our perception. Along the same lines, the section "The Roots of Human Behavior" addresses the fundamental concept of human behavior and how our emotions were built by our genes, helping us understand basic and complex human choices. Finally, the section "The Neuroscience of Creativity" discusses the neural basis of creativity and its relation to intelligence by dissecting what neuroscience already knows about the development of creativity and how the work environment could foster creativity. The discussion of these topics in this chapter aims to enlighten readers of the importance of neuroscience knowledge in the curriculum and how the arts and design practices can benefit society to become more tolerant.

INTRODUCTION

Although many leading concepts in modern neuroscience can be traced back to the speculation of ancient Greek philosophers (Crivellato & Ribatti, 2007), the neuroscience was not established as a unified discipline until 1971, date of the first meeting of the Society for Neuroscience. Neuroscience as a

DOI: 10.4018/978-1-5225-5478-3.ch002

discipline is considered the rebirth of the mind study and one of the first truly transdisciplinary fields, giving researchers a large conceptual umbrella under which they could posit hypotheses about the neural basis of thinking at all levels (Tokuhama-Espinosa, 2010).

The first section of this chapter presents an event that catapulted neuroscience to stardom as a pop phenomenon, when on the 26th of February 2015 the photo known as #TheDress became a viral Internet Meme with over 10 million people from around the world astonishingly arguing over the dichotomy in color perception caused by its ambiguity: #whiteandgold or #blueandblack. This image taught the general public that color perception can be as relative as human perception in general. The brain is equipped with a mechanism called perceptual constancy, responsible for bringing some stability to our already troubled lives. In the specific case of color, this mechanism is constantly compensating for changes in lighting in order to aid the color appearance of objects to remain stable. Without the color constancy, we would perceive objects changing color constantly because the light emitted by them actually changes according to the change in lighting. The full understanding of the reasons behind people perceiving #TheDress as either white-and-gold or blue-and-black and what they have in common is still a scientific mystery to be solved, but the most important lesson from this viral phenomenon is that neuroscience principles can be easily grasped by the general public and that color perception can be as subjective as a political opinion or a purchase decision.

The second section presents humans as a congeries of genes. These genes are responsible for humans greatly appreciating their own well being, and greatly dealing with their own pain. They are also responsible for mutual assistance among relatives. The closer the relationship the more likely they are to help a family member: a mother is always willing to sacrifice more for her son than her nephew, and the nephew more than the son of a neighbor, and so on. What lies behind all of this is the selfish gene. Humans don't think and act on their own need to increase their chances of being replicated. The selfish gene is a replicator and an almost immortal replicator lasting millions of years. The mortal lasting only a few years is the human. The selfish gene is responsible for humans fighting so much and almost never helping a stranger. But the selfish gene doesn't control everything and humans might be the only animals on the planet who are aware of being the result of selfish genes as well as having the capacity to transcend the selfishness of this very gene. Only humans can change the rules of the game so that they can become real humans.

The third section discusses the neuroscience of creativity and the ability of thinking outside the box, that is; thinking differently from the norm. Creativity is a combination of genetics and environment. However, neuroscience studies still cannot explain how exactly these two factors are combined in order to enhance the development of creativity. The increase in size of the human brain happened during two different periods of our evolution and may have generated a sophistication in human memory that provided humans with a better mental understanding of the environment and with greater creativity on how to socially behave. This section also discusses the neural basis of creativity and its relation to intelligence and also tackles the importance of enhancing the working environment in order to foster creativity.

CONTEXT IS EVERYTHING: LESSONS FROM #THEDRESS

Color vision scientists around the world went to sleep just to wake up on Friday the 27th of February, 2015 with several emails and messages questioning them about the color of #TheDress. At first glance, most scientists concluded that they were all seeing the same picture on different screens and that different screens producing the emission of different lights in turn ended up influencing the perception of the observer. However, the color difference was stark and even when people were looking at the same screen they were divided between gold-and-white and blue-and-black (Mahler, 2015; Rogers, 2015).

The general principle behind this dichotomy is "how we think we see". But to understand this sentence is necessary to understand at least some basic concepts of how the human mind works. Millions of people around the world were really passionate about the color of #TheDress and when someone perceived a different color it was like an affront, a threat to their own self-image (Rogers, 2016).

Color perception is as relative as human perception in general. A better way to understand this notion would be to use price as a measurement. For instance, the dress was priced at $77. Is this expensive or cheap? It depends. To a Hollywood celebrity is close to nothing but to a homeless person is quite a bit. So what is the relationship between price and color? The same way the notion of being expensive or cheap is related to a bank account, the color of the dress also depends on how each person's brain works. In the case of #TheDress, people are divided into two main categories, #whiteandgold or #blackandblue. The color of the original dress is black and blue (pictured below). Who is right? The answer is: everyone.

A brief testing was conducted by this author on February 27, 2015 with 24 observers using the same device: 12 individuals perceived gold-and-white and 12 others perceived blue-and-black. At the same time, across the town of Chicago, a student tested 29 co-workers: 11 perceived gold-and-white and 18 perceived blue-and-black. The conclusion pointed out the fact that the ratio of gold-and-white and blue-and-black could only be safely set after rigorously tested in a bigger sample. It also confirmed that the photo is actually controversial and people are divided between gold-and-white and blue-and-black. In general, color perception differences occur frequently but are so subtle that people usually don't pay much attention to them. However, #TheDress caused a stark color difference and that is why it turned viral.

The brain is equipped with a mechanism called perceptual constancy and that brings some stability to our already troubled lives. Among others, both the color constancy and the constancy of size or shape aim to ease the perceptual instability of our daily lives. In dealing with size, if a person is very close to you, their image projected onto your retina is different from the projected image when they are far away but your brain has no problem understanding that it is perceiving the same person with the same size, no more no less (Goldstein, 2011).

In the specific case of color, this mechanism is constantly compensating for changes in lighting in order to aid the color appearance of objects to remain stable. Without the color constancy, we would perceive objects changing color constantly because the light emitted by them actually changes according to the change in lighting - whether natural or artificial (Shevell, 2006).

The first known report on color constancy dates back to 1694 by Philippe De La Hire who addressed the fact that we don't realize that the colors are different under different lightings. Later on, a week before the French Revolution (1789) Gaspard Monge made a brilliant statement about color constancy, drawing attention to the phenomenon to the Royal Academy of Sciences in Paris. Clad in a red knit, he asked his colleagues to observe the clothing through red lenses. The result was an awestruck audience surprised that they had the feeling the mesh had a very whitish red tone - almost white (Feitosa-Santana et al., 2006; Mollon, 2006).

In the case of #TheDress, some brains assume that lighting is yellow and discount this lighting realizing the dress as blue-and-black while others assume that the lighting is blue, thus perceiving a gold-and-white dress.

Before the digital era, taking pictures required us to choose between Kodak and Fuji films – the former offering warmer tones and the latter cooler ones. Analog cameras were not equipped as our brain and were unable to discount the room lighting. So we had to do what we called white-balancing before taking a picture which is similar to what our brain does with color constancy in order to stabilize our perception of the colors.

But if we were aware of this phenomenon for so long, why is that the world is only addressing this mechanism now? It so happens that the color difference caused by #TheDress is brutal and in order to realize why, we need to understand how we build relationships between colors (color space) within our brain.

Just over a hundred years ago, Ewald Hering proposed that the experience of color results from the analysis of colors in opposite pairs: green opposing red and blue opposing yellow. Thus explaining why we are unable to perceive greenish-red/reddish-green or bluish-yellow/yellowish-blue. He also used examples of afterimages that we realize after setting a straight stare for about 30 seconds on the same image (Feitosa-Santana et al., 2006).

As explained by Hering, our perception of color is based on two chromatic channels working in opposition: the blue-and-yellow and green-and-red. And what does that mean, exactly? Blue is opposite to yellow; they do not co-exist and so we cannot see a bluish-yellow or yellowish-blue (figure 1). The same goes for green and red. It is important to stress that our vision is based on light-colors but not pigment-colors – the latter being easily understood by observing the process of mixing pigment-colors used by printers and painters (Feitosa-Santana et al., 2006).

Figure 1. #TheDress

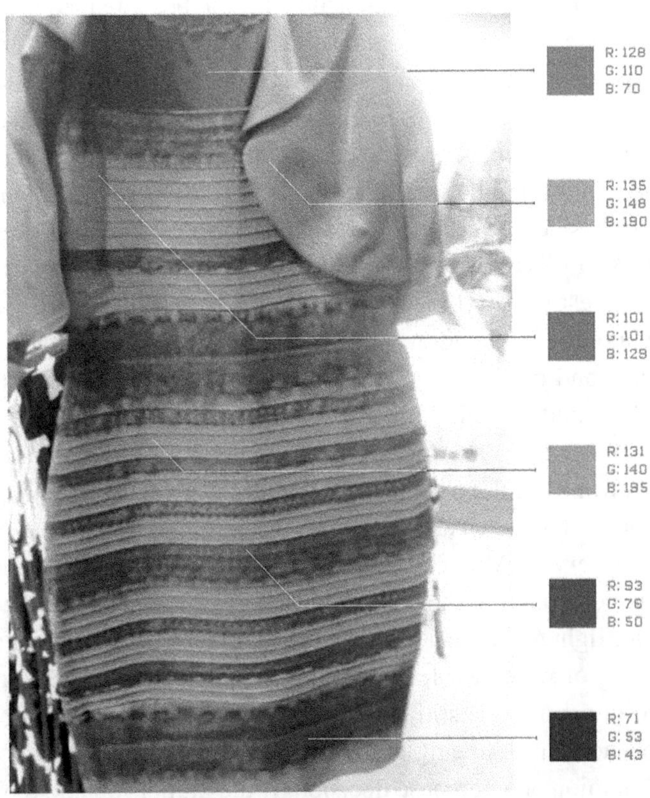

**For a more accurate representation see the electronic version.*

The original dress emits lights that usually cause the color perception of blue-and-black. The dress-image, however, came out very differently. The photo analysis (picture above) tells us that the dress should be perceived as blue-and-gold but the image is unstable and itself full of lights - requiring an arduous task from the brain in order to decide in milliseconds which is the predominant color of the lighting, so that it can then discount and generate the color perception: in this particular picture either blue lighting or yellow lighting.

If it were the case that our brains had to decide between green or blue lighting, the dress most likely would have not become a viral phenomenon, since blue and green are not opponents and perceptual differences would have passed unnoticed. Blue lighting creates a perception predominantly yellow, and green lighting generates a perception dominantly red. Yellow and red are two different colors but between each other there is a plethora of colors such as yellowish-red, yellowish-orange, orangish-red and many other similar descriptions that we are very used to, so it would not cause any real controversy.

In the case of blue and yellow lighting, which are opponents, the result is completely different depending on which of the two the brain decides to perceive. The blue lighting creates a perception predominantly yellow and soon people identify the color as gold or yellow. Discounting yellow lighting generates a perception predominantly blue. Between the blue and yellow there is no chromatic intermediary, creating the stark difference in colors of #TheDress.

The study conducted by this author and colleagues (Feitosa-Santana et al., 2016) in a sample of 52 subjects indicated that the distribution of color perception of #TheDress was approximately of 30% perceiving as blue-and-black, 40% as white-and-gold and 30% as blue-and-gold. The study conducted by Lafer-Sousa et al. (2015) reported different proportions for their 53 subjects: approximately 50% reporting blue-and-black, 40% reporting white-and-gold, and 10% reporting blue-and-gold. The difference in the color perception distribution between these studies could be explained by the fact that most of the subjects in the study conducted by Feitosa-Santana and co-authors were art students very interested in the discussion about #TheDress and may be influenced by the knowledge that the colored stripes of the dress-image were actually blue (and not white) and gold (or orange, but not black) when evaluated by the Photoshop RGB system, not considering the color constancy mechanism in action.

Still, it is unclear the relationship between the brains of those perceiving white-and-gold and those perceiving blue-and-black besides the fact that the former attains its observation in shadows and the latter under daylight (Brainard & Hurlbert, 2015; Gegenfurtner et al., 2015; Winkler et al., 2015). Lafer-Sousa and colleagues (2015) suggest that it all depends on people's daily schedules: morning people would perceive the dress as white-and-gold and evening people would perceive the dress as blue-and-black. However, Feitosa-Santana and colleagues (2016) did not find a correlation between the chronotype (morning/evening schedule) and the color of #TheDress.

Feitosa-Santana and colleagues (2016) found that the perception of the color of #TheDress was associated to the white perception and to color preferences. White-and-gold observers added more blue than the other groups to determine a white patch as well as they preferred more light colors than blue-and-black as well as blue-and-gold observers. Moreover, this study found that the preference for light green can be used as a predictor to the perception of the color of #TheDress, meaning that #whiteandgold observers had higher preferences for light green than others, blue-and-black or blue-and-gold.

The fully understanding of what people perceiving #TheDress in the same colors (white-and-gold or blue-and-black) have in common is still a scientific mystery to be solved, but the most important lesson from this viral phenomenon is being grasped by the general public. Color perception can be as subjective as a political opinion or a purchase decision. If the brain is unable to perceive the world as it is, what exactly does it do? It creates a world that can be useful to us and, therefore, becomes our reality. What we see or what we think is just a point of view, sometimes shared by many but sometimes shared by few or none. As renowned Brazilian Poet Carlos Drummond de Andrade once said "Each one opted in accordance with their whim, illusion or myopia". According to Wallisch (2015), just because we see something in a certain way it doesn't mean that everyone else will see it in the same way. Having this statement in mind helps us to built better relationships as well as a more tolerant society.

THE ROOTS OF HUMAN BEHAVIOR

Whether a denied favor or a broken deal, the reality is that usually the benefited person of the equation is a relative and the harmed one is not. In most cases, the excuse given for denying a favor or breaking a deal is directly related to the personal need of someone or that of a relative. Behind this attitude lies the selfish gene setting a scale of priorities according to the percentage of genes that they have in common with their relatives. The more genes in common the greater the benefit. The lesser degree of relation, the smaller the possibility of a benefit. In the calculation that defines the grantee, the winner is the one with the highest percentage of genes in common. The logic: to benefit the one with more genes in common increases the chances of preserving its own gene (Pinker, 1997).

This is a simplified summary of how the selfish gene, the central unit of the theory of evolution accepted by most scientists today, which has its roots in the association between the natural selection of Charles R. Darwin (1809-1882) and the genetics of Gregor J. Mendel (1822-1882) and was synthesized nearly half a century after the death of both. Since then it has been extensively studied in almost every branch of science. The selfish gene shows that animals were not adapted to preserve their species, but to preserve their genes. As Darwin noted, natural selection favors genes that replicate better, the selfish ones. Examples of the selfish gene in action are found in both the animal and plant kingdom, and with very little effort, you can identify them in the human society. From a simple broken deal to the killing of a stepson, there are many examples of the selfish gene speaking louder in *Homo sapiens* (Coyne, 2009; Dawkins, 2006; Trivers, 1971).

The term selfish gene was coined by Richard Dawkins, and despite some criticism, it is still widely used in discussing the evolution of species. Some authors prefer to call it the immortal gene or selfish genetic element and Dawkins regrets the fact that he didn't chose the term 'immortal gene' as strongly advised by Tom Maschler (Ridley, 2016). Anyway, it is worth noting that all authors are unanimous in saying that, although termed as the selfish gene, these genes have no conscience and therefore just do what needs to be done for the gene to be replicated. All authors, including the author of this article, agree that when the gene is selfish or, for example, the creator of something – it does not imply motivation or moral connotation (Dawkins, 2006).

The selfish gene is the one that needs to replicate, and the more replicators it gets, the greater the chances of success. To understand what is the selfish gene, it is necessary to differentiate body from gene. In the case of human beings, we must distinguish the person from the genes that compose them. See: Robert and Elisabeth had one son, Bobby, who is not a clone of the parents but a package of genes composed of genetic material from both sides – half from the father and half from the mother. So what Bobby has in common with Robert and Elisabeth are genes. In this gene package, Bobby has one in two chances of having genes in common with both his father and his mother. Robert and Elisabeth didn't want Bobby as an only child, and because they don't live in China they could increase their family. Therefore,

Bobby now has a sister, Beth, and again he has a one in two chances of having genes in common with his sister because they are children of the same father and same mother. Bobby and Beth grew. He had a daughter and she had a son. So Bobby has a one in four chances of having genes in common with his nephew, and Beth also has a one in four chances of having genes in common with her niece. So, the nephew and niece are cousins and have one in eight chances of having genes in common with each other. This probability in the percentage of genes in common creates a kinship scale as the number of genes in common increases. In this genetic relatedness there is no novelty, no mystery, and it is difficult to find someone who does not have this knowledge.

What is not new, but many still do not know, is that the scale of relatedness is directly related to the predisposition to cooperate (Hamilton, 1964). The more genes in common, the greater the chances are of cooperation. The less genes in common, the lower the chances of cooperation. It is within the motivation scale to make the largest investment in the selfish gene, and there are few biologists who do not accept this theory: genes are selfish because they need to ensure their own replication. The bodies do not replicate and therefore should not be selfish. Thus the selfish genes need to create neural connections of pleasure and pain to regulate the actions of the animal and thus increase their chances of replication. That's why animals appreciate their own welfare and suffer horrors with their own pain. On the issue of welfare, humans never get tired of self-benefit: shopping, traveling or any other action with no apparent purpose but to benefit oneself. "Never is enough" is the best expression for it. Thus, selfish genes need to create circuits in which humans can appreciate and feel good around each other. The more genes you have in common with someone the better you feel – increasing the chances of cooperation, and therefore, creating greater chances of replicating the selfish gene. Remember, the selfish gene has no conscience (Dawkins, 2006; Ridley, 2016).

Natural selection is not synonymous with preservation of the species and to think this way is a mistake. Darwin knew this, but was and still is often misunderstood. If the overarching goal were the species, the blood ties would be irrelevant. And what we see in the animal kingdom, including mankind, is just the opposite. In practice, it is easy to see that every mother is always more willing to sacrifice herself for her children than for her nephews, and she is more willing to help her nephews than the nephews of her husband and is less willing to help the nephews of her neighbor and so on (Pinker, 1997).

The truth is not nice and animals do not care about what happens to the species or ecosystem. In the film *March of the Penguins,* the penguin whose baby did not survive, suffers from his loss while simultaneously not caring one bit about the puppy beside him being dragged by a sea lion. This same penguin is afraid to dive for food because he does not know if there is a sea lion ready to attack him. So, what does he do? He waits for a much hungrier penguin to dive or he even tries to push the penguin neighbor in order to check whether the local water is free of sea lions and, therefore, safe to get food. Penguins, as much as men, are not concerned about their species or the ecosystem. Think in terms of humans; how many people do you know that skip their vacation to donate the money of their planned travel to benefit a stranger? How many couples do you know who chose to be foster parents instead of biological parents? By contrast, it is very common to find people who do not like their stepchildren, or worse, stepchildren who were obliged to leave home because they are victims of abuse by their stepparent. In general, animals behave in a selfish way depending on how the emotional circuits are designed, and selfish genes designed these circuits.

As stated earlier, it is necessary to separate the selfishness of the gene from the selfishness of the person, but because selfish genes built circuits in which people have more pleasure when they act in accordance to the wishes of the selfish gene, they end up being selfish too. Return to the exercise presented in the beginning of this paper. Most people are governed by their neural circuits, when in situations where they must choose between benefiting a relative or a non-relative. These neural circuits usually cause them to benefit the parent because of the well-being they feel from cooperating with the relative is much stronger when compared to the well-being they feel cooperating with a non-relative. For your friend, to benefit you at the expense of a relative would have much higher cost compared to the benefit given to a relative at your expense. The choice is often automatically and unconsciously based on cost-effectiveness (Hamilton, 1964; Pinker, 1997).

Now, imagine a change in scenery and your friend's dilemma is not to choose between his relative and you, but between his relative and his boss. If your friend benefits his relative, he runs the risk of being retaliated against by his boss. If your friend benefits his boss, he runs the risk of being retaliated against by the family. He probably will choose to benefit his boss and the chances of not being retaliated against by his family are enormous. In this dilemma, almost all human beings would benefit the boss. Why? When choosing a relative, the person is putting himself and his job in jeopardy, and he has more genes in common with himself than he has in common with a relative.

No one gets as sad about the pain of the neighbor as with his own pain. No one is as happy with the happiness of the neighbor as one's own happiness. But if that neighbor is a relative, the more genes in common, the more likely he is to have empathy and, therefore, the more likely he is to feel sadness or happiness for that family member with more intensity. And that is the essence of love, to feel pleasure with the pleasure of the other, and feel pain with the pain of the other. Today, knowing the selfish gene, it is possible to understand that cooperation between relatives or descendants of people with common genes is equivalent to helping themselves, and it is called kin selection. So if the love that exists between those who have genes in common is a mere reflection of kin selection, the question arises: love is nothing more than a neural circuit designed to work in favor of its creator, the gene selfish? Apparently, Yes!

But cooperation between people with common genes that corresponds to genes helping themselves is not the only type of cooperation in the animal kingdom, and especially in the human society. Cooperation between unrelated people exists, and friendship is an example (Trivers, 1971). The more you cooperate with your friend, the greater are your chances that they cooperate with you. This is mutual cooperation, also known as reciprocal altruism. From the point of view of the selfish gene, it may be good that you have the ability to cooperate with a non-relative. In practice, if you cooperate with your friend and your friend cooperates equally with you, you are even. But in the mathematics of friendship, no one wants to be abused and, therefore, you must be careful with the dodgers. The cheater is the one who wants to sell its cooperation for more than it's worth. Thus, he might be better benefited than those with whom he cooperates. A very clever dodger is one who can sell, successively, its cooperation for a little more than it is really worth. According to many evolutionists, cooperation between non-kin seems to have an important role in human evolution, and Robert Trivers (2011), in particular, suggests that humans are machines that have been adapted to cheat, to detect cheaters and avoid being seen as cheaters. In practice, it is easy to check, just look at the behavior of any child... It's no wonder that human society has its economy based on money, and also is full of laws, regulations, records, certificates, contracts, taxes, fines, etc. All examples that were made to prevent cheaters from having success.

Nobody wants to be cheated and there are impressive tactics developed through evolution to protect against a cheater. In much of the animal kingdom, females, unlike males, have much more certainty of having their genes in their offspring. Thus, males of several species have developed skills to ensure that they do not invest in offspring of other males and thus are not cheated. As an example, male mice secrete a chemical when a pregnant female smells him in order to check if he is indeed the father of the baby she is carrying. If she is not pregnant with him, the smell of the substance secreted causes abortion. Thus, it destroys their potential male stepson and the female is rendered to breed just for him. Think of the human race: the list of atrocities carried out by man to control the loyalty of "their" women is enormous. On the other hand, from the point of view of the female, most of them do not want a partner who is only interested in mating but do not want to collaborate in the creation of offspring. Therefore, in several species, the female requires a long courtship in order to check how much the male is really interested in investing in the future family. Recently, in modern society, another technique was developed to detect cheaters, man or woman, the DNA test.

In the eighties, the world followed a fairy tale that ended in tragedy. The story was a hoax, and many were the swindlers involved in it. Opinions are divided, and nearly all sided with Lady Di. According to the typical protocol, Prince Charles, who needed an heir to the English throne, could not marry Camilla and therefore sought the hand of his girlfriend, Diana, in marriage. Lady Di then discovered that she was not in a fairy tale and that her contribution to the marriage was only the "nobility" of her genes. Thus, she made no effort to hide from the world that she had been cheated. Human beings do not like cheaters. Lady Di was and still is admired by millions and millions of people worldwide. Recently, the couple's first child, William repeated the fairy tale and Kate was the chosen one. The royal family seems to have learned something and not encouraged, again, the realization of a marriage based only on genes. William and Kate were motivated to date for many years, trying to avoid another scandal that could, once again, upset the fans of the British royal family.

In the modern society, especially urban and virtual, there is a new type of cheating. Think of a friend who is a friend of all. They are the ones who please as many people as possible and rarely put themselves in conflict with others and they never take sides. They take sides only when they have no choice, and for sure their choice is based on who they see most likely to repay for their cooperation in the future. Whether consciously or unconsciously, their goal is always the same: self-benefit. Humans calculate the cost-benefit of any choice to cooperate in the future, always seeking chances to benefit from the cooperation of the chosen - short, medium or long term. Their motto: invest in their social network. You've just been introduced to one of the newest products of the selfish gene, the contemporary dodger. They are social chameleons, wearing the personality necessary to suit the environment, but you never really know who they are, much less what they think.

Genes have no morals, those who have morals are the animals - especially humans. Genes are not aware; those who are aware are people. The genes are potentially immortal while humans are not. Genes live for millions and millions and millions of years, people for just a few. Genes replicate, humans die. The gene does not learn, it just replicates. Humans learn, but take their knowledge to the grave. Not anymore. In oral society, humans lived an eternal cordless phone and knowledge could be lost easily. But humans, unlike other animals, invented writing and have since avenged death. Therefore, with death avenged, the knowledge built by Darwin and Mendel is immortalized and enabled the development of the evolutionary theory presented here. Darwin descendants and some others have also contributed to the development of this theory. Mendel, in turn, had no children. His genes are gone, but his knowledge remained present. When his studies were rediscovered, many years after his death, they allowed the development of hundreds of studies that led to the understanding, treatment and often the cure of many genetic diseases (Heligman, 2009).

Humans are the only animals on earth that can break the totalitarianism of their genes. Genes created humans and they are the only ones that can turn against their creator. By rebelling, they can transcend their selfish nature and use free will to re-design the future of the human society in this planet.

THE NEUROSCIENCE OF CREATIVITY

Creativity is the ability to think differently from the norm, regardless of common sense. In other words, thinking "outside the box". What makes us creative? Like almost everything from personality traits to disease development, creativity is also a combination of genetic (nature/biology) and environment (nurture/culture), but the neuroscience studies still can not say how exactly these two factors are combined to the development of creativity.

Despite human lineage starting 6 million years ago, the first signs of human creativity are merely over 2 million years old – when *Homo habilis* was able to produce the first stone tools. Between *Homo habilis* and humans today, we have seen an increase in size of our brain that is associated with the improvement of our creativity, intellectual and technological productivity. The first increase in size of the head was noted in the *Homo erectus* (1.8 mi years ago) and coincides with the presence of creative thinking – the same that has adapted to climate differences, development of more sophisticated tools and long distance hunting abilities. The second significant increase in the brain has approximately 600 to 150 thousand years ago with a series of inventions like glue, protective mix of mosquitoes and fire, culminating in a giant leap of human creativity. Between 90 and 30 thousand years ago, it occurs the emergence of syntax in language along with the arts, sciences, and politics. Brain increase in these two moments may have generated a sophistication in human memory that enabled better mental representation (understanding) of the environment and greater skill in social behavior (Gabora & DiPaola, 2012).

Creative people in various fields like music, visual arts, poetry, etc. present creative processes with very similar standards such as the ease in creating metaphors. Besides the ease in creating metaphors, *Synesthetes* – people with synesthesia, also show 7 times more likelihood to pursue creative careers. Synesthesia is a consequence of cortical formation where 3 regions – temporal, parietal and occipital are not separated completely and the overlap of them is responsible for this perceptual phenomenon where two distinct senses are perceived simultaneously (such as seeing a color while at the same time listening to a certain musical note or tasting a flavor while at the same time feeling a certain texture of touch) (Ramachandram & Hubbard, 2003).

The study conducted by Jung and colleagues (2010) suggests that in neuroscience, creativity is correlated with increased cortical thickness in some regions of the brain and reduced cortical thickness in other regions of the brain. These findings indicate that creativity is a complex activity that requires excitement in some regions of the brain as well as inhibition in others – e.g., the improvisation arts require synaptic inhibition in a certain region of the brain that is associated with a power giant concentration. Lhommée and colleagues (2014) showed how the dopaminergic circuits (dependent dopamine) are important for stimulating creativity.

Creativity and intelligence require distributed neural participation in a complex combination of brain activities, although both work with partially different circuits, and it is suggested that creativity is not merely the result of a high intelligence but a circuit that has evolved parallel to intelligence (Jung et al., 2013). Insights are more likely to happen after a good night's sleep: those who sleep well are more likely to have a light bulb moment (Eureka!) than those who are sleep deprived (Wagner et al., 2004).

Humans avenged death when they invented writing, making the transmission of knowledge after one's passing a reality. This revenge is the basic difference between human culture and the culture of other species as we tend to develop culture faster. Chimpanzees teach their children how to exterminate ants but can't refine this technique adding something else because they are unable to pass on improvements in this behavior for future generations as humans do.

So, what exactly is needed to develop one's creativity? For children, a good diet, many hours of sleep and diverse stimulation (motor, visual and intellectual) because the ability of our brain must create new neural connections and our neural plasticity is much stronger when we are young. The creative process involves many simultaneous connections between different brain regions. For adults, in addition to a good diet and good night's sleep (and the amount of time varies from person to person), time, attention and a physical activity that results in well-being. In this context, time translates into *quality* and attention into *concentration* and *focus*. Some studies show that the only way to enable creative solutions in a short period of time is through attention (dedication or exclusive concentration towards the solution of the problem). Too much time without quality or little time without attention are harmful ways to stimulate creativity. Two great examples of time in the creative process are Darwin and Einstein, both worked years and years to synthesize their theories.

How does the work environment foster creativity? Most workplaces do not encourage creativity. One needs a flexible work environment, respecting the employee's personality so that in turn this can exercise their creativity, since contrary to popular belief, stressful environments work against creativity.

CONCLUSION

The primary goal of neuroscience is to understand how the human mind works in its various capabilities. Its main barrier is finding ways to dialogue with the general public and in the classroom. #TheDress Meme gave neuroscience the opportunity to become pop culture material and took scientific material to the daily discussions of millions around the globe. Despite a nostalgia constantly evoking the past, there are plenty of scientific evidence (Pinker, 2011) proving that humanity has never been better, and that we live in a much safer and compassionated society than our ancestors. Our biology may be still the same for thousands of years but our cultural changes are responsible for all this present positive change. Humans are the only animals on earth that can break the totalitarianism of their genes. And we hope to keep on going until we are done with that job.

REFERENCES

Axelrod, R. (2011). Launching "The Evolution of Cooperation". *Journal of Theoretical Biology*, *299*(2012), 21-24.

Baldassari, D., & Grossman, G. (2011). Centralized sanctioning and legitimate authority promote cooperation in humans. *PNAS Early Edition*, *108*(27), 11023–11027. doi:10.1073/pnas.1105456108 PMID:21690401

Brainard, D. H., & Hurlbert, A. C. (2015). Colour vision: Understanding #TheDress. *Current Biology*, *25*(13), R549–R568. doi:10.1016/j.cub.2015.05.020 PMID:26126278

Burt, A., & Trivers, R. (2011). *Genes in Conflits: the biology of selfish genetic elements*. Cambridge, London: The Belknap Press of Harvard University Press.

Coyne, J. A. (2009). *Why Evolution is True?* New York: Penguin Group.

Crivellato, E., & Ribatti, D. (2007). Soul, Mind, Brain: Greek philosophy and the birth of neuroscience. *Brain Research Bulletin*, *71*(4), 327–336. doi:10.1016/j.brainresbull.2006.09.020 PMID:17208648

Dawkins, R. (2006). The Selfish Gene (30th ann. Ed.). New York: Oxford.

Feitosa-Santan, C., Lutze, M., Barrionuevo P., & Cao, D. (submitted). *The Perceived Color of #TheDress is Associated with Individual Color Preferences.*

Feitosa-Santana, C., Oiwa, N., Costa, M. F., Tiedemann, K. B., Silveira, L. C. L., & Ventura, D. F. (2006). Espaço de Cores. *Psicologia USP*, *17*(4), 35–62. doi:10.1590/S0103-65642006000400003

Gabora, L., & DiPaola, S. (2012). How did humans become so creative? *A computational approach. International Conference on Computational Creativity.*

Gegenfutner, K. R., Bloj, M., & Toscani, M. (2015). The many colours of 'the dress'. *Current Biology*, *25*, R1–R2. PMID:25562290

Goldstein, S. E. (2011). *Cognitive Psychology: Connecting mind, research, and everyday life experience* (3rd ed.). Belmont: Wadsworth Cengage Learning.

Grossberg, S. (2014in press). The visual world as illusion: The ones we know and the ones we don't. In A. Shapiro & D. Todorovic (Eds.), *Oxford Compendium of Visual Illusions*. Oxford, UK: Oxford University Press.

Hamilton, W. (1964a). The genetical evolution of social behaviour I. *Journal of Theoretical Biology*, *7*(1), 1–16. doi:10.1016/0022-5193(64)90038-4 PMID:5875341

Hamilton, W. (1964b). The genetical evolution of social behaviour II. *Journal of Theoretical Biology*, *7*(1), 17–52. doi:10.1016/0022-5193(64)90039-6 PMID:5875340

Heligman, D. (2009). *Charles and Emma – The Darwin's Leap of Faith*. New York: Harry Holt and Company, LLC.

Jung, R. E., Mead, B. S., Carrasco, J., & Flores, R. A. (2013). The structure of creative cognition in the human Brain. *Frontiers in Human Neuroscience*, *7*(330), 1–13. PMID:23847503

Jung, R. E., Segall, J. M., Bockholt, H. J., Flores, R. A., Smith, S. M., Chavez, R. S., & Haier, R. J. (2010). Neuroanatomy of Creativity. *Human Brain Mapping*, *31*(3), 398–409. PMID:19722171

Lafer-Sousa, R., Hermann, K., & Conway, B. R. (2015). Striking individual differences in color perception uncovered by 'the dress' photograph. *Current Biology*, *25*(13), pR545–R546. doi:10.1016/j.cub.2015.04.053 PMID:25981795

Lhommée, E., Batir, A., Quesada, J.-L., Ardouin, C., Fraix, V., Seigneuret, E., & Krack, P. et al. (2014). Dopamine and the biology of creativity: Lessons from Parkinson's disease. *Frontiers in Neurology*, *5*, 55. PMID:24795692

Mahler, J. (2015, February 27). The White and Gold (No, Blue and Black!) Dress that Melted the Internet. *New York Times*.

Mollon, J. (2006). Monge: The Verriest Lecture, Lyon, July 2005. *Visual Neuroscience*, *23*(3-4), 297–309. doi:10.1017/S0952523806233479 PMID:16961961

Pinker, S. (1997). *How the Mind Works*. New York: W. W. Norton & Company.

Pinker, S. (2011). *The Better Angels of Our Nature: Why Violence Has Declined*. New York: Viking Books.

Raihani, N. J., & Bshary, R. (2011). Resolving the iterated prisoner's dilemma: Theory and reality. *Journal of Evolutionary Biology*, *24*(8), 1628–1639. doi:10.1111/j.1420-9101.2011.02307.x PMID:21599777

Ramachandran, V. S., & Hubbard, E. M. (2003). The phenomenology of synaesthesia. *Journal of Consciousness Studies*, *10*, 49–57.

Rogers, A. (2015, February 26). The Science of why no one agrees on the color of this dress. *Wired*.

Rogers, A. (2016, February 26). A year ago, a Dress murdered the idea of objective color. *Wired*.

Shevell, S. K. (2003). Color appearance. In S. K. Shevell (Ed.), *The Science of Color* (2nd ed.; pp. 149–190). Oxford, UK: Elsevier. doi:10.1016/B978-044451251-2/50005-2

Tokuhama-Espinosa, T. (2010). *Mind, Brain, and Education Science: The new brain-based learning*. New York: W. W. Norton & Company.

Trivers, R. (1971). The evolution of reciprocal altruism. *The Quarterly Review of Biology*, *46*(1), 35–57. doi:10.1086/406755

Trivers, R. (2011). *The Folly of Fools: The Logic of Deceit and Self-Deception in Human Life*. New York: Basic Books.

Trivers, R. L. (1972). Parental investment and sexual selection. In B. Campbell (Ed.), *Sexual selection and the descent of man, 1871-1971* (pp. 136–179). Chicago: Aldine.

Wagner, U., Gals, S., Halder, H., Verleger, R., & Born, J. (2004). Sleep inspires insight. *Nature*, *427*(6972), 352–355. doi:10.1038/nature02223 PMID:14737168

Wallisch, P. (2015, February 27). Lessons from the Dress – Vision is fundamentally ambiguous but usually not this ambiguous. *Slate*.

Winkler, A. D., Spillmann, L., Werner, J. S., & Webster, M. A. (2015). Asymmetries in blue–yellow color perception and in the color of 'the dress'. *Current Biology*, *25*(13), R547–R548. doi:10.1016/j.cub.2015.05.004 PMID:25981792

KEY TERMS AND DEFINITIONS

#TheDress: A photo that became a viral meme on 26 February 2015 also associated with the hashtags #whiteandgold and #blackandblue.

Color Constancy: One type of perceptual constancy in which our brain compensates for changes in lighting in order to aid the color appearance of objects to remain stable.

Color Perception: A perceptual phenomenon determined by neural processes in the brain.

Context: The frame of reference, circumstances, factors or conditions involved in a scenario or event, background.

Creativity: The ability to think differently from the norm, regardless of common sense. In other words, thinking "outside the box".

Illusion: The deception of the senses, changing the actual appearance for a fake one.

Selfish Gene: The immortal gene that is the core view of evolution theory.

This research was previously published in Projective Processes and Neuroscience in Art and Design edited by Rachel Zuanon, pages 239-252, copyright year 2017 by Information Science Reference (an imprint of IGI Global).

Chapter 3
A Study on the Interface Between Arts and Sciences:
Neuroesthetics and Cognitive Neuroscience of Art

Alexandre Siqueira de Freitas
Universidade Federal do Sul da Bahia, Brazil

ABSTRACT

This chapter discusses issues related to two fields of knowledge: neuroesthetics and cognitive neuroscience of art. These two fields represent areas that link historically dichotomic instances: nature and culture. In the first section, the author introduces a brief discussion on this dichotomy, reified here as science and art/aesthetics. Based on a preliminary analysis of these fields, as well as potential interfaces and articulations, the author then situates neuroesthetics and cognitive science of art. In both cases, the main definitions, usual criticisms, and comments on potential expectations regarding the future of these two areas will be presented.

INTRODUCTION

Arts and sciences are, in their most basic senses, "two great engines of culture", as coined by Stephen Wilson (2002, p. 5). According to Wilson, they consist of *stimuli*, sources of creativity and identity marks. However, the ways through which these engines converge, diverge, and sometimes merge present substantial and complex issues. Even though it's possible to trace analogies (this is our intent here), relating the creative processes from both terrains (arts and sciences) or the visual aspects of its products, some qualities attributed to each field often seem to remain distinct.

In order to propose a brief illustration, the author would like to recuperate the term "engine", used by Wilson. On one hand, we have an art object whose movement of axes, gear wheels and clock hands are the main attraction to its public. On the other hand, we have an engine that presents the exact same components, but, in this case, the public's main interest is focused on the movement of clock hands, that indicate time. These two pictures are, at once, similar and distinct. Regarding their semantic values

DOI: 10.4018/978-1-5225-5478-3.ch003

and cultural symbolism, they are quantitatively identical and qualitatively different. The first object is primarily aesthetical and ideally gratuitous, if we think from a Kantian perspective. The second one, by its turn, belongs to a primarily cognitive sphere, from all the scientific fields. However, both objects present, to varying degrees, aesthetic and cognitive elements. This example unveils some of the potential tensions and ways through which one might notice the divergences and convergences between art and science products.

In order to raise epistemological questions related to fields of knowledge focused on art and sciences – neuroesthetics and cognitive neuroscience of art –, the author believes it's important to analyze, even if briefly, the duality between arts and sciences, including thoughts on aesthetics, a central issue to the subject discussed here. These "two cultures" divided by an abyss of mutual incomprehension – as C. P. Snow (1990) presented the dichotomy between sciences and humanities in 1954 –, unveil, at the present day, signs of change in this landscape. These signs consist of small cracks in humanistic and scientific epistemologic strucutures, in their modern conception. Cracks that may indicate major ruptures, future transformations. And they tend to get more and more intense in our current historic moment, which some would call postmodernity. Therefore, this work aims at discussing these transformations, which are present in certain researches and new theoretical fields. The author will firstly describe some of the elements that were (and still are) used in the construction of fields we call science, arts and aesthetics – all of them embedded with signification and history. Secondly, the author will address the neuroesthetics and the cognitive neuroscience of art separately, with a brief history of each, critiques and expectations.

BACKGROUND: SCIENCE, ART AND AESTHETICS

Science, as we know it today, bears great expectations about "how" and "why" general *phenomena* occur. It's primarily focused on the cognition of the "natural world" and reflects the necessity of extracting order structures from a vast, chaotic background from time to time. According to Júlio Plaza (2003, p. 38), influenced by Ortega's ideas, knowledge is the mental effort that extracts information, or language, from disorder. Plaza also argues that in order to actually know something we must reach its very "being". This "being" is not achieved through perception, but through an intellectual model. The production of knowledge is thus the construction of intellectual models that results from a work process based on complex information (Srour *apud* Plaza, 2003, p. 38). Scientific knowledge is grounded on the faith in empirical information, with which scientists usually create massive theoretical structures to link observation and reality (Wilson, 2001, p. 19). Science pursues ways of effectively acting upon reality and, therefore, "builds conceptual models or representations that reflect, with some isomorphy, aspects of the world's objective organization" (Vieira, 2006, p. 48).

We are facing, now, one of the great pillars of science: objectivity. In its core resides a cumulative expansion movement driven by the specificities of the operations that guide observation. Concepts, principles, and codes are combined in the quest for "objective and universal truths", as defined by neuroscientist Jean-Pierre Changeux (2013, p. 97). Even though the history of science proves these truths are transitory, the driving forces behind scientific knowledge are not as dynamic as common sense believes – maybe due to the troubled distinction between science and technology (which is not the purpose of this work). According to Thomas Kuhn's classic and essential *The Structure of Scientific Revolutions* (1962), even truths, or better yet, scientific paradigms, are constantly changing. Such paradigms, models created to address the biggest number of observations in inside a coherent landscape, suffer the tension of what Kuhn called anomalies. That goes on until a new conceptual landmark – a potential rival to the preceding one – is created. Therefore, objectivity (in its mobility) is linked to two inseparable instances: man's inventiveness and the enhancement of instruments, the latter frequently associated with mathematics. As for subjectivity, its influence is usually avoided by the scientific method through control mechanisms and, posteriorly, by the critique and comments on the results published. This movement reveals an inescapable intersubjectivity, which is also inherent to the science experiment. In such a wide scope, empirists underscore the importance of observation, rationalists focus on the logical process of construction and derivation, and critical theorists, among other branches of knowledge, see science as the modern disillusionment (Wilson, 2001, p. 12). Theorists in this last branch point out to underestimated social forces and metanarratives, though they structure questions and paradigms in the scope of scientific investigations. Bruno Latour, author of *Science in Action* (1987) is one of these theorists. The persuasive influence of gender, class, national identity, and history are also underscored. Besides these three groups, Wilson mentions the constructivists, who question our ability to find truths that are applicable to all times and cultures. Objectivity is always questioned and conditioned in different ways, but, despite suspicion, it endures in the core of Western civilization.

The relation between science and the natural world (usually in opposition to the cultural) is another essential issue for its characterization. Science emphasizes an observation of the natural world as reality, implying an essentially ordered character of nature. The author says "essentially" because, when analyzed through perception, nature is usually seen as chaotic. As argued above, scientists are responsible for the objective organization of the world. And science fiercely believes in the potential discovery/organization of the world.

The scientific thinking, as Plaza (2003, p. 38) puts it, is a productive process that aspires universality. Still according to Plaza and following Bense's precepts, knowledge lies on the intervention of intelligent beings who identify things in the world in order to be cognizant of it. Such identification happens in the natural world (according to the principle of exteriority) and functions according to different rules, in order to establish and understand things and *phenomena*.

Ideally, science continually tests and refines its hypotheses, enhancing its instruments, defining its concepts and intellectually interfering in symbolic objects, such as representations, intuitions, and observations of all kinds.

Art and Aesthetics

It's natural and maybe fundamental that a concept as polissemic as "art", carrying so many different meanings through history, especially in the "posthistoric" time in which we live[1]. "There really is no such thing as Art. There are only artists" – wrote Gombrich in his *The Story of Art* (1951, p. 12). Taking art as a kind of social combination, the author of this celebrated work chooses to present art history as a story about artists. Gombrich's choice is certainly appropriate in terms of historic dimension, since art objects results from the encounter between the artists and their desires, conceptions, cultural and social circumstances, selected materials and media. However, for the purposes of an epistemologic study, one must look at art as a subject, in spite of its inseparability from environment. After all, every product ever made by men is also the result of social encounters and combinations.

In order to clarify a few ideas on the concept of art, the author turns to Italian thinker Luigi Pareyson, whose conception is structured on three historically recurrent definitions: art as action, as knowledge and as expression (2001, p. 21). All along Western history, these definitions opposed, excluded and dialogued with each other in different ways. In Acient history, action was art's predominant attribute, embedded in its etymology: *ars*, from the Latin, and *techné*, from the Greek. In the Renaissance, the cognoscitive element was highlighted, and art became a way of knowing and interpreting the world. Finally, Romanticism underscored art as expression, and the canon shifted from beauty as a model to the beauty of expression.

Regarding the executive aspect of art, Pareyson argues that:

(...) art consists of intensive, eminent, absolute production and realization, to the point that it was actually called creation. It produces, though, not only organisms that are autonomous and independent (as in nature), but also radically new objects, which are reality's very own evolution, an onthological innovation. (Pareyson, 2001, p. 25).

Through knowledge, embodied in congnoscitive elements, art takes on "functions from other human activities, i.e., from science, philosophy, religion, or morality" (*Pareyson, 2001*, p. 24). But the relation between "art-as-knowledge" and "art-as-action" is not necessarily peaceful.

Art doesn't have, per se, a revealing or cogniscitive role, and it's not even limited to knowledge, especially if we think of a contemplative kind of knowledge. The emphasis on art's cognoscitive, visible, contemplative, and theoretical attributes weakened its most essential and fundamental aspects – its executive and accomplishing dimension – and did a disservice to art theory and practice. (Pareyson, 2001).

Art-as-expression highlights, unsurprisingly, art's expressive aspect, which, by its turn, allows us to think of art as language, and even to establish semantic theories. However, while art is certainly expression, this is not its essence.

(...) art also has, alongside other attributes, an expressive character. At most, one could add that, without it [the expressive character], it could not even be art, once it would lack that sense of finished humanity, which is the condition to any human production. (Pareyson, p. 22).

Therefore, while such conceptions (art-as-action, art-as-knowledge, art-as-expression) bear essencial artistic marks, they should not be essentialized or isolated, because all three of them are based upon vigorous historic engines.

Another interesting way of understanding the art phenomenon is presented by semiotician Max Bense (1971) and summed up by Júlio Plaza (2003, p. 39): the definition of art as a view of the world as "something made", as part of the creative structure. This is different from seeing the world as "meaning and signification", a semantic and communicative perspective, or "something given" (the physical dimension), according to the causal model. While the latter is highly determined, the semantic and communicative model is defined in a conventional way, and the model of the world as a "given entity" – the aesthetic state one – is singularly and ambiguously determined. Science aims at determining and establishing ideas, by identifying, transposing, translating and ordering physical states. In the case of aesthetic states, such determination happens in a slight way, as in the distribution of colors seen in a painting by Ticiano (*Plaza, 2003*, p. 39).

Even though art's expressive dimension approximates it to language in a certain way, aesthetical information transcends semantic information, once it is open to much higher unpredictability and improbability of signs, due to its slight determination and low degree of codification (*Plaza, 2003*). This is a game of possibilities of perception, more than a game with rules, which is the case of one of the most important dimensions of science.

As for the art-as-knowledge modality presented above, Plaza places it according to a perspective that diverges from science. While science addresses knowledge from an abstract perspective and applies it universally, at any time and place, art addresses it in a concret plan made of individual, irreplaceable, and unique objects. Even though semantic information might subsist, art is presented to the world primarily as aesthetic information.

It's also important to mention that the idea of art bears certain contradictory elements such as the two definitions presented by Étienne Souriau in his *Vocabulary of Aesthetics* (2010, p. 178). On one hand, art is defined as a set of precepts that ensure a good artistic realization – possibly associated with the notion of *beaux-arts* and the rise and consolidation of academies in the 17th and 18th centuries, a movement in line with the Enlightenment ideals. On the other hand, art is considered as knowledge that escapes rules and guidelines – the *savoir-faire* goes beyond information and procedures acquired. Still according to Souriau, some claim that a work of art reaches its higher aesthetic quality when rules are applied in a way that the technical *savoir-faire* is no longer apparent (*Souriau, 2010*).

After exploring the artistic terrain and the potentiality of its multiple definitions, we must address the concept of aesthetics (even if briefly) in order to rectify some usual misconceptions. These come not only from the common sense, but are also closely related to issues to be discussed in this work, especially those rising from neuroesthetics.

Aesthetics, as most of the disciplines, is unique and multiple. It might be possible to offer a relatively clear definition of it, but hardly an exact one (Freitas, 2015, p. 48). Seeing it as art contemplation might be a correct, yet vague assumption. However, the more precise we get, the more subject to contradictions and errors we will be, once Aesthetic's meaning lies on its uses, which usually are very diverse. The problem of considering it a simple contemplation of art resides in the fact that other instances or disciplines, such as Poetics, Criticism, and Art Theory, might interfere and create conceptual misunderstandings and a few methodology problems.

Aesthetics, as we call it today and as a field pursuing a space as discipline, was introduced by the German philosopher Alexander Gottlieb Baumgarten (1714-1762), as the "science of beauty" in 1735 (Souriau, 2010, p. 725). Opposing to *noêta*, facts of intelligence, and to *aistheta*, facts of sensibility, Baumgarten took the sensible as a central subject and thus made the discipline's autonomy possible. Aesthetics was only possible thanks to an epistemological turn, occurred in the 16th and 17th centuries, that changed the way we conceive the world and how it presents itself to our sensations. The emancipation of the visual arts and its attempt to join the liberal arts during the Renaissance contributed to the rise of the modern concept of art, or at least initially, of the arts.

In the 20th century Aesthetics faced countless obstacles, and many of them were related to changes in the way art was seen. Nothing seems classifiable and the art object, as a sensible and clearly delimited totality, is frequently set aside. In the beginning of the last century, Aesthetics influenced the work of philosophers as Benjamin and Adorno. Benjamin introduces the "loss of the aura" in the work of art in "the age of its reproduction", while Adorno claims that the art objects don't have an intrinsic essence[2] (Talon-Hugon, 2010, p. 89-92).

In the second half of the 20th century, Étienne Souriau (2010, p. 728-729) distinguished five variations of aesthetics, which we summarize as follows:

1. **Philosophical Aesthetics:** The earliest one, introduced by Plato and consisting of a reflexive and analytic method.
2. **Psychological Aesthetics:** Created in the second half of the 19th century and focused on understanding how art and ideals of beauty influence thought. It includes psychoanalytical aesthetics.
3. **Sociological Aesthetics:** Its main theorists are Marx, Taine, and Guayau, and it's focused on establishing a place for art in society.
4. **Comparative Aesthetics:** Remains exclusively focused on strictly aesthetic issues. If focuses on similarities, transpositions and mutual influences of different kinds of art.
5. **Morphological Aesthetics:** Uses the analytic and even the mathematic method. It's prospective and aims at unveiling inventive principles and ways of building structures in the interior of works of art.

Even though all these variations are, to a certain degree, imbricated, psychological aesthetics (addressing aspects related to cognitive dimensions and to "art-as-knowledge") and morphological aesthetics (focused on understanding form, in its broad and specific senses) seem to be the closest to issues regarding the interfaces between sciences (neurosciences) and artistic activities.

To conclude our thoughts on aesthetics as discipline, the author reproduces an introduction by Carole Talon-Hugon, whose definition is based on similarities, convergences and filiations of different concepts of aesthetics: "Aesthetics is the analysis of a certain field of objects where the terms 'beautiful', 'sensible' and 'art' are dominant." (2010, p. 4). Even though it sounds slightly vague, this assertion is interesting because it allows us to historically analyze Aesthetics from this triple base and unveils an interchange of these poles, as in Pareyson's definition of art. Thus, the core shifts from one concept to another: art, the sensible, and beauty. These three instances are always present, but never in balance. For example, the idea of beauty, as we'll see further in this work, hasn't been central since the 20th century, when ideas of art and the sensible became more important.

After these relatively long yet necessary considerations on issues regarding science, art, and aesthetics, we turn to relatively recent fields situated in this disciplinary "tension zone". We will attempt to establish connections between disciplines and to approach the expansion of the taxonomy of fields that were fairly stable until recently.

MAIN FOCUS: NEUROESTHETICS AND COGNITIVE NEUROSCIENCE OF ART

The name "neuroesthetics" reveals an attempt to include different theoretical fields which, since the decline of the Renaissance, became autonomous and tended to diverge due to epistemological issues. In the present day, these interdisciplinary branches are included in a dynamic center of thought focused on renewing traditional views and highlighting new understandings and perspectives about all kinds of *phenomena*.

The knowledges related to the application of scientific and biologic methods to the investigation of artistic works and processes are relatively old, even though not always under the name of "neuroesthetics". Neuroesthetics is the confirmation of this history and founded a few specific guidelines and nuances. It's hard to define its historic origins, due to the coexistence, in the last centuries, of scientific currents and punctual observations simultaneously focused on the analysis of aesthetic perspectives and biologic facts.

Neuroesthetic's recent origins were located by Jean-Pierre Changeux (2008, p. 97) in his own investigations, as well as in Alexander Luria's. Luria, a Russian neuropsychologist who was very active during the mid-20th century, conducted a series of investigations attempting to reveal details of the interaction between mind and brain in cinema, articulating film theory, psychology, biology and philosophy (Valisseva, 2013). Luria contributed to the rise of studies focused on the interfaces between arts and sciences and influenced important scientists and researchers, such as neurologist and writer Oliver Sacks.

We could also mention I. A. Richards, an important English literary critic, as one of the main precursors of neuroesthetics or even of neurosciences applied to arts. Richards advocates, in his 1926 essay "Science and Poetry", the use of psychology of instincts and emotions as an effective yet underdeveloped tool for textual interpretation (Ashton, 2011). The psychology of instincts is strongly related to evolutive biology and, therefore, to ideas that originated the neurosciences applied to arts.

Changeux defines neuroesthetics as a terrain focused on "raising questions on the neural bases of the contemplation and the creation of the work of art" (2013, p. 97). The author believes that congnitive activities, specially conscience and artistic practices, are related to a better development of the brain organization. Changeux also reflects on "neurohistory", including investigations on color, beauty, empathy, synesthesia, among others themes. The field called neuroesthetics includes all of these ideas.

However, it's not possible to discuss neuroesthetics without mentioning the British neurobiologist Semir Zeki, who coined the concept. Zeki's concept of "visual brain", defined as a specific region of the frontal cortex where the aesthetic experience primarily takes place, was established in the last decades with his particular interest on visual mechanisms. In the 1970s, Zeki conducted a pioneer study on specialized and overlapped "visual channels", located in multiple areas, in different regions of the cerebral cortex (Changeux, 2013, p. 118). The neurobiologist thus proposes two major laws of the visual brain: *constancy* and *abstraction*. The first one is related to the brain's ability to retain the object's constant and essential elements, while abstraction liberates it from memory limitations (Zeki, 2001, p. 51-52).

Besides the mentioned general definitions, which are related to an understanding of biologic and neurobiologic bases of the aesthetic experience, Zeki places art as the extension of an essential brain function: knowledge acquisition. In the description of neuroesthetics provided in his website[3], the scientist does not refuse subjectivity, but expresses his wish to articulate it with a common neural organization that allows us to communicate through art[4]. His investigations are guided by questions on how ordinary visual activations can provoke emotions. Besides that, Zeki also argues, in his website, that celebrated artists could be, to a certain extent, successful neuroscientists who instinctively understand common emotional and visual organizations, as well as how the brain works. Zeki's investigations often refer to themes as beauty, neutrality and ugliness. Based on the registry of brain activity through advanced methods, he analyzes responses, their intensity, and the regions in the cerebral cortex where they take place. Different painting categories are analyzed and reactions (high and low frequencies of brain activity) to those considered "beautiful" or "ugly" are registered. High brain activity in response to beautiful images would represent a reward reaction. The scientist also explains, in his introduction to his website, the great expectations usually attributed to neurosciences, even more than to arts. Zeki believes that the neurosciences are the appropriate field for the achievement of a proper understanding of human nature. He also believes that, by understanding neural laws we can also understand all kinds of human activity (morality, religion, law, etc.).

Regarding the idea of beauty, critics usually claim that neuroesthetics is related to the ways by which scientists handle this concept. This theme is also related to a need for defining certain parameters and, for that reason, specialists from this field are constantly looking for the neural bases of beauty. In an attempt to dialogue with aesthetic and philosophical terrain, they focus on a discourse about beauty that is no longer popular among art and philosophy thinkers. Semir Zeki, for example, tries to find in Kant a perspective that would be appropriate to his own scientific investigation: a more stable idea of beauty[5]. Beauty as discussed by Kant – something that pleases universally – can easily fit into neuroesthetic purposes. However, a brief look at the history of this concept reveals otherwise.

Aesthetic ideals vary, alternate, change, and coexist from time to time. Western art history clearly shows us the diversity in this concept. Beauty has been associated with symmetry, reason, light, color, romantic ideals, technologic ideals, consumism, etc. And all these associations are still partially true. The pursue of neural bases of beauty, in an attempt to universality, usually tend to become a failed investigation, if we consider the wide range of types of "beauty" created by historic and geographic situations. The search for universal principles in the work of great masters of the past (like some neuroesthetics theorists do) seem to be the search for an universal ideal or some static entity in the interior of an extremely slippery, fluid concept. Therefore, it might be problematic to address established beauty as empirical experiment. Early in the 19th century Gustav Fechner – somehow precursor of neuroesthetics with his experimental psychology – already knew that the appreciation of beauty was inherently subjective and that such analysis should serve as background to all other investigations (Conway; Redhing, 2013). A "discovery" of universal beauty, by means of the intensive use of neuroimaging technologies, could become a normative and fruitless idea to artistic appreciation itself. According to Conway and Rehding, scanners that reveal areas of higher activation in a certain region of the cerebral cortex (especially the medial orbitofrontal cortex, mOFC) are actually measuring preferences, and not the beauty of the objects observed (*idem*). Futhermore, studies reveal the existence of highly creative people who suffered damages in this region, often regarded as essential to artistic perception (*ibidem*). The technology applied to the comprehension of neural mechanisms would, then, bear certain analytic pitfalls. Even aesthetic components in color images of brain processes can, many times, suggest particular interpretations that, under the influence of the scientific method, could succumb to the desire for regulation.

Still regarding beauty, science usually tends to relate it to the idea of art. As we already argued, this idea has long been rejected by the art field and is almost inexistent in the current artistic discourse. "Beauty" is definitely not an issue for the aesthetics developed in the last decades inasmuch philosophical analysis. For example, neuroesthetics theorists erroneously identify artistic appreciation as beauty appreciation. This common association tends to corroborate the idea (usually revealing some prejudice from the arts and humanities) that scientists are naive and conservative. A brief look at the choices made by some of the most representative theorists from neuroesthetics, or neuroscientists focused on art, in terms of the type of art analyzed would show otherwise. Ramachandran often focuses on Indian art, Zeki on European modernists and Kandel on the study of Vienna Secession, an art movement from his hometown. While it might be natural that people are guided by familiar patterns, subjective convictions might also influence the results of their investigations. This could lead to a universalization of particular value attributes, according to Conway and Rehding (2013).

As we analyze the history of definitions from neuroesthetics (in the work of its precursors or not), we realize that this terrain is more hybrid than its name suggests. It combines principles from perception psychology, evolutive biology, phenomenology, anatomy and brain function (in all its branches), radiology applied to neuroimaging, and obviously philosophy, aesthetics, and art criticism and theory. Likewise, we can notice that even though some researchers working in the interfaces between art and science could call filiate themselves to neuroesthetics, they don't. Such posture is probably due to their opposition to epistemological concepts from neuroesthetics that are often criticized.

In order to discredit neuroscientific studies applied to art (or neuroesthetic studies), John Carey, a famous literary critic, compared neuroscientist Vilaynur Ramachandran and his colleague, philosopher of mind William Hirstein, to comedians Laurel and Hardy (Hirstein, 2012). Carey believes that neurosciences could offer nothing but a limited perspective of art[6]. One might question the radical character of the British professor's criticism regarding scientific studies focused on art (in this case, poetry). However, we can't deny that it reveals a certain discomfort in the pretense encounter of neurosciences and art practices. The reductionist perspective mentioned by Carey results from the fact that science does not admit the coexistence of given truths, while art can (and should) allow the existence of different truths in terms of values, practices and appreciation.

The belief in science, with a limiting perspective of the artistic phenomenon, can also be explained from other perspectives. Science generally reflects a certain obsession humans have for mastering and controlling objects. We must standardize and, therefore, reduce. A mountain turns into a cone; the moon turns into a sphere. Pleasure, whether in appreciation of art or drug use, turns into a reward in the limbic system. But the degrees of complexity of human beings, of things and their processes tend to keep scientists at a certain distance from the object analyzed. "Science manipulates things and refuses to habit them" is the first sentence in Marleau-Ponty's *Eye and Mind* (2004, p. 13). The phenomenologist questions the fact that science sees all beings as "general objects", "as if they meant nothing for us and were, thus, subject to our schemes".

Scientific reductionism can also turn into the common scientific practice of applying similar research models to different situations. "When a model is successful in a set of questions, science applies it to all other situations" (Merleau-Ponty, *idem*). That explains modern science's tendency to follow trends and the proliferation of "neurodisciplines", such as neuroethics, neurosemiotics, neurophenomenology, neuroesthetics, and neuromarketing.

But how can we practice science without establishing standards, without pointing what's right and what's wrong? The refusal of judgement might imply subjection and, therefore, the risk of discreditation. It's hard, in science, to avoid the right-and-wrong pitfall. It's easy to succumb to the desire for standards. Since this desire is almost a rule, standardization tends to create some tension when it approaches fields like art, which doesn't even have clear limits.

Even though artists usually work according to a wide range of rules and techniques, the regulation of practices and artistic values and goals can disturb them immensely. Phillip Ball (2013), writer and former editor for the *Nature* magazine, recalls, for example, Paul Klee and his interpretation of a study on the use of color, written by Wilhelm Ostwald, amateur painter and Nobel prize winner in Chemistry. Ostwald praises the possibility of creating harmony with balanced tones and criticizes the "incorrect use" of blue in a Ticiano's painting. For Klee, Bauhaus professor and famous colorist, the scientist's normative thoughts represent the absolute abandon of the artist's rich and complex soulfulness.

Another important issue that might represent a critique to neuroesthetics, or to neurosciences applied to art, is the misunderstandings regarding causes, effects and meanings of the work of art. These scientific investigations fail to distinguish the effects of art from its meaning. There is a tendency to deny the interpretative dimension of these researches, because they believe they have a better understanding of brain processes, evolution mechanisms and the ways through which environment and body format these processes (Ashton, 2011). Neurosciences are relatively successful in their analysis of causes and effects of the work of art. However, the interpretations, the broader understandings, depend on responses that are more closely related to meanings and senses. In other words, most of the neuroscientists believe, equivocally, that the discovery of causes and effects equals the discovery of what the artwork means.

Furthermore, according to Jennifer Ashton (*idem*), neuroscientists also tend to mistake meaning for feelings inspired by the artwork, during its conception or its appreciation. For example, different paintings can have similar effects and very distinct meanings. The identification of how and where in our brains the aesthetic responses take place does not tell us what they actually are or mean. This is, hence, a background, epistemological question that neurosciences address with difficulty. Thus, the field gets subject to fierce and justifiable criticism.

The initial arguments regarding neuroscience's limited perspective towards art, especially the neuroesthetic's, is appropriately validated by arguments related to epistemologic and conceptual questions that rise from the tension between scientific and artistic/humanistic fields. The practice of neuroesthetics, at least as conducted by the most frequently mentioned theorists, can easily provide us, almost naturally, with arguments that question the field's epistemologic foundation. By reading the texts that founded the field one can quickly identify an apparently exaggerated expectation regarding discoveries and the future of the discipline. For example, Semir Zeki highlights, in the introduction to his definition of neuroesthetics in his website, the belief in the possibility of defining art. According to him, that would happen if we could incorporate references to our brains (which is responsible for art creation, execution and appreciation). However, the polissemy inherent to the concept of art is the condition to its very existence. And we also must highlight the fact that when neuroesthetics theorists discuss art, they are actually limited to visual arts. This is another example of how exaggerated science's expectations probably are: In order to respond to the criticism aimed at neuroesthetics, William Hirstein claimed that it's possible to understand the processes of artistic creation due to the fact that the brain is an entirely physical system[7]. Even though we can reveal with higher precision the plasticity of neural networks and complex brain processes through sophisticated technology and sagacious methods, we are still not able to ensure the real apprehension of what the work of art might mean.

Therefore, we notice that the research field focused here can be subject to endless criticism, which is justified by certain authors' perspectives which we tried to include in this text. Most of these comments are basically an understanding of the simplification with which scientists approach the art field. While, on one hand, the demonstrations and understandings of brain processes related to art are hugely complex, on the other hand they tend to a reductionist perspective of the background concerning the field's own epistemologic basis.

However, it's important to notice that this epistemologic tension and the uncomfortable situation and conceptual problems caused do not annul the importance of results and discoveries produced by neurosciences (especially neuroesthetics) in its dialogue with artistic facts.

Cognitive Neuroscience of Art

In the beginning of this millennium, the neurosciences have become broader and established dialogues with different knowledge instances, whether hard sciences or humanities. Not surprisingly, the art field faces a series of tensions and problems (as those mentioned here and related to neuroesthetics) in its dialogue with sciences. The criticism that denounces neurosciences' reductionist perspective regarding art is not limited to neuroesthetics. The same happens with the mutual mistrust between, on one hand, artists, philosophers, and art theorists and, on the other hand, neuroscientists. This criticism aimed by philosophers and art theorists at reductionist perspectives frequently rises from methodological questions proposed by the neurosciences and its attempts to explain aesthetic dimensions of art. These neuroscientific explanations about art are rarely sufficient to address all the aesthetic properties of an object and the experiences related to it (Seeley, 2013). However, we must remember that any approach to an artwork (or to most objects and events) in a specific disciplinary field is necessarily limited to a mere perspective of a multifaceted object.

Now the author would like to turn to some definitions and discussions on the cognitive neuroscience of art, a field that, despite a few similarities, represents an alternative to neuroesthetics.

According to William P. Seeley's insightful definition (2011), cognitive neuroscience of art is a branch from empiric aesthetics that is focused on applying neuroscientific methods in order to study our relations with artworks. These methods come from a broad research scope that approaches all of our psychologic relations with the world. Through its contact with humanities (especially philosophy), neurosciences can, among other things, dialogue with philosophical theories that approach the entirety of our relations with artworks and artistic practices.

Empirical aesthetics, which, according to Seeley (*2011*), created the cognitive neuroscience of art, was originated in the 19th century with Gustav Fechner, who, in 1871, wrote "On experimental aesthetics"[8]. Fechner was a rigorous scientist, filiated to the monist philosophical current. He also wrote poems and comedy plays, such as "*Vergleichende Anatomie der Engel*" (Comparative anatomy of angels). In this play from 1825, Fechner gives us a detailed description of men and then elaborates a series of analogies relating them to angels' spiritual states. Fechner's creativity, empiric perspective, methods, and experiments later made him the father of future cognitive neurosciences of art.

Even though art is not limited to rules and guidelines, artistic develop glossaries related to specificities and general subjects from their backgrounds, along with composition strategies that produce behavioral responses in their public (or consumers, according to cognitive neuroscientists)[9]. Artworks are not limited to guidelines, but most of them follow internal rules which are inherent to the singularity of the object or to certain styles and genres. These are rhythmic patterns in a broad sense, internal narrative agreements, ways of handling space. Ultimately, they are elements which are inherent to the poetics of a given artwork or a certain group of artworks.

Cognitive sciences consist of a group of scientific disciplines that focus on the description, explanation and, in some cases, simulation of mechanisms of though as well as the whole system of information processing, regarding the acquisition, retention, use, and transmission of knowledge (Miller, 2003, p. 144). According to Miller, cognitive sciences include six disciplines: philosophy, neurosciences, informatics, psychology, linguistics, and anthropology.

Hence, cognitive neuroscience of art can be defined as the study of how art consumers acquire, represent and manipulate the information that is present in the structure of an artwork. It is the set of *stimuli* intentionally projected in order to produce emotional, perceptive, and cognitive responses (Seeley, 2011). In this context, the work of art is seen as a form. This idea is related to the work of philosopher Luigi Pareyson, mentioned earlier in this work. According to Pareyson, the concept of form is appropriate to the work of art, and "formativity" qualifies the artistic activity[10]. Therefore, artworks consist of forms. However, they are live forms, "organisms that are autonomous and independent, like those from nature" (as previously quoted in this text).

Thus, cognitive neuroscience of art is seen as a tool for modeling the artwork consumer's processes and behaviors (Leote; Oliveira; Baraúna, 2015; Seeley, 2011). So, when we understand the analysis of the work of art as an attempt to explain how it works, the definition of this neurosicence as a tool is very appropriate. This modality of cognitive science is, then, presented as a potential "key" for understanding the nature of art and the aesthetic experience. And the author believes this key is consciously situated with others that already exist and have long been applied. Such is the case of those implicitly aggregated to art theories, music theories, film analysis, discourse analysis, phenomenology, etc.

In comparison to neuroesthetics, cognitive neuroscience of art tends to include a wider range of possibilities, due to its desire to comprehend the art phenomenon and its multiple expressions. Meanwhile, neuroesthetics, following the paths created by precursor Semir Seki, is usually limited to visual perception. Besides that, the epistemologic positioning of the cognitive neuroscience of art solves an important problem of neuroesthetics. It's specifically focused on art and not on aesthetics, while the latter "does not encompass non-aesthetic semantic or ontological questions about the nature of art or our engagement with artworks" (Seeley, 2011). We get, then, to the distinction between art philosophy and aesthetics. Issues related to the nature and onthology of art belong to art philosophy, while aesthetics, as the science of sensible knowledge, includes other questions than those related to art. Since philosophy is considered a cognitive science, it must naturally dialogue (in order to approach the art field) with neurosciences and other disciplines that might be involved (psychology, linguistics, anthropology, among others). We should also notice that semantic elements of the artwork have regulating functions in our neurophysiologic systems, and sometimes it defines the aesthetic character of our relation with specific artworks (*Seeley, 2011*). Even though the problem of signification is inherent to art (and that's why it's so difficult to define it as language), the semantic question plays a central role in the relation between the artwork and its public.

Mark Rollins suggests the notion of artworks as entities which are full with exogenous forces that are preponderant in the process of information transmission (2004, p. 101). Artworks must be seen as "atentional strategies" – an idea that, somehow, is in line with Pareyson's concept of artwork as form and of the artistic activity as formativity. At the same time, the expresive/communicative aspect of art is highlighted, with emphasis on the artwork, and not on the artist. Somehow, art becomes something that communicates and informs. In this context and with the rise and consolidation of new technologies, we can imagine, as does Wilson (2001, p. xxi), that the arts inform and enrich scientific investigations, and vice versa.

Once the basis of this new and specific modality of neuroscience is well established, we will obviously face methodological issues, due to the vastness and the complexity that are inherent to the relation between artwork and consumer. This problem becomes even more complex when we refer to the art definitions presented in this text and realize that we can only define what is sufficiently determined – which is not the case of the works of art, that consist of aesthetic states. However, isn't science the practice through which one extracts order structures from chaos? Then, we must find ways (μέθοδος, methods, *hodos* = way) that allow us to deal with the relations between semantic contents and aesthetic states presented in artworks. Besides that, it's necessary to achieve certain theoretical models or pertinent explanations of the nature and the onthology of art.

Other researchers besides Seeley see the cognitive neuroscience of art as a more appropriate alternative than neuroesthetics for the comprehension of perceptive processes in art (Leote et al., 2015). The first one seems to nurture more realistic expectations towards the future of this field of investigation. Cognitive neuroscience of art could, for example, confirm our best theories about artworks and enrich our philosophical practices (Seeley, 2011). It could also, maybe, help us understand the mechanisms for art perception and reveal onthological aspects of the artistic activity.

CONCLUSION

The analysis of the criticism aimed at the neurosciences with focus on art, presented in this text, especially in the section that discusses neuroesthetics, reveals a series of tensions in the contact between arts and sciences. These tensions rise, in most of the cases, from mutual ignorance regarding background issues that guide each field. Terminological issues and conceptual clashes represent a substantial problem in Zeki's neuroesthetics. In this discourse, two different fields with distinct ideologies clash, and one of them becomes subordinated to the other. Anyway, the comprehension of the creative processes and functions/functioning processes of our brains necessarily encompasses the comprehension of perception and reward processes, as well as neurologic and psychologic questions, among others. And these might be interesting questions for theoretical fields and art making.

Rosangella Leote argues that the art field needs to understand natural processes frequently investigated by science, "and use this situation in favor of actions focused on art, in any kind of experience" (2012, p. 164). The artistic production can be enriched by investigations developed by the neuroesthetics, the cognitive neuroscience of art and, more generally speaking, by the neurosciences focused on art. It can also contribute to these theoretical fields – not only through the attenuation of the borders between art and science, but also through a fruitful dialogue of mutual enrichment.

The consolidation of these new research fields as fertile terrains for knowledge and action requires the development of a comprehension of how art consumers generally acquire, represent, and manipulate the information produced by the formal structure of *stimuli*. That would allow a more comprehensive understanding of perception mechanisms and the modeling of processes and behaviors (if pertinent). That means to keep tracking the paths created by Gustav Fechner back in the 19th century, suggesting readings and models with scientific rigor, avoiding static categories and, besides that, remaining aware that clarifications about the artwork's signification processes do not represent the meaning itself. The comprehension of the ways through which mind works, of evolutive developments and of restrictions imposed by the environment and the human body on these processes allows us to elaborate a rich panorama of our relation with art (Ashton, 2011).

The results of efforts made by researchers from the neuroesthetics and the cognitive neuroscience of art will probably be better understood and received by the academy once they offer serious investigation of our aesthetic relations with works of art. They also must offer effective instruments for the analysis of artworks and styles, so that it becomes possible to understand how different kinds of art work from the biologic perspective.

Despite the epistemological differences between sciences and humanities/arts, one might say, from a scientific point of view, that the discovery of perceptive processes and the comprehension of the structural functioning of works of art offer interesting possibilities. Several results of investigations on senses and arts are emerging and will continue to be published in the future. Consequently, several fields of knowledge will fructify. From the perspective of humanities – especially the aesthetics – and arts, the dialogue with science could offer incitement and new ways of seeing, hearing and experiencing art.

REFERENCES

Ashton, J. (2011). *Two problems with Neuroesthetic Theory of Interpretation.* Emory College of Arts and Sciences: nonsite.org. Available from http://nonsite.org/issues/issue-2/two-problems-with-a-neuro-aesthetic-theory-of-interpretation

Ball, P. (2013). *Neuroaesthetics is Killing Your Soul: Can Brain Scans Ever Tell Us Why We Like Art?* Nature Publishing Group. Available from http://www.nature.com/news/neuroaesthetics-is-killing-your-soul-1.12640

Bergeron, V. (2011). *What Should We Expect from the New Aesthetic Sciences?* Aesthetics: American Society for Aesthetics. Available from http://www.aesthetics-online.org/articles/index.php?articles_id=55

Carey, J. (2010). *What Good Are the Arts?* Oxford, UK: Oxford University Press.

Changeaux, J.-P. (2013). *O verdadeiro, o belo e o bem.* Rio de Janeiro: Civilização Brasileira.

Conway, B. R., & Rehding, A. (2013). Neuroaesthetics and the Trouble with Beauty. *PLoS Biology, 11*(3), e1001504. doi:10.1371/journal.pbio.1001504 PMID:23526878

Danto, A. C. (2014). *After the End of Art: Contemporary Art and the Pale of History.* Princeton, NJ: Princeton University Press.

Freitas, A. S. (2015). *Rencontre des arts: correspondances entre œuvres sonores et visuelles au xxe siècle.* Paris: Harmattan.

Gombrich, E. H. (1951). *The History of Art.* New York, NY: Phaidon Publishers Inc.

Hirstein, W. (2012). Neuroaesthetics: Responding to the Critics. *Psychology Today.* Available from https://www.psychologytoday.com/blog/mindmelding/201212/neuroaesthetics-responding-the-critics

Kuhn, T. S. (1996). *The Structure of Scientific Revolutions.* Chicago, DC: The University of Chicago Press. doi:10.7208/chicago/9780226458106.001.0001

Latour, B. (1987). *Science in Action.* Cambridge, MA: Harvard University Press.

Leote, R. (2012). Multisensorialidade e Sinestesia: Poéticas Possíveis? *Proceedings of 6th International Conference on Digital Arts, ARTECH.*

Leote, R., Oliveira, H. C., & Baraúna, D. (2014). *Teoria neurológica da experiência estética de Ramachandran e Hirstein: apresentação, comentários e proposta de aplicação.* Paper presented at the *5º* Encontro Internacional de Grupos de Pesquisa: Realidades Mistas & Convergências entre Arte, Ciência e Tecnologia. Escola de Artes e Comunicações ECA-USP, São Paulo.

Miller, G. A. (2003, March). The Cognitive Revolution: A Historical Perspective. *Trends in Cognitive Sciences, 7*(3), 141–144. doi:10.1016/S1364-6613(03)00029-9 PMID:12639696

Pareyson, L. (2001). *Problemas da Estética.* São Paulo: Martins Fontes.

Plaza, J. (2003). *Arte/ciência: uma consciência.* Revista Ars, Universidade de São Paulo.

Rollins, M. (2003). Pictorial Strategies and Perceptual Content. In H. Hecht, R. Schwartz, & M. Atherton (Eds.), *Looking Into Pictures: An Interdisciplinary Approach to Pictorial Space* (pp. 99–112). Cambridge, MA: MIT Press.

Seeley, W. P. (2011). *What is the Cognitive Neuroscience of Art ... and Why Should We Care?* American Society for Aesthetics – Aesthetics on-line. Available from http://www.aesthetics-online.org/newsletter/31.2.pdf

Snow, C. P. (2012). *The two cultures*. Cambridge, MA: University Press. doi:10.1017/CBO9781139196949

Souriau, E. (2010). *Vocabulaire d'esthétique*. Paris: Presses Universitaires de France.

Talon-Hugon, C. (2010). *L'Esthétique*. Paris: Presses Universitaires de France.

Vassilieva, J. (2013). Eisenstein/Vygotsky/Luria's project: Cinematic Thinking and the Integrative Science of Mind and Brain. *Screening the Past Publications, 38*. Available from http://www.screeningthepast.com/2013/12/eisenstein-vygotsky-luria%E2%80%99s-project-cinematic-thinking-and-the-integrative-science-of-mind-and brain/

Vieira, J. A. (2007). *Teoria do conhecimento e arte*. Fortaleza: Lab. Estudos da Oralidade.

Wilson, S. (2002). *Information Arts: Intersections of Art, Science and Technology*. Cambridge, MA: The MIT Press.

Zeki, S. (2001). Artistic Creativity and the Brain. *Science Magazine, Essays on Science and Society, 293*(5527). Available from http://www. sciencemag.org/content/293/5527/51.full

KEY TERMS AND DEFINITIONS

Art: The main definition of art used in this chapter is by Luigi Pareyson, whereby the concept of art is structured on three historically recurrent definitions: art as action, as knowledge and as expression.

Cognitive Sciences: A group of scientific disciplines that focus on the description, explanation and, in some cases, simulation of mechanisms of though as well as the whole system of information processing, regarding the acquisition, retention, use, and transmission of knowledge.

Critical Theorists: Theorists that see science as the modern disillusionment. They point out to underestimated social forces and metanarratives, though they structure questions and paradigms in the scope of scientific investigations.

Empirical Aesthetics: Empirical study of esthetical proprieties of art.

Epistemology: A branch of philosophy concerned with the nature and scope of knowledge.

Formativity: Term coined by Luigi Pareyson to qualifies the artistic activity.

Science: Science pursues ways of effectively acting upon reality and, therefore, builds conceptual models or representations that reflect, with some isomorphy, aspects of the world's objective organization.

Visual Brain: Defined by Semir Zeki as a specific region of the frontal cortex where the aesthetic experience primarily takes place.

ENDNOTES

1 "Posthistoric" is the term used by US phylosopher and critic Arthur Danto (2014) to discuss the "end of art", which is actually the end of a master narrative that guides Western art and embodies certain homogeneity.

2 This is a reference to Walter Benjamin's famous text: "The Work of Art in the Age of Mechanical Reproduction", published in 1936.

3 <http://neuroesthetics.org/statement-on-neuroesthetics.php>

4 No discipline can actually be clearly and statically delimited. However, art seems to question disciplinary borders even more fiercely, especially in the present day.

5 The reference to Kant in presented in his website and also in an interview conceded to Brazilian newspaper *O Globo*, in the section *Prosa e Verso* of the edition from 12/30/2006, available at: http://www.umsl.edu/~carrolljc/Documents%20linked%20to%20indiex/darwin4%20%281%29.pdf

6 John Carey's critique can be found in: Carey, John. *What Good Are the Arts?* Oxford: Oxford University Press, 2010.

7 See https://www.psychologytoday.com/blog/mindmelding/201212/neuroesthetics-responding-the-critics.

8 Vincent Bergeron also mentions Gustav Fechner, but associates him to neuroesthetics, as an activity that investigates where and how different components of our aesthetic responses are implemented in our brains (2011, p. 6). Bergeron's idea of neuroesthetics is very similar to the definitions of cognitive neuroscience of art mentioned in this text.

9 The author believes the term "consumer", often mentioned by William Seeley, can also be applied to artistic situations of interactivity. The author prefers the terms "spectator" and "interator".

10 "Formativity", according to Pareyson, asserts the process, the dialog between the artist and raw materials, as well as the work's driving force over the artist himself/herself.

This research was previously published in Projective Processes and Neuroscience in Art and Design edited by Rachel Zuanon, pages 71-86, copyright year 2017 by Information Science Reference (an imprint of IGI Global).

Chapter 4
Understanding the Interdisciplinary Meaning of Beauty to Neuroscience:
Designing Beauty to Neuroscience

Bruno H.S. Araujo
Universidade Federal de São Paulo, Brazil

Ibrahim Elias Nasseh
Universidade de São Paulo (USP), Brazil

ABSTRACT

The search for the source of beauty has exercised the speculation of philosophers and writers throughout the ages. The classical conception is that beauty consists of an arrangement of integral parts into a coherent whole, according to proportion, harmony, symmetry, and similar notions. The opposing view is that art is a fully subjective enterprise and our preferences are shaped by our values and experiences. To study beauty, neuroscience evokes the juxtaposition of concepts such as symmetry, geometry, proportion, judgment, affection, grotesque, erotic, sensitive with certain loci in the human brain. However, to explain and understand the aesthetic experiences at the network level, neuroscience have to incorporate the ambiguity in their research field. And with the continuous increasing of knowledge and comprehension of how beauty is perceived by the human brain, the world in its wholeness will be highly impacted, since we will start to live in pleasant societies.

INTRODUCTION

To deliberate about beauty, we first need to contemplate different types of considerations and approaches that, in history, were indorsed on this subject. Therefore, in an epistemic census, the key is to establish the possible heuristics endeavor and to elect the "epistemic figures"[1] of sense and possibilities for a new understanding.

DOI: 10.4018/978-1-5225-5478-3.ch004

The theory of beauty with universal principles was attempted at the first time by Plato (Panofsky, 2000), whose established separation between art and beauty, in opposition rules. To Plato, the beauty, has a completely existence, independent and higher than the physical reality, however it also has a universal true value, what make it undeniable. In this way, beauty cannot be defined or held in any touchable manifestation. On the other hand, for Plato, art was *mimesis*, imitation and therefore unchangeable. It had, consequently, the same calling for misunderstanding and confusion that the sensations have on us. The true beauty could only be contemplated by the intellect. It is the Plato's theory of Ideas, a conception of the universe in its wholeness, which was mobilized to give logical basis to the concept of beauty.

Some reconciliation between beauty and art only came up from Cicero who was the first to allow the artist the ability to create superiors forms more than simple copying or *mimesis* the nature. The true artist's model is inside, by the contemplation of his own intellect. He can thus correct the "mistakes" of nature and create a perfect model of beauty. We need to keep in mind that such reconciliation takes place, however, through a distortion of the platonic concept of art and idea (Panofisky, 2000). In Cicero and over the antiquity, the transcendental Plato's doctrine of idea, as harmonic and changeless principle of the universe, and the art as a simple sensitive and objective figuration (in the sense that its reality come true in physical objects like paints or sculptures, so transients and mortal) gave raises to opposite paradigms which art releases itself from his eminently sensible externality to became the achievement of a mental interiority. In this way, the idea falls from its pedestal of transcendental essence to be a thought or internal concept in human immanent consciousness.

In Panofky (2000), the Plato's concept of idea and its development over the time is the main topic to explain the split of the historiography of beauty into certain ages. But although we can see the history of art and beauty through this prism as like evolutionary, which could invoke an Apollonian ideal, harmonic or positivist - whether we would take a most appropriate term to some philosophy of science - we also can see the history of beauty in an opposite sense. This sense is expressed as Dionisiac, chaotic and open.

The categorizing effort, which historiography is a lessee, perhaps can be understood through the key of the critique of historical culture on which Nietzsche weaves fierce criticism.

In this sense the historiographical anachronism does not recognize the becoming[2] as a stream in which all meanings transit. In this sense, the historiography demand to find the same or the identity under the otherness. This goal in such classificatory approach which in the same time excludes and imprisons the hallmarks of a unique kind of art.

Thus,

... the historiography is bogged down in the furrows that it knocks: the scholar is surprised when comes across the reference to "medieval´s" authors in Renaissance texts; so the reviews Ghiberti, dealing with optics, brings authors such as Al-Hazen ... (Novaes,1994,p.10).

Meanwhile, from this concept of genealogy of Nietzsche - which term he uses to conceptualize the authentic "historical spirit" – what happens is the reverse of historiography offers.

(genealogy) will pretend, in first place, to determine the differences instead of to forge identities, it will be attentive to mutations of meanings and suspicious of concepts supposedly unambiguous. Thus, it will investigate the history with no claim to find there the realization of any eternal ideal, thus it will not put our present there at the origin, as if there was a predestination to be performed in a prior sense

to be deployed. It shows that history is a succession of disparate meanings, without any predetermined unification, a succession of interpretations, dominations that alternate. (Moura, 2005, p 144).

Thereby it is the dispute of power by itself that moves the historical development and there is no supra-historical morality that determine it, least of all retain it in a certain context that eternize in some closed periods, being much more like a coming and going of non-perennial and dialectical realities. This concept of game was precisely a main subject to Heraclitus - appropriated by Nietzsche - which decouples the reason as a necessity in history and starts to think on your own contests and your own disparity.

Placed this view of concepts of art and beauty in a problematizing outlook, the new conceptual field that neuroscience brings can invoke the juxtaposition of concepts such as symmetry, geometry, proportion, judgement, affection, grotesque, erotic, sensitive - and many other significant concepts to the genesis and development of what is meant as beauty - with certain *loci* in the human brain, or to certain mathematical relationships, or even to certain universality. Neuroscience not only does, but also settles the method and the necessary conditions, without which the science would never reach the necessary rigor of the study of its object nor the success of your enterprise. We shall see how this happened in neuroscience in the forward paragraphs.

It is part of the modus operandi of science reduce and abstract objects, decomposing them - just like recommending Descartes in his Discourse on the Method (Descartes, 1973). Thus, topics such as art and beauty will be more resisted to this kind of approach. This resistance may benefit science if it remains opens to the knowledge from other disciplines such as philosophy and conceptual ideas created by the artists themselves. This resistance increases progressively throughout history and set requirements for the universal concept of beauty becomes progressively difficult. This may occurs most acutely after modernity and avant-garde art where what is considered beautiful seems to collapse and the very term beauty seems to be emptied within the philosophical thought itself as seen in Dewey, Collungwood and Lyotar (Jonhson, 2010).

But, what in art and beauty may seem a particular case of knowing, in several other scopes it has been a rule. In the core of the interdisciplinarity, we will testimony the increasing emerge of this new-old way of knowing, as a sort of comprehensive and genealogical epistemology.

How many different areas and cognitive resources, like geographic, economic, political, diplomatic, physical, engineering, were vital to mobilize for the construction of the European Council for Nuclear Research (CERN), the largest particle collider in the world?

Substantiating in the whole idea of interdisciplinarity and even transdisciplinarity, many of the concepts about the beauty that science employ, were provided by philosophy and even the art itself. In this way, the art as it was understood in its classical period provided to science epistemic figures of which beauty was comprehended as proportion and harmony, that furnish the basis of the prime researches on the beauty issues (Umberto, 2004).

THE CLASSIC CONCEPTION OF BEAUTY

During an aesthetic appreciation, in general, what is analyzed is the physical properties of the object. However, as we know, appreciation cannot only be reduced to this quantitative criterion, since holistic contemplation attributed to the object is also important (Walker, 2006). A biologist should instead suggest that the objective[3] side of beauty stems from built-in predispositions for certain features, colours, shapes or proportions. The opposing view is that art is a fully subjective enterprise and our preferences are shaped by our values and experiences. The real answer is likely to lie somewhere in the middle – after all, new art apprentices learn basic common skills such as proportion, perspective and symmetry before represent their overall vision of the world (Bayles & Orland, 2007). In this account, beauty is at least in part objective[3].

The classical conception is that beauty consists of an arrangement of integral parts into a coherent whole, according to proportion, harmony, symmetry, and similar notions. This is a primordial Western conception of beauty, and is embodied in classical and neo-classical architecture, sculpture, literature, and music wherever they appear. Aristotle says in the Poetics that

to be beautiful, a living creature, and every whole made up of parts, must … present a certain order in its arrangement of parts" (Aristotle, volume 2, 2322 [1450b34]) and in the Metaphysics: "The chief forms of beauty are order and symmetry and definiteness, which the mathematical sciences demonstrate in a special degree (Aristotle, volume 2 1705 [1078a36]), (Sartwell, 2014).

This view, as Aristotle implies, is sometimes boiled down to a mathematical formula, such as the golden ratio.

It is important to note that the idea of 'symmetry' in ancient texts is richer than its current implication of bilateral similarity, though it incorporates that as well. It also refers to the sorts of harmonious proportions characteristic of objects that are beautiful in a classical sense.

Vitruvius, the ancient Roman architect, gives as good a characterization of the classical conception as any, both in its complexities and in its underlying unity: Architecture consists of Order, which in Greek is called taxis, and arrangement, which the Greeks name diathesis, and of Proportion and Symmetry and Decor and Distribution, which in the Greeks is called oeconomia. (Vitruvius, 1970).

Order is the balanced adjustment of the details of the work separately, and, as to the whole, the arrangement of the proportion with a view to a symmetrical result.

Proportion implies a graceful semblance: the suitable display of details in their context. This is attained when the details of the work are of a height suitable to their breadth, of a breadth suitable to their length; in a word, when everything has a symmetrical correspondence.

Symmetry also is the appropriate harmony arising out of the details of the work itself: the correspondence of each given detail to the form of the design as a whole. As in the human body, from cubit, foot, palm, inch and other small parts come the symmetric quality of eurhythmy. (Vitruvius, 1970)

The golden ratio (1.618) has been studied widely and has attracted interest like no other number in mathematics. It is believed by many to be the basis of aesthetic quality in the visual and auditory arts and has been employed in architecture, painting, music, and industrial design (Livio, 2003).

In one of the oldest empirical and most classic studies in psychology, Fechner (1876) presented a range of rectangles whose width-to-height ratios varied from 1:1 to 2:5. and he found that observers preferred those near the golden ratio. On the order hand, Davis and Jahnke (1991) used squares, rectangles, circles, ellipses, trapezoids, and parallelograms and they found no preference for the golden ratio. Instead, they observed a strong preference for shapes whose sides were in a 1:1 ratio, even when these were pitted directly against golden ratio shapes. They conclude that symmetry is the driving factor behind preference of geometric shape. Other studies support this conclusion. Höge (1995) had participants draw "beautiful" quadrangles and found more of them corresponded to square-like proportions. Boselie (1984) created a variety of different geometric patterns, including kite shapes and star-like shapes. In one version of these patterns there were multiple 1:1 ratios of side lengths. These ratios were absent in corresponding shapes. In the majority of cases, the shapes with more unity ratios among their side lengths were preferred.

In addition to the golden ratio, several other numbers have been proposed to have special status with regard to aesthetic judgments. One of these is the plastic number (1.321). This number emerges from the Padovan series created by three ones and then generating each additional digit by adding those that come two or three digits back (1, 1, 1, 2, 2, 3, 4, 5, 7, 9, 12, 16, 21, etc.). The ratios of adjacent number pairs in this sequence converge on the plastic number. Stewart (1996) argues that the plastic number should be used as a design principle. Another ratio very close in value is the Pythagorean number (1.333) formed by the ratio 4:3, sometimes known as the Pythagorean triple. Shortess et al (1997) took measurements of art paintings, art supplies, scientific figures, television and movie screens, custom frames, and other objects, all of which had median ratios close to 1.3. The prevalence of this number among these materials suggests that it may also be considered attractive.

It has been shown that groups of neurons in the brain are able to respond to different characteristics, such as proportion and when the signals from these individual processing centers are interpreted, it generates a positive feeling – a sense of beauty (Di, Macaluso, & Rizzolatti, 2007). Briefly, Di et al. (2007), showed images of 15 masterpiece sculptures from the Classical and Renaissance periods, to volunteers who had no experience in art theory. As they rated the pictures, the scientists scanned their brains. Some of the images were subtly altered. In their original form, the sculptures were proportioned according to the golden ratio. In the altered versions, the images had different proportions but were the same in every other respect. The viewers' evaluation of the stimuli, as expressed in the aesthetic judgment condition, showed that the original images were mostly evaluated positively (76%), whereas the modified images were generally scored with a negative rating (63%). In summary, their findings shown that the presence of a parameter, like the golden ratio, can elicit a specific neural pattern underlying the sense of beauty in the observer.

The Other Side of the Coin

When we start to passing through the modern and contemporary art and their conceptual fields, what seems uniform and unambiguous seems to undo in mirages. When reference thinkers like Merleau-Ponty, Derrida, Husserl, Heidegger, Adorno and Benjamin are faced with the representatives of a new conceptual field in the art such as Picasso, Braque, Matisse, Cezanne, Mallarmé, Stravinsky, Duchamp and Schöenberg, we find ourselves dealing with an increasingly complexity and astonished ambiguous in the semantic field. This does not happen by a simple increase of knowledge as a whole, but through the recognition of the very ambiguity inherent in the cognitive objects of art and beauty and endurance that the very historical time with cultural barriers, economic, social, scientific and linguistic. In summary, the subject of art demands a continuous updating to make sense (Figure 1). Baudelaire recognizes this resistance in his notes:

This is in fact an excellent opportunity to establish a rational and historical theory of beauty, in contrast to the academic theory of an unique and absolute beauty; to show that beauty is always and inevitably of a double composition, although the impression that it produces is single – for the fact that it is difficult to discern the variable elements of beauty within the unity of the impression invalidates in no way the necessity of variety in its composition. Beauty is made up of an eternal, invariable element, whose quantity it is excessively difficult to determine, and of a relative, circumstantial element, which will be, if you like, whether severally or all at once, the age, its fashions, its morals, its emotions. Without this second element, which might be described as the amusing, enticing, appetizing icing on the divine cake, the first element would be beyond our powers of digestion or appreciation, neither adapted nor suitable to human nature. (Baudelaire, 1995, p.3)

A golden thinker who denounces the incompleteness and ambiguity that lies to the conceptual universe of modern and contemporary was Merleau-Ponty.

In his aesthetic work and within a general theory of knowledge as is evident in the Phenomenology of Perception (Merleau-Ponty, 2012), the author, through a dialogue with modern art, science and psychology, sees aesthetics as a perfect sensitive knowledge in a certain way. The object priority of Merleau-Ponty, is not the field of art or beauty, but the sensation as aisthesis and, in a comprehensive way, all forms of appearance, at the same time, complementary and distinct of a logical framework. (Abbagnano, 2007)

Merleau-Ponty criticizes science not to draw its boundaries or describe its objects but to point its incompleteness. For the author, modern art is thought in a phenomenological perspective. In phenomenology the objects have an existence only as related and co-existing with a particular consciousness. In this way the objects are captured in a global manner in which cultural determinants, social, poetic, sentimental and sensorial aspects are a unit that is not perceived only conceptually as the example of lemon and molasses (Merleau-Ponty, 2002), but with real world objects that emergence as unitary and indivisible in its physical embodiment in its all-round relationship to human consciousness (Figure 1).

Despite millions of nuances that can weave about the phenomenal object it appears to us as a complete informational complex of consciousness. I feel the object, I realize its temperature, odor and etc... Groping their recesses, we immediately connect memories, awake feelings before even think about it. Everything about it tells me in every thing, all that is in it for me. This perception of reality implies the obvious ambiguity of values that can exist in complex object as art. So we see the realization of the modern artist more like a reality experimentation field and its many missteps by the intricate universe of human consciousness than just the enjoyment of the beauty's site. This impression can be generalized to other areas of knowledge. Thus, sciences also reviews its concepts and relativize its limits, relaxes the margins of what you can see. To deal with new observations of the empirical world, creates mathematical models that imply a reinterpretation of reality in its radical nature. Creates concepts such as non-spatiality (since the same object can be in two different places) and not causality (as the simultaneity of events does not establish a causal link even in a basic temporal time-line), like in Quantum Theory.

Returning to the initial proposal pointed in the beginning of this chapter, neuroscience can also benefit itself from the same procedures achieved in others areas of knowledge. Rising questions and not resolving some dubiousness as if they do that exist, and not necessarily doing that seeking the extinction of ambiguity. Perhaps neuroscience should incorporate the ambiguity in its conceptual horizon, and not extinguishing it. To achieve this goal, neuroscience has philosophy, art and science itself (in its most paradigmatic representative – Physics) as a conceptual body of knowledge or epistemic pictures, pioneers in the treatment of this paradox. Although this objective appears extremely hard to achieve, we cannot doubt about this possibility, in the future horizon of events. Citing Gaston Bachelar;

Missing, rectify, diversify are types of dynamic thinking fleeing the confidence of the unity, and they finding in homogeneous systems more obstacles than stimulus. In short, the man moved by the scientific spirit wonders, but for once, the better questioning.......The Epistemologist must take the facts as if they

Figure 1. The emphasis on symmetry, proportion and geometry is put into practice by Vitruvius and many architects and architectural authors in the Renaissance. They went out of their way to build according to rules and methods that were considered necessary for beautiful and good architecture during the classical period. Proportion, as golden ratio, which is commonly incorporated in classical architecture, might be universally pleasing to the mind, however, these proportions are completely absent from successful modern buildings such as Frank Gehry's Guggenheim museum in Bilbao, Spain (B). This is an expressive example of how is beauty evolving over time with a sequence of variation of its interpretations throughout history. A: Example of Doric project with notes of geometry and proportion in classical architecture Adapted from Tavola IX from Vignola (1562). B: Frank Gehry's Guggenheim museum in Bilbao, Spain.

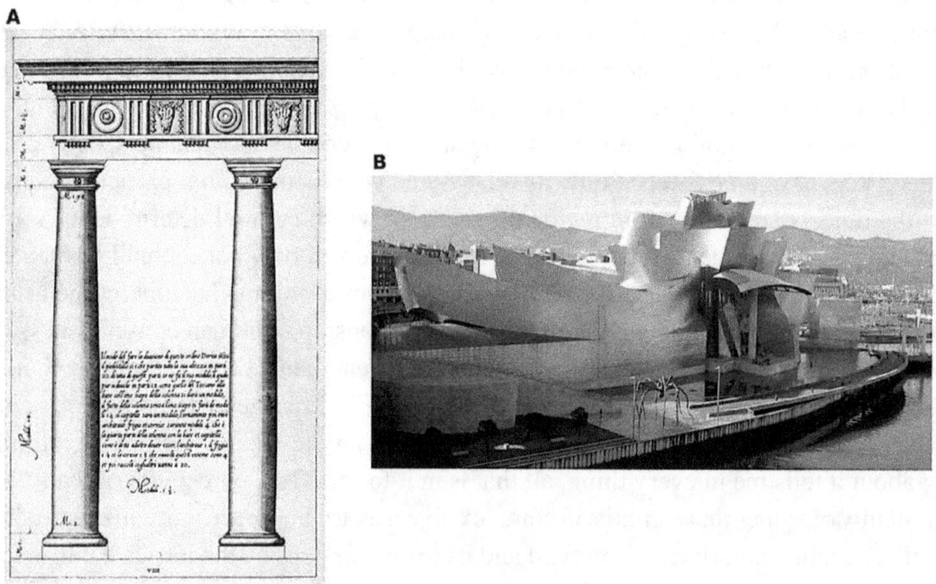

were ideas by entering them into a system of thought. A fact misunderstood by a time remains for the historian, a fact. For the epistemologist, it is an obstacle, a counter-thought. (Bachelar, 1996).

Even the empirical world in which neuroscience is based, points to this ambiguity and multiplicity of possibility, as we shall see.

THE MULTIPLICITY AND AMBIGUITY OF BEAUTY IN NEUROSCIENCE

Neuroesthetics received its formal definition in 2002 as the scientific study of the neural bases for the contemplation and creation of a work of art (Nalbantian, 2008). This field has emerged at the intersection of psychology, neuroscience and aesthetics, and its diverse aims and methods reflect its multidisciplinary nature. (Nadal & Pearce, 2011) Since our minds are part of the world, and how we think and experience and act has been molded by the world over years of evolution and the experience of beauty comes from the interactions between our minds and the world (Chatterjee, 2013).

In spite of our brain my be evolved to find some objects universally beautiful and even if experiences of beauty maybe shared universally, cultural influences and context affect these experiences (Chatterjee, 2013)

The majority bibliographic data of neuroscience suggests that we do not have specific brain network for beauty (Muñoz & Martín-Loeches, 2015). In other words, it is proposed that beauty emerges from cognitive integration and processing of different information by the brain of the subject who perceive it, like: visceral reaction, perceptual analysis, imagery, implicit memory, reward and judgment. Such cognitive process might have at least three components. The perceptual processing of the object itself, the emotional response to the object and, when relevant, an explicit judgment about the object's beauty. Today, the use of technics like neuroimaging, has been contributing in advancing our understanding of the cognitive and neural underpinnings of aesthetic appreciation, since it can provide a new analytical level of evidence in the understanding of cognitive processes (Nadal, & Skov, 2013). Researchers undertook human brain imaging experiment, using functional magnetic resonance imaging (fMRI). fMRI is a technique for measuring and mapping brain activity that is noninvasive. It is being used in many studies to better understand how the healthy brain works, and in a growing number of studies it is being applied to understand how that normal brain function is disrupted in disease.

In the simplest fMRI experiment a subject alternate between periods of doing a particular task and a control state, for example subjects are asked to view pictures of paintings and listen to brief musical excerpts, during a short period of time, and then rate as beautiful, neutral or ugly, while the activity of their brains are imaged. With this type of method is possible to identify a single area or set of areas whose activity would correlate with the experience of beauty or ugly, regardless of whether it was derived from an auditory or visual source.

A review of the neurobiology of beauty reveals that the experience of visual, musical, and moral beauty all correlate with activity in different brain areas (Figure 2). More recently, fMRI studies found that when consumers are exposed to a higher level of beautiful products, brain activations are similar to getting rewards; in contrast, exposure to a lower level of beauty results in similar brain activations as feelings of nausea or unfairness (Chatterjee, 2011). Studies have reported that attractiveness activate areas within the orbitofrontal cortex, the nucleus accumbens or the ventral striatum (Aharon et al., 2001; Ishai, 2007; Kampe, Frith, Dolan, & Frith, 2001; Kranz & Ishai, 2006; O'Doherty et al., 2003) and that the amygdala has a nonlinear relationship to it (Figure 2) (Winston, O'Doherty, Kilner, Perrett, & Dolan, 2007).

These regional activations, within neural circuitry dedicated to reward systems, are interpreted as reflecting the emotional valence attached to attractiveness (Senior, 2003). The particular emotional valences are those involved in the expectation of rewards and satisfaction. The judgment of beauty, as distinct from its emotional evocations, involves parts of the prefrontal córtex (Nakamura et al., 1998).

Many, if not all studies that have addressed the neural correlates of the experience of beauty using fMRI have found high activity in medial orbitofrontal cortex (mOFC) (Zeki, Romaya, Benincasa, & Atiyah, 2014). The orbitofrontal cortex occupies the ventral surface of the frontal part of the brain. In terms of its neuroanatomical connectivity, the orbitofrontal cortex is uniquely placed to integrate sensory and visceral motor information to modulate behavior through both visceral and motor systems. This has led to the proposal that the orbitofrontal cortex is an important part of the networks that are involved in emotional processing, reward, decision-making and hedonic experience (Kringelbach., 2005).

The results for many studies shown that the mOFC was the only common cortical area activated by stimuli that were judged to be beautiful, regardless of whether the origin of the stimulus (example: visual

Figure 2. Image showing different brain areas: A: Lateral view of the brain showing in blue the Frontal Cortex. B: Sagital view of the brain showing in pink the location of the medial orbitofrontal cortex. C: Limbic system – The structures of the limbic system are involved in motivation, emotion, learning, and memory. The two most studied areas in the limbic system are highlighted, in red the amygdala and in grey the hippocampus. D: Figure showing the functional regions of the cerebral cortex - the motor areas are highlighted.

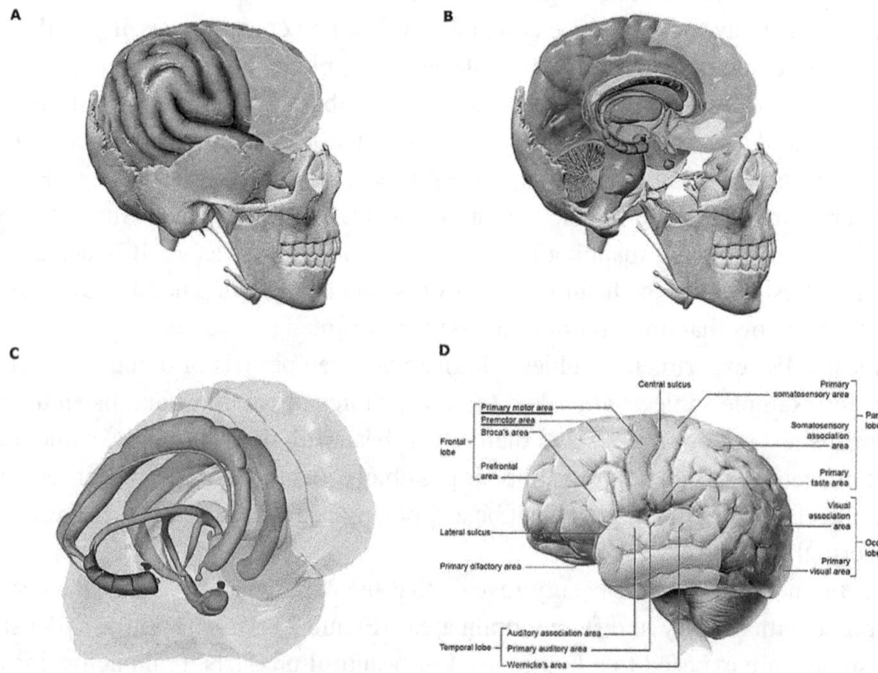

**For a more accurate representation see the electronic version.*

or auditory) (Ishizu & Zeki, 2011). For example, it is know that mOFC is activated in response to facial attractiveness (Tsukiura & Cabeza, 2011; Cloutier, Heatherton, Whalen, & Kelley, 2008; O'Doherty et al., 2003) and even in studies that involved judgments of beauty vs. symmetry of geometric shapes, the activation of mOFC was reported (Jacobsen, Schubotz, Höfel, & Cramon, 2006).

Probably, the most amazing experiment done on this field is from Zeki et al. (2014). What made this research so interesting and unusual was the different type of methodology employed to explore beauty. Instead of do analysis of shapes, art objects or human faces, Zekis et al. (2014) used mathematical formulae to try to understand the neurobiological basis of beauty. In this work, the authors used fMRI to image the brain activity of 15 mathematicians when they viewed mathematical formulae that they had previously rated as beautiful, neutral or ugly. The results showed that the experience of mathematical beauty also correlates with activity in the mOFC.

For the general people who are not mathematician the mathematical formulae appear dry and inaccessible but to a mathematician an equation can be beauty. What make this so interesting is that the even abstract source as mathematics could be correlated with something beauty, and moreover this work also shown that bases of beauty is not sustained only in the geometric shape of the object (on the object itself), like symmetry, it is more than these, it is something that is also related to your life history.

With this in mind, the integration of the processes of perceptual analysis, implicit memory, explicit classification, cognitive evaluation (Leder, 2013) and the induced emotion by the object in the individual (Brattico, Bogert, & Jacobsen, 2013) is the base of the evaluative process to judge if something is beautiful or not.

With the accumulative scientific progress on the understanding of neuroanatomy and the bibliography collection available on various experimental trials about the neural correlates of different physiological conditions and neuroimaging. Researchers start to design experiments to known if it is possible to modulate brain activity without the use pharmacological treatment or any invasive technique such as deep brain stimulation. Using non-invasive stimulation techniques Chib et al. (2013) produced behavioral changes in volunteers, making them to judge faces as more attractive than before their brains were stimulated, in other words they were able to manipulate the beauty judgment of the volunteers.

The two predominant means of non-invasively stimulating the brain are transcranial magnetic stimulation (TMS) and transcranial direct current stimulation (tDCS) (George & Aston-Jones, 2009). Both of these stimulation methods result in changes in brain function by causing neurons resting membrane potential to depolarize ("activating") or hyperpolarize ("deactivating"). Positive stimulation (as in the cases of TMS and anodal tDCS) increases the neuronal excitability and more spontaneous cell firing ("activating"). Negative stimulation (in the cases of cathodal tDCS) decreases the neuronal excitability and decreased spontaneous cell firing ("deactivating") (Chib et al., 2013). The only bottleneck of these stimulation techniques is the problem of accessing the deep brain regions such as medial areas in frontal cortex.

A number of studies have associated increases in ventromedial prefrontal cortex (VMPFC) activity (Au-Young, Shen, & Yang, 1999; Tzschentke, & Schmidt, 2000) and decreases in dorsolateral prefrontal cortex (DLPFC) activity (Bertolino et al., 2000; Pycock, Kerwin, & Carter, 1990), with increases in midbrain activity. With these relationships in mind, the researchers hypothesized that excitatory/anodal and inhibitory/cathodal tDCS electrode placement over VMPFC and DLPFC, respectively, would result in the remote activation of the ventral midbrain and that remote activation would manifest behaviorally as increases in participants' rewarding appraisals.

To test their hypothesis, the researchers asked volunteers to judge the attractiveness of groups of faces both before and after the volunteers' brains had been stimulated with tDCS. The study participants rated the faces while inside a fMRI scanner, which allowed the researchers to evaluate any changes in brain activity caused by the stimulation.

Those in the main group rated the faces presented after stimulation as more attractive than those they saw before stimulation. There were no differences in the ratings from the control groups. This change in ratings in the main group suggests that the researchers were able to activate the ventral midbrain, and that result in changes in brain activity in this deep-brain region associated with rewarding appraisals.

While it looks trivial in our daily lives, the complexity in the acquisition and interpretation of a certain stimulus to generate a response, whether beautiful or ugly, is huge and involves several brain areas. As incredible as it may seem, researchers have shown that both beautiful and ugly stimuli modulate activity in the same brain areas and it seems that the difference between judging opposite situations as beauty and ugliness, appears to be the activation intensity of these brain areas. The findings suggests that the orbitofrontal cortex is activated more strongly for things considered beautiful than those seen as ugly, while in the motor cortex the response appears to be opposite (strongly in ugly) (Kawabata & Zeki, 2004).

The activation of motor cortex is of special interest, since it is not expected activation of this region in tasks that are not related to motor activity, such as during aesthetic judgment. With this in mind, brain activity seems to support that beauty and ugliness are not independent esthetic categories, sharing most, if not all, of the same brain areas. The activation of the same areas, during the judgment of beauty and ugliness, may explain, at least in part, why some people are attractive to ugliness. Such considerable similarity might result as problematic for a plain elucidation of whether beauty and ugliness are actually related or independent in the human brain (Martín-Loeches, Hernández-Tamames, Martín, Urrutia, 2014).

CONCLUSION

The Learn'd Astronomer

When I heard the learn'd astronomer,/
When the proofs, the figures, were ranged in columns before me,/
When I was shown the charts and diagrams, to add, divide, and measure them,/
When I sitting heard the astronomer where he lectured with much applause in the lecture-room,/
How soon unaccountable I became tired and sick,/
Till rising and gliding out I wander'd off by myself,/
In the mystical moist night-air, and from time to time,/
Look'd up in perfect silence at the stars.
(Walt Whitman)

REFERENCE

Abbagnano, N. (2007). *Dicionário de filosofia.* Martins Fontes.

Aharon, I., Etcoff, N., Ariely, D., Chabris, C., O'Connor, E., & Breiter, H. (2001). Beautiful faces have variable reward value: FMRI and behavioral evidence. *Neuron, 32*(3), 537–551. doi:10.1016/S0896-6273(01)00491-3 PMID:11709163

Alberti, L. B. (1989). *Da Pintura.* Campinas, Brazil: Editora da UNICAMP.

Aristotle, . (1984). *The Complete Works of Aristotle, in two volumes* (J. Barnes, Ed.). Princeton, NJ: Princeton University Press.

Au-Young, S. M., Shen, H., & Yang, C. R. (1999). Medial prefrontal cortical output neurons to the ventral tegmental area (VTA) and their responses to burst-patterned stimulation of the VTA: Neuroanatomical and in vivo electrophysiological analyses. *Synapse (New York, N.Y.), 34*(4), 245–255. doi:10.1002/(SICI)1098-2396(19991215)34:4<245::AID-SYN1>3.0.CO;2-D PMID:10529719

Bachelar, G. (1996). *A Formação do Espirito Científico: Contribuição para uma psicanálise do conhecimento.* Rio de Janeiro, Brazil: Contraponto.

Baudelaire, C. (1995). *The Painter of Modern Life and Other Essays*. London: Phaidon Press.

Bayles, D., & Orland, T. (2007). *Art and Fear. Observations on the perils (and rewards) of artmaking*. Eugene, OR: Image Continuum Press Edition.

Bertolino, A., Breier, A., Callicott, J. H., Adler, C., Mattay, V. S., Shapiro, M., & Weinberger, D. R. et al. (2000). The relationship between dorsolateral prefrontal neuronal N-acetylaspartate and evoked release of striatal dopamine in schizophrenia. *Neuropsychopharmacology*, *22*(2), 125–132. doi:10.1016/S0893-133X(99)00096-2 PMID:10649825

Boselie, F. (1984). Complex and simple proportions and the aesthetic attractivity of visual patterns. *Perception*, *13*(2), 91–96. doi:10.1068/p130091 PMID:6504684

Brattico, E., Bogert, B., & Jacobsen, T. (2013). Toward a neural chronometry for the aesthetic experience of music. *Frontiers in Psychology*, *4*, 206. doi:10.3389/fpsyg.2013.00206 PMID:23641223

Chatterjee, A. (2011). Neuroaesthetics: A coming of age story. *Journal of Cognitive Neuroscience*, *23*(1), 53–62. doi:10.1162/jocn.2010.21457 PMID:20175677

Chatterjee, A. (2013). The illogic of beauty. In *The Aesthetic Brain: How We Evolved to Desire Beauty and Enjoy Art*. Oxford, UK: Oxford University Press. doi:10.1093/acprof:oso/9780199811809.003.0010

Chib, V. S., Yun, K., Takahashi, H., & Shimojo, S. (2013). Noninvasive remote activation of the ventral midbrain by transcranial direct current stimulation of prefrontal cortex. *Translational Psychiatry*, *11*(3), e268. doi:10.1038/tp.2013.44 PMID:23756377

Cloutier, J., Heatherton, T. F., Whalen, P. J., & Kelley, W. M. (2008). Are attractive people rewarding? Sex differences in the neural substrates of facial attractiveness. *Journal of Cognitive Neuroscience*, *20*(6), 941–951. doi:10.1162/jocn.2008.20062 PMID:18211242

Cunningham, M., Barbee, A., & Philhower, C. (2002). Dimensions of facial physical attractiveness: The intersection of biology and culture. In G. Rhodes & L. Zebrowitz (Eds.), *Facial attractiveness. Evolutionary, cognitive, and social perspectives* (pp. 193–238). Westport, CT: Ablex.

Davis, S. T., & Jahnke, J. C. (1991). Unity and the golden section: Rules for aesthetic choice? *The American Journal of Psychology*, *104*(2), 257–257. doi:10.2307/1423158

Descartes, R. (1973). Discurso do método. In *Coleção Os Pensadores* (J. Guisnburg & P. B. Júnior, Trans.). Abril Cultural.

Di Dio, C., Macaluso, E., & Rizzolatti, G. (2007). The golden beauty: Brain response to classical and renaissance sculptures. *PLoS ONE*, *2*(11), e1201. doi:10.1371/journal.pone.0001201 PMID:18030335

Dutton, D. (2010). *The Art Instinct: Beauty, Pleasure, and Human Evolution*. Oxford, UK: Oxford University Press.

Fechner, G. T. (1876). *Vorschule der Aesthetic*. Berlin, Germany: Breitkopf & Härtel.

George, M. S., & Aston-Jones, G. (2009). Noninvasive techniques for probing neurocircuitry and treating illness: Vagus nerve stimulation (VNS), transcranial magnetic stimulation (TMS) and transcranial direct current stimulation (tDCS). *Neuropsychopharmacology*, *35*(1), 301–316. doi:10.1038/npp.2009.87 PMID:19693003

Höge, H. (1995). Fechner's experimental aesthetics and the golden section hypothesis today. *Empirical Studies of the Arts*, *13*(2), 131–131. doi:10.2190/UHTQ-CFVD-CAU2-WY1C

Ishai, A. (2007). Sex, beauty and the orbitofrontal cortex. *International Journal of Psychophysiology*, *63*(2), 181–185. doi:10.1016/j.ijpsycho.2006.03.010 PMID:16759727

Ishizu, T., & Zeki, S. (2011). Toward a brain-based theory of beauty. *PLoS ONE*, *6*(7), 21852. doi:10.1371/journal.pone.0021852 PMID:21755004

Jacobsen, T., Schubotz, R. I., Hofel, L., & Cramon, D. Y. (2006). Brain correlates of aesthetic judgment of beauty. *NeuroImage*, *29*(1), 276–285. doi:10.1016/j.neuroimage.2005.07.010 PMID:16087351

Jonhson, G. A. (2010). The retrieval of beautiful: thinking though Merleau- Ponty's aesthetics. Evanston, IL: Northwestern University Press.

Kampe, K., Frith, C., Dolan, R., & Frith, U. (2001). Reward value of attractiveness and gaze. *Nature*, *413*(5856), 589. doi:10.1038/35098149 PMID:11595937

Kawabata, H., & Zeki, S. (2004). Neural correlates of beauty. *Journal of Neurophysiology*, *91*(4), 1699–1705. doi:10.1152/jn.00696.2003 PMID:15010496

Kossovith, L. (1996). Permanência e Renovação nas Artes. *Discurso*, *26*, 83–92.

Kossovith, L. (2006). *Tradição Clássica. Desígnio: revista de história de arquitetura e do urbanismo. nº5*. São Paulo, Brazil: Ed. Annablume/FAU-USP.

Kranz, F., & Ishai, A. (2006). Face perception is modulated by sexual preference. *Current Biology*, *16*(1), 63–68. doi:10.1016/j.cub.2005.10.070 PMID:16401423

Kringelbach, M. L. (2005). The human orbitofrontal cortex: Linking reward to hedonic experience. *Nature Reviews. Neuroscience*, *6*(9), 691–702. doi:10.1038/nrn1747 PMID:16136173

Kumar, M., & Garg, N. (2010). Aesthetic principles and cognitive emotion appraisals: How much of the beauty lies in the eye of the beholder? *Journal of Consumer Psychology*, *20*(4), 485–494. doi:10.1016/j.jcps.2010.06.015

Leder, H. (2013). Next steps in neuroaesthetics: Which processes and processing stages to study? *Psychology of Aesthetics, Creativity, and the Arts*, *7*(1), 27–37. doi:10.1037/a0031585

Livio, M. (2003). *The Golden Ratio: The Story of Phi, the World's Most Astonishing Number*. New York, NY: Broadway Publishing.

Martín-Loeches, M., Hernández-Tamames, J. A., Martín, A., & Urrutia, M. (2014). Beauty and ugliness in the bodies and faces of others: An fMRI study of person esthetic judgement. *Neuroscience, 26*(277), 486–497. doi:10.1016/j.neuroscience.2014.07.040 PMID:25086316

Merleau-Ponty, M. (2002). *Causeries. 1948, établies et annotées par S. Ménasé.* Paris: Seuil.

Merleau-Ponty, M. (2012). *Phenomenology of Perception* (D. Landes, Trans.). London: Routledge.

Moura, C. A. R. (2005). *Nietzsche: civilização e cultura.* São Paulo, Brazil: Martins Fontes.

Muñoz, F., & Martín-Loeches, M. (2015). Electrophysiological brain dynamics during the esthetic judgment of human bodies and faces. *Brain Research, 1594*, 154–164. doi:10.1016/j.brainres.2014.10.061 PMID:25451119

Nadal, M., & Pearce, M. T. (2011). The Copenhagen Neuroaesthetics conference: Prospects and pitfalls for an emerging field. *Brain and Cognition, 76*(1), 172–183. doi:10.1016/j.bandc.2011.01.009 PMID:21334125

Nadal, M., & Skov, M. (2013). Introduction to the special issue: Toward an interdisciplinary neuroaesthetics. *Psychology of Aesthetics, Creativity, and the Arts, 7*(1), 1–12. doi:10.1037/a0031842

Nakamura, K., Kawashima, R., Nagumo, S., Ito, K., Sugiura, M., Kato, T., & Kojima, S. et al. (1998). Neuroanatomical correlates of the assessment of facial attractiveness. *Neuroreport, 9*(4), 753–757. doi:10.1097/00001756-199803090-00035 PMID:9559951

Nalbantian, S. (2008). Neuroaesthetics: Neuroscientific theory and illustration from the arts. *Interdisciplinary Science Reviews, 33*(4), 357–368. doi:10.1179/174327908X392906

Novaes, A. (1994). *Artepensamento.* São Paulo, Brazil: Cia das Letras.

O'Doherty, J., Winston, J., Critchley, H., Perrett, D., Burt, D. M., & Dolan, R. J. (2003). Beauty in a smile: The role of medial orbitofrontal cortex in facial attractiveness. *Neuropsychologia, 41*(2), 147–155. doi:10.1016/S0028-3932(02)00145-8 PMID:12459213

Panofsky, E. (2000). Ideia: a evolução do conceito de belo. *Nature, 286*, 74–76.

Sartwell, C. (2014). *Beauty.* The Stanford Encyclopedia of Philosophy. Retrieved from http://plato.stanford.edu/archives/spr2014/entries/beauty/

Senior, C. (2003). Beauty in the brain of the beholder. *Neuron, 38*(4), 525–528. doi:10.1016/S0896-6273(03)00293-9 PMID:12765605

Shortess, G. K., Clarke, J. C., & Shannon, I. (1997). The shape of things: But not the golden section. *Empirical Studies of the Arts, 15*(2), 165–165. doi:10.2190/3UBV-2LJU-T2T9-D7XN

Stewart, I. (1996). Tales of a neglected number. *Scientific American, 274*, 102–102.

Symons, D. (1979). *The Evolution of Human Sexuality.* Oxford University Press.

Tsukiura, T., & Cabeza, R. (2011). Shared brain activity for aesthetic and moral judgments: Implications for the Beauty-is-Good stereotype. *Social Cognitive and Affective Neuroscience*, *6*(1), 138–148. doi:10.1093/scan/nsq025 PMID:20231177

Tzschentke, T. M., & Schmidt, W. J. (2000). Functional relationship among medial prefrontal cortex, nucleus accumbens, and ventral tegmental area in locomotion and reward. *Critical Reviews in Neurobiology*, *14*(2), 131–142. doi:10.1615/CritRevNeurobiol.v14.i2.20 PMID:11513242

Umberto, E. (2004). *História da beleza* (E. Aguiar, Trans.). Rio de Janeiro, Brazil: Editora Record.

Vignola, J. (1562). Regola delli cinque ordini d'architettura. Rome: n.p. Facsimile.

Vitruvius, . (1970). *On Architecture, Frank Granger, trans.* Cambridge, MA: Harvard University Press.

Walker, S. (2006). Sustainable by Design: Explorations in Theory and Practice. London: Earthscan.

Winston, J., O'Doherty, J., Kilner, J., Perrett, D., & Dolan, R. (2007). Brain systems for assessing facial attractiveness. *Neuropsychologia*, *45*(1), 195–206. doi:10.1016/j.neuropsychologia.2006.05.009 PMID:16828125

Zeki, S., Romaya, J. P., Benincasa, D. M., & Atiyah, M. F. (2014). The experience of mathematical beauty and its neural correlates. *Frontiers in Human Neuroscience*, *8*(68), 1–12. PMID:24592230

KEY TERMS AND DEFINITIONS

Aesthetic: The branch of philosophy dealing with the nature of art, beauty, and taste, with the creation and appreciation of beauty.

Art: The expression or application of human creative skill and imagination, typically in a visual form such as painting or sculpture, producing works to be appreciated primarily for their beauty or emotional power.

Beauty: A combination of qualities that pleases the aesthetic senses.

Deep Brain Stimulation (DBS): DBS is a neurosurgical procedure, involving the implantation of a medical device called a neurostimulator, which sends electrical impulses, through implanted electrodes, to specific parts of the brain.

Functional Magnetic Resonance Imaging: Functional neuroimaging procedure using magnetic resonance imaging technology that measures brain activity by detecting changes associated with blood flow.

Genealogy: In philosophy is a historical technique in which one questions the commonly understood emergence of various philosophical and social beliefs.

Historiography: The study of the methodology of historians in developing history as an academic discipline, and by extension is any body of historical work on a particular subject.

Medial Orbitofrontal Cortex: Prefrontal cortex region situated in the ventral medial surface of the frontal lobes of the brain.

Neuroesthetic: Branch of the neuroscience that study the neural bases for the contemplation and creation of a work of art.

Prefrontal Cortex: In mammalian brain anatomy, the prefrontal cortex is the cerebral cortex which covers the front part of the frontal lobe. This brain region has been implicated in planning complex cognitive behavior, personality expression, decision making, and moderating social behavior.

Transcranial Magnetic Stimulation: Noninvasive procedure that uses magnetic fields to stimulate nerve cells in the brain.

ENDNOTES

1 Images that endures during history and have the capacity to change his concept during time.

2 Understood as Nietzsche organized.

3 In this context objective means features of the object.

This research was previously published in Projective Processes and Neuroscience in Art and Design edited by Rachel Zuanon, pages 103-118, copyright year 2017 by Information Science Reference (an imprint of IGI Global).

Chapter 5
Perceptual Processes and Multisensoriality:
Understanding Multimodal Art from Neuroscientific Concepts

Rosangella Leote
Universidade Estadual Paulista (UNESP), Brazil

ABSTRACT

Perception seems to be an overly discussed subject in theories of Art, giving us the impression that there is nothing new to add. So much research and so many conceptions have been developed on the subject. Nevertheless, many holes can be perceived in these theories with regard to the mental process that operates in the perception phenomena. We have chosen to look to neuroscience for possible answers to these holes. In this paper, based on the knowledge of Vilayanur S. Ramachandran and António Damásio, among others, we focus on that which we are emphasizing as "perceptual processes". We will restrict ourselves herein to the perceptual processes of a multisensory nature that take place in the perceivers' relationship with the artworks, which contain multimodal stimuli, promoted by physical and digital interfaces of an assistive nature. Said perceivers, however, are both people with severe motor and vocal disabilities as well as those without these restrictions.

WHY?

In this paper we present the evolution of a part of our research that is being developed within the research group under our coordination - GIIP[1] in one of its lines of research, namely "Physical and digital interfaces for the arts: from diffusion to inclusion"[2].

The approach assumes the existence of poetic, not merely artistic, produced and/or perceived by people with special needs, but also by those who are "exempt" from this condition.

DOI: 10.4018/978-1-5225-5478-3.ch005

We believe that the perceptual experience involves both the enjoyment and the making of Art by any person, despite the diversity of physical, emotional and cognitive qualities that individuals possess. Based on that, we demonstrate that it is the relationship between self-perception, stimulus and sensing that qualifies the "perceptual process" [3].

Aspects of our perception are clearly important for the construction of our consciousness and hence of our world and understanding of Art.

As artists, our understanding of the perception phenomenon is important in many ways and for different reasons. To understand it, artist-researchers who objectively seek understanding come across and develop several methodologies, generally empirical in nature. Like them, we have followed a similar path based on a variety of fundamentals and reasons. Until now, our strategy has been to develop low-cost assistive interfaces that can be used by disabled and non-disabled people to develop and/or teach Arts.

The poetics, in our study, are those that present multisensoriality and multimodality. However, we seek to analyse and substantiate them through contributions from neuroscience. This contribution does not propose the exclusion of other analysis models. On the contrary, it aims to verify which contributions from neuroscience lead to a deeper understanding of perception, and can be associated with some of the better known theories for this purpose, used by the arts in general, namely, Semiotics and the Theory of Complex Systems. We have added phenomenology and Gestalt psychology to the scope of our work. Nevertheless, the relationship between theories is not discussed in this text, and the reason for this omission will be dealt with in this text.

In this study we favour those works qualifiable as artistic, i.e. aesthetic manifestations, in diverse poetics and produced by a multimodality of stimuli.

It became necessary for us to study neuroscience when we began trying to understand the processes taking place in the individual's mind when his or her perception is triggered, so that they can be verified through brain mapping technologies such as, for example, magnetic resonance (fMRI) and tomography (PET scan).

Although no scientific field has as yet been able to shed light on the cerebral operation whose occurrence generates these mental processes, is it through neuroscience that we can come closest to the quality of these processes.

It is clear to us that multimodality leads to multisensoriality, but the latter varies according to the acuity of the perceiver's natural sensors, or in consonance with the physical nature of the multimodal stimuli. We will later deal with the multimodal amplitude[4].

The myth that external or internal stimulus is only understood, qualitatively, by only one specific meaning according to the nature of the channel, has been weakened. The perceptual process experience is a continuous one. We live because we perceive. By identifying certain events that we give emphasis to during a significant experience in our daily living, we are revealing world events that we were forced or decided to experiment, in varying intensities.

The stimuli provide elements that trigger consciousness of sensations, as explained by Damásio (2000), captured by the various senses in a simultaneous operation, in the construction of the brain map that enables the stimulus that is highlighted by the perceiver to be identified in each perceptual experience.

All perceptual processes occur selectively. But those whose selectivity can be easily tested in humans, through several assessment tests, are hearing, sight, smell and taste.

We are selective up to the limit of our core consciousness (Damásio: 2000, 2004 e 2011). There are many other perceptual states that occur in a manner that is unintelligible to the core consciousness, i.e.

they are in the extended consciousness. Neuropsychology has endeavoured to shed light on this fact. However, we still know more about the reasons that control what we can see even if by optical illusion or another psychophysical conditioning, than what we cannot see, even though it is in our perceptual field. It is highly unlikely that we will develop instruments or our extended consciousness enough, to the point of being able to take in perceptual events in their entirety. If this occurred, it would be an anomaly. Most of the perceptual choices we make, i.e. selecting from among the stimuli offered to us, become lodged forever or for quite some time in the less accessible layers of our consciousness (in the extended consciousness).

Despite being choices, our measuring instruments are not able to present the mind process that experiences the event intensely and makes such choices. We know about the existence of parts "that have been suppressed by the memory" when, for whatever reason, unintentionally or induced (i.e. therapy, hypnosis, medications and other drugs) we re-examine the events and retrieve certain suppressed memories of things we did not realize at the time of the experience. The nature of trauma is like this, and also of subliminal messages

Hi-Tech has made it possible to capture information with significant details about the areas of the brain[5] that are activated depending on the stimulus provided. However, this information, of a physiological order, is not by itself efficient at showing WHAT thoughts, feelings and emotions are and hence the perception of the world and HOW they are built. Science understands a large part of the "WHERE" this takes place. For now, we will approach the first two because it is likely that a long time will pass before scientists can take us to the innermost recesses of the human mind.

EXPOUNDING ON THE WHAT

The concept of multisensoriality is explained by the simultaneous activation of multiple natural sensors in the interactor's various levels of perception. This concept evokes that of multimodality.

This term generated our concept of "interactive multimodality" used to describe artistic and non-artistic works, where multiple modes of stimuli, together, trigger this perceiver/interactor's various natural sensors within the testing space, so that the perceiver is conscious of this multisensoriality.

We can extend the notion of multisensoriality to comprise both natural and artificial sensors, (which are extensions of the natural ones). In any case, consciousness is at the same time a filter and organizer of multisensory states.

Although multisensoriality is present in any event experienced in our daily lives, the examination of the event, within an exhibition space, requires from the artist or proponent of the work, a better understanding of how it happens. It is also necessary to guide the user's cognition towards an interaction that is more coherent with the Multisensoriality model that the work intends to achieve.

When, in the perceptual event involving a work of art, it does not occur to us to pay attention to the technological device, or even to the scientific knowledge used, and these take second place, it is the poetic that is highlighted. This is what the majority of artists that produce in this field desires. If this happens it is often said that the technology became invisible. Although technologically impossible in our era, it is a plausible qualification. Even if our technology becomes literally invisible, there will still be the perceptual processing that attracts our attention, including the technological structure that contributes, with its share, to the work.

We would venture to say that there is a straight and horizontal line leading multimodality to multisensoriality. But the same cannot be said between stimuli and sensing. This is because one stimulus may trigger several sensors simultaneously.

Our focus is purely a resource to understand how multimodals operate and how multisensors respond.

In fact, scientists are increasingly unravelling how human senses function and have been convinced for a long time that the division into five senses, based on Aristotle, is limited. For present-day Sciences, and depending on the theoretical line, more than 30 senses may be listed. However, if we were to take into account the natural sensors that we know of, including those of animals, the number would increase greatly (i.e. thermal, photo, mechanical, chemical, etc.). And, we haven't even touched on the so-called "extra-sensory senses." If we accept the thinking of António Damásio (2000, 2011), although he did not discuss this matter, we infer that perception of an "extrasensory" nature is also the result of the organization of the brain maps, which reflect different states of consciousness, therefore, everyday occurrences in the brain that the individual may be made more or less aware of. But this is a controversial issue, as we cannot measure the stimulus that generates the extra-sensory perceptual processes.

We, of course, agree with the idea that the division into any number of senses is limited, but we do not have sufficient knowledge to concur with any specific number of human senses or, likewise, to discuss a subdivision about which not even scientists have come to a consensus.

As this is not our aim, what interests us, for now, is to examine the multisensoriality that reveals itself to perception at some heightened level of consciousness, produced by various modes of stimuli.

Multimodality can be understood as an immeasurable abundance of stimuli, which are offered by the world we create, to our consciousness.

Nevertheless, it occurs to us that as the worldview and the construction of reality is the work of each individual, in line with his or her Umwelt[6], it would be impossible to quantify both multimodality and multisensoriality.

However, as a work methodology, we reduced the breadth of meaning of multimodal stimuli to focus on those who are both easier to be recreated and also perceived with our natural sensors. Similarly, we study the natural sensors that can be simulated by the technologies available in our contemporaneity.

Thus, we distinguish stimuli related to certain natural sensors, with or without the intervention of artificial ones. The choice is linked to the modal range we can observe in the hybridized field of Art/Science/Technology. Therefore, to us sight, hearing, smell, touch, taste, proprioception and kinaesthesia are relevant.

In this part of the research, we highlight sound and visual perceptions.

The Deaf Perceive Sound

Several studies and devices that were created to replace senses that are barely or not in the least functional have been around for a long time. We have been acquainted with some of these devices for so long that we barely pay attention to their purpose linked to perception, such as eyeglasses and hearing aids.

Although our society is "moved" by images, it is sounds that reach us, in most cases, without invitation. So that we do not hear them we need to carry out an effort of psychic disconnection that is greater than the rapid action of closing our eyes, which we do so as not to see.

Sound information is a type of signal that is converted by ear canals into a auditory brain map and captured by our consciousness as a code that can only be decoded in greater detail by those who hear it. But it is actually received in greater intensity by the sense of touch than by the sense of hearing. This invalidates the notion of deafness, for the deaf perceive sound.

The range of sound awareness that people have is greatly reduced in comparison to the sound frequencies that we know through physics, mathematics and chemistry. Each person develops his or her own abilities to perceive in the space of frequencies combined with amplitude, between the bass (about 20 Hz) and treble frequencies (about 20,000 Hz) that are not solely responsible for our understanding of sound.

The condition of the environment where the sound source and the perceiver are located, contributes both to the sound's qualification, and to a higher acuity of its perception. By the route taken by the sound to reach our body we can identify its course and origin, as well as balance and locate ourselves in the space in relation to it. Sound can be perceived without the person's ear being able to catch it. Sound frequencies are received competing with other stimuli witnessed and activating the natural sensors available, promoting understanding of "the existence of sound", although without qualification and importance beyond the auditory fact. It is the ear that allows the body to balance itself.

Therefore, a malfunctioning of the ear may affect an individual's proprioception, the body's notion of balance, without this having a direct bearing on his or her ability to hear. We say direct because, as we have seen, other aspects cause sound to be perceived.

We have heard about deaf people's ability to develop rhythmic sensitivity to dance or music, through the tactile perception of sound (SACKS, 2003). On a large-scale, everyone perceives sound's vibration in his or her body, when it is high enough. Tolerance to pitch also varies from individual to individual.

Sound establishes a path that touches the whole body by the impact of the sound wave. But it only becomes relevant when this impact reaches the most appropriate channel for its conversion into a auditory brain map, which is the ear. We are barely aware, however, of the tactile role that the hearing process plays, with the simple reverberation of sound in the eardrum.

This is one of the reasons why some people tend to listen to music with the headphone volume turned up too high, despite the decibel control system installed in the sound reproducing equipment. People, when using headphones, fail to perceive the sound's reverberation throughout their entire body, therefore they tend to compensate for the lack of this tactile experience by increasing the volume.

Research and artistic work done by the Spaniard Josep Cerdá[7] use what he calls inaudible sounds to show other sound frequencies, not perceived by our senses. He argues that no sound is repeatable, however controlled the conditions are for its repetition. That is, the frequencies, both those inaudible to human senses and those not perceived by our attention at the time of the experience, but which are constructive parts of the listening experience, underlie the sound event. Indeed, we agree that the soundscape underlying any perceptual sound experience will possess, unconditionally, unique and competent matrices to make the landscape specific. Here we are using two concepts jointly.

The first concerning soundscape in the way identified by Raymond Murray Schaefer (1992; 1997), can be summarized as the specificities of the sound conditions that can be found in that specific location, the space surrounding the perceiver. The idea comes from the theory of Acoustic Ecology, founded by him and his research group in the 70s.

The second refers to the concept of sound matrix, developed by Lucia Santaella (2001), who identifies it through semiotics in firstness state. According to her, matrices are elements that, by semiosis and hybridizations, will become sounds, per se, and may become the language of sound (musical or otherwise). The same would happen with the visual and verbal languages. All these matrices are potentially capable of developing into hypermedia language.

The reception of auditory information, in a different way from those who are not able to hear, only occurs after pressure occurs in the eardrum with each pulse of the captured frequency. Only after this pressure/tact information travels the entire path by a mechanical vibration process of the ossicles does it reach the cochlea, where it is converted into electrical signals that are directed through the auditory nerve to the brain. Other physical conditions of the perceiver exert pressure before this moment, but those shall be left aside for the purpose of simplifying the concept. Next, the encoded sound information becomes perceivable by the so-called auditory brain, the brain mapping area that predominantly interprets and brings the auditory perception to the core consciousness. To reinforce the previous idea, we could say that we would be "hearing with the ears".

But everyone's auditory acuity is not identical. Each individual following his or her own auditory route will always be "a little deaf" to certain sound conditions. Perceptual acuity changes with the variations of our bodily state that is directly influenced by external factors such as altitude, air pressure, temperature, number of people etc. Internal factors create interferences that are often determinant to the rest of a person's life.

There are also the transitory factors such as pain, fever, allergies, flu and emotional states that also alter not only our auditory perception, but also interfere with all our natural sensors.

This collection of characteristics that is formed individually and varies from moment to moment, is what we use to build our world and our aesthetic sense. With it we also complete the meaning of the works of art, in our case, the multimodal ones.

The soundscape is only one part of the multimodal spectrum. It is the soundscape's specificity and originality that induces the perceiver to multisensoriality. He or she is immersed in the event, even if unconsciously in part. It is in this immersion that we locate the qualification of the relationship between multimodality and multisensoriality. Note that there will never be control over how or how many of the interactive modes will affect the perceiver, in order to imprint on him or her the consciousness of multisensoriality.

We Reinterpret the Image

The "visual brain," as described by António Damásio (2000) and Margareth Livingstone (2008)[8] builds the intelligibility of the visual stimuli that reach the retina in the form of light pulses. Part of the process is optical; another is mechanical and the other neurological, until the image map is formed in the mind. This investigation also focuses on the visual map. Just as with sound, it is not only what is "available" to sight that is interpreted in the layer of core consciousness. This visual map resides only in the mind. In fact, most of what is absorbed in the perceptual process is unknown to the interpreter.

Our vision is very selective. The area that we see is limited to the arc that the eye draws when scanning the environment and to our ability to approach the viewed object. This arc varies according the anatomy of the perceiver's eye and face.

Apart from the perceiver's physical conditions, all the properties of odour, taste, volume, temperature and light reflected from the object that are available to sight, come together to build the perceptual dynamic. But it is not enough to perceive the physical information to establish judgement about them, as this occurs in the core consciousness. So we say that the individual is conscious of the perception made. If there is any perceived light, a visual perceptual process is going on, even if it does not allow the person to identify the general qualities of the stimuli at work in this sliver of light. Ultimately total blindness is the extreme inability to perceive light.

NÃO SE APROXIME / DO NOT COME CLOSER

We have aimed for transdisciplinarity, by using theoretical bases and studying our artistic production.

We are interested in how interactors deal with contemporary art installations. This issue is crucial to us and demands constant investigation that includes analysis of the relationship between art, science and technology and the perception of these interactors in these transduced and contaminating fields.

In order to demonstrate a part of what we perceive and use in the field we shall describe the interactive installation called "Do not come closer" (Figure 1), which we developed in 2013. It is a work that stemmed both from conceptual research and artistic production. The work aims to promote interaction with two video clips that overlap on the same projection area, The progression and fusion between them is computer instructed by the contact of the interactor's hand on a surface sensorized by ultrasounds. We used low-cost technology and open source programming. Although we are the authors of the work, it was developed alongside some of our research partners[9].

We changed the project in 2014, when minor alterations were made to the program and in the arrangement of the ultrasound sensors used to detect the interactor's presence and actions.

The performer's (this author) image is projected on video, on a narrow and vertical screen. There a fusion occurs between the clothed body that invites interaction and the naked body, whose hands are suspended and cuffed. In the two videos the body turns 360 degrees, slowly ever looping. The plot entails a performer wearing a long black dress, who with gestures invites the interactor to come closer. If the interactor does not move closer, the image is modified using ultrasound sensors applied to a demarcated area at the entrance of the installation, where the interactor places his or her hand. Otherwise, the video image of the clothed body is maintained. Therefore, it is only at this spot, away from the screen, that the interactor visualizes the body's transformation when disrobing, in the fusion programmed between the two videos. In other words, when trying to get closer to the screen, i.e., near the naked body, the interactor is obliged to remove his or her hand from the sensorized area, causing the first video to keep on playing. In summary, drawing closer prevents the vision of the naked form.

Symbolically, the interactor's bold proximity reduces the possibility of the shackled creature's release. Any approach results in the destruction of the original meaning, creating ambiguity about the installation's intent. The fruition zone[10] is not limited to the voyeur aspect.

The approach that "Do not Come Closer" introduces is in the context involving the interactor's participation at the installation, where our gaze is turned to the problem of visual and sound perception.

Obviously, the engagement of the visual sensors is the most prominent mode of stimulation. Without any sound included, the work leads to immersion and sound perception, both because of the "silence", and the space. It encourages spatial perception, due to the size of the dark hallway (where, at the end, the bluish image is projected) and the invitation to the visitor's journey to the screen; olfactory perception regarding the materials (wood, fabric, paint, metals, electronic components) and bodily odours of other interactors and/or the interactors themselves; tactile perception, resulting from the necessary action of touching the sensorized area, as well as other perceptions whose nature depends on the interactor.

The installation "Do not Come Closer" is, in fact, soundless. The purpose is to make the interactor immerse himself in the image, while at the same time realizing that the lack of a sound source is not silence. On the contrary, on noting the lack of a sound source in the work's syntax, the auditory sense's perception is highlighted; the impression of space and the relationship with the image is changed. Despite the prominence that the visuality has in this work, the image is not treated as the main component. What interests us is the relationship that the individual who perceives it has with the image, the space, the silence and the prompting to a particular type of interaction, even if it does not occur (i.e. placing a hand on the sensorized area).

With the continuous reduction of equipment and components, investment in these researches can be more rewarding, from a marketing perspective, accelerating the accessibility of these devices to the needy and interest parties. The price of the technology developed by large corporations is still very high, but the bottom up movement, permitted by this miniaturization and, similarly, by the popularity of open source, has reverted their knowledge to another field of application, one with lower cost equipment. The proposed production of devices for technology focused on the arts that we are developing fits the second (but not only) opportunity. With open source and hardware with low-cost, easily customizable components, we can reach a wider and less privileged social group in economic and social terms (Figure 1).

As in this work, an important feature that we perceive in the artworks using emerging media is the ability to conceive an expanded constitution of the work's identity. We describe this at another time (2008) when we discuss the expansion of the work's identity in greater detail. However, to summarize the thinking, we say that the work that has technological aesthetic characteristics can have its body or structure expanded to the point of involving electronic spaces (virtual) and physical spaces (lived). By experiencing theses spaces (virtual and lived) we immerse ourselves in them.

We must differentiate these spaces (virtual and lived) and consider that while together they compose the same sensation of lived space, relative to places that we hold dear, separately they represent different levels in the form of experiencing. With Bollnow (2008) we learn that living and experiencing a space are different things. This difference is very important because it is experience that promotes the understanding of the concept of place. The more we experience a space, the closer it comes to the concept of place.

We believe that a work moves toward concretization when it can direct the interactor into making the installation space a place of his or her own.

Figure 1. Não se aproxime (Do not Come Closer)

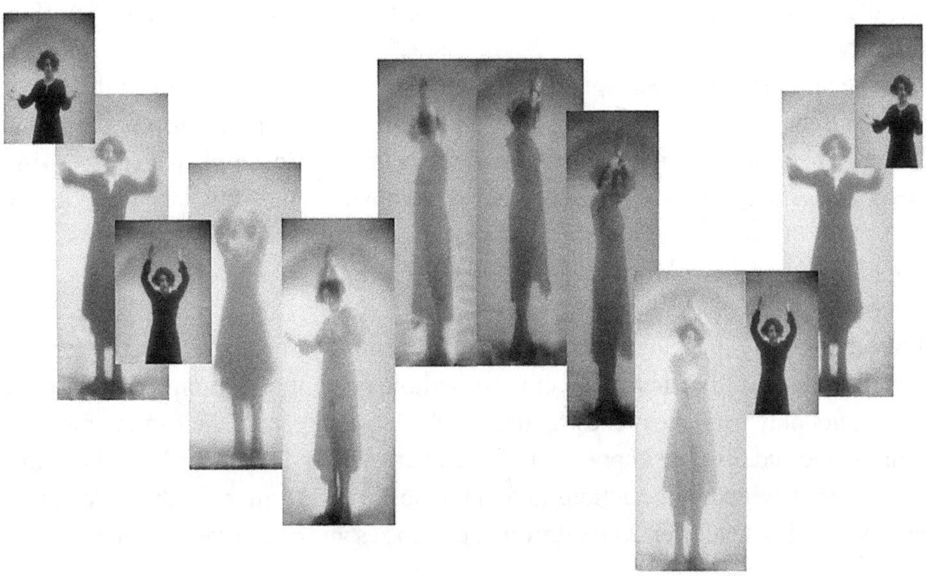

ABOUT THE ANSWERS

Therefore, what is perception? We do not need to criticize or propose another referral to the logical model of interpretation of perceptual phenomenon made by Peirce's semiotics. We are fully confident that it is competent and even spectacular. As for Gestalt psychology (Wertheimer, Koffka, Köhler) although very coherent in its structure, it poses certain questions about the verification models of the experiments that led to the conclusions that gave rise to the theory. These experiments are still used today. Our criticisms dwell precisely in the somewhat hazy interstitial space between the modal stimulus and the brain map created. The theory was developed in the early 20th century; long before the technological systems of brain reading achieved the breakdown of the "concrete" brain operations available nowadays. While such devices are still far from providing full understanding of perception - and we will probably never reach that level with the kinds of technologies we know about - they can already give us indications of certain brain action specificities that cannot be measured because they occur only in the mind.

Many of the conclusions from Gestalt analyses were obtained more through inference than the organization of solid data that could be placed in isolated conditions with the possibility of test reproduction and whose analysis would be predominantly objective.

Another factor that compels us to consider the need to carefully examine Gestalt theory is the concept of consciousness. Arno Engelmann (2002) argues about the existence of two types of consciousness: "immediate-consciousness" (true fact) and the "other's-mediate-consciousness" (probabilistic). This idea that there is an outside and an inside world does not reconcile with what we understand as consciousness based on our neuroscience studies.

Another incompatibility with our scientific basis of both semiotics and neuroscience is the concept of percept, introduced by some Gestaltists, which is understood as "characterizing the type of perceptual content"[11] of the consciousness (Engelmann, 2002).

According to our understanding, the difference between percept and perception in Gesltalt literature is not clear. We take percept as firstness while percept, as extracted from Gestalt Theory, would be a complete perceptual process.

We hope this research will help us to seek clarity and the connection between what we have already learned from the theories we use in art, especially the two aforementioned theories (Semiotics and Neuroscience).

As we have seen, in standard perceptual conditions, the senses, interdependently, provide our core consciousness with the assurance that we exist in full connection with our surroundings and create our world. But what is the difference in the amplitude of this guarantee in people who possess limited conditions of perception? Although at this stage of our research we are far from answering this question, we dare to propose that the very awareness of the lack of acuity or severe limitation of some sense guarantees the creation of one's own world, thanks to the action of the core consciousness, based precisely on these limitations.

Silent films, that strictly speaking never were so, can exemplify this. In the space of meaning, between the image and the lack of sound, there were endless audible perceptions coming from stimuli existing in the environment, not only from the orchestral music which accompanied the film, in most of the screenings. The soundscape and also the impression of sound created in the viewers' minds, coming from the sounds promised or imagined in the actions and dialogues in text form, built their listening experience. The film was not heard. The viewers knew, through memories and cognitive experiences, that in the film would reside sounds whose quality was intuited by each individual. In other words, each viewer created auditory brain maps, based on memory.

We understand that it is in this "intuitive" capacity, for lack of another word, that resides one of the origins of the qualia of sound, its absolute firstness. If we listen to a dialogue without images, preferably within a cinema's spatial condition, the same will occur with the image: we will create it, through brain maps, in relation to that sound. Touch, smell and taste are conducted in the same manner. Proprioception, on the other hand, receives less interference from external stimuli of this nature.

People who have completely lost the tactile perception of their body, or parts of it, due to trauma, may not feel it when one of their limbs or their entire body is touched, but be aware of the fact of the contact. Memory will retrieve the *qualia* in the experience. It can be said, then, that the existence of the contact is perceived to some degree. Touch, by its very nature, depends on the functioning of specific receptors (corpuscles and terminations).[12]

But perception, at least to the core consciousness, will always be "mistaken". We hope that António Damásio and Oliver Sacks can agree with this. As perception is selective, we can either not perceive or overperceive. This applies to the perception of pain or touch in limbs that have been amputated. These are the so-called phantom limbs, discussed at length by Sacks (2007, 1998) and Ramachandran (2005, 2012).

A perceptual experience can be carried out with people whose arms have a fully functioning tactile capacity, even though they haven't suffered an amputation.

Children usually do it as a game. Someone is asked to close his or her eyes tight and to stretch out either arm to the person applying the test. The tester slowly travels up the arm with his or her fingers, as if walking, gently, from the middle of the hand towards the anterior elbow crease (it can also be done with the tip of a pencil). The person whose eyes are closed must inform the other when his or her fingers reach the crease line. Interestingly most people indicate the contact on the crease long before or after it is reached. This has to do with the tactile receptors in this area because, on the forearm, relative to other areas of the arm, the number of these receptors is lower and its distribution is more dispersed. The back of the knee is even less sensitive than this area of the arm. In this case, identification of the touch on the crease depends not only on tactile perception. At the very least, temporal, spatial, auditory and also visual perception contribute to the experience. This is because, even with his or her eyes closed, a person can sense the amount of light received, i.e. the shadow of the person applying the test can interfere with the emission of light rays, indicating the other's approach; both the opening of the fingers and this "walk" can intuit the path travelled. The duration of the movement will also supply information about this path and the warmth of another person's hand provides indicators to the senses. But all this only takes place if the perceiver's consciousness of the limb corresponds to its physical quality and if the perceiver's cognition of such stimuli is profound. Now some of these perceptual details may not reach the core consciousness, and even so, the perceiver is able to locate the exact point of the touch on the crook of his arm. This means that the expanded consciousness has perceived the qualias of the experience.

But if there is no cognition, based on the body's memories, not even these qualias will be perceived. The body's lack of cognition occurs under special conditions, where the person has been deprived, even before birth, of said cognitive experiences involving these perceptions. The reasons are several, but those we have paid most attention to are related to congenital development problems and paralysis.

How We Can Conclude This?

In short, we are always ascending from one perceptual state to another (or interweaving, transducing them) until reaching the core consciousness.

That which is learned about the perceptual states (which occurs though imprecise states of involvement with what is being perceived) is what an individual uses to build his or her world, which is always under construction.

For now, semiotics provides us with us the logical fundamentals for the subject.

Both neuroscience, and neurotechnology and biotechnology present results that can be understood as predominantly "verifiable", regarding the problem of perception. As we know that none of these fields of study can as yet give us the definitive answer to what we could consider an understanding of the perceptual phenomenon, in the current phase of our work we chose to study perception using theoretical instruments from neuroscience. A future step will be to organize these results in comparison with the theories indicated above and, if possible, focus our work on the development of an analytical model combining all of them. This is still a hypothesis.

REFERENCES

Bollnow, O. F. (2008). *O homem e o espaço*. Curitiba: Editora UFPR.

Cairns-Smith, A. G. (1996). *Evolving the mind on the nature of matter and the origin of consciousness*. Cambridge, UK: Cambridge University Press.

Damásio, A. (1996). *O erro de Descartes*. São Paulo: Cia das Letras.

Damásio, A. (2000). *O mistério da consciência*. São Paulo: Cia das Letras.

Damásio, A. (2004). *Em Busca de Espinosa: prazer e dor na ciência dos sentimentos*. São Paulo: Companhia da Letras.

Damásio, A. (2011). *E o cérebro criou o homem*. São Paulo: Companhia das Letras.

Engelmann, A. (2012). A Psicologia da Gestalt e a ciência empírica contemporânea. *Psicologia: Teoria e Pesquisa, 18*(1). 10.1590/S0102-37722002000100002

Koffka, K. (1957). *Princípios da Psicologia da Gestalt*. São Paulo: Cultrix.

Köhler, W. (1980). Psicologia da Gestalt. Belo Horizonte: Itatiaia.

Livingstone, M. (2002). Vision and art: The biology of seeing. New York: HNA Ed.

Peirce, C. S. (1977). *Semiótica*. São Paulo: Editora Perspectiva.

Ramachandran, V.S., & Blakeslee, S. (2012). *Fantasmas da mente: uma investigação dos mistérios da mente humana*. Editora Record.

Ramachandran, V. S., & Hubbard, E. M. (2005). The emergence of the human mind: Some clues from synesthesia. In *Synesthesia - Perspectives from Cognitive Neuroscience. Oxford-NewYork* (pp. 147–190). Oxford University Press.

Sacks, O. (1996). *An Anthropologist on Mars*. New York: Vintage.

Sacks, O. (1998). *The Man Who Mistook His Wife for a Hat: and other Clinical Tales*. New York: Touchstone.

Sacks, O. (2007). *Alucinações Musicais*. São Paulo: Companhia das Letras.

Santaella, L. (2001). *Matrizes da Linguagem e Pensamento*. São Paulo: Iluminuras.

Schafer, M. R. (2001). *A Afinação do Mundo*. São Paulo: UNESP.

Schafer, M. R. (2003). *O Ouvido Pensante*. São Paulo: UNESP.

Uexküll, T. (2014). A teoria da *Umwelt* de Jakob von Uexküll. *Revista Galáxia, 7*, 19-48. Retrieved from: http://revistas.pucsp.br/index. php/galaxia/article/viewFile/1369/852

Wertheimer, M. (1971). O fenômeno Phi como um exemplo de nativismo na percepção. In *Textos básicos de História da Psicologia*. São Paulo: Herder e EDUSP.

KEY TERMS AND DEFINITIONS

Art and Neuroscience: Connection between art and neuroscience that involves artistic and/or scientific production and study, whose benefits can be shared between the two areas of knowledge even if only one is emphasized.

Multimodality: Simultaneous stimuli that are given to the individual, and that can lead to perceptual processes.

Multisensoriality: Simultaneous perception, through more than one of the individual's sensory receptors, whose reception may not be conscious in its totality.

Perception: Selective and adaptive capacity of beings endowed with neuronal functions, to gather information from the surrounding environment and the organism's internal functions through nerve endings. This phenomenon depends on the mental and biophysical qualities of the organism, when applicable.

Perceptual Processes: Event during which the organism's internal or external signals are configured as a perception.

Open Source: Open software technology, which allows different individuals to modify significant parts of it, without onus in relation to law and patent software. However, this type of distribution can establish copyright, on key parts of the software, which can not be changed without granting of rights from the author.

Qualia: First instant of the perceptual process that is triggered by mental processes which qualify in firstness, the emergence of a perception.

ENDNOTES

1 GIIP (International and Inter-institutional Research Group on Convergences between Art, Science and Technology) is composed of Research Group Leaders and the researchers linked to them as collaborators. Currently, 36 people participate in the group including researchers, students and technicians, all registered in the CNPq system, in addition to students that are teaching assistants and Scientific Initiation (IC) students without a scholarship. Researchers both nationally and internationally are also part of the network. GIIP's lines of research are 1 - Convergence between Art, Science and Technology; 2 - Physical and Digital interfaces for the arts: from diffusion to inclusion; 3 - Digital and Mobile Technologies and their Applications; 4 - Nanoart. The group is certified by the UNESP and by the CNPq.

2 This line of research has five sub-projects. These results refer to this author's subproject entitled "Neuroscience and Perception: Multisensoriality and Multimodality'. It was developed as a Post doctorate research at the Universidade Aberta (Open University) in Lisbon, Portugal, under the supervision of Dr. Adérito Marcos, and research grant from the Fundação de Amparo à Pesquisa do Estado de São Paulo - FAPESP (Research Support Foundation of the State of São Paulo) in Brazil.

3 As explained in our paper/publication "Multisensoriality and Synesthesia: Possible Poetics?" In: Proceedings of the 6th International Conference on Digital Arts, ARTECH. Lisbon: 2012 and in the Revista Ars. Ano 2, nº 24, 43-61. São Paulo: USP.

4 By modalities we understand types of stimuli.

5 Here there is no concept of separation between brain and body. Our approach follows that of Damásio (2000: 61), for whom the brain is part of an integrated, live and complex organism.

6 Jacob Von Uexküll, a German biologist, developed this concept in 1909. He created the building blocks for the development of the theory of biosemiotics.

7 Cerdá is one of our research collaborators, a GIIP member. The artist's website: www.artesonoro.org. Access in January 2014.

8 The reader may notice that we are quoting Margaret Livingstone instead of Semir Zeki. We suggest consulting our publication "Multisensoriality and Synesthesia: Possible Poetics?" In: Proceedings of the 6th International Conference on Digital Arts, ARTECH. Lisbon: 2012, where we criticized the author's text and, as explained in another publication, although we accept his statements about the qualification of the visual brain, we disagree with some of his theoretical applications in the context of art.

9 Team: Rosangella Leote - Project, design and development; Rec(O)rganize (Rodrigo Rezende and Fernanda Duarte) - hardware, software, programming and video editing; Miguel Alonso - Installation Assembly Assistant. We used ultrasonic sensors, an Arduino microcontroller and Pure Data software on the Windows system.

10 "Zonas de Compensação" is both the name of the extension project and also of the annual exhibition held by the same project. It is organized by the GIIP, under our coordination, and is where the work was shown.

11 Emphasis added.

12 Expanding the concept, Somaesthesia can be treated as a perceptual group where the submodalities known as touch, pain, temperature and pressure are found. Somaesthetic perception occurs thanks to receptors that are known as Meissner's corpuscles, Krause's corpuscle, Pacinian corpuscle, Merkel's corpuscle, hair follicle endings, Ruffini endings and Free Nerve Endings. This is a scientific classification, which we are not concerned with here, understanding that so far it has seemed, first of all, unnecessary to invest in this breakdown into minute detail, since we prefer to focus on multisensoriality with stimulus modalities that can be simulated by technologies. Secondly, because the research sources we have encountered up to now have not provided us with enough clarity to get into this issue, as we have noticed excessive incompatibilities, especially in the classification of the senses. It seems, however, that there is consistency in the recognition of three kinds of perception: the exteroception (stimuli from outside the body), proprioception (the body's own stimuli - also called kinaesthetic) and visceroception, also known as interoception (stimuli of the viscera), although the latter is also treated as a specific sensing. The somaesthetic system as a whole includes sight and hearing (containing the information receptors that are outside the body). Despite smell and taste also carrying information from outside, these are not considered part of exteroception but rather visceroception, which once again, puts us in an uncomfortable position to accept this or that classification.

This research was previously published in Projective Processes and Neuroscience in Art and Design edited by Rachel Zuanon, pages 1-14, copyright year 2017 by Information Science Reference (an imprint of IGI Global).

Chapter 6
Enabling Creativity:
Using Garden Exploration as a Vehicle for Creative Expression and Analysis

Becky Boesch
Portland State University, USA

ABSTRACT

This chapter uses current developments in cognitive neuroscience to explore the notion that educational activities should involve the whole person. To that end, the author explains in depth an undergraduate college assignment that allows for learning through the coupling of creativity (divergent) and analysis (convergent) thinking in an integrative learning task. First, students explore the concept of metaphor which provides both mental association and ambiguity. With this underpinning, students experience three very different types of gardens and try to uncover the metaphors of nature lying within their design. Students record a journal and take images of the gardens and later create a photographic montage of each garden which reflects the metaphor that the students saw emerging in the garden itself. Accompanying the visual image is a written reflection which discusses the metaphor they experienced sensually in the garden and how it is represented in their images.

INTRODUCTION

Recent developments in cognitive neuroscience (CN) particularly as it relates to creativity should challenge the educational community to reassess their teaching practices. The understanding of creativity has expanded through CN research. New studies indicate creative thinking involves the whole brain and is not localized. Creativity utilizes both divergent (the entertaining of multiple ideas) and convergent (the coalescing of this multiplicity) thinking. These new findings hint at the potential benefits of developing creativity in students. Based on this new information, the development or encouragement of creative thinking should exist in all educational classes not just those in such artistic fields of art, music, dance and creative writing. Educators who are committed to the development of students who can examine diverse and seemingly unrelated ideas in order to create a "new" understanding should make creative learning

DOI: 10.4018/978-1-5225-5478-3.ch006

tasks the center of their pedagogy. By including creative learning tasks, educators are encouraging two important learning concepts, critical thinking and integrative learning.

This chapter will first examine relevant CN research as it relates to creativity and creative thinking. Of particular interest will be the use of representational systems such as metaphor in the creative task and how divergent and convergent thinking work together in the act of creation. These key findings will be discussed in the ways they connect to two educational constructs: critical thinking and integrative learning. The author will provide a specific assignment using gardens which takes both CN research and educational pedagogy into account as a model for enabling and developing creativity within students at the undergraduate college level and will conclude with ways that the research between CN, creativity and educational pedagogy need to be continued and clarified.

COGNITIVE NEUROSCIENCE AND CREATIVITY

What is creativity? The most accepted definition of creativity was proposed by Sternberg & Lubart in 1999 and is composed of two components. Creativity is "the ability to produce work that is both novel (i.e. original, unexpected) and appropriate (i.e. useful, adaptive concerning task constraints)" (p.677). While this definition is well accepted, its understanding and application to CN is still in its infancy. "There are few neuroscience studies of creativity or of the creative process. This is most likely due to the difficulties of defining creativity and the lack of psychometric means of assessing it....Nonetheless, there may well be a neural basis for creativity" (Haier & Jung, 2008, p. 172). Even though much more research needs to be done, initial understandings have emerged. Whereas creativity was initially thought to reside primarily in the right hemisphere, that idea has been debunked and now it is clear that creativity is much more complex in its processing and involves the whole brain (Sawyer, 2011, Dietrich, 2007). Many regions of the brain, in both hemispheres are active during creative tasks. In fact, to take it a step further, Immordino-Yang (2011) and Immordino-Yang & Damasio (2007) stress that "affective neuroscience is revealing that the mind is influenced by an interdependency of the body and brain; both the body and brain are involved, therefore, in learning" (p. 99). So, not only is the whole brain involved in creative acts but the body as well. These new realizations have import for education. If educators are committed to maximizing learning, then learning activities should be designed that allow for whole brain and body learning in order to encourage and develop creativity.

Studies have also revealed in more detail the mental processes involved in creative acts. Initially creativity was thought to involve only divergent thinking. "Divergent thinking refers to unbound ideational searching or open-ended thinking that is typically evoked in creativity tasks where solutions need to be generated for problem situations that do not have right or wrong answers" (Abraham & Windmann, 2006, p. 39). Divergent thinking has often been paired with two other "prerequisites" to creative acts, defocused attention or disinhibition. However, research shows that creative thinking also potentially involves convergent thinking. Convergent thinking has long been associated with intelligence and has been connected with the concepts of focused attention and inhibition. The distinction between these two is that "convergent cognitive processes...arrive at one correct answer" whereas in divergent thinking "multiple correct responses are plausible" (Haier & Jung, 2008, p. 175). Gorborz and Necka (2003) stated, "these two processes – generation of ideas and evaluation – seem to require intellectual operations that are contradictory in nature" (p. 183) but feel that both are necessary for a creative act. "That is, inhibition, focused attention and controlled processing as well as disinhibition, defocused attention and automatic processing may be relevant for producing creative ideas" (Edl, Benedk, Papousek, Weiss & Fink, 2014, p. 39). In fact, researchers are beginning to recognize that intelligence and creativity play complementary roles in the brain.

Cognitive activities ensuing from an ill-defined task can be categorized as convergent thinking (which is associated with intelligence) and divergent thinking (which is associated with creativity) establishing that creative processes contain both modes of thinking intertwined. In the creative process... convergent thinking assesses, evaluates, and integrates generated information. (Jaarsveld, fink, Rinner, Schwab, Benedek, Lachmann, 2015, p. 172)

Simonton (2004) as quoted in Dietrich (2007) makes a case for the notion that creativity is a Darwinian process in that it involves variation and selection. Ideas are generated all the time but require a selection process to deem whether they are truly creative. In fact, Edl, Benedek, Papousek, Weiss, and Fink (2014) proposed that divergent and convergent thinking followed a particular process in creative work. In the earlier stages when the problem or idea is ill defined, divergent thinking plays a larger role because the mind can pull in associated and seemingly irrelevant ideas. However as the task continues, the thinking becomes more convergent as specific ideas or courses of action develop (p.38). Abraham and Windmann (2007) further elaborate this process by stating that various creative processes "all involve the initial generation of potential ideas, or 'preinventive' structures, such as the formation of associations between stored conceptual structures in memory and the analogical transfer of information from one domain to another" (p. 39).

One example which demonstrates this process of how divergent and convergent thinking are utilized is in the making of art. "Works of art can be tentatively viewed as elements of a human-specific nonverbal communication, distinct from language 'halfway between scientific knowledge and mythical thought'" (Changeaux, 2011, p. 3). The common view of art and art making is that it is completely divergent in its process but it too uses convergent thought to coalesce artistic expression. Changeaux (2011) states that artistic expression involves subjectivity, conscious access of schema and application of rules that shape that artistic expression. "These *regles of l'art* hypothetically viewed as acquired patterns of connections or scaffoldings...include, among others, novelty, the coherence of parts with the whole, parsimony or the most frugal route of expression, the tension between bottom-up realism and top-down abstraction, the search for societal recognition, and the artist's conception of the world" (Changeaux, 2001, p. 4). CN research has shown that multiple brain systems are utilized in the creation of art.

These ideas shed light on educational practices in a number of ways. If the goal of education is to develop and encourage intelligence then this new research would advocate that intelligence can be further enhanced through allowing and developing creativity. Teaching in a way that only employs convergent thinking is limiting the student's learning. Perhaps students shouldn't only be taught the "right answer."

Another insight that develops from this research maps how educators, who want to develop creativity, can construct their assignments so that students successfully participate in creative acts. First, any assignment that wants to employ creativity should allow students time initially to associate the issue/problem with their previous understandings, but it is important what previous associations are activated. Some evidence suggests that allowing individuals to access previous concepts inhibits creativity. Abraham and Windmann (2007) found that subjects produced ideas for new toys that were more conforming to the norm after they were shown exemplars of novel toys. In other words, their creativity decreased (p. 41). Snyder (2007) agrees in that "mindsets" often make individuals blind to alternative interpretations (p.

9). In both of these previous examples, the researchers are referring to schemas. Schemas are ideal representations of an object. For example, when someone uses the word "dog," individuals evoke an image of a dog or the concept of a dog that best approximates the ideal dog. This can indeed become limiting because then all dogs are unconsciously measured against the schema of an ideal dog. For example, a German shepherd would, for many, be a closer approximation to our schema of a dog then a Chihuahua. The schema of dog restricts associations rather than expands them. However, there are associations that can potentially expand our creativity. Sawyer (2011) stresses that research shows that creativity is enhanced through association and, in particular, distant association (p. 149). This implies that these associations need to move beyond schema to broader mental associations that encourage metacognition. This is represented in metaphorical thinking.

The nature of metaphorical thinking allows for greater dissociative thought through broad diffuse associations. Metaphor initially was seen as merely a literary device but Lakoff and Johnson (1980) in their landmark work revealed how metaphor is the guiding mechanism by which we understand the world. "Our ordinary conceptual system, in terms of which we both think and act, is fundamentally metaphorical in nature" (Lakoff & Johnson, 1980, p. 3). One of Lakoff and Johnson's examples is the metaphor that money is time. Rather than limiting the concept of money, the metaphor allows for greater and wider associations because two seemingly disparate concepts of time and money interact. Thus we have such common taken for granted statements as:

- You're running out of time.
- He's living on borrowed time.
- You don't use your time profitably.
- You need to budget your time. (Lakoff & Johnson, 1980, p. 8).

Gruber (1981) in studying creative individuals found that "the creative individual also pursues (or is pursued by) a number of dominant metaphors. These figures are images of wide scope, rich, and susceptible to considerable exploration, exposing the investigator to aspects of phenomena that might otherwise remain invisible to him." (Gardner, 1982, p. 354). The creative role of metaphor surfaced in Darwin's evolutionary idea of the branching tree, William James' image of ideas as a stream or river, and John Locke's focus on symbolism within the falconer (Gardner, 1982, p. 354).

So the key here for the educator is to allow students to make associations that expand their thinking rather than limit it. Two well-known education concepts that encourage these diverse associations is critical thinking and integrative learning. The American Association of Colleges and Universities (AAC&U) has recognized the need for reconceptualizing what we do in higher education by identifying four essential areas that need to be addressed in a college education in which both critical thinking and integrative learning are named (Hovland & Scneider, 2011).

Critical thinking is in essence the process of challenging taken for granted assumptions and coming up with novel ways to readdress or understand a situation or problem. In other words, a critical thinker is able to challenge existing ideas and explore new ones in order to arrive at a new understanding. The most common accepted definition of critical thinking was arrived at by a Delphi consensus panel. The Delphi technique of data collection utilizes questionnaire and reiterative feedback among experts in the field in order to develop a wider and more inclusive understanding of specific concepts or issues. The consensus reached by the panel states that *[w]e understand critical thinking to be purposeful, self-regulatory judgment which results in interpretation, analysis, evaluation, and inference, as well as explanation of the evidential, conceptual, methodological, criteriological, or contextual considerations upon which that judgement is based. (Facione, 1990, p. 2)*

Thus, in order to practice critical thinking a student looks at possible meanings and associations (divergent thinking, defocused attention, and disinhibition) in order to achieve a new novel understanding (convergent thinking, focused attention and inhibition).

Another related area in education is integrative learning. The Association of American Colleges and Universities define integrative learning as "an understanding and a disposition that a student builds across the curriculum and co-curriculum, from making simple connections among ideas and experiences to synthesizing and transferring learning to new, complex situations within and beyond the campus" (Rhodes, 2010, p. 51). Mary Taylor Huber (2005) perhaps says it best when she states that integrative learning is

[f]ostering the intellectual art of making, recognizing, and evaluating sound, meaningful connections across different concepts, cases, or experience....[w]hen people single out integrative learning as deserving special attention, they are usually talking about the larger leaps of imagination rather than little ones, about mastering intellectual arts that can be used to make productive, provocative, and memorable connections between domains that have usually been—in one's personal, academic, or cultural experience as isolated, separated, or kept apart. (p.3-4)

Integrative learning coincides with creative thought as does critical thinking but perhaps on a larger more disparate scale. "Integrative learning is, at its core, a process for synthesizing learning across multiple experiences, coalescing meaning, and also creating new learning and meaning" (Taylor, 2011, p.14). In some respects, integrative learning could be argued to be the most creative of the two processes. Dietrich (2007) speculates that "[a]t a minimum, we can assume that the more integrative the neural structure involved in the computation, the more combinational novelty might occur" (p. 25). Both of these types of learning hold promise for bringing new connections together to arrive at a new "creative" understanding.

In conclusion, new findings in CN and creativity studies have perhaps strengthened what some have already been advocating in education in terms of encouraging critical thinking and integrative learning in the classroom. This is because both critical thinking, on a smaller scale, and integrative learning are in themselves creative thought because they first encourage students to entertain divergent, dissociative thoughts and then converge those ideas into a new, novel understanding. In addition, educators now know that there seems to be a connection between creativity and intelligence. If one wishes to develop intelligence in students then attention should be paid to creativity as well. Shouldn't learning, particularly higher level learning be creative? In other words, students should be presented with disparate but related ideas and must cogitate on the relationship of those ideas to arrive at a new creative understanding and expression.

GARDEN EXPLORATION AS A MEANS FOR DEVELOPING CREATIVITY

The above developments and understandings underpin an assignment which was created for undergraduate freshman students. This assignment is part of a year-long freshman general education course which has four learning goals: communication, human diversity, critical thinking and social/ethical responsibility.

Each of these courses are created jointly by faculty from disparate disciplines. The theme of the course for this particular assignment was Human/Nature. The course description is as follows:

The human animal is considered to be both a part of and yet distinct from nature. This relationship between our human selves and the natural world we inhabit is complicated and perplexing. This theme explores the complex connections between humans and nature. In what ways are we humans "natural"? Is there such a thing as human nature, and if so, what is it? How are we related to nature and our larger natural surrounds? How have we described and represented nature to ourselves? How have humans over the course of time understood and interacted with the natural world? How have our understandings of nature changed? Do humans have unique responsibilities toward the natural world and if so, what are they? Over the course of the year we will attempt to answer these questions, drawing on the resources of the social and biological sciences, history, literature and the arts.

These freshman interdisciplinary undergraduate courses encourage diverse and varied associations, thus making this course a perfect setting for the development of critical and integrative thought through creative activities. Such disciplinary areas as evolutionary biology, anthropology, linguistics, visual art, poetry, environmental science, and history are just a few of the lenses through which we tackle this theme. The assignment addressed in this chapter occurs in the middle of the course when we are exploring the questions of how humans have understood the natural world and their role in it. The focus of this particular term involves lots of creative "stretching" for the students as they visit art museums, write poetry and, as in this assignment, visit gardens. The entire project begins with an initial look at the use of metaphor as it applies to landscape. To help prime the students, they read the chapter, "The Beholding Eye: Ten Versions of the Same Scene" by D. W. Meinig, (1979). In terms of landscape, Meinig (1979) states that

even though we gather together and look in the same direction at the same instant, we will—we cannot—see the same landscape. We may certainly agree that we will see many of the same elements—houses, roads, tree, hills—in terms of such denotations as number, form, dimension, and color, but such facts take on meaning only through association; they must be fitted together according to some coherent body of ideas. Thus we confront the central problem: any landscape is composed of not only what lies before our eyes but what lies within our heads (p. 34).

He goes on to illustrate and explain ten potential ways that someone could "define" landscape, i.e., landscape as nature, habitat, artifact, system, problem, wealth, ideology, history, place and aesthetic. The initial discussion from this article challenges students to increase their associations, connections and understanding of landscape. They begin to realize that there are many ways to "see" landscape. The goal here is to encourage divergent thought "...unbound ideational searching or open-ended thinking that is typically evoked in creativity tasks" (Abraham & Windmann, 2006, p. 39). Besides the discussion of the article, the class is shown various images of landscape and the students are asked initially to decide individually what they see as the most prominent metaphor emerging from the image. This is then opened up for a class-wide discussion and students are confronted with the fact that many of their peers did not see the same metaphor or metaphors in the image further forcing them to see that there are multiple ways to understand (divergent thinking) rather than one correct answer (convergent thinking).

Once the students have been exposed to the multiplicity of interpretations (metaphors), which can be brought to bear when viewing landscape, the class turns its attention to one of Meinig's metaphors, in particular, landscape as aesthetic. As the term implies, this metaphor sees landscape as a work of art. While Meinig talks about this metaphor in terms of landscape painting, the same ideas apply to designed gardens/parks. Both express many meanings (metaphors) through their art by paying specific attention to things such as color, texture, mass, line, position, symmetry, balance and tensions. Landscape as aesthetic

...rests upon the belief that there is something close to the essence, to beauty and truth, in the landscape. Landscape becomes a mystery holding meanings we strive to grasp but cannot reach, and the artist is a kind of gnostic delving into these mysteries in his own private ways but trying to take us with him and to show what he has found. (Meinig, 1979, p. 46)

Hopefully, by this time the students have increased the complexity of their thought, first in the realization that landscape can be "seen" through many metaphors, and second in understanding landscape and its meaning incorporates such artistic aspects as color, texture, mass, symmetry. At this point, students should have begun to implement divergent thinking which is the initial step needed in critical and integrative thinking as well as creativity. This is done in the hope that the more disparate the associations, the greater the creativity will be later on in the project.

Once their thinking has been primed, the students are given the assignment (Appendix). Students are asked to visit one garden in each of the provided three categories. The reason for providing categories is for students to have the most diverse experience in the types of gardens/parks that are available in the area. Upon visiting the gardens, the students are asked to create a written journal of the experience as they are moving through the gardens. The purpose of this journal is twofold. The first is to have the students record their visceral experience of the garden. How did it touch their senses such as sound, smell, touch, sight? This is done with the understanding that creativity is a complex mental process that uses many different portions of the brain and that an activity that utilizes as many of the senses as possible has a greater chance of engaging a student's creativity. The second goal is to have the students hone their observational skills by noticing such things as shapes, colors, textures, sizes, surfaces as they connect to their experience in the garden. This exercise reflects Wilhelm Wundt's process of introspection which recorded thoughts and sensations and then organized them into categories. This also addresses what Immordino-Yang (2011) states as whole body learning. "Affective and social neuroscience findings suggest…that emotion and cognition, body and mind, work together in students of all ages" (p. 102). In addition to the written journal, they are asked to take photographic images of the garden in order to document their experience. Not only will these photos be used later in the assignment but they also provide a visual record of their experience for them to draw on in their reflection.

Once the students have visited the gardens and have their initial journal and images, they are then asked to begin the process of analyzing and interpreting the garden. These two artifacts should present them with a number of random thoughts, ideas, feelings, and senses. From this morass, a larger and more coalesced meaning begins to emerge. Some of the ideas they should grapple with from their journal are things such as:

Where does your eye wander (or is it led?) as you enter the space? What dominates the space? In other words, what "happens" when the elements of the garden are situated as they are? Think about the relationships between the shapes, colors, textures, sizes, surfaces and you, the visitor to the garden.

Once, they have started to make "sense" of their experience, students are then asked to determine what metaphor emerges for them in each garden. What is the overall meaning emerging from the aesthetic of the garden? It is through this process that they move from the most diverse of experiences (divergent thinking) to greater clarity and convergence of thought.

Once they have determined the most significant metaphor for them that emerges out of each of the gardens, they then take this understanding and construct their own creative expression of it through utilizing the photos they have taken and representing them through visual software. In other words, they are asked to visually represent the metaphor through the images they took at each garden. Students come to the class with various levels of skill in terms of using appropriate software. Some can only utilize basic software provided through Microsoft Office but others are skilled enough to use such increasingly complex systems like Photoshop. They are given no set instructions on what each of these images should look like other than the construction should reflect the metaphor represented for them in the garden. They also include a short written explanation of the rationale as to why they visually presented the images as they did.

To illustrate the range of possible visual interpretations, the author includes examples of two students' work (see Figures 1-6).

These two students' works reflect the incredible creativity that can potentially emerge from this assignment. The students represented their understandings quite differently but successfully. Students in their end of term reflections often mention this assignment and how enjoyable it was. This is because this

Figure 1. Student 1 example 1

Figure 2. Student 1 example 2

Figure 3. Student 1 example 3

Figure 4. Student 2 example 1

project allows for learning that is expansive and creative. Students are able to engage in both intellectual and experiential learning. The project allows them to move through the creative process successfully by

1. Presenting the concept of metaphor as a construct (i.e. providing a larger association for the project).
2. Complicating and expanding the notion of metaphor thus allowing for the beginning of divergent thought.
3. Allowing for a seemingly random collection of experiences, observations, and insights from their garden visits.
4. Providing mechanisms (journal, photos) for students to reconstruct their understanding of the gardens and of metaphor (convergent thought).
5. Encouraging a final project which allows for creative expression (i.e., work that is both novel and appropriate) both through written and visual expression.

At the same time that this assignment allows for creative expression from the students, it also helps develop both critical thinking and integrative learning. With critical thinking the student has to begin by questioning his/her understanding of metaphor and then must develop a new understanding from new knowledge and experiences. The student participates in integrative learning because he/she has to synthesize learning from multiple experiences (reading and discussion on metaphor, garden visitations,

Figure 5. Student 2 example 2

journal writing, and photographic images) and then must bring those together for a new understanding of metaphor, gardens, and art in his/her artistic expression of the metaphors emerging from the gardens.

SOLUTIONS AND RECOMMENDATIONS

Implementing this kind of project is not without challenge. Students are initially resistant to completing the project for a number of reasons. First, many question their own ability to be creative. This is because our educational system has, in essence, severely limited creativity in students or at least their perception of their own creativity. Sir Ken Robinson's example (2006) of the longitudinal study involving paper clips in his famous TED talk comes to mind. In this longitudinal study, children were asked at various ages starting from an early age to give as many possible uses for a paper clip. As their schooling progressed, the number of uses for a paperclip dramatically declined as the children grew older. This is reflective of how our schools limit rather than develop creativity in students. By the time students get to college, many no longer believe that they are creative or have the potential to be creative. They feel much more comfortable with the standard modes of learning which involve a midterm, final and perhaps a paper. They have been conditioned to see learning as convergent (one right answer) rather than divergent (multiple possible answers). To alleviate this anxiety, the instructors attempt to create a climate in the class for students so that they feel comfortable taking creative risks. Fortunately, because this project

Figure 6. Student 2 example 3

occurs midway in a year-long course, the instructor already knows the students quite well and can offer encouragement and support risk.

Another resistance which surfaced also ties into students' expectations of what learning is. Many complained about having to "leave the classroom." Again, they have been conditioned to think that learning occurs within the classroom not outside of it. For many of them, learning is sitting at tables or desks where the mind is exercised but nothing else. This resistance can partially be alleviated by talking about the nature of learning itself and how learning is essentially experiential and lifelong. People learn by doing. Another hurdle encountered with this assignment was the different level of computer knowledge represented in the class. Some students are very well acquainted with image manipulation software and others are not. In evaluating this project, the instructor needs to take this disparity into account.

If education is truly committed to enhancing creativity within our students, then much of the paradigm of what learning is and how it works needs to be reconsidered. Educators from kindergarten to college need to incorporate a much wider range of learning activities that allows students to engage holistically in their learning. There has been far too much emphasis placed on the development of convergent thought without understanding that divergent thought is equally important. Part of this stems from myths in education that continue to persist. Many educators still believe that creativity is a separate skill that occurs primarily in the right hemisphere of the brain and is only employed in certain subject areas such as art, music, and drama. It is clear now that creativity is needed in all subjects.

FUTURE RESEARCH DIRECTIONS

There are a number of areas where much more research needs to be done in order for us to understand creativity and implement it effectively in education. In terms of cognitive research into creativity, researchers have barely scratched the surface of how the brain actually works. To be able to fully understand this is a daunting task. As mentioned at the beginning of this chapter, Haier and Jung, 2008, state that clarity still needs to emerge as to what creativity is and, in addition, how to psychometrically measure it. Most researchers agree with Sternberg & Lubart's (1999) definition of creativity but how does that translate into how the brain functions? So, two significant questions need to be answered in cognitive neuroscience research. What is creativity in the brain? How do you measure it? This is no small task. Researchers have provided tantalizing evidence but our understanding is still fragmented and incomplete.

Besides tackling these overarching questions, there are current areas of CN research that hold promise for education. For example, in terms of Immodino-Yang's research, they discuss the current findings in how both the body and mind work in learning. "Interestingly, evidence from social and affective neuroscience is shedding new light on the neural underpinnings of such social processing, affective responses and their relation to learning" (Immordino-Yang, 2011, p. 98). This research should be expanded to explore how the mind and body work together in the creative act. Researchers suspect there is a connection but they need more specificity. If we can answer this question, then educators can redesign the pedagogy of their classrooms to allow for this. Experiential learning would potentially become much more significant within educational circles.

One of the more promising areas of CN research for education is how divergent and convergent thought work together in the creative act. (Perhaps divergent and convergent thought is involved in all higher order thinking?) Preliminary research indicates that divergent thinking precedes convergent thinking in the creative act but these findings are still tentative. In connection with this is the need to clarify what kind of divergent thinking, i.e. what kind of associations are helpful in encouraging creativity since the research is conflicting in this area. Some researchers state that preexisting ideas limit creativity, others state that these associations can be helpful. Further clarity here could potentially have significant impact in what educators do. First, if divergent thinking plays a critical role in convergent thinking then educators would need to rethink how they teach and what they teach. Rather than initially presenting information as right/wrong (convergent thinking), the educator would need to present information in a way that allows and encourages problem solving (divergent thinking) which can then subsequently result in convergent thinking. Changes like this in educational pedagogy could result in significant learning gains for students. Not allowing for enough divergent thinking in the teaching process might explain why students often forget what they have learned after taking a test. Perhaps their learning processes were truncated and therefore not long-lasting because they were not allowed to explore the ideas divergently initially.

Another area of research that is desperately needed is to connect current educational concepts such as critical thinking and integrative learning with brain research into creativity. There seems to be tantalizing parallels between what we currently know about creative thought and what critical thinking and integrative learning attempt to accomplish. In fact, both critical thinking and integrative learning seem to follow the divergent/convergent thinking process involved with creativity. However, this researcher was not able to find any CN studies that have tried to map how either of these processes work in the brain. If critical thinking and integrative learning can be shown to follow similar patterns in the brain as creativity, it would validate the notion of those educators who are advocating and/or utilizing the incorporation of these pedagogies in their classrooms. In conclusion, educators need a more complete picture of creativity from CN researchers to be able to align educational practices so that creativity is encouraged and developed in students.

CONCLUSION

Although we have just begun to explore the complexities of creative processing in the brain, the initial results have been both illuminating and encouraging for those involved in education. This research has revealed that creativity is a whole brain process rather than a predominantly right brain activity. In fact, learning, of which creativity is a part, involves the whole body. Creativity also employs not only divergent thinking but convergent thinking as well. Creativity is facilitated by allowing for diverse associations and seemingly random explorations of ideas which then enable a convergence of new ideas. Creativity and intelligence seemed to be linked. These are just some of the findings that have surfaced in the early stages of research.

These initial findings are exciting and hold the potential for revolutionizing the way educators teach. "For education to truly benefit from these neuroscientific findings in a durable, deep way, for the full implications to become apparent, educators must examine closely the theory on which good practice is built, to reconcile new and exciting evidence with established educational models and philosophies" (Immordino-Yang, 2011, p. 102). The paradigm of education as a banking concept (Freire, 1976) where students' minds are repositories of information which they memorize and regurgitate no longer seems a viable philosophy or at the very least it is one that needs to be significantly restructured. If these initial findings hold true, it would imply that teachers need to involve students in experiential learning where all of the senses are employed rather than treat them like passive objects in the classroom. Teachers need to construct learning activities that allow for varied exploration of divergent ideas before students are asked to arrive at a conclusion. These potential findings lend support to the notions of critical thinking and integrative learning where students are asked to question and challenge assumptions and pull new ideas from a multiplicity of areas. This garden project was presented as a tangible example of how educators can begin to incorporate what we currently know about creativity into college teaching through the use of metaphorical thinking, experiential learning, and artistic representation of ideas. Education, in light of this new research will need to become more exploratory in nature by allowing students to take creative risks on a regular basis.

To borrow Sir Ken Robinson's example of the paper clip study, educators need to help students find more uses for paper clips over the course of their education rather than fewer. For it is in the act of creativity that some of our greatest accomplishments as a species have come about whether as a symphony by Mozart or the Theory of Natural Selection by Darwin. In this rapidly changing world, our society needs individuals that are more creative rather than less to tackle the problems that are facing our world. Educators need to encourage intelligence that can take in and process the seemingly random multiplicity of ideas/issues confronting our world. Teaching or promoting creativity should no longer be the purview of certain educational disciplines but needs to be embraced by all in order to equip students for the challenges that lay ahead.

REFERENCES

Abraham, A., & Windmann, S. (2007). Creative cognition: The diverse operations and the prospect of applying a cognitive neuroscience perspective. *Methods (San Diego, Calif.), 42*(1), 38–48. doi:10.1016/j.ymeth.2006.12.007 PMID:17434414

Changeux, J. P. (2011). The neuroscience of art: A research program for the next decade? *Mind. Brain and Education, 5*(1), 3–4. doi:10.1111/j.1751-228X.2011.01102.x

Dietrich, A. (2007). Who's afraid of a cognitive neuroscience of creativity? *Methods (San Diego, Calif.), 42*(1), 22–27. doi:10.1016/j.ymeth.2006.12.009 PMID:17434412

Edl, S., Benedek, M., Papousek, I., Weiss, E. M., & Fink, A. (2014). Creativity and the Stroop interference effect. *Personality and Individual Differences, 69*, 38–42. doi:10.1016/j.paid.2014.05.009

Facione, P. A. (1990). Critical thinking: A statement of expert consensus for purposes of educational assessment and instruction. Research Findings and Recommendations. Newark, DE: American Philosophical Association. (ERIC Document Reproduction Service No. ED315423).

Freire, P. (1976). *The Pedagogy of the Oppressed*. New York, NY: Continuum Publishing Co.

Gardner, H. (1982). *Art, Mind, and the Brain: A Cognitive Approach to Creativity*. New York: Basic Books, Inc.

Groborz, N., & Necka, E. (2003). Creativity and cognitive control: Explorations of generation and evaluation skills. *Creativity Research Journal, 15*(2-3), 183–197. doi:10.1080/10400419.2003.9651411

Haier, R. J., & Jung, R. E. (2008). Brain imaging studies of intelligence and creativity: What is the picture for education? *Roeper Review, 30*(3), 171–180. doi:10.1080/02783190802199347

Huber, M. T. (2005). *Integrative learning as an intellectual art*. Paper presented at the meeting of the AAC&U Network Conference on Integrative Learning, Denver, CO.

Immordino-Yang, M. H. (2011). Implications of affective and social neuroscience for educational theory. *Educational Philosophy and Theory, 43*(1), 98–103. doi:10.1111/j.1469-5812.2010.00713.x

Jaarsveld, S., Fink, A., Rinner, M., Schwab, D., Benedek, M., & Lachmann, T. (2015, March). Intelligence in creative processes; An EEG study. *Intelligence, 49*, 171–178. doi:10.1016/j.intell.2015.01.012

Lakoff, G., & Johnson, M. (1980). *Metaphors We Live By*. Chicago, IL: The University of Chicago Press.

Meinig, D. W. (1979). The beholding eye: Ten versions of the same scene. In D. W. Meinig (Ed.), *The Interpretation of Ordinary Landscapes* (pp. 33–48). New York: Oxford University Press.

Rhodes, T. L. (Ed.). (2010). *Assessing Outcomes and Improving Achievement: Tips and Tools for Using Rubrics*. Washington, DC: American Association of Colleges and Universities.

Robinson, K. (2006). *Do Schools Kill Creativity?* Retrieved June 23, 2015, from https://www.ted.com/talks/ken_robinson_says_schools_kill_creativity

Sawyer, K. (2011). The cognitive neuroscience of creativity: A critical review. *Creativity Research Journal, 23*(2), 137–154. doi:10.1080/10400419.2011.571191

Snyder, A. (2007). Learning and creativity—accelerated by suppressing or circumventing certain brain regions. In M. Tokoro & K. Mogi (Eds.), *Creativity and the Brain* (pp. 3–16). Singapore: World Scientific Publishing Co.

Sternberg, R. J., & Lubart, T. I. (1996). Investing in creativity. *The American Psychologist, 7*(7), 677–688. doi:10.1037/0003-066X.51.7.677

Taylor, S. H. (2011). Engendering habits of mind and heart through integrative learning. *About Campus: Enriching the Student Learning Experience, 16*(5), 13–20. doi:10.1002/abc.20076

KEY TERMS AND DEFINITIONS

Cognitive Neuroscience: The scientific study of brain functioning through the use of EEG, PET and fMRI.

Convergent Thinking: The process of thinking which assesses, evaluates, and integrates information.

Critical Thinking: The process of thinking which examines taken-for-granted assumptions in the light of new information and develops a new understanding from that information.

Divergent Thinking: The process of thinking which allows for open-ending thinking and broad associations.

Experiential Learning: A methodology which utilizes all the senses in learning.

Gardens: A piece of ground or landscape intentionally designed for use as a park of public recreation area.

Integrative Learning: Interdisciplinary study across knowledge domains or a creative co-mingling of curricular and co-curricular to broaden learning.

Metaphor: A figure of speech in which a word or phrase is applied to an object or action that allows for broad and more metacognitive associations.

Schema: A cognitive framework or concept that organizes information into conceptual categories.

This research was previously published in Exploring the Benefits of Creativity in Education, Media, and the Arts edited by Nava R. Silton, pages 117-135, copyright year 2017 by Information Science Reference (an imprint of IGI Global).

APPENDIX: HUMAN/NATURE LANDSCAPE PROJECT

For this project, you will visit three gardens in the Portland area—one garden from each of the categories below. When you visit these gardens, record your thoughts and experiences and take as many digital photos as possible to capture the essence, i.e., the metaphor, of the garden.

Garden Categories (Choose 1 From Each Section)

Note! These gardens have free admission unless otherwise noted.

Category A

We will visit the Japanese garden as a class. If you cannot make it you can go to either on your own

1. Classical Chinese Garden (requires admission: see http://www.portlandchinesegarden.org/).
2. Japanese Garden in Washington Park (requires admission: see http://www.japanesegarden.com/visiting/) $6.50 admission with student ID.

Category B

1. Peninsula Park Rose Garden, 700 N Rosa Parks Way. Directions, etc.: http://www.rosegardenstore.org/peninsulagardens.cfm.
2. International Rose Test Gardens in Washington Park (just below the Japanese Garden). Directions, etc.: http://www.rosegardenstore.org/thegardens.cfm
3. Crystal Springs Rhododendron Garden, SE 28th and SE Woodstock.

Category C

1. Laurelhurst Park (between SE 39th and SE Stark).
2. Westmoreland Park (SE McLaughlin Blvd and SE Bybee).
3. Gabriel Park (on SW Vermont, in Raleigh Hills towards Beaverton).

NOTE: For each category, I've selected at least one garden that has easy public transit access. *If you cannot get to any of the gardens in one of the categories, see me and we'll see if we can find an alternative.*

Guidelines for Visiting the Gardens/Parks

- As you are visiting these gardens and trying to decide what to journal and what to photograph, feel free to use the following questions as a guide. In order to help you address the questions below, it is important that while you are in the garden, you jot down notes, impressions, and ideas. Describe what you see. Using all of your senses, what are the characteristics of the garden? Be as neutral as possible as you describe the garden initially: look at the shapes, colors, spaces, volumes, objects. Be as specific as you can. What kinds of plants are in the garden? (If you don't know the names, describe them.) What are the other elements that make up the garden? Imagine that you are describing the garden to someone who is not there with you.

- Do some research on the history of the garden, and if relevant, the cultural context. Compare and contrast what is important to the original creators. Is that in line with your experience or different?
- Once you have enough raw material from your visit, address the following two aspects in your written response.

Written Brainstorming Guidelines

1. **Analyze the Garden:** *Here are some tips that might help.* Try to describe the relationships among the things you see. Where does your eye wander (or is it led?) as you enter the space? What dominates the space? Is there anything you didn't notice at first but noticed later? In other words, what "happens" when the elements of the garden are situated as they are? (Are you encouraged to look at particular things, wander in particular directions, do particular things?) Think about the relationships between shapes, colors, textures, sizes, surfaces, smells and you, the visitor to the garden. (What do you like? What do you relate to? Why?) If it is too much to analyze the whole garden, focus on just one part of the garden and describe its location within the whole.
2. **Interpret the Garden:** Now explore what the garden "means." This section is particularly getting at Meinig's idea of landscape. This garden was constructed for a specific aesthetic purpose. Again, you can focus on just one part of the garden as you do this. What emotions did you feel as you wandered in the garden? What questions came up? (and did you come with any ideas/answers?) What were the most meaningful parts of the garden for you?

Image Construction Guidelines

1. Once you have visited all three gardens/parks and have an ample collection of photos, use Photoshop or some other appropriate software to arrange the photos for each garden in a way that represents the metaphor that you feel emerges from each garden. In other words, you will create three digital collages, one for each garden, paying special attention to the images you choose and how you arrange them so that you can clearly communicate the metaphor of nature/landscape that is emerging from the garden.
2. Once you have completed your Photoshop images, write a one page explanation addressing questions 1 and 2 in the written response guidelines as they are represented in your images. Explain the metaphor that you have chosen for that garden and how it emerges in the your images and design.

Chapter 7
Cognitive and Neural Mechanisms Involved in Interactions Between Touch and Emotion

Zhilin Zhang
Beijing Institute of Technology, China

Tianyi Yan
Beijing Institute of Technology, China

Qiang Huang
Beijing Institute of Technology, China

Jinglong Wu
Beijing Institute of Technology, China

ABSTRACT

Touch has been described as the most fundamental means of contact with the world and the most primitive modality among all sensory systems. In the past, the study of emotional communication has focused almost exclusively on facial and vocal channels but has ignored the channel for the sense of touch. However, the latest studies have documented that the sense of touch can convey at least six emotions, and its accuracy rate is comparable to that of facial expressions and vocal communication. Moreover, there is also mounting evidence indicating that the modality of touch encompasses two dimensions, which provide not only its well-recognized discriminative input from glabrous skin to sensory cortex but also an affective input from hairy skin to the insular cortex because a type of low-threshold mechanosensitive receptor that innervates hairy skin has been shown to convey emotions via C fibers. In light of recent advances in our research, this chapter aims to illustrate the cognitive and neural mechanisms that underlie interactions between touch and emotion.

DOI: 10.4018/978-1-5225-5478-3.ch007

INTRODUCTION

Recently, research has indicated that the sense of touch can convey at least six emotions and that its accuracy rate is comparable to that of facial expressions and vocal communications. Here, I summarize the current state of knowledge on the elaborate mechanisms that facilitate the formation of communication via touch. First, I describe behavioral results related to touch-communicated emotions. Next, I show that in agreement with physiological evidence, there is a class of low-threshold mechanosensitive C fibers that innervate hairy skin and that these neurons represent the neurobiological substrate for affective touch. Then, I present evidence that touch communicates distinct emotions and that this process requires several neural mechanisms, some of which are concerned to emotional processing and others that are involve in a specialized tactile perception system. Finally, I highlight factors that I believe are related to the neural bases of interactions between touch and emotion.

BACKGROUND

With the rapid aging of our population, human-robot emotional interactions are becoming a topical issue related to the process of helping the elderly and sick and performing medical rehabilitation as well as social services (Figure 1). However, humans would rather choose a more emotional animal than a cold machine companion. The emotional interactions between robots and humans that occur via the sense of touch seem to be an impenetrable barrier not only in the field of brain science but also in the field of humanoid robots. Recently, a growing amount of evidence has indicated that the sense of touch has another dimension and that it provides not only its well-recognized discriminative input to the brain but also an affective input (McGlone, Wessberg, & Olausson, 2014). This result will hopefully promote the development of methods to integrate the senses and new research in the brain sciences, provide theoretical and technological support for research into emotional interactions between human beings and robots, and drive the brain sciences and other industries related to advanced robots to thrive in the world.

Figure 1. (a) The aging population of China, including the increasing proportion of elderly people. (b) Humanoid robots can communicate emotions to people via face, voice, and touch modalities.

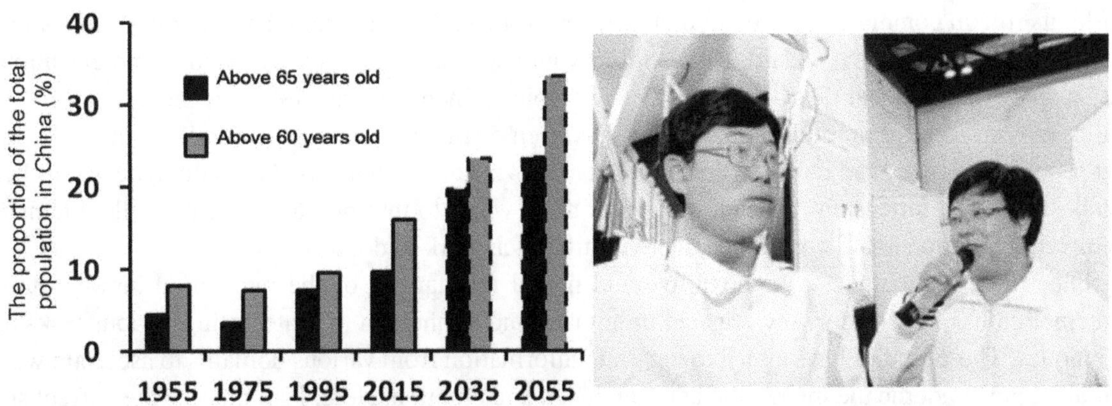

(a) (b)

Touch and Emotion

Happy, sad, fearful, anxious, elated, smitten, disappointed, angry, pleased, disgusted, excited, shameful, guilty, and infatuated are just some of the terms we use to describe our emotions. Emotions can be roughly described by the process of evaluating such an event, which includes the process of becoming aware of a feeling. Emotions are often intertwined with mood, temperament, personality, disposition, and motivation (Damasio, 1994). However, emotions are complex. There is currently no unified and strict definition for this term that could be used around the world. Previously, researchers adopted two primary approaches to unifying our definition of emotion. First, basic emotions attempt to examine the universality of facial expression and define a finite set of human emotions (Ekman & Friesen, 1971). However, another way to approach the categorization of emotions is to describe them not as discrete states but as reactions to events in the world that vary along a continuum. Obviously, these basic and dimensional approaches make it difficult to converge all of our emotional experiences. The question of how we can scientifically research "emotion" has been plaguing us for many years.

Since the development of the field of neuroscience, humans have obtained a new understanding of the brain and emotions. However, the notion that a network of brain structures underlies emotional behavior has been very influential (Damasio, 2000). Since James Papez (1937) first proposed a circuit theory of the brain and emotions, MacLean's theory that the limbic system is the "emotional" brain has been generally accepted (Maclean, 1949). However, it remains difficult to determine criteria to use to define which structures and pathways should be included in the limbic system. For example, the hippocampus may be part of both the limbic system and systems involved in non-emotional processes (Scoville & Milner, 1957). Hence, the view that emotion is a unitary concept or that it is governed by a single neural system is inaccurate. When an individual performs different emotional tasks or scenes, we expect different neural systems to be involved. These systems include brain regions that may only be responsible for emotional processes and also a number of brain regions with multiple functions. Understanding the involvement of these different regions is crucial to our investigation into the cognitive neuroscience of emotion.

Early attempts at investigating the existence and localization of cortical circuits that govern emotion in humans have relied exclusively on experiments involving the facial and vocal channels. These studies have ignored the sense of touch. In these experiments, subjects have been scanned either while watching affective pictures or while listening to an emotional voice. Despite the sense of touch being the first sense we acquire and the fact that it is described as our most fundamental means of contact with the world, its role in communicating emotion has received much less attention than the roles of facial and vocal displays of emotion (Geldard, 1960). Some would even say that touch is thought to merely amplify the intensity of emotional displays by the face and voice (Knapp & Hall, 1997). Although probing our ability to nonverbally communicate emotions has been difficult to accomplish until recently, the idea that people can impart and interpret emotional content via another nonverbal modality, touch, has been demonstrated by Hertenstein. Indeed, we appear to be wired to interpret touch by our fellow humans. However, in relation to affective touch, there is still much work to do.

Since these early studies, the cognitive and neural mechanism of the interaction between touch and emotion has been widely investigated in humans and nonhuman primates using various tasks and techniques. These studies have aimed to integrate information from various domains to ascertain which neural circuits underlie the interaction between touch and emotion. Here, I summarize the current state of knowledge relating to the elaborate mechanisms that facilitate the development of communication via touch. First, I describe the behavioral results of touch that communicates emotions. Next, I show that in agreement with this physiological evidence, a class of low-threshold mechanosensitive C fibers innervates hairy skin, and these fibers represent the neurobiological substrate for affective emotion. Then, I present evidence showing that touch communicates distinct emotions and that this process requires several neural mechanisms, some of which are concerned with emotional processing and others that involve a specialized tactile perception system. Finally, I suggest factors that I believe are likely to be related to the neural bases of the interaction between touch and emotion.

The Behaviors of Interactions Between Emotion and Touch

The sense of touch helps us to discriminate the location of a stimulus on the skin surface, to explore objects haptically and to identify and manipulate objects. It also contributes to an integrated sense of our own body (Thompson & Hampton, 2011). However, research into the wide-ranging functions of touch leave out a very essential fact: touch can also be emotional. Hertenstein's early work documented that touch plays a role in emotional communication among humans (Hertenstein, Verkamp, Kerestes, & Holmes, 2006). As they stated, the game pat-a-cake, which is composed solely of tactile stimulation, is capable of generating positive emotions (Wolff, 1961). Subsequently, Gewirtz (1992) used two sets of stimuli to investigate infant preferences and showed that young infants who received touch displayed more smiles and vocalizations and spent less time crying than did infants who received no touch (Gewirtz & Peláez-Nogueras, 1992). Although touch is capable of generating positive emotions and modulating negative emotions, it is also capable of generating negative emotions (Peláez-Nogueras, Field, Hossain, & Pickens, 1996). An earlier study indicated that static touch (simply laying a hand on an infant's abdomens with little pressure) was less reinforcing and generated more negative emotional displays than other stimuli that involved dynamic tactile stimulation (Brossard & Decarie, 1968). Touch is clearly central to the communication of emotion and particularly the hedonics of emotion. Therefore, we need to construct reliable coding systems to further investigate the qualities and parameters of tactile stimulation.

The study of how touch communicates emotions was advanced by Hertenstein's landmark work on emotion and behavior coding (Hertenstein, Keltner, App, Bulleit, & Jaskollka, 2006). When encoders were asked to think and make contact with the decoder's bare arm from the elbow to the end of the hand to signal each emotion, all of the tactile displays were recorded by the researchers, including the specific type of touch and the intensity and duration of each emotion. Of the well-studied emotions, anger, fear, and disgust were decoded more significantly than happiness and surprise. Participants were more inclined to interpret attempts to communicate prosocial emotions (love, gratitude, sympathy) than self-focused emotions (embarrassment, envy, pride). However, these observations could be linked to those that were obtained in behavioral studies, in which different types of touch have been shown to indicate responses to different emotions in tests involving unacquainted dyads that were performed without visual cues. In another study, the subjects were divided into female-female, male-male, and male-female groups. Anger was communicated at greater-than-chance levels only when a male comprised at least one member of a communicating dyad. Sympathy was communicated at greater-than-chance levels only when a female comprised at least one member of the dyad. Happiness was communicated only if females comprised the entire dyad. Therefore, gender asymmetries also play a role in the accuracy of communicating distinct emotions via touch (Hertenstein & Keltner, 2011).

Another study provided more degrees of freedom to communicate various emotions by allowing encoders to touch the other member of the dyad anywhere on the body that was appropriate. This study also advanced our knowledge by developing precise descriptions of emotion-specific signals (Hertenstein, Holmes, McCullough, & Keltner, 2009). It provided evidence showing that touch communicates at least eight emotions: anger, fear, happiness, sadness, disgust, love, gratitude, and sympathy with accuracy rates that ranged from 50% to 70%. Touch communications showed more negatively valenced (anger, fear, sadness, disgust) and positive emotions (happiness, gratitude, sympathy, love) than communications inferred by the face and voice. Moreover, the authors also documented specific tactile behaviors that were associated with each emotion and located where and what percentage of the time the decoder's body was contacted by the encoder when emotions were accurately decoded (Table 1). For example, fear was communicated by holding the other, squeezing, and contact without movement, whereas sympathy was communicated by holding the other, patting, and rubbing.

A more recent study investigated touch's ability to increase emotion in communal relationships (Simao & Seibt, 2015). The authors proposed that a brief, friendly touch to the shoulder increases feelings of gratitude. One possible reason that was suggested was that a friendly touch on the shoulder was probably perceived as a responsive gesture because the receiver may have perceived it as signaling for attention, care, and consideration (Capraiello & Reis, 2010). Another similarity to a responsive gesture is intentionally responding to a partner's needs by benefitting him or her (Algoe, Haidt, & Gable, 2008). Nevertheless, the link between touch and gratitude was mediated by communal feelings rather than by liking or by positive effects. Therefore, touch signals the intention to initiate or maintain a communal relationship, thus confirming and reinforcing the existence of a communal feeling (Okun, et al, 2015). Although these experiments indicate that some features of communication between emotion and touch are similar in humans, the specific cognitive mechanism that underlies these processes remained unclear. Next, I propose a set of underlying neural mechanisms that are involved in the communication of distinct motions by touch. I then provide suggestions for a more careful and detailed interpretation of these proposals.

Table 1. Behaviors, durations, intensities and locations of tactile displays that were associated with each of the emotions

Emotions	Behaviors	Duration Time (s)	Intensity of Tactile Displays	Location of Decoder's Body
Anger	Shake, Push	4.5±5.1	Strong	Arm
Fear	Shake, Squeeze	7.6±8.5	Moderate	Forearm
Happiness	Shake, Swing	5.8±3.7	Strong	Forearm, Hand
Sadness	Contact, Hug, Nuzzle	6.4±4.3	Light	Arm, Hand
Disgust	Push	5.8±7.0	Moderate	Forearm
Love	Hug	5.3±4.7	Moderate	Back
Gratitude	Shake	6.0±4.6	Moderate	Hand
Sympathy	Pat, Hug, Rub	6.4±5.6	Light	Arm, Forearm, Hand, Back

The Neuropsychology of the Interaction Between Touch and Emotion

Since it was initially shown that there is a link between touch and emotions, the neuronal pathway underlying the interaction between emotion and touch has been a topic of investigation both at the level of peripheral neurophysiology and at the level of the brain. Four sub-modalities are often described to involve low-pressure mechanoreceptors (LTMs), including pressure/vibration, temperature, itch, and pain. These modalities share their anatomical locations in the skin with receptors that encode thermal and chemical stimuli (Figure 2) (Griffin, McArthur, & Polydefkis, 2001). Each channel generates distinct sensory/perceptual qualities that are performed by classes of stimulus-specific neurons that project in defined anatomical pathways to the cerebral cortex (Mountcastle, 2005).

Historically, researchers have viewed touch as generally subserving a primarily discriminative role (Johansson, Landstrom, & Lundstrom, 1982). However, microneurography experiments have demonstrated that there is a population of unmyelinated C-tactile afferents that respond optimally to emotional touch. The high degree of arborization of these C fibers is distributed in the infraorbital nerve, the supraorbital nerve, the arm and the leg (Vallbo, Olausson, & Wessberg, 1999). Interestingly, C-tactile afferents have never been recorded from nerves innervating the palmar skin of the hands. Information from neurophysiological experiments concluded that C-tactile afferents respond to very low indentation forces in the range of 0.3-2.5 mN with high frequency responses (50-100 impulses/s^{-1}) to innocuous stimuli, such as gentle stroking with a soft brush (Wessberg, Olausson, Fernstrom, & Vallbo, 2003). As has been shown in humans, C-tactile afferents showed an inverted U-shaped relationship between brushing velocity and firing frequency, with the highest responses recorded between 1 and 10 cm/s^{-1} (Cole, et al., 2006). Another fundamental process during temperature modulation involves the transformation of C-tactile afferents. These afferents discharge preferentially in response to slowly moving stimuli at a neutral (typical skin) temperature, rather than at cooler or warmer stimulus temperatures (Ackerley, et al., 2014). In fact, C-tactile afferents are not only velocity-tuned but also temperature-tuned.

Figure 2. Four types of mechanoreceptors are located in the skin

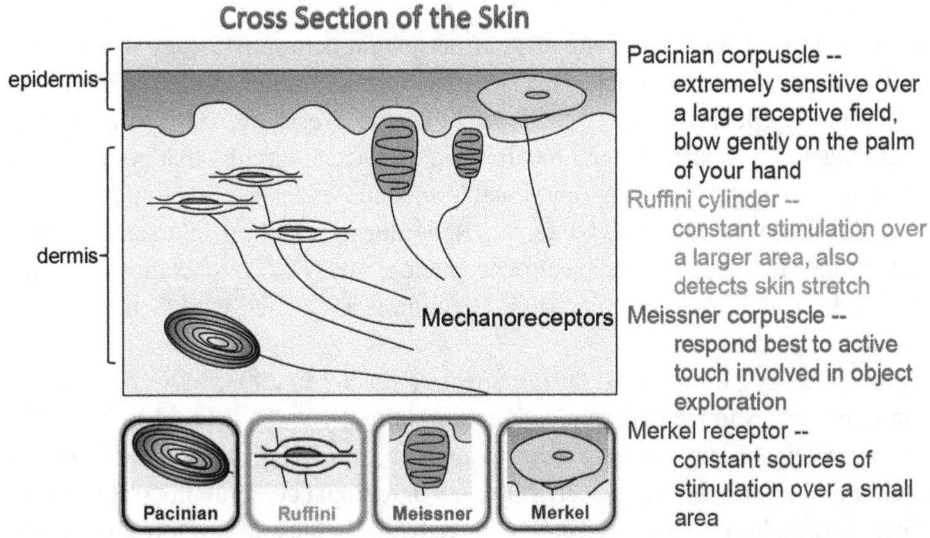

**For a more accurate representation see the electronic version.*

Another study focused on animals during handling or maternal contact. These contacts resulted in more adaptive stress responses that reduced the effects of impaired cognitive ability or other stress-related diseases (Meaney, et al., 1991). An abundant amount of evidence indicates that handling and maternal contact increases glucocorticoid receptor expression, leading animals to exhibit adaptive and appropriate responses rather than prolonged responses to stress (Liu, et al., 1997; Meaney, Aitken, van-Berkel, Bhatnagar, & Sapolsky, 1988). In addition, handling appears to increase sensitivity to negative feedback from glucocorticoids in rats, perhaps because it causes an increase in receptor sites. Increased sensitivity to glucocorticoids is necessary to inhibit the release of more glucocorticoids, which exacerbates the stress response (Meaney, Aitken, Bodnoff, Bhatnagar, & Sapolsky, 2013; Francis, Caldji, Champagne, Plotsky, & Meaney, 1999).

Although microneurography has provided valuable insights into the electrophysiological properties of C-tactile afferents, little is known about what stimulus would adequately replicate the gentle, low-force stroking of hairy skin, which is a stimulus that is not in the nociceptor range. Depending on the principles of the Darwinian theory of evolution, any stimulus that is associated with a reward or punishment should be linked to the simple purpose of survival. The functional significance of a nociceptive system to survival remains unequivocal. Therefore, McGlone and colleagues (2010) proposed a similar duality for touch in which discriminative touch is mediated by myelinated fast-conducting Aβ nerves, while affective touch is mediated by C-tactile afferents that are activated only when touching stimulus has the appropriate properties for an adequate stimulus (Björnsdotter, Morrison, & Olausson, 2010). A more direct molecular neurobiology result was shown in electrode recordings that were obtained from the nerves of awake human volunteers while their tactile receptive fields were stimulated using a soft brush. However, this result also supports the notion that intact C-tactile afferent pathways are likely to be necessary for the normal evaluation of touch pleasantness on hairy skin, in addition to the notion that C-tactile afferents may operate as selectors that are activated in parallel with Aβ afferents. Importantly, these two systems are not separate, despite being at least partly dissociable. There are therefore undoubtedly sensory discriminative and motivational affective pathways for touch interactions (Morrison & Downing, 2010).

The Neuroimaging of Interaction Between Touch and Emotion

It is clear from recent investigations at the level of peripheral neurophysiology that C-tactile afferents transmit the affective properties of touch. This is also a topic for studies of the brain (Loken, Wessberg, Morrison, McGlone, & Olausson 2009; McGlone, Loken, & Wessberg, 2007). Brain imaging experiments have investigated the existence and localization of cortical circuits that communicate emotions by touch in humans. Some studies using functional Magnetic Resonance Imaging (fMRI) have focused on selected neuroanatomical regions of interest (ROI) during emotion interactions that involve the sense of touch. Other studies have used electroencephalograms (EEGs) to examine activity in cortical processing networks, and these have found that activation is associated with tactile-guided emotion in many additional areas.

Which neural pathways are involved during affective touch? In 2008, McCabe et al. used fMRI to investigate brain responses to the communication of emotion by touch (McCabe, Rolls, Bilderbeck, & McGlone, 2008). When participants were asked to apply moisturizing lotion to the hairy skin but not to the glabrous skin of the palm, increases were observed in regional cerebral blood flow in many cortical and subcortical areas, including motor and premotor areas, and in the mid-orbitofrontal cortex. In another study, subjects underwent fMRI to compare the effects of stroking at a C-tactile afferent optimal velocity (3 cm/s) to stroking at a non-C-tactile afferent optimal velocity (30 cm/s). C-tactile afferent stroking resulted in higher rates of regional cerebral blood flow in extensive regions of the posterior insula that are associated with somatic and multimodal responses than the rates provoked by non C-tactile afferent touching in the same areas (Bolognini, Rossetti, Fusaro, Vallar, & Miniussi, 2014).

Early attempts to use brain imaging to identify the functional anatomy underlying the generation of goal-directed arm movements during spontaneous facial self-touch gestures in humans used EEG. However, the changing dynamics of theta waves indicated that spontaneous self-touch gestures play a role in emotional processes, primarily in stressful situations (Grunwald, Weiss, Mueller, & Rall, 2014). Moreover, being touched by another person influences our readiness to empathize with and support that person. Event-related potential experiments have indicated that touch sensitizes ongoing cognitive and emotional processes and that this sensitization is mediated by bottom-up somatosensory processing. Moreover, touch seems to be perceived as a stronger sensory cue than tone (Gallese, 2003). However, when using single trial classification techniques, beta-power significantly distinguished between pleasant and unpleasant stimulation on a single trial basis with high accuracy (Singh, et al., 2014).

Other studies have focused on brain responses to the unpleasant emotion of pain, and the results of these studies have suggested that feeling and seeing pain may share neural processes (Morrison, Lloyd, di Pellegrino, & Roberts, 2004). When participants were asked to observe the painful experiences of others, brain areas coding the unpleasant aspects of pain contributed to an observer's representation of the other's pain (Jackson, Meltzoff, & Decety, 2005). Compared to the control group, pain-related motivational affective regions were more activated in the anterior and mid-cingulate cortex, anterior insula, motor and sensory cortex (Avenanti, Minio-Paluello, Bufalari, & Aglioti, 2006; Cheng, et al., 2007). However, there is currently no consensus regarding the exact location and existence of overlapping cingulate activity between felt and seen pain, which vary between individuals (Morrison, Loken, & Olausson, 2010).

Soft brush stroking on hairy skin activates the somatosensory areas SI and SII in addition to the posterior contralateral insular cortex. The insular cortex is classically considered to be a specific region for affective processes, and it is considered a gateway from sensory systems to emotional systems, especially in the posterior contralateral insular cortex (Craig, 2008). Olausson also demonstrated that when similar soft brushing stimuli were applied to Aβ-deafferented subjects, no activity was observed in somatosensory areas, whereas the posterior insular region was activated (Olausson, et al., 2002). Importantly, when brain activity was compared in response to touching either C-tactile afferent on hairy skin or glabrous skin, higher levels of cortical activity were observed in the posterior insular cortex and mid-anterior orbitofrontal cortex when the C-tactile afferents were touched than when the glabrous skin was touched (Gordon, et al., 2013). In fact, stimulation of unmyelinated C-tactile afferents evoked significant fMRI deactivation in the somatosensory cortex, probably because more emotional areas, mainly those related to emotion-related paralimbic cortical systems, were activated. In experiments addressing this issue, the posterior superior temporal sulcus, the medial prefrontal cortex and the dorso-anterior cingulate cortex have also been implicated in processing C-tactile afferent-targeted touch (Kramer, et al., 2007).

Although the contributions of tactile Aβ afferents and C-tactile afferents to emotional processing have not been widely explored, there is some evidence to support the notion that Aβ afferents may well underpin pleasant sensations. Studies in which similar touch stimuli were applied to the palm, where C-tactile afferents are lacking, showed fMRI responses in a target area for insular afferents in the orbitofrontal cortex, which is involved in complex emotional evaluations (Rolls, et al., 2003). Therefore, we hypothesize that glabrous skin can communicate emotions and that it is involved in secondary reinforcement mechanisms that involve low-threshold mechanoreceptors signaling in a pattern that is typical of tactile processing (McGlone, et al., 2012).

A growing amount of neurobiological evidence indicates that touch is more than a sensory input for discriminating what is on the skin. Instead, the gentle touch that is often observed during nurturing and social interactions reflects the presence of an evolutionary mechanism that is mediated via C-tactile afferents that project strongly to emotional systems but little or not at all to the discriminative cognitive systems that promote physical contact in specific contexts (Figure 3). In fact, growing up cannot occur without physical contact with others, and touching provides us with much more of a benefit than we appreciate throughout life.

FUTURE RESEARCH DIRECTIONS

Because there is an urgent demand for improving human-robot emotional interactions that can help the elderly and sick and assist in medical rehabilitation and social services, some pivotal problems strike us as particularly important for understanding how the human brain actively facilitates multisensory integration and its attention for visual, audio and touch perception and how this information can be used in human-robot emotional interaction. How do we allocate limited attention? How do we integrate multisensory information (Figure 4)?

The results of this research will hopefully promote the development of methods to integrate the senses, improve research into brain science, provide theoretical and technological support for research into the emotional interactions between humans and robots, and drive brain science and related industries to develop advanced robots that can thrive in the world.

Figure 3. Schematic model of affective and discriminative pathways involved in touch in the skin

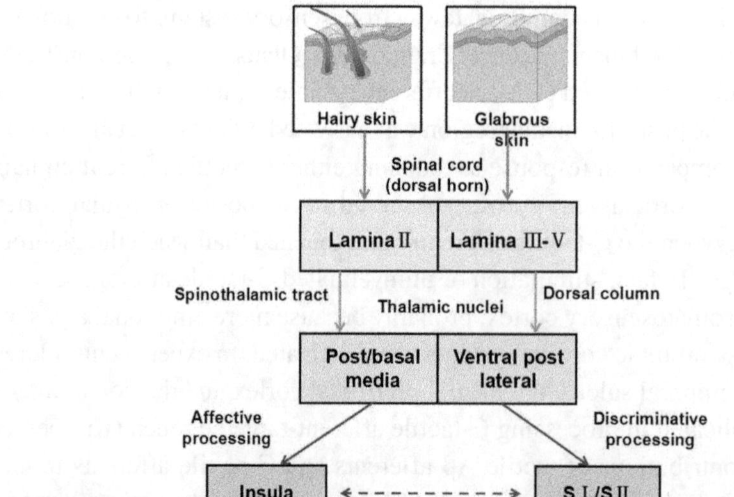

Figure 4. Multisensory brain cognitive mechanisms involved in multisensory integration and attention

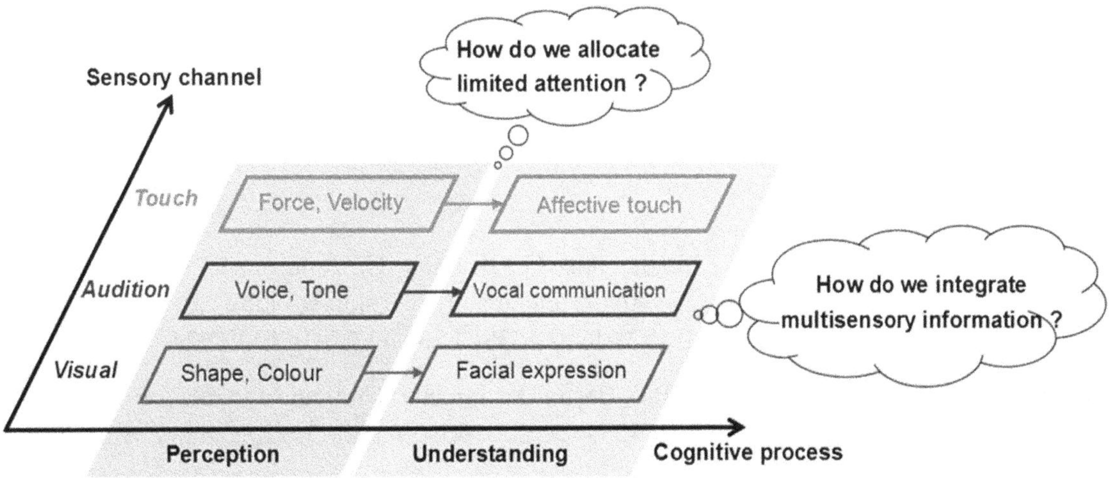

CONCLUSION

The development of fMRI has coincided with improvements in designs that may help researchers to address many of the problems involving in conducting emotion-centered studies. Earlier research and theories tended to view emotion as separate from cognition, implying that they could be studied and understood separately. However, all emotions have unique and defining characteristics that can be studied along with more "cognitive" behaviors. Ideally, a coordinated series of neuroimaging experiment should be performed to explore the interaction between emotion and touch. In addition, MRI-compatible systems now make it possible to perform electrophysiological and possibly kinematic recordings while subjects perform various tasks during scanning.

Some progress has been made in characterizing the cognitive neuroscience of emotion and the neural substrates that underlie it. Nonetheless, much remains unknown, and many important issues have yet to be addressed. The "affective touch hypothesis" implies that the essential role of the C-tactile afferents system is to provide or support emotional, hormonal, and behavioral responses to skin-to-skin contact with conspecifics. However, it is too difficult to quantitatively define the kind of behavior that is related to each emotion. In addition, the intensity and even the quality of the emotional response that is evoked by a particular stimulus is highly dependent on contextual factors. Therefore, an emerging shift in our approach to studying the cognitive neuroscience of emotion is movement toward an emphasis on the study of neural structures to investigations of neural systems.

Finally, we have learned that the sense of touch makes unique contributions to emotional communication and affective styles. Only through careful and thoughtful experimentation that uses converging techniques involving both the brain and behavior might we come to understand the neural systems underlying the interactions between emotion and touch.

ACKNOWLEDGMENT

This work was supported by the Chinese National Natural Science Foundation (Grants 61473043 and 61305142).

REFERENCES

Ackerley, R., Backlund-Wasling, H., Liljencrantz, J., Olausson, H., Johnson, R. D., & Wessberg, J. (2014). Human C-tactile afferents are tuned to the temperature of a skin-stroking caress. *The Journal of Neuroscience*, *34*(8), 2879–2883. doi:10.1523/JNEUROSCI.2847-13.2014 PMID:24553929

Algoe, S. B., Haidt, J., & Gable, S. L. (2008). Beyond reciprocity: Gratitude and relationships in everyday life. *Emotion (Washington, D.C.)*, *8*(3), 425–429. doi:10.1037/1528-3542.8.3.425 PMID:18540759

Avenanti, A., Minio-Paluello, I., Bufalari, I., & Aglioti, S. M. (2006). Stimulus-driven modulation of motor-evoked potentials during observation of others' pain. *NeuroImage*, *32*(1), 316–324. doi:10.1016/j.neuroimage.2006.03.010 PMID:16675270

Björnsdotter, M., Morrison, I., & Olausson, H. (2010). Feeling good: On the role of C fiber mediated touch in interoception. *Experimental Brain Research*, *207*(3-4), 149–155. doi:10.1007/s00221-010-2408-y PMID:20963582

Bolognini, N., Rossetti, A., Fusaro, M., Vallar, G., & Miniussi, C. (2014). Sharing social touch in the primary somatosensory cortex. *Current Biology*, *24*(13), 1513–1517. doi:10.1016/j.cub.2014.05.025 PMID:24954046

Brossard, L., & Decarie, T. (1968). Comparative reinforcing effects of eight stimulations on the smiling responses of infants. *Journal of Child Psychology and Psychiatry, and Allied Disciplines*, *9*(1), 51–59. doi:10.1111/j.1469-7610.1968.tb02207.x PMID:5729746

Capraiello, P. A., & Reis, H. T. (2010). Perceived partner responsiveness minimizes defensive reactions to failure. *Social Psychological & Personality Science*, *2*(4), 365–372. doi:10.1177/1948550610391914

Cheng, Y., Lin, C. P., Liu, H. L., Hsu, Y. Y., Lim, K. E., Hung, D., & Decety, J. (2007). Expertise modulates the perception of pain in others. *Current Biology*, *17*(19), 1708–1713. doi:10.1016/j.cub.2007.09.020 PMID:17900903

Cole, J., Bushnell, M. C., McGlone, F., Elam, M., Lamarre, Y., Vallbo, A., & Olausson, H. (2006). Unmyelinated tactile afferents underpin detection of low-force monofilaments. *Muscle & Nerve*, *34*(1), 105–107. doi:10.1002/mus.20534 PMID:16518853

Craig, A. D. (2008). *In handbook of emotions* (3rd ed.). New York, NY: Guilford Press.

Damasio, A. R. (Ed.). (1994). *Descartes' error: emotion, reason, and the human brain*. New York, NY: Vintage Press.

Damasio, A. R. (2000). Eighth C.U. Ariëns Kappers Lecture. The fabric of the mind: A neurobiological perspective. *Progress in Brain Research, 126*, 457–467. doi:10.1016/S0079-6123(00)26029-9 PMID:11105662

Ekman, P., & Friesen, W. V. (1971). Constants across cultures in the face and emotion. *Journal of Personality and Social Psychology, 17*(2), 124–129. doi:10.1037/h0030377 PMID:5542557

Francis, D. D., Caldji, C., Champagne, F., Plotsky, P. M., & Meaney, M. J. (1999). The role of corticotropin-releasing factor--norepinephrine systems in mediating the effects of early experience on the development of behavioral and endocrine responses to stress. *Biological Psychiatry, 46*(9), 1153–1166. doi:10.1016/S0006-3223(99)00237-1 PMID:10560022

Gallese, V. (2003). The manifold nature of interpersonal relations: The quest for a common mechanism. *Philosophical Transactions of the Royal Society of London. Series B, Biological Sciences, 358*(1431), 517–528. doi:10.1098/rstb.2002.1234 PMID:12689377

Geldard, F. A. (1960). Some neglected possibilities of communication. *Science, 131*(3413), 1583–1588. doi:10.1126/science.131.3413.1583 PMID:13827173

Gewirtz, J. L., & Peláez-Nogueras, M. B. (1992). Skinner's legacy to human infant behavior and development. *The American Psychologist, 47*(11), 1411–1422. doi:10.1037/0003-066X.47.11.1411 PMID:1482005

Gordon, I., Voos, A. C., Bennett, R. H., Bolling, D. Z., Pelphrey, K. A., & Kaiser, M. D. (2013). Brain mechanisms for processing affective touch. *Human Brain Mapping, 34*(4), 914–922. doi:10.1002/hbm.21480 PMID:22125232

Griffin, J. W., McArthur, J. C., & Polydefkis, M. (2001). Assessment of cutaneous innervation by skin biopsies. *Current Opinion in Neurology, 14*(5), 655–659. doi:10.1097/00019052-200110000-00016 PMID:11562579

Grunwald, M., Weiss, T., Mueller, S., & Rall, L. (2014). EEG changes caused by spontaneous facial self-touch may represent emotion regulating processes and working memory maintenance. *Brain Research, 1557*, 111–126. doi:10.1016/j.brainres.2014.02.002 PMID:24530432

Hertenstein, M. J., Holmes, R., McCullough, M., & Keltner, D. (2009). The communication of emotion via touch. *Emotion (Washington, D.C.), 9*(4), 566–573. doi:10.1037/a0016108 PMID:19653781

Hertenstein, M. J., & Keltner, D. (2011). Gender and the communication of emotion via touch. *Sex Roles, 64*(1-2), 70–80. doi:10.1007/s11199-010-9842-y PMID:21297854

Hertenstein, M. J., Keltner, D., App, B., Bulleit, B. A., & Jaskollka, A. R. (2006). Touch communicates distinct emotions. *Emotion (Washington, D.C.), 6*(3), 528–533. doi:10.1037/1528-3542.6.3.528 PMID:16938094

Hertenstein, M. J., Verkamp, J. M., Kerestes, A. M., & Holmes, R. M. (2006). The communicative functions of touch in humans, nonhuman primates, and rats: A review and synthesis of the empirical research. *Genetic, Social, and General Psychology Monographs*, *132*(1), 5–94. doi:10.3200/MONO.132.1.5-94 PMID:17345871

Jackson, P. L., Meltzoff, A. N., & Decety, J. (2005). How do we perceive the pain of others? A window into the neural processes involved in empathy. *NeuroImage*, *24*(3), 771–779. doi:10.1016/j.neuroimage.2004.09.006 PMID:15652312

Johansson, R. S., Landstrom, U., & Lundstrom, R. (1982). Responses of mechanoreceptive afferent units in the glabrous skin of the human hand to sinusoidal skin displacements. *Brain Research*, *244*(1), 17–25. doi:10.1016/0006-8993(82)90899-X PMID:6288178

Knapp, M. L., & Hall, J. A. (Eds.). (1997). *Nonverbal communication in human interaction*. Boston, MA: Harcourt Brace College Press.

Kramer, H. H., Lundblad, L., Birklein, F., Linde, M., Karlsson, T., Elam, M., & Olausson, H. (2007). Activation of the cortical pain network by soft tactile stimulation after injection of sumatriptan. *Pain*, *133*(1-3), 72–78. doi:10.1016/j.pain.2007.03.001 PMID:17449179

Liu, D., Diorio, J., Tannenbaum, B., Caldji, C., Francis, D., & Freedman, A. et al.. (1997). Maternal care, hippocampal glucocorticoid receptors, and hypothalamic-pituitary-adrenal responses to stress. *Science*, *277*(5332), 1659–1662. doi:10.1126/science.277.5332.1659 PMID:9287218

Loken, L. S., Wessberg, J., Morrison, I., McGlone, F., & Olausson, H. (2009). Coding of pleasant touch by unmyelinated afferents in humans. *Nature Neuroscience*, *12*(5), 547–548. doi:10.1038/nn.2312 PMID:19363489

Maclean, P. D. (1949). Psychosomatic disease and the visceral brain; recent developments bearing on the Papez theory of emotion. *Psychosomatic Medicine*, *11*(6), 338–353. doi:10.1097/00006842-194911000-00003 PMID:15410445

McCabe, C., Rolls, E. T., Bilderbeck, A., & McGlone, F. (2008). Cognitive influences on the affective representation of touch and the sight of touch in the human brain. *Social Cognitive and Affective Neuroscience*, *3*(2), 97–108. doi:10.1093/scan/nsn005 PMID:19015100

McGlone, F., Olausson, H., Boyle, J. A., Jones-Gotman, M., Dancer, C., Guest, S., & Essick, G. (2012). Touching and feeling: Differences in pleasant touch processing between glabrous and hairy skin in humans. *The European Journal of Neuroscience*, *35*(11), 1782–1788. doi:10.1111/j.1460-9568.2012.08092.x PMID:22594914

McGlone, F., Vallbo, A. B., Olausson, H., Loken, L., & Wessberg, J. (2007). Discriminative touch and emotional touch. *Journal of Experimental Social Psychology*, *61*(3), 173–183. PMID:17974312

McGlone, F., Wessberg, J., & Olausson, H. (2014). Discriminative and affective touch: Sensing and feeling. *Neuron*, *82*(4), 737–755. doi:10.1016/j.neuron.2014.05.001 PMID:24853935

Meaney, M. J., Aitken, D. H., Bodnoff, S. R., Bhatnagar, S., & Sapolsky, R. M. (2013). Early postnatal handling alters glucocorticoid receptor concentrations in selected brain regions. *Behavioral Neuroscience, 127*(5), 637–641. doi:10.1037/a0034187 PMID:24128353

Meaney, M. J., Aitken, D. H., van-Berkel, C., Bhatnagar, S., & Sapolsky, R. M. (1988). Effect of neonatal handling on age-related impairments associated with the hippocampus. *Science, 239*(4841 Pt 1), 766–768. doi:10.1126/science.3340858 PMID:3340858

Meaney, M. J., Mitchell, J. B., Aitken, D., Bhatnagar, S., Bodnoff, S. R., Iny, L. J., & Sarrieau, A. (1991). The effects of neonatal handling on the development of the adrenocortical response to stress: Implications for neuropathology and cognitive deficits in later life. *Psychoneuroendocrinology, 16*(1-3), 85–103. doi:10.1016/0306-4530(91)90072-2 PMID:1961847

Morrison, I., & Downing, P. E. (2007). Organization of felt and seen pain responses in anterior cingulate cortex. *NeuroImage, 37*(2), 642–651. doi:10.1016/j.neuroimage.2007.03.079 PMID:17588777

Morrison, I., Lloyd, D., di Pellegrino, G., & Roberts, N. (2004). Vicarious responses to pain in anterior cingulate cortex: Is empathy a multisensory issue? *Cognitive, Affective & Behavioral Neuroscience, 4*(2), 270–278. doi:10.3758/CABN.4.2.270 PMID:15460933

Morrison, I., Loken, L. S., & Olausson, H. (2010). The skin as a social organ. *Experimental Brain Research, 204*(3), 305–314. doi:10.1007/s00221-009-2007-y PMID:19771420

Mountcastle, V. B. (2005). *The sensory hand.* Cambridge, MA: Harvard University Press.

Okun, M., Steinmetz, N. A., Cossell, L., Iacaruso, M. F., Ko, H., Barthó, P., & Harris, K. D. et al. (2015). Diverse coupling of neurons to populations in sensory cortex. *Nature, 521*(7553), 511–515. doi:10.1038/nature14273 PMID:25849776

Olausson, H., Lamarre, Y., Backlund, H., Morin, C., Wallin, B. G., Starck, G., & Bushnell, M. C. et al. (2002). Unmyelinated tactile afferents signal touch and project to insular cortex. *Nature Neuroscience, 5*(9), 900–904. doi:10.1038/nn896 PMID:12145636

Peláez-Nogueras, M., Field, T. M., Hossain, Z., & Pickens, J. (1996). Depressed mothers' touching increases infants' positive affect and attention in still-face interactions. *Child Development, 67*(4), 1780–1792. doi:10.2307/1131731 PMID:8890507

Rolls, E. T., O' Doherty, J., Kringelbach, M. L., Francis, S., Bowtell, R., & McGlone, F. (2003). Representations of pleasant and painful touch in the human orbitofrontal and cingulate cortices. *Cerebral Cortex, 13*(3), 308–317. doi:10.1093/cercor/13.3.308 PMID:12571120

Scoville, W. B., & Milner, B. (1957). Loss of recent memory after bilateral hippocampal lesions. *Journal of Neurology, Neurosurgery, and Psychiatry, 20*(1), 11–21. doi:10.1136/jnnp.20.1.11 PMID:13406589

Simao, C., & Seibt, B. (2015). Friendly touch increases gratitude by inducing communal feelings. *Frontiers in Psychology, 6*, 815. doi:10.3389/fpsyg.2015.00815 PMID:26124737

Singh, H., Bauer, M., Chowanski, W., Sui, Y., Atkinson, D., Baurley, S., & Bianchi-Berthouze, N. et al. (2014). The brain's response to pleasant touch: An EEG investigation of tactile caressing. *Frontiers in Human Neuroscience*, *8*, 893. doi:10.3389/fnhum.2014.00893 PMID:25426047

Thompson, E. H., & Hampton, J. A. (2011). The effect of relationship status on communicating emotions through touch. *Cognition and Emotion*, *25*(2), 295–306. doi:10.1080/02699931.2010.492957 PMID:21432672

Vallbo, A. B., Olausson, H., & Wessberg, J. (1999). Unmyelinated afferents constitute a second system coding tactile stimuli of the human hairy skin. *Journal of Neurophysiology*, *81*(6), 2753–2763. PMID:10368395

Wessberg, J., Olausson, H., Fernstrom, K. W., & Vallbo, A. B. (2003). Receptive field properties of unmyelinated tactile afferents in the human skin. *Journal of Neurophysiology*, *89*(3), 1567–1575. doi:10.1152/jn.00256.2002 PMID:12626628

Wolff, P. H. (1961). Observations on the early development of smiling. *Determinants of Infant Behavior, 2,* 113-138.

ADDITIONAL READING

Bloom, F., Nelson, C. A., & Lazerson, A. (2001). *Brain, mind and behavior* (3rd ed.). New York, NY: Worth Publishers.

Bullock, T. H. (1984). Comparative neurosceince holds promise for a quiet revolution. *Science*, *225*(4661), 473–478. doi:10.1126/science.6740319 PMID:6740319

Churchland, P. S. (1986). *Neurophilosophy: Toward a unified science of mind/brain.* Cambridge, MA: MIT Press.

D'esposito, M., Zarahn, E., & Aguirre, G. K. (1999). Event-related functional MRI: Implications for cognitive psychology. *Psychological bulletion, 125, 155-164.*

Gazzaniga, M. S. (2000). Cerebral specialization and interhemispheric communication: Does the corpus callosum enable the human condition? Bran, 123, 1293-1326.

Gazzaniga, M. S. (2005). *The ethical brain.* New York, NY: Dana Press.

Georgopoulos, A. P. (1995). *The Cognitive Neurosciences*. Cambridge, MA: MIT Press.

Hellige, J. B. (1993). *Hemispheric asymmetry: What' s right and what' left*. Cambridge, MA: Harvard University Press.

Hillyard, S. A. (1993). Electrical and magnetic brain recordings: Contributions to cognitive neuroscience. *Current Opinion in Neurobiology*, *3*(2), 710–717. doi:10.1016/0959-4388(93)90213-I PMID:8513235

Ivry, R. B., & Rovertson, L. C. (1998). *The two sides of perception*. Cambridge, MA: MIT Press.

Jacob, F. (1977). Evolution and tinkering. *Science*, *196*(4295), 1161–1166. doi:10.1126/science.860134 PMID:860134

Jeannerod, M. (1997). *The cognitive neuroscience of action*. Cambridge, MA: Blackwell Science Publishers.

Johnson, M. H. (1997). *Developmental cognitive neuroscience*. Cambridge, MA: Blackwall Science Press.

Kandel, E. R., Schwartz, J. H., & Jessell, T. M. (1995). *Essentials of neural science and behavior*. Norwalk, CT: Appleton and Lange Publishers.

Kass-Simon, G., & Farnes, P. (1990). *Women of science: righting the record*. Bloomington, IN: Indiana University Press.

Lindzey, G. (Ed.). (1936). *History of psychology in autobiography* (Vol. III). Worcester, MA: Clark University Press.

Merzenich, M. M., & Kaas, J. H. (1980). *Principles of organization of sensory- perceptual systems of mammals*. New York, NY: Academic Press.

Michael, S. G., Richard, B. I., & George, R. M. (2013). *Cognitive neuroscience: The biology of the mind* (4th ed.). New York, NY: W. W. Norton & Company Ltd. Press.

Passingham, R. (1993). *The friontal lobes and voluntary action*. New York, NY: Oxford University Press.

Pinkers, S. (1997). *How the mind works*. New York, NY: W. W. Norton Publishers.

Ponser, M. I., & Raichle, M. E. (1994). *Images of mind*. New York: W. H. Freeman Publishers.

Raichle, M. E. (1998). Behind the scenes of functional brain imaging: A historical and physiological perspective. *Proceedings of the National Academy of Sciences of the United States of America*, *95*(3), 765–772. doi:10.1073/pnas.95.3.765 PMID:9448239

Rapp, B. (2001). *The handbook of cognitive neuropsychology: What deficits reveal about the human mind. Philadephia*. Psychology Press.

Rosenbaum, D. A. (1991). *Human motor control*. San Diego, CA: Academic Press.

Shaw, P., Greenstein, D., Lerch, J., Clasen, L., Lenroot, R., Gogtay, N., & Giedd, J. et al. (2006). Intellectual ability and cortical development in children and adolescents. *Nature*, *440*(7084), 676–679. doi:10.1038/nature04513 PMID:16572172

Shepherd, G. M. (1988). *Neurobiology* (2nd ed.). New York, NY: Oxford University Press.

Shepherd, G. M. (1992). *Fundations of the neuron doctrine*. New York, NY: Oxford University Press.

Wilson, E. O. (1975). *Sociobiology, the new synthesis*. Cambridge, MA: Belknap Press of Harvard University Press.

KEY TERMS AND DEFINITIONS

Affective Touch: The affective dimension of touch pertains to the hedonic valence of tactile sensation.

Brain Cognitive Mechanism: A set of informational processes that occurs in the brain.

Discriminative Touch: Discriminative touch is a sensory modality that allows a subject to sense and localize touch.

EEG: An electrophysiological monitoring method to record electrical activity in the brain.

Emotional Interaction: The emotional communication between a human and a computer/robot.

fMRI: A functional neuroimaging procedure that uses MRI technology to measure brain activity by detecting changes associated with blood flow.

Multisensory Integration: The study of how information from different sensory modalities, such as sight, sound, touch, smell, self-motion and taste, are integrated by the nervous system.

This research was previously published in Improving the Quality of Life for Dementia Patients through Progressive Detection, Treatment, and Care edited by Jinglong Wu, pages 166-186, copyright year 2017 by Medical Information Science Reference (an imprint of IGI Global).

Chapter 8

Integrating Interpersonal Neurobiology Into the Play Therapy Process:

Advancing Adlerian Play Therapy

Dalena L. Dillman Taylor
University of Central Florida, USA

Naomi Joy Wheeler
University of Central Florida, USA

ABSTRACT

Advancements in neuroscience provide theoretical support for principles and practices of counseling and play therapy intervention (Badenoch & Kestley, 2015; Siegel, 2012). Likewise, AdPT shares many conceptual similarities with IPNB and neuroscience research including emphasis on childhood experience, social interest and relationships, purposefulness of behavior, holism, and the internalized narrative or self-schema. This chapter per the authors will (a) develop connections from neuroscience and IPNB to play and play therapy, (b) review the IPNB model's domains of integration and their relationship to AdPT, and (c) provide an updated AdPT framework that encompasses the most recent neuroscience empirical support.

NEUROSCIENCE, IPNB AND PLAY: ADVANCING ADLERIAN PLAY THERAPY

The first six years of life are the most influential period of neuronal growth, as well as, the most vulnerable to maltreatment (National Scientific Council on the Developing Child, 2010; U.S. Department of Health and Human Services, 2013). This period of vulnerability makes early childhood a critical and influential time in development and learning. Recent advancements in neuroscience provide evidence that our environment, early life experiences, and relationships contribute to physical and psychological health (Kandel, 1998; Panksepp & Biven, 2012; Porges, 2011; Siegel, 2012). Negative and positive experiences exert a differential impact to how our brains are formed, function, and integrate information.

DOI: 10.4018/978-1-5225-5478-3.ch008

The Interpersonal Neurobiology model (IPNB; Siegel, 2012) incorporates several theories from neuroscience that address (a) memory encoding, (b) the polyvagal theory (Porges, 2011), (c) neuroplasticity (Kandel, 1998), and (d) the midbrain's affective emotional-motivational systems (Panksepp & Biven, 2012). These theories underscore the value of new experiences and interaction in our social environment to reshape and restructure both brain and behavior. Therefore, this introduction begins with a review of neuroscience theory within the framework of the IPNB model, play, and play therapy, specifically Adlerian play therapy.

IPNB

In the Interpersonal Neurobiology (IPNB) model, Siegel (2012) postulates that health is influenced by the integration of the brain, mind, and in interactions with others in our social environment. All relationships have the potential to inhibit or contribute to neural integration. IPNB addresses nine domains of neural integration: *consciousness, bilateral, vertical, memory, narrative, state, interpersonal, temporal,* and *transpirational* or *identity integration* (refer to IPNB domains of integration and Adlerian Play Therapy section below). Each domain contributes to overall health and requires attention in therapeutic and educational intervention (Siegel, 2012), because each domain also produces energy and information flow through the nervous system and body. We are constantly receiving information, interpreting and incorporating messages into conscious and unconscious responses. Likewise, energy and information are shared between nervous systems with others in the social environment. We are impacted by those around us and respond in accordance to the integrated messages experienced from within (internal cues) and between (our system in interaction with others in the environment).

Similarly, Alfred Adler (1998) identified early childhood influences to personality and lifestyle development from the social system (e.g. family constellation, gender roles, parental attitudes, and family economic position) and inherited qualities (i.e. health and appearance). Adler noted the importance of a person's perceptions of childhood experiences for adult life and the value of social interest for well-being. He theorized that our beliefs and logic about self, others, and the world are created in childhood and serve as the foundation for future attitudes and behavior. We are wired to seek integration among brain, mind, and in interactions, which contributes to growth, health, and restoration. Principles emerging from IPNB provide theoretical support for many prevention and intervention strategies with adults as well as children.

Play

Children are able to work through experiences and make meaning of their life events, environment, and relationships through play. In fact, play contributes to neurobiological development as well as cognitive, socio-emotional, physical development, acquisition of social skills, and capacity for self-regulation (Berk, 2012; Erickson, 1963; Greenspan, 1993; Ray, 2011). Play is the first, natural language for children whereby needs, emotions, and thoughts are communicated (Landreth, 2012). Through play, children interpret their experiences and relationships to cultivate social skills, self-regulation, and a nervous system responsive to life. Play contributes to secure relationships and social connection that provides safety for exploration and engagement with the world. However, when a child experiences trauma or poor attachment to significant caregivers, development can be interrupted resulting in heightened stress responses, immobilization, and developmental delays (Badenoch, 2008; Kestley, 2015). Therefore, when challenging or traumatic events occur in early childhood, appropriate assessment and intervention may be crucial to influence maladaptive health trajectories and schemas initiated.

When untreated, posttraumatic stress symptoms persist at six-month follow-up in young children exposed to a traumatic event (Meiser-Stedman, Smith, Glucksman, Yule, & Dalgleish, 2008). Children exposed to trauma seem to be more sensitive to subsequent stressors. This heightened sensitivity contributes to an increased risk for problematic internalizing and externalizing behaviors (Grasso, Ford, & Briggs-Gowan, 2013). When poor attachments exist, education regarding the neurological underpinnings of disrupted attachment may be helpful to parents and the child alike. Neuroeducation may help to facilitate understanding between family members and contribute to responses that promote safety and facilitate healthy neurological growth. Further, many theories of play therapy, such as Adlerian play therapy, involve the parent/guardian in the therapeutic process of a child.

Play Therapy and Adlerian Play Therapy

Play therapy is a developmentally responsive approach to therapeutic work with children based in empirical support (Bratton, Ray, Rhine, & Jones, 2005; Landreth, 2012; Lin & Bratton, 2015; Ray, 2011). Children are able to express themselves symbolically and through play, supporting their needs while in therapy (Landreth, 2012; Ray, 2011). Research supports play therapy as an efficacious intervention for children with problematic or disruptive behaviors (Bratton et al., 2005; Lin & Bratton, 2015). Additionally, play therapy is an effective intervention for treatment of several different presenting issues: (a) externalizing or disruptive behaviors (i.e., aggression, ADHD symptoms, anger, impulsivity; Flahive & Ray, 2007; Garza & Bratton, 2005; Ojiambo & Bratton, 2014; Packman & Bratton, 2003; Ray, Blanco, Sullivan, & Holliman, 2009; Ray, Schottelkorb, & Tsai, 2007), (b) academically at-risk (Blanco & Ray; 2011; Blanco, Ray, & Holliman; 2012), and (c) functional impairment (Ray, Stulmaker, Lee, & Silverman, 2013). The therapeutic relationship between therapist and child can facilitate greater levels of awareness and regulation.

Adlerian Play Therapy (AdPT) shares many conceptual similarities with IPNB and neuroscience research including emphasis on childhood experience, social interest and relationships, purposefulness of behavior, holism, and the internalized narrative or self-schema. AdPT is empirically supported as effective (e.g., Meany-Walen, Bratton, & Kottman, 2014; Dillman Taylor & Meany-Walen, 2015; Meany-Walen, Bullis, Kottman, & Dillman Taylor, 2015). Additionally, AdPT is one of the leading approaches to play therapy (Lambert et al., 2007), emphasizing social interest and connection, subjective perspectives, a holistic approach to working with children, and the purposefulness of behavior – adaptive or maladaptive – to meet one's needs (Kottman & Meany-Walen, 2016). Due to the influence of early childhood experiences, children develop subjective perspectives about self, others, and the world (e.g., life style) and how to form relationships through this lens. Children enter therapy with mistaken beliefs in their lifestyle, and behaviors that once helped them to feel safe, secure, and significant are no longer feasible. Additionally, relationships with significant others – parents, caregivers, and/or teachers – are strained. Adlerian play therapists holistically conceptualize children that enter therapy, considering the child's (a) assets, (b) Crucial C's (e.g. perception of self as connected, capable, count as a human being, courageous; Lew & Bettner, 2000), (c) life tasks, (d) family constellation, (e) family atmosphere, (f) early recollections, (g) goal of misbehavior, (h) personality priorities, (i) lifestyle convictions, (j) private logic, and (k) play themes. AdPT "is further distinguished from other models of play therapy by specific techniques used in the play process, critical role of encouragement, emphasis on collaboration with caregivers, and type of information gathered through play themes" (Dillman Taylor & Bratton, 2014, p. 208).

Advancements in neuroscience provide theoretical support for the principles and practices of counseling and play therapy intervention (Badenoch & Kestley, 2015; Siegel, 2012). However, in spite of the theoretical connections between play therapy and neuroscience (Wheeler & Dillman Taylor, 2016), further research is needed to examine this relationship. Additionally, outside of the scope of this chapter, applications of IPNB informed play therapy intervention with diverse and/or special needs populations may be a useful next step. This chapter will (a) develop connections from neuroscience and IPNB to play and play therapy, (b) review the IPNB model's domains of integration and their relationship to AdPT, and (c) provide an updated AdPT framework that encompasses the most recent neuroscience empirical support.

INTEGRATING PLAY THERAPY AND NEUROSCIENCE

The interpersonal neurobiology (IPNB) model incorporates several theories from neuroscience to explain the connections among the brain, mind, and relationships with others (Siegel, 2012). Also, in the IPNB model, integration is based in neuroscience research regarding how memories are developed and the influence of biological instinctual responses for emotion and survival. Basic summaries for neuroscience theories including (a) memory encoding, (b) polyvagal theory, and (c) affective emotional-motivational systems will be provided; however, this review is not exhaustive, nor replacement for review of the original scholarly works (See Kandel, 1998; Panksepp & Biven, 2012; Porges, 2011; Siegel, 2012 for a detailed explanation of the aforementioned theories).

Memory Encoding

Encoding is a biological process whereby memories are created (Siegel, 2012). Events perceived by our senses are given attention and create or strengthen neural pathways. Implicit memory refers to unconscious recalled mental models of emotions, images, or behaviors that influence our reactions. Implicit memories begin to develop prenatally; and once born, the caregiver attachment influences neural development and the child's view of self. According to Adler (1998), attachment and the parent-child relationship are influenced by parental attitudes such as rejecting, democratic, or overprotective. The parent-child dynamic contributes to feelings of equality, inferiority, or superiority.

Conversely, explicit memory appears around age two. Explicit memory requires focused attention for conscious encoding of events and to be able to add a sense of time to the memory. The development of explicit memory contributes to autobiographical recollection where storytelling can help children learn and attach meaning. Autobiographical memories are often more easily recalled due to their strong connection to enhanced emotional states and neural activity.

At the core, an implicit and nonverbal process often guides play. We are wired at a primitive level motivated for play (Panksepp, 2010). This natural process occurs repeatedly in play therapy where preverbal implicit memories may emerge (Kestly, 2014). In response, the play therapist attunes to the child's needs and experiences through the restoration and reconsolidation process of the implicit. Implicit interpretations of the world and its rules are often reflected in play behaviors (Marks-Tarlow, 2012). Similarly, children are provided the opportunity for implicit learning through play.

Implicit learning occurs through engagement in the implicit world and by making sense of the implicit explicitly (McGilchrist, 2009). Children engage in metaphorical play that emerges naturally in the play therapy process, thus grounding this experience in nonverbal storytelling. Additionally, in play therapy, the therapist verbally communicates the child's process, thereby connecting the metaphoric and verbal worlds. In so doing, autographical and metaphorical understanding of the child's embodied story increases (Kestly, 2014; McGilchrist, 2009). When facilitated in a safe environment and safe attachment, play moves from implicit to explicit and contributes to autonomy and social engagement (Marks-Tarlow, 2012). Through the safety of the therapeutic relationship, children engage in the right-left-right progression of integration (McGilchrist, 2009). The progression from processing between the right and left hemispheres of the brain (i.e. right-left-right) facilitates integration of implicit memories and narrative into the explicit, conscious world.

Polyvagal Theory

According to the polyvagal theory (Porges, 2011), our neural systems are continuously assessing our experience and environment to match a response as a means of survival. Of the three neural system responses, the most primitive is 'immobilization' or more commonly called the 'freeze' response where an individual feigns death for survival. Second, the 'fight-or-flight' response allows us to take action or exit a perceived challenge or threatening situation. Finally, higher order processes, such as social engagement or self-soothing behaviors, can only be employed once the sympathetic nervous system (SNS) is calm. The SNS facilitates the 'fight-or-flight' response, whereas the parasympathetic nervous system stimulates 'rest-and-digest' and a return to homeostasis. Our physiological state influences our options for behavioral response because we are wired to seek safety. Additionally, older circuits are engaged sequentially to regain a sense of safety, responding from more developed circuits first. As a result, in counseling, the polyvagal response allows for development of safety and trust in the therapeutic relationship once both systems are receptive.

To experience the benefits of play discussed to this point, the child's SNS must first enter a calm state. As Porges (2011) outlines in the polyvagal theory, every human is wired to seek out a sense of safety, limiting behavioral options depending on which neural system is engaged. In play therapy, the polyvagal response allows for development of safety and trust within the playroom as well as in the therapeutic relationship.

The playroom should be designed and maintained in a way to promote stability and predictability for the child (Axline, 1964; Kestley, 2014; Landreth, 2012; Ray, 2011). Kestly (2014) refers to this room as the "play sanctuary," in which the environment is designed to create a calm, safe space that is inviting and welcoming of the child. Therapist considerations for creation of a play sanctuary outlined by Kestly (2014) include that it: (a) is inviting, (b) is organized and predictable, (c) includes special play areas for nurturing, aggressive, make-believe, reality, and creative toys, and (d) presents flexible boundaries. However, the child must *perceive* the environment as safe in order to calm his or her SNS and engage in the circuitry of *PLAY* (Panksepp & Biven, 2012). The perception of a threat to safety may be influenced by an individual's style of life including attitudes towards self, others, the world and challenges encountered in any of those domains. As the *PLAY* circuit is engaged and safety is felt, children are capable of accessing their implicit memories in play for integration into their autobiographical story (Kestly, 2014)

Affective Emotional-Motivational Systems

Instinctual drives for survival also contribute to the basic emotional networks in the brain or affective neuroscience (Panksepp, 2010). These primitive networks are located in the subcortical regions of the brain, yet inform higher order psychological processes and motivate behavior. There are seven basic emotional processes including *SEEKING, RAGE, FEAR, LUST* (seen in adolescence), *CARE, PANIC/ GRIEF*, and *PLAY*. The emotional processes are indicated with capital letters to designate the associated systems fundamentally engaged across species without conditioning. These emotional responses allow us to interpret safety or survival concerns in the environment.

Several of these systems, including *SEEKING, PLAY*, and *CARE*, underscore the significance of play and play-based interventions. Oftentimes when children are in internal distress, their disruptive behaviors can create challenges for parents/caregivers to recognize the underlying need for connection and safety. By engaging the circuitry of play, children can explore their world and internal interpretation through the safety of a symbolic medium. Likewise, in play therapy children can begin to "widen the window of tolerance" by experiencing safety in the playroom and within the therapeutic relationship while processing and integrating disturbances.

Neuroplasticity

New neural connections are developed across the lifespan through exposure to novel stimuli. Also known as neuroplasticity (Kandel, 1998), the brain continues to adapt and grow in response to experience. Similarly, reparative experiences can enhance neural development contributing to hope for change.

New ways of thinking, feeling, and experiencing prior memories or beliefs can be created through novel and repetitive experiences (Kandel, 1998; Siegel, 2012). In play therapy, children explore their world and experiences through play while maintaining safety in both the physical space and relationship with the therapist. AdPT (Kottman & Meany-Walen, 2016) posits the experience of safety while exploring past disturbances can challenge previous beliefs of self or others. Over time, these experiences in the playroom can alter beliefs about self, others, and the world. Ultimately, such schema modification creates a more accurate and complex depiction of the world that incorporates both positive and negative components.

Interpersonal Neurobiology (IPNB)

Relationships, experiences, and interaction with the environment exert bi-directional influences, each informing the next experience throughout our lifetime. In IPNB the focus is on how this interplay contributes to the overall development of the nervous system. Regulation of information and energy between the brain, mind, and interpersonal experience is called integration (Siegel, 2012). Oftentimes, the greater the integration between these systems, the more functional and healthy the individual. Therefore, all relationships and experiences have the potential to inhibit or contribute to integration and health. Furthermore, therapeutic relationships can facilitate integration by seeking to enhance awareness and build from experiences of co-regulation.

IPNB DOMAINS OF INTEGRATION AND ADLERIAN PLAY THERAPY

The IPNB model (Siegel, 2012) addresses nine domains of neural integration including: *consciousness, bilateral, vertical, memory, narrative, state, interpersonal, temporal,* and *transpirational* or *identity integration.* We will briefly outline each domain of integration, as well as, related neuroscience theory with application to AdPT below.

Consciousness Integration

Integration of consciousness is a teachable skill that aims to increase intentional attention to each of the senses. Consciousness can be practiced through awareness to the present moment and each of the four areas of sensation (Siegel, 2014). In the 'wheel of awareness' Siegel addresses (a) the five senses (touch, taste, smell, sight, hearing); (b) the sixth sense or the sensations of the interior of the body; (c) the seventh sense or mental activities including thoughts, feelings and beliefs; and (d) the eighth sense, or interconnectedness to others, the earth, the past, and future. Although just one form of consciousness integration, the wheel of awareness underscores the linkage between sensations, thoughts, feelings, and connection as contributors to mindfully experiencing the here-and-now. Present moment awareness can only be accomplished as the SNS is calmed and diffuse physiological arousal is reduced. In play therapy, intentional coordination and organization of the 'play sanctuary' can convey comfort, safety and invite children to explore the space and toys (Kestley, 2014; Landreth, 2012). Similarly, the therapeutic presence and relationship can enable the connection between self and others (Crenshaw & Kenney-Noziska, 2014; Geller & Porges, 2014). The first phase of AdPT focuses on development of the therapeutic and egalitarian relationship.

In AdPT, the therapeutic relationship between the therapist and child creates a feeling of safety and security in the playroom, thus opening up the child to experience and begin integration of consciousness. The play therapist will direct attention to the current moment through reflections of sensory experiences including the sights, sounds, and tactile sensations. In the Adlerian playroom, toys and materials are displayed in an open arrangement that is inviting to children. The structure of an Adlerian playroom allows the child to engage all five senses. Kottman and Meany-Walen (2016) identified appropriate materials and toys that encourage exploration. To engage touch, children can use water, sand, paints, and others. Children can engage their sound sense by playing instruments, popping the bubble wrap, etc. Visual stimuli in a playroom include the lights and the variety of colors of the toys and materials. Children can also engage in the sense of smell (i.e., the smell of paints, markers) and taste (e.g. baby bottle, water, or other toys). See Kottman and Meany-Walen (2016) for a complete list of playroom materials.

Bilateral Integration

Bilateral integration refers to the connection between the left and right hemispheres of the brain. Although each hemisphere performs similar functions, each attends to information in unique ways. Autobiographical memory, symbols, and creativity are primarily held in the right hemisphere (Siegel, 2010); whereas, the left hemisphere adds logical, linear, and literal processing including the addition of words and context to memories. When bilateral integration has not occurred, the individual may be challenged to incorporate emotions and cognitions in communication. Therefore, Siegel (2010) proposes connecting with other's on the emotional level and then redirecting cognitions to facilitate bilateral integration – connect to redirect.

Adlerian play therapists use specific skills to encourage bilateral integration. A unique Adlerian skill in conducting this level of integration is metacommunication. Kottman and Meany-Walen (2016) defined metacommunication as "the counselor step[ping] outside the interaction and communicat[ing] about the communication taking place in the relationship" (p. 111). In other words, the therapist notices the emotion, underlying meaning, or pattern being expressed by the child then reflects that information to the child. Through this interaction, the therapist is helping the child bring the emotional, abstract level to the forefront in a cognitive manner (i.e., metacommunication, reflection). When the child recognizes or "owns" this reflection, the child has established insight into his or her own patterns of viewing self, others, and the world (e.g., lifestyle) and thus, bilateral integration.

Vertical Integration

Vertical integration refers to the incorporation of higher (prefrontal cortex) brain functions with lower (subcortical) brain functions and body sensations. An individual responds to a triggering event by activating the emotional circuitry, FEAR and SEEKING. In response to the sensory input, the limbic region ("downstairs brain") is engaged with the goal of re-establishing safety. However, the individual can also engage higher order thinking from the prefrontal cortex ("upstairs brain"). In so doing, integration supports a shift from a state of reactivity to receptivity and openness to new information and experience. The play therapist can provide opportunities for learning and incorporate vertical integration by teaching children about their upstairs and downstairs brain through the *hand model of the brain* (Siegel & Payne Bryson, 2012).

Play therapists integrate many interventions into the play therapy modality to best meet each child's needs. In AdPT, Kottman and Meany-Walen (2016) outlined several metaphoric techniques such as (a) designing therapeutic metaphors, (b) mutual storytelling, and (c) directing role-plays; all that can help children retell experiences that were challenging or created anxiety. The therapist initiates storytelling to help the child connect his or her limbic area to the prefrontal cortex. When these two areas are not integrated, they may impede the child's progress in play therapy. Through storytelling, the therapist uses a custom-designed therapeutic metaphor that relates to the child's own experiences but creates distance for the child to establish safety as needed (Kottman & Meany-Walen, 2016). In turn, the use of metaphoric storytelling engages the *PLAY* circuitry (Panksepp & Biven, 2012). Storytelling includes a beginning, middle, and end – in that the main character overcomes the problem in a clear, concrete process. The child witnesses the therapist telling the story first, engaging the child's "upstairs brain" (i.e., prefrontal cortex). For younger children, their developmental level may not warrant this level of cognitive processing; however, repetition of the story can still encourage vertical integration. For older children, this process can be more explicit. By reading a child's non-verbal signals to assess his or her level of readiness to process, the therapist can purposefully discuss the obstacles and resolution in the story to enhance vertical integration. The therapist can also invite the child to tell/re-tell the story depending on his or her expressed interest in participating. When the child retells the story, this narrative directly relates to the process of vertical integration – using the prefrontal cortex to make sense of a triggering/challenging event.

Memory Integration

Memory integration distinguishes explicit from implicit and incorporates factual with autobiographical memory to create an awareness of self across time. Recall that implicit memories are based on experience and contribute unconsciously to options in behavioral response. Memory integration supports differentiation between past and present, so that an individual can avoid flooding responses and experience the past as the past. Additionally, through memory integration, an individual can become more aware when expectations for the world or mental models are triggered and attach factual reasoning before responding (Siegel & Payne Bryson, 2012).

Adlerian play therapists can utilize metacommunications to increase awareness of these mental models. Small nuances (i.e, body language – shrugs, downward glances; body position – toward or away from the therapist; verbal noises – sighs, heavy/shallow breathing) that the child communicates during session tend to relate to an internal struggle that he or she is experiencing and cannot cognitively articulate. Through use of metacommunications, the therapist is drawing the child's attention toward these nuances. Once the child is aware, the therapist can begin using interventions to make the implicit, explicit.

Further, Adlerian play therapists can also implement a specific intervention to recall emotions associated with memories – early recollections (Kottman & Meany-Walen, 2016).

Each early recollection elicited by the counselor should be a single, specific incident preferably occurring before the age of 10. Early memories are not coincidences; they are often projections. In large measure, what we selectively attend to from the past is reflective of what we believe and how we behave in the present, and what we anticipate from the future (Watts, 2013, p. 464).

Early recollections provide insight into the child's lifestyle – view of self, others, and the world. Therapists may find it more challenging to engage younger children in the memory process necessary for this activity. Therefore, the child needs to be at least six or seven years old when using early recollections in play therapy (Kottman & Meany-Walen, 2016). The therapist works with the child to develop his or her autobiographical memory, recalling up to seven early life memories. This activity should be conducted over several sessions versus the one session used for adults given the developmental level of children who are appropriate for play therapy (e.g., 3-10 years).

Narrative Integration

Narrative integration supports an individual's meaning-making process of their lived experiences. Violated expectations influence the adaptation of inaccurate and skewed schemas of the self, other, or the world, especially where trauma may have occurred. Storytelling can integrate thoughts, sensations, and feelings (Siegel & Payne Bryson, 2012). Likewise, narrative integration provides context for our experience, allowing negative events to remain in the past instead of intruding on the present (Collie, Backos, Malchiodi, & Spiegel, 2006; Siegel, 2012). However, storytelling and development of a self-narrative begins before language development (Badenoch, 2008). Therefore, storytelling activities can enable the externalization of internal challenges (Butler, Gutterman, & Rudes, 2009). Art, dance, sandplay, puppetry, music or free play may all be useful play therapy interventions that facilitate storytelling, depending on the developmental level of the child and his or her readiness to engage.

Unlike the custom-designed metaphor described earlier, the Adlerian therapist can also implement mutual storytelling, an intervention that allows *the child* to direct the story. Kottman and Meany-Walen (2016) suggested encouraging:

...the child [to] invent a story that has not happened in real life or been the plot of a television show, movie, video game, or book. This keeps the story in the 'made-up' realm, which lets children process their experiences without evoking their defenses (p. 237).

During the storytelling process, the therapist carefully listens for how the story is similar or what aspects of the story are similar to the child's presenting issues. The therapist then tells the child that he or she will tell another story using the same characters. The therapist retells the story using the same characters, same setting, and same beginning; however, she changes the middle and end to demonstrate more socially appropriate ways to handle the situation with a more positive outlook on life. This activity allows for narrative integration in a developmentally appropriate manner. Children feel safer when their internal struggles are explored in metaphor. Through the use of mutual storytelling, they can begin to see their own strengths, identify positive ways to build relationships, and learn strategies to enhance their own feelings of connectedness, capability, significance, and courageousness.

State Integration

State integration addresses the temporary nature of feelings experienced in life, relationships, and external events that can influence how we view or identify ourselves internally. Personality, character, and a sense of identity are generally persistent over time; however, experiences of a particular state may create internal conflict seeming incongruent or at odds with how an individual views him/herself. In response, play therapists can support resolution of this internal tension by bringing awareness to other choices or truths within the current moment.

The therapist can develop the client's understanding of state transience throughout the AdPT process (e.g. across phases) by validating a child's emotional state with reflection of feeling or metacommunications and acknowledging the fluidity of feelings about our experiences. In the fourth phase of Adlerian play therapy (i.e. reeducation/reorientation), the therapist teaches the child new, socially appropriate behaviors and provides opportunities for the child to try out new or learned feelings, thoughts, and behaviors that he or she has developed over the course of therapy. Within this phase, the therapist can teach children how to differentiate between thoughts that "I feel angry" (state) and "I am angry" (trait). Distinguishing between state and trait can contribute to more adaptive narrative for the self as described above, thus, promoting the reorientation aspect of therapy.

Interpersonal Integration

Human brains are wired for connection and interaction with others in the environment (Siegel & Payne Bryson, 2012). Therefore, once one system becomes integrated and differentiated within, interpersonal integration can occur between, through the introduction of another system. Interpersonal integration links the individual to a bigger purpose and degree of intimacy by incorporating input from others – movement from a 'me' to a 'we' perspective. We are attuned when we respond to the perceived and interpreted emotional states of those around us accurately. Mirror neurons allow us to predict and respond to inferred patterns in others and demonstrate connections to the human capacity for empathy. In response to the 'felt connection', neural firing becomes more coherent. In play therapy, attunement and coherence with another person contribute to the development of self-regulation within the therapeutic relationship.

In AdPT, the interaction between child and therapist encourages the natural healing that occurs when two systems resonate with one another. Facilitative responses communicate a trust and overall belief in the potential of the child to be *inter*dependent, learning to collaborate and work with the therapist to achieve therapeutic goals (Kottman & Meany-Walen, 2016). The therapeutic relationship is defined as a shared partnership, in that both the therapist and child share the responsibility and power in the play sessions. This nurturing, collaborative relationship is essential for the growing mind and child. AdPT interventions such as returning responsibility and allowing the child to take the lead empower interpersonal integration within the playroom. These skills communicate to the child that he or she is capable, competent, counts, and is courageous (four C's, Lew & Bettner, 1996) to try new tasks. By encouraging the child's four C's, the therapist builds the child's confidence in oneself and in the therapeutic relationship. Unlike other play therapy theories/approaches, AdPT emphasizes the importance of others in the process of change. Kottman and Meany-Walen (2016) stressed the critical nature of not only building an egalitarian relationship with the child, but also with the parents/caregivers and teachers. This systemic, interpersonal approach has been deemed most useful for integration (i.e., mental health). For instance, AdPT techniques applied in family therapy seem to not only increase involvement by the child, but also enhance positive emotional experiences within the family during family therapy sessions (Willis, Walters, & Crane, 2014). Related, children participating in group AdPT demonstrate improvements in socio-emotional competence and skills through interactions and cooperation between members (Chinekesh, Kamalian, Eltemasi, Chinekesh, & Alavi, 2014).

Temporal and Transpirational Integration

Developmentally, young children may be limited in their experience of the last two domains of integration – temporal and transpirational integration. Temporal integration examines the existential tensions between our desire for certainty and immortality versus the reality of uncertainty and death. Similarly, transpirational integration focuses on the larger belonging, connection, and purpose at the social, community, or planetary level. The capacity for abstract and hypothetical thought does not develop until adolescence (Berk, 2012; Piaget, 1951). Additionally, young children lack the cognitive ability to conceptualize personalized understandings of death (Himebauch, Arndol, & May, 2008). However, temporal integration also incorporates the connection of self in time related to the past, present and anticipated future (Siegel, 2012). Self-knowledge is increased by experiences and interactions with others that create our life story.

Children can communicate stories (as discussed with narrative integration), developing a stronger sense of their experiences as having a beginning, middle, and end, thus developing a stronger sense of their autobiographical memory. Adlerian play therapists provide a safe, nonjudgmental environment *with* the child; this environment is co-created. In this space, children can play out events from their past as a way to process and integrate those events into their life story (Siegel, 2012). Children involved in play therapy can use the facilitative responses of an Adlerian play therapist (including tracking, reflection of content, metacommunication) to make sense of both internal and external experiences (Kottman & Meany-Walen, 2016).

PUTTING IT ALL TOGETHER: NEUROSCIENCE-INFORMED ADLERIAN PLAY THERAPY

AdPT consists of four unique phases, although not rigid or linear. At any time, the therapist can transition to an early phase to meet the needs of the child client. Kottman and Meany-Walen (2016) outlined the four phases as (a) building an egalitarian relationship, (b) investigating the lifestyle, (c) gaining insight, and (d) reorientation/reeducation. The therapist role is fluid in this process, alternating between directive and non-directive approaches (Dillman Taylor & Bratton, 2014). Adlerian play therapists consider the therapeutic relationship to be a critical element in the change process for the child client. The purpose of this section is to provide an updated AdPT framework that is inclusive of neuroscience research (Table 1). We outline the nature of people, therapeutic goals, the role of the therapist, and the four therapeutic phases of Adlerian theory through a neuroscience lens.

Nature of People

Adlerian theorists view individuals as "indivisible, social, creative, decision-making beings whose beliefs and behavior have a purpose" (Carlson & Slavik, 1997, p. xi). Current neuroscience research supports this view of individuals – highlighting the importance of holism, social connection, purposefulness of behavior, and creativity (Miller & Dillman Taylor, 2016). Individuals strive to belong, gain significance, and develop social interest within their family of origin first, then generalizing this approach to other important relationships. This process of conceptualization is defined as the lifestyle – "the use of the personality, traits, temperament, and psychological and biological processes in order to find a place in the social matrix of life" (Maniacci, Sackett-Maniacci, & Mosak, 2014, p. 66). Lifestyles create predictability and emotional safety, thus, paving the path for involvement in the social engagement system.

Table 1. Integrating Adlerian play therapy and interpersonal neurobiology

	AdPT	AdPT and IPNB
Nature of People	Indivisible, social, creative, purposeful behavior, decision-makers	Holistic, social, purposeful behavior, creative, subjectivity
Therapeutic Goals	Primary: Increase social interest	Primary: Striving for well-being, mental health, and social interest (i.e., neural integrational, differentiation and linking)
Role of the Therapist	Fluid and fluctuating according to the need of the client and phase of therapy	Fluid and fluctuating according to the need of the client and phase of therapy
Phase I	Building an egalitarian relationship between therapist and client	Building an egalitarian relationship between therapist and client with a focus on conscious, interpersonal, and bilateral levels of integration
Phase II	Investigating the lifestyle	Investigating the lifestyle with a focus on memory and narrative levels of integration
Phase III	Gaining insight	Gaining insight with a focus on vertical, narrative, and memory levels of integration
Phase IV	Reeducation/Reorientation	Reeducation/Reorientation with a focus on interpersonal integration and neuroeducation

Even if a child is responding in a negative manner (e.g., one lacking in social interest), this behavior provides a sense of safety and manages to calm the SNS due to the predictability of others' reactions to this behavior. Adlerians' theorize that family members often develop a fixed view for who the child is and how he or she will behave by the age of six or seven. This view of the child acts as a 'psychology box' that can encapsulate the child. However, current neuroscience research supports Adlerians' view that these boxes can be modified through a trusting, collaborative therapeutic relationship with the child client and his or her family.

Further, social interest, a critical piece to Adlerian theory, "describes a person's connection to and appreciation of others' contributions to the betterment of the entire community" (Kottman and Meany-Walen, 2016, p. 21). "Social interest is consistent with a neurobiological view of optimal wellbeing" (Miller & Dillman Taylor, 2016, p. 113). Reflective of neural integration is the selfless response of having concern for and acting in the benefit of others (Siegel, 2010).

Adlerian therapists view individuals as goal-directed and their behaviors as achievement towards one's desire for significance and belonging. Clients in therapy, in collaboration with the therapist, transition from goals of misbehavior (inadequacy, revenge, power, and attention) to positive goals of behavior (involvement, independence, fairness, and competence) that can aid as a barometer of mental health and increased improvement (Kottman & Meany-Walen, 2016). Neuroscience defines goals of human growth and development as survival and belonging (Baumeister & Leary, 1995; Cozolino, 2010; Immordino-Yang & Damasio, 2007; Siegel, 2012; Spielberg et al., 2012). Individuals consistently assess risk and engage protective instincts of fight, flight, or freeze. In early childhood, children learn to adapt to their environment utilizing key behaviors that may create obstacles in adulthood when bids for connection are made; thus, the individual struggles to connect with others in an emotionally supportive, interpersonal relationship – another important goal. Therefore, to remain consistent with current literature, we advise Adlerian therapists to embrace the goal of survival in the conceptualization of clients due to the hierarchical relationship between establishing safety and the goals of belonging and significance.

In addition, people view life subjectively – this perspective aids in the development of the client's worldview and how he or she strives to achieve the goals as listed above. Rudolf Dreikurs is credited for once stating, children are keen observers, but poor interpreters. They tend to develop inaccurate beliefs about events and interactions. These inaccurate interpretations and goals of misbehavior contribute to the development of mistaken beliefs – a belief based on one's biased perception of self, others, and the world. People are also considered creative. The concept of neuroplasticity validates individuals' ability to adapt and change over time (Siegel, 2012). Adlerian therapists view the client holistically, focusing on the big picture based on the information gathered, and not singling out one aspect of his or her worldview. Therapists, within session and through the conceptualization process, focus on encouraging an individual's assets, uniqueness; exploring one's Crucial c's, goals, personality priorities, private logic, thought, emotions, and behaviors through a holistic manner; and examining the client's uniqueness across the five life tasks (work/school, friends, love, finding meaning in life, and coming to terms with oneself [self - self-regulation]; Kottman & Meany-Walen, 2016). This approach of viewing the client as indivisible is supported by neuroscience (e.g., interpersonal neurobiology; Siegel, 2012).

Goals of Therapy

The overarching goal of Adlerian therapy is to increase social interest. Given that social interest is a marker for mental health, this goal ties directly to IPNB (Miller & Dillman Taylor, 2016). For example,

Fredrickson et al. (2013) found a correlation between individuals' healthier gene expression and meaningful happiness (e.g., eudemonia). Therefore, preliminary neuroscience research supports that social interest and mental well-being go hand-in-hand, acknowledging the need to strive for social interest as priority in therapy. However, Miller and Dillman Taylor (2016) recommended that the Adlerian definition of mental health (e.g., degrees of social interest and discouragement) should be expanded to include brain-based understanding of mental health, particularly the definition proposed by Siegel (2012). From the IPNB perspective, "the mind as a complex, self-organizing, non-linear system that achieves wellbeing when aspects of the system (e.g., parts of the embodied brain, interpersonal relationships) are allowed to sufficiently differentiate (i.e., specialize) and then link with other aspects of the system" (Miller & Dillman Taylor, 2016, p. 122-123). In keeping this more holistic view of mental health in mind when working towards achievement of goals in AdPT, therapists will be able to address the whole person and his or her system as Adler first proposed.

According to Kottman and Meany-Walen (2016), secondary goals within AdPT aim to continue advancing the client's mental health. Some of these goals include: (a) decrease feelings of inferiority; (b) overcome feelings of discouragement while recognizing and using personal resources; (c) make changes in life goals and mistaken beliefs about self, others, and the world; (d) alter negative motivation to achieve significance and belonging; (e) gain a sense of equality with others; and (f) become cooperative, contributing members of society. Neuroplasticity adds support grounded in neuroscience for the possibility of change and achievement towards these goals within a supportive, encouraging therapeutic relationship.

Role of the Therapist

In AdPT, the therapist role is fluid, fluctuating according to phase of therapy. Within the first phase of therapy, the therapist responds to the child in a non-directive manner and power is shared between the therapist and client (Dinkmeyer, Dinkmeyer, & Sperry, 1987; Kottman & Meany-Walen, 2016). The therapist's role in phase two transitions to a more directive approach, acting as a detective to gain information regarding the client's lifestyle (Kottman, 2011). In phase three, the therapist's role is more fluid alternating between non-directive and directive approaches to challenge the child's mistaken beliefs and to increase insight (Kottman & Meany-Walen, 2016). During the last phase of therapy, the therapist shifts to a more active role, teaching and encouraging the child to integrate, practice, and incorporate new skills and perceptions about self, others, and the world.

Four Phases of Adlerian Play Therapy

Adlerian theory consists of four phases of therapy: (a) building an egalitarian relationship, (b) investigating the lifestyle, (c) gaining insight, and (d) reeducation/reorientation. Each of the four phases builds on the former, allowing the therapist to revisit a previous phase as needed. To advance Adlerian theory to current day practice, Miller and Dillman Taylor (2016) recommended therapists incorporate a developmental framework (outlined in phase two below) and neuroscience informed interventions (across the four phases, with emphasis in phases three and four).

Phase I: Building an Egalitarian Relationship

A strong, collaborative relationship is the foundation for change to occur and for the child to move towards *conscious integration* within the therapeutic process. This phase lays the foundation for the other three; thus, the therapist is consistently working on the maintenance of the therapeutic relationship and *interpersonal integration* for the duration of therapy (Kottman & Meany-Walen, 2016). To develop safety in the playroom, Kottman and Meany-Walen (2016) stated that the therapist treats the child client as an equal from the first meeting by: (a) getting on the child's level and introducing self to child client prior to the parent/caregiver in the waiting room, (b) asking the child client what he or she was told regarding play therapy, (c) demystifying the counseling process, (d) giving the child a business card, and (e) communicating in a developmentally appropriate manner the nature of confidentiality. Once safety is established, the child's play is free to move from the implicit to explicit, engaging in the right-left-right progression of integration. Further, the therapist utilizes several skills in order for *conscious integration* to occur – awareness of non-verbal behavior – such as tracking behavior, restating content, metacommunicating, reflecting feelings, answering and asking questions, returning responsibility to the child, and interacting actively with the child. Metacommunication is a particularly important skill in AdPT. This skill is interwoven into all four phases to help the client gain insight into his or her lifestyle. The use of metacommunication aids in *bilateral integration*, connecting the right hemisphere (dominated by emotions) to the left hemisphere (dominated by language).

Phase II: Investigating the Lifestyle

Early childhood experiences contribute to one's overall physical and psychological health (Kandel, 1998; Panskeep & Biven, 2012). Therefore, the therapist works with caregivers, teachers, and the child to understand early life experiences and their influence on the child's development of lifestyle. Further, as Miller and Dillman Taylor (2016) recommended, therapists should take inventory of the child's early life experiences through a developmental theory such as attachment theory. Through the integration of attachment theory into AdPT, therapists will be better equipped to conceptualize the whole child and his or her relational influences to one's worldview as Adler intended. Therefore, the therapist takes a detailed developmental history (Kottman & Meany-Walen, 2016; Ray, 2011). The child's lifestyle is multifaceted, consisting of several components that together formulate the overall conceptualization of how the child views self, others, and the world. The therapist explores with the child and significant others the following aspects: (a) assets, (b) functioning at life tasks, (c) play themes, (d) psychological birth order, (e) family constellation, (f) family atmosphere, (g) early recollections, (h) goals of misbehavior, (i) crucial c's, and (j) personality priorities. In particular, early recollections aid in *memory integration*. As the client is able to describe and piece together early childhood memories, he or she links these early recollections into an autobiographical narrative, thus increasing *memory* and *narrative integration*. In this phase, the therapist gathers information to formulate tentative hypotheses regarding the client's worldview. In the next phase, the client begins to gain insight into his or her lifestyle, thus aiding further integration.

Phase III: Gaining Insight

A child begins to develop an internalized narrative or mistaken beliefs about self, others, and the world from early experiences. Therapists, in this phase, work to provide the client with insight into this internalized framework. As noted earlier in the chapter, Kottman and Meany-Walen (2016) outlined several metaphoric techniques to aid in *vertical* and *narrative integration* such as (a) designing therapeutic metaphors, (b) mutual storytelling, and (c) directing role-plays; all that can help the child retell experiences that were challenging or created anxiety. Further, the therapist uses the collection of early memories from phase two to provide the client with tentative hypotheses regarding his or her worldview and increase the client's *memory* and *narrative integration*. The therapist uses retelling of early memories and/or storytelling techniques to attach meaning to memories and help further implicit learning through play, the child's natural language.

Phase IV: Reorientation/Reeducation

The child learns to develop new ways to think, behave, feel, and experience prior memories and beliefs through novel and repetitive experiences. In this phase, the therapist collaborates with the client to modify mistaken beliefs and private logic to more socially useful ways of seeing self, others, and the world. Through repetition of practicing new behaviors and thoughts, the client can begin to internalize and experience his or her world more accurately. The therapist uses this phase to educate the client on more socially useful approaches to engaging and interacting with others – both to increase in social interest and *interpersonal integration*. Further, as Miller and Dillman Taylor (2016) recommended, inclusion of *neuroeducation* is impactful for clients to begin understanding how their behaviors relate to brain development. Siegel and Payne Bryson (2011) found that teaching children the hand model of the brain was helpful in providing insight, increasing recognition of cues preceding "flipping their lid," and implementing mindfulness strategies to regain control. Similarly, Siegel and Payne Bryson (2011) suggested education to teach calming skills, skills to redirect attention, and skills to integrate states, feelings, and desires as methods of state integration. Adlerian play therapists can integrate various techniques to help the child focus on his or her breath as it fits with the child's presenting issue, such as (a) use of pinwheels to demonstrate deep breathing, (b) bubbles for slow controlled breaths, or (c) deep belly breaths while balancing a toy on a child's stomach. These activities can be generalized for use in other settings including home or school.

SUMMARY

Current neuroscience research provides theoretical support for the principles and practices of play therapy (Badenoch & Kestley, 2015; Siegel, 2012). Specifically, AdPT shares many conceptual similarities with IPNB, including an emphasis on childhood experience, social interest and relationships, purposefulness of behavior, holism, and the internalized narrative or self-schema. The connections between neuroscience principles and play therapy were demonstrated, reviewed the IPNB model's domains of integration and their relationship to AdPT, and provided an updated AdPT framework that encompassed current neuroscience empirical support. Therefore, the integration of AdPT with IPNB demonstrates significant promise given the ease of connection between these two therapeutic models. It is likely that therapists who integrate AdPT with IPNB can help facilitate healthy brain development and interpersonal and intrapersonal relationships; although, future research is needed on the efficacy of this approach.

REFERENCES

Adler, A. (1998). *Understanding human nature*. London: Oneworld Publications. (Original work published 1927)

Axline, V. M. (1964). *Dibs: In search of self - personality development in play therapy*. Boston, MA: Houghton Mifflin.

Badenoch, B. (2008). *Being a brain-wise therapist: A practical guide to interpersonal neurobiology*. New York, NY: Norton.

Badenoch, B., & Kestly, T. (2015). Exploring the neuroscience of healing play at every age. In D. Crenshaw & A. Stewart (Eds.), *Play therapy: A comprehensive guide to theory and practice* (pp. 524–538). New York, NY: Guildford Press.

Baumeister, R. F., & Leary, M. R. (1995). The need to belong: Desire for interpersonal attachments as a fundamental human motivation. *Psychological Bulletin*, *117*(3), 497–529. doi:10.1037/0033-2909.117.3.497 PMID:7777651

Berk, L. (2012). *Infants and children: Prenatal through middle childhood* (7th ed.). Boston, MA: Pearson Allyn and Bacon.

Blanco, P. J., & Ray, D. C. (2011). Play therapy in elementary schools: A best practice for improving academic achievement. *Journal of Counseling and Development*, *89*(2), 235–243. doi:10.1002/j.1556-6678.2011.tb00083.x

Blanco, P. J., Ray, D. C., & Holliman, R. (2012). Long-term child centered play therapy and academic achievement of children: A follow-up study. *International Journal of Play Therapy*, *21*(1), 1–13. doi:10.1037/a0026932

Bratton, S. C., Ray, D., Rhine, T., & Jones, L. (2005). The efficacy of play therapy with children: A meta-analytic review of treatment outcomes. *Professional Psychology, Research and Practice*, *36*(4), 376–390. doi:10.1037/0735-7028.36.4.376

Butler, S., Guterman, J. T., & Rudes, J. (2009). Using puppets with children in narrative therapy to externalize the problem. *Journal of Mental Health Counseling*, *31*(3), 225–233. doi:10.17744/mehc.31.3.f255m86472577522

Carlson, J., & Slavik, S. (1997). *Techniques in Adlerian Psychology*. New York, NY: Routledge.

Chinekesh, A., Kamalian, M., Eltemasi, M., Chinekesh, S., & Alavi, M. (2014). The effect of group play therapy on socio-emotional skills in pre-school children. *Global Journal of Health Science*, *6*(2), 163–167. PMID:24576376

Collie, K., Backos, A., Malchiodi, C., & Spiegel, D. (2006). Art therapy for combat-related PTSD: Recommendations for research and practice. *Art Therapy: Journal of the American Art Therapy Association*, *23*(4), 157–164. doi:10.1080/07421656.2006.10129335

Cozolino, L. (2010). *The neuroscience of psychotherapy: Healing the social brain* (2nd ed.). New York, NY: W. W. Norton & Co.

Crenshaw, D. A., & Kenney-Noziska, S. (2014). Therapeutic presence in play therapy. *International Journal of Play Therapy*, *23*(1), 31–43. doi:10.1037/a0035480

Dillman Taylor, D., & Bratton, S. C. (2014). Developmentally appropriate practice: Adlerian play therapy with preschool children. *Journal of Individual Psychology*, *70*(3), 205–219. doi:10.1353/jip.2014.0019

Dillman Taylor, D., & Meany-Walen, K. (2015). Investigating the effectiveness of Adlerian Play Therapy (AdPT) with children with disruptive behaviors: A single-case research design. *Journal of Child and Adolescent Counseling*, *1*(2), 81–99. doi:10.1080/23727810.2015.1081519

Dinkmeyer, D., Dinkmeyer, D. Jr, & Sperry, L. (1987). *Adlerian counseling and psychotherapy* (2nd ed.). Columbus, OH: Merrill.

Erickson, F. (1963). Viewpoints on children in hospitals. *Hospitals*, *1*, 37–47.

Erickson, F. (1963). Viewpoints on children in hospitals. *Hospitals*, *1*, 37–47.

Flahive, M. W., & Ray, D. (2007). Effect of group sandtray therapy with preadolescents. *Journal for Specialists in Group Work*, *32*(4), 362–382. doi:10.1080/01933920701476706

Fredrickson, B. L., Grewen, K. M., Coffey, K. A., Algoe, S. B., Firestine, A. M., Arevalo, J. G., & Cole, S. W. et al. (2013). A functional genomic perspective on human well-being. *Proceedings of the National Academy of Sciences of the United States of America*, *110*(33), 13684–13689. doi:10.1073/pnas.1305419110 PMID:23898182

Garza, Y., & Bratton, S. C. (2005). School-based child-centered play therapy with Hispanic children: Outcomes and cultural consideration. *International Journal of Play Therapy*, *14*(1), 51–80. doi:10.1037/h0088896

Geller, S., & Porges, S. (2014). Therpeutic presence: Neurophysiological mechanisms mediating feeling safe in therapeutic relationships. *Journal of Psychotherapy Integration*, *24*(3), 178–192. doi:10.1037/a0037511

Grasso, D., Ford, J., & Briggs-Gowan, M. (2013). Early life trauma exposure and stress sensitivity in young children. *Journal of Pediatric Psychology*, *38*(1), 94–103. doi:10.1093/jpepsy/jss101 PMID:23008502

Greenspan, S. (1993). *Playground politics: Understanding the emotional life of your school-aged child.* Reading, MA: Harvard University Press.

Himebauch, A., Arnold, R., & May, C. (2008). Grief in children and developmental concepts of death #138. *Journal of Palliative Medicine*, *11*(2), 242–243. doi:10.1089/jpm.2008.9973 PMID:18333739

Immordino-Yang, M., & Damasio, A. (2007). We feel, therefore we learn: The relevance of affective and social neuroscience to education. *Mind, Brain, and Education*, *1*(1), 3–10. doi:10.1111/j.1751-228X.2007.00004.x

Kandel, E. (1998). A new intellectual framework for psychiatry. *The American Journal of Psychiatry*, *155*(4), 457–469. doi:10.1176/ajp.155.4.457 PMID:9545989

Kestly, T. (2015). Sandtray and storytelling in play therapy. In D. A. Crenshaw & A. L. Stewart (Eds.), *Play therapy: A comprehensive guide to theory and practice* (pp. 156–170). New York, NY: Guilford Press.

Kestly, T. A. (2014). *The interpersonal neurobiology of play: Brain-building interventions for emotional well-being*. New York, NY: W.W. Norton & Company.

Kottman, T. (2011). *Play therapy: Basics and beyond* (2nd ed.). Alexandria, VA: American Counseling Association.

Kottman, T., & Meany-Walen, K. K. (2016). *Partners in play* (3rd ed.). Alexandria, VA: American Counseling Association.

Lambert, S. F., LeBlanc, M., Mullen, J. A., Ray, D., Baggerly, J., White, J., & Kaplan, D. (2007). Learning more about those who play in session: The National Play Therapy in Counseling Practices Project (Phase 1). *Journal of Counseling and Development*, *85*(1), 42–46. doi:10.1002/j.1556-6678.2007.tb00442.x

Landreth, G. (2012). *Play therapy: The art of the relationship* (3rd ed.). New York, NY: Brunner Routledge.

Lew, A., & Bettner, B. L. (1996). *Responsibility in the classroom: A teacher's guide to understanding motivating students.* Connexions Press.

Lew, A., & Bettner, B. S. (2000). *A parent's guide to motivating children.* Connexions Press.

Lin, Y., & Bratton, S. C. (2015). A meta-analytic review of Child-Centered Play Therapy approaches. *Journal of Counseling and Development*, *93*(1), 45–58. doi:10.1002/j.1556-6676.2015.00180.x

Maniacci, M., Sackett-Maniacci, L., & Mosak, H. (2014). Adlerian psychotherapy. In D. Wedding & R. J. Corsini (Eds.), *Current psychotherapies* (10th ed.; pp. 55–94). Belmont, CA: Thomson Brooks/Cole.

Marks-Tarlow, T. (2012). The play of psychotherapy. *American Journal of Play*, *4*(3), 352–377.

McGilchrist, I. (2009). *The master and his emissary: The divided brain and the making of the western world*. New Haven, CT: Yale University Press.

Meany-Walen, K., Bratton, S., & Kottman, T. (2014). Effects of Adlerian play therapy on reducing students disruptive behaviors. *Journal of Counseling and Development*, *92*(1), 47–56. doi:10.1002/j.1556-6676.2014.00129.x

Meany-Walen, K., Kottman, T., Bullis, Q., & Dillman Taylor, D. (2015). Effects of Adlerian play therapy on childrens externalizing behavior. *Journal of Counseling and Development*, *93*(4), 418–428. doi:10.1002/jcad.12040

Meiser-Stedman, R., Smith, P., Glucksman, E., Yule, W., & Dalgleish, T. (2008). The posttraumatic stress disorder diagnosis in preschool- and elementary school-age children exposed to motor vehicle accidents. *The American Journal of Psychiatry*, *165*(10), 1326–1337. doi:10.1176/appi.ajp.2008.07081282 PMID:18676592

Miller, R., & Dillman Taylor, D. (2016). Does Adlerian theory stand the test of time? Examining individual psychology from a neuroscience perspective. *Journal of Humanistic Counseling*, *55*(2), 111–128. doi:10.1002/johc.12028

National Scientific Council on the Developing Child. (2010). *Early experiences can alter gene expression and affect longterm development* (Working paper 10). Retrieved from http://developingchild.harvard.edu/resources/reports_and_working_papers/working_papers/wp10/

Ojiambo, D., & Bratton, S. C. (2014). Effects of group activity play therapy on problem behaviors of preadolescent Ugandan orphans. *Journal of Counseling and Development*, *92*(3), 355–365. doi:10.1002/j.1556-6676.2014.00163.x

Packman, J., & Bratton, S. C. (2003). A school-based group play/activity therapy intervention with learning disabled preadolescents exhibiting behavior problems. *International Journal of Play Therapy*, *12*(2), 7–29. doi:10.1037/h0088876

Panksepp, J. (2010). Affective neuroscience of the emotional BrainMind: Evolutionary perspectives and implications for understanding depression. *Dialogues in Clinical Neuroscience*, *12*(4), 533–545. Retrieved from http://www.ncbi.nlm.nih.gov/pmc/articles/PMC3181986/pdf/DialoguesClinNeurosci-12-533.pdf PMID:21319497

Panksepp, J., & Biven, L. (2012). *The archaeology of mind: Neuroevolutionary origins of human emotions*. New York, NY: Norton.

Piaget, J. (1951). *Play, dreams, and imitation in childhood*. New York, NY: Norton.

Porges, S. (2011). *The polyvagal theory: Neurophysiological foundations of emotions, attachment, communication, and self-regulation*. New York, NY: Norton.

Ray, D. C. (2011). *Advanced play therapy: Essential conditions, knowledge, and skills for child practice*. New York, NY: Routledge/Taylor & Francis Group.

Ray, D. C., Blanco, P. J., Sullivan, J. M., & Holliman, R. (2009). An exploratory study of child-centered play therapy with aggressive children. *International Journal of Play Therapy*, *18*(3), 162–175. doi:10.1037/a0014742

Ray, D. C., Schottelkorb, A., & Tsai, M. (2007). Play therapy with children exhibiting symptoms of attention deficit hyperactivity disorder. *International Journal of Play Therapy*, *16*(2), 95–111. doi:10.1037/1555-6824.16.2.95

Ray, D. C., Stulmaker, H. L., Lee, K. R., & Silverman, W. K. (2013). Child-centered play therapy and impairment: Exploring relationships and constructs. *International Journal of Play Therapy*, *22*(1), 13–27. doi:10.1037/a0030403

Siegel, D., & Payne Bryson, T. (2012). *The whole-brain child: Twelve revolutionary strategies to nurture your child's developing mind.* New York, NY: Delacorte Press.

Siegel, D. J. (2010). Commentary on integrating interpersonal neurobiology with group psychotherapy: Reflections on mind, brain, and relationships in group psychotherapy. *International Journal of Group Psychotherapy, 60*(4), 483–485. doi:10.1521/ijgp.2010.60.4.483 PMID:21028973

Siegel, D. J. (2012). *The developing mind: How relationships and the brain interact to shape who we are* (2nd ed.). New York, NY, US: Guilford Press.

Siegel, D. J. (2014, August 1). *Mindfulness as integration* [blog post]. Retrieved from http://www.drdansiegel.com/blog/2014/08/01/mindfulness-as-integration/

Spielberg, J. M., Miller, G. A., Warren, S. L., Engels, A. S., Crocker, L. D., Sutton, B. P., & Heller, W. (2012). Trait motivation moderates neural activation associated with goal pursuit. *Cognitive, Affective & Behavioral Neuroscience, 12*(2), 308–322. doi:10.3758/s13415-012-0088-8 PMID:22460723

U.S. Department of Health and Human Services, Administration on Children, Youth, and Families, Children's Bureau. (2013). *Child Maltreatment 2012.* Retrieved from http://www.acf.hhs.gov/sites/default/files/cb/cm2012.pdf#page=31

Watts, R. E. (2013). Adlerian counseling. In B. J. Irby, G. Brown, R. Lara-Alecio, S. Jackson, B. J. Irby, G. Brown, & S. Jackson (Eds.), *The handbook of educational theories* (pp. 459–472). Charlotte, NC: IAP Information Age Publishing.

Wheeler, N., & Dillman Taylor, D. (2016). Integrating interpersonal neurobiology with play therapy. *International Journal of Play Therapy, 25*(1), 24–34. doi:10.1037/pla0000018

Willis, A., Walters, L., & Crane, D. (2014). Assessing play-based activities, child talk, and single session outcome in family therapy with young children. *Journal of Marital and Family Therapy, 40*(3), 287–301. doi:10.1111/jmft.12048 PMID:24118088

KEY TERMS AND DEFINITIONS

Adlerian Play Therapy (AdPT): A developmentally responsive approach to therapeutic intervention with children based upon the tenants and ideas of Alfred Adler. AdPT follows four phases of intervention including: (a) building an egalitarian relationship, (b) investigating the lifestyle, (c) gaining insight, and (d) reeducation/reorientation.

Affective Emotional Motivational Systems: The seven emotional processes (i.e. SEEKING, RAGE, FEAR, LUST, CARE, PANIC/GRIEF, PLAY) that motivate behavior across species and as a component of survival.

Interpersonal Neurobiology (IPNB): A model for neural integration developed by Daniel Siegel that addresses the associations between the brain, mind, and interpersonal interactions for an individual's health and well-being.

Memory Encoding: The physiological process of memory formation whereby experiences contribute to neural pathways for (a) implicit or unconscious and (b) explicit and conscious responses.

Neural Integration: The process of connection for the parts of the brain and mind within an individual as well as between others in the environment that contribute to a healthy mind.

Neuroplasticity: The ability of the brain to continue neural growth and development across the lifespan through new experiences.

Play: The first natural language for human beings that (a) is a primitive emotion that motivates behavior when safety is perceived, (b) may reveal implicit and nonverbal interpretations of the world, (c) allows for exploration of the world, and (d) facilitates learning, neural development and skill acquisition.

Polyvagal Theory: A component of human beings' survival response whereby the sympathetic and parasympathetic nervous systems are engaged to respond to a perceived threat (i.e. fight, flight, or freeze) or return to homeostasis respectively, as a means to assess and obtain safety.

This research was previously published in Emerging Research in Play Therapy, Child Counseling, and Consultation edited by Rheta LeAnne Steen, pages 60-80, copyright year 2017 by Information Science Reference (an imprint of IGI Global).

Section 2

Computer Science and Information Systems

Chapter 9
Cognitive Neuroscience in Information Systems Research

Yeli Zhao
Chinese University of Petroleum–Beijing, China

Keng Siau
Missouri University of Science and Technology, USA

ABSTRACT

This paper reviews cognitive neuroscience and several neurophysiological tools (e.g., fMRI, PET, EEG, MEG, and eye tracking). The strengths and weaknesses of such tools for information systems research are presented. The paper provides examples of existing cognitive neuroscience studies in varies areas, such as neuroeconomics, neuromarketing, and eye tracking. In addition, this paper provides an overview of brain areas that response to various mental processes, and discusses the localization and functionality of each brain area. Because of the popularity of eye-tracking research in information systems, measurements and metrics related and derived from eye-tracking technique (e.g., fixation, saccades and scanpath) are described and discussed in this paper. Opportunities for applying cognitive neuroscience techniques to IS research as well as future research directions are also discussed.

INTRODUCTION

Neuroscience is the scientific study of the nervous system. As a branch of science, neuroscience covers the fields of anatomy, physiology, biochemistry, or molecular biology of nerves and nervous tissue. Cognitive neuroscience spans over and overlaps many disciplines such as neuroscience, physiological psychology, cognitive psychology, and neuropsychology.

Cognitive neuroscience aims to find the causal relationship between neural circuits in the brain and their corresponding psychological or cognitive functions. The use of neuroimaging functional tools (e.g., fMRI) enables the examination of brain mechanisms and brain activations while the subject performs mental tasks. In order to identify the brain area activated by a specific mental task, researchers have studied different research topics such as consciousness, cognition, memory, attention, emotion, decision making and others.

DOI: 10.4018/978-1-5225-5478-3.ch009

Basics of Cognitive Neuroscience

The human brain contains about one hundred billion (10^{11}) neurons and 100 trillion (10^{14}) synapses, making it the most complex organ in human body. There are two major brain systems that control and influence our daily behaviors, namely the prefrontal cortex (PFC) and the limbic system. The prefrontal cortex is the brain's outer layer (cerebral cortex) that covers the front part of the prefrontal lobe. It is associated with higher cognitive processing such as reasoning, planning, problem solving, decision-making, and movements. The limbic system deals mainly with emotions and memories. Specific brain areas and their subdivisions are responsible for different behaviors. Table 1 summarizes the existing understanding of different parts of the brain and their associated cognitive behaviors.

OVERVIEW OF NEUROPHYSIOLOGICAL TOOLS

Functional Neuroimaging Techniques

In order to interpret the association between brain activation and corresponding stimulus, we need to identify which area is activated when the subject performs a particular task. Usually, there are four functional neuroimaging techniques that are commonly used to observe brain activation and they are fMRI, PET, EEG and MEG. fMRI and PET detect changes associated with cerebral blood flow whilst EEG and MEG are electrophysiological techniques that monitor electromagnetic brain activity.

Functional Magnetic Resonance Imaging or Functional MRI (fMRI)

fMRI is probably the most commonly used functional neuroimaging procedure using MRI technology. Based on the fact that regional cerebral blood flow always associate with neuronal activation during cognition, fMRI is able to localize and track changes in blood oxygenation, which is a reasonable proxy for neural activity. fMRI studies range from the examination of familiar cognitive processes (e.g., human

Table 1. Brain systems and responsive cognitive behaviors

Brain system and Functions	Key Area	Cognitive Behaviors
Prefrontal cortex - Problem solving - Personality expression - Calculation - Short-term memory - Moderating acceptable behavior - Decision making	Dorsal prefrontal cortex	Attention, cognition and action (Goldman, Rokic, 1988) Working memory (Braver et al., 1997; Cohen et al., 1997) Cognitive effort (Owen et al., 2005; Van der Linden et al., 2003)
	Ventral prefrontal cortex	Emotion (Price, 1999)
	Medial prefrontal cortex	Slow-wave sleep (SWS) (Mander et al., 2013)
Limbic System - Emotion - Behavior - Motivation - Long-term memory - Olfaction	Hypothalamus	Motivation, emotion, learning, and memory
	Hippocampus	- Spatial memory (Kheirbeck & Hen, 2011) - Learning (CurlikShors & Shors, 2012)
	Amygdala	- Episodic-autobiographical memory (EAM) networks (Markowitsch & Staniloiu, 2011) - Attention and emotional process, social processing

memory and language processing) to novel studies (e.g., neurofunctional components of humor) (Geol and Dolan 2001). One of the reasons fMRI is frequently used is that it is convenient because it does not involve ingesting substances, taking shots, or exposure to radiation. fMRI is accepted as a major non-invasive technique in Cognitive Science.

ElectroEncephaloGraphy (EEG)

EEG is a technique to record electrical activity along the scalp and to measure voltage fluctuations resulting from the ionic current, which are synchronized to behavioral responses. During an experiment, an electrode will be attached to the subject's skin to detect the summed potentials that are generated by a large number of neurons. EEG is primarily used for temporal resolution.

According to Debener et al. (2006), EEG and fMRI have complementary advantages with regard to the spatial and temporal resolution. For instance, EEG measures the brain's electrical activity directly, so it is good in identifying the temporal response of a brain, but weak in identifying the exact location of the activated brain area. fMRI measures changes in blood flow and identifies the activated spot. New applications are being developed to acquire EEG and fMRI data simultaneously.

Positron Emission Topography (PET)

PET is a functional imaging technique that produces a three-dimensional image to observe metabolic processes in the body. PET measures blood flow in the brain, which is a reasonable proxy for neural activity. By injecting certain radioactive neurotransmitters, specific neurochemical changes (e.g., spatial distribution) in the brain can be visualized with a PET scanner.

MagnetoEncephaloGraphy (MEG)

MEG is a functional neuroimaging technique to map brain activity. By using very sensitive magnetometers, MEG records magnetic fields produced by electrical currents generated in the brain. MEG can also be used for temporal resolution as well as detect activities deeper in the brain structures.

Psychophysiological Tools

Eye Tracking

Eye tracking is a technique used to measure an individual's eye movements, usually with the aid of an eye tracker. For example, a camera above the monitor will record where a person is looking at, moments when the eyes are relatively stationary (fixation) and the sequence in which the person's eyes shift from one point to another (saccades). Fixations and saccades are two of the primary measurements used in eye-tracking research. Other metrics used in measuring are eye gaze, scanpath, pupil size, and blink rate. Applications of these measurements will be discussed in later sections.

Emotion Measurement Technology

Emotion measurement technology enables computers to recognize human emotions based on facial cues or physiological responses. Using a webcam to detect smirks, smiles, frowns and furrow, it can measure a user's level of surprise, amusement or confusion. It can also measure heart rate without the need for the user to wear any special devices. This is done by using the reflection of the person's face which shows blood flow. Examples of such products are Affdex and Q Sensor.

Other measurements of psychophysiological indicators such as heart rate, blood pressure, galvanic skin response (GSR, sweating in the palms) and pupil dilation are also used to detect a subject's reaction to a stimulus.

Table 2 depicts the neurophysiological tools and measurements.

MAJOR STRENGTHS OF NEUROPHYSIOLOGICAL TOOLS

The primary advantage of neurophysiological tools is that subjects cannot consciously manipulate their responses because all of the data tracked and retrieved from the brain or eyes are exposed to these neurophysiological tools before they can be consciously processed or modified. Consequently, physiological and brain data are relatively objective and reliable. The data is not susceptible to subjectivity bias, social desirability bias, and demand effects. Therefore, neuroscience techniques can serve as a complement to existing research tools by providing reliable data that are difficult or impossible to obtain with traditional tools such as self-reported or archival data.

Another benefit of neurophysiological data is the real-time measurement. It enables continuous data collection when a subject is executing a task or responding to a specific stimulus. Continuous data collection is difficult to achieve using techniques such as self-reports.

Further, subjects may feel uncomfortable or are unwilling to participate in self-reporting studies. These may be attributed to various reasons. For example, sensitive issues regarding value proposition (e.g., gender, race, religion, culture, and taboos), personal issues (e.g., fear, resist, and other passive moods), deep emotions or motivations (e.g., guilt) (Jan et al., 2014). Subjects may also be exposed to

Table 2. Neurophysiological Tools and Measurements

Neurophysiological Tools	Measurement
Neuroimaging Tools	
Functional Magnetic Resonance Imaging (fMRI)	Neural activities by changes in blood flow
Positron Emission Tomography (PET)	Metabolic processes by gamma rays
ElectroEncephaloGraphy (EEG)	Electrical brain activities by voltage fluctuation
MagnetoEncephaloGraphy (MEG)	Magnetic fields by electronic current
Psychophysiological Tool	
Eye Tracking	Fixation (gaze) and Saccades (movements)
Others	
Emotion Measurement Technology	Facial and physiological responses; heart rate

complex cognitive processes that overwhelm them or encounter issues related to social cognition and moral judgment. Neurophysiological tools can help to alleviate these problems and thereby providing researchers valuable research opportunities that were not available before.

MAJOR WEAKNESSES OF NEUROPHYSIOLOGICAL TOOLS

A primary weakness of neurophysiological tools is the cost. The cost of neuroimaging tools can be very substantial. In addition to the huge equipment cost, personnel costs need to be considered when there is a need to hire technicians with specialized knowledge to operate, interpret, and analyze the results.

The artificial environment is another issue. Some researchers question the validity of results derived from such studies because results may be affected as subjects may feel stressed or uncomfortable when attached with sensors from neurophysiological tools. Another issue is the difficulty to account for body movement in the data collected. Preparing subjects for proper recordings (e.g., calibration for eye tracking) requires trained and experienced researchers and technicians. The analysis of vast amounts of neurophysiological data can be an extremely time-consuming and sophisticated task. Extracting the right segments of data for the behavior of interest requires skills and experience.

Finally, in comparison to traditional data sources, neurophysiological results may be difficult to analyze and interpret given the current state of research and understanding of human brain. For example, the meanings of eye-tracking measurements are still under debate (Rayner 1998). Our understanding of the operations and functions of each part of the human brain is still far from complete.

COGNITIVE NEUROSCIENCE STUDIES IN VARIOUS DISCIPLINES

Cognitive neuroscience has been exploring the practical application in social science, such as neuroeconomics, neuromarketing, and psychology. Economists, marketers, and psychologists have started to exploit the potential of cognitive neuroscience and functional neuroimaging to enhance knowledge in their fields.

Neuroeconomics

Neuroeconomics analyzes and understands economic-related behaviors from the cognitive neuroscience perspective by using functional neuroimaging techniques (Zak, 2004). The neuroeconomics literature focuses mainly on examining how subjects develop strategic decision-making and how to balance various tradeoffs when they perform experimental games under different scenarios (Rustichini, 2005).

Regarding the scenarios wherein strategic decision-making is anticipated, researchers have developed the following four aspects: cognitive and emotional aspects of decision making, decision making under uncertainty and ambiguity, anticipation of gains and losses, and evaluating payoffs and outcomes (Dimoka, 2010). Examples of related literature and arguments are shown in Table 3.

Generally, the neuroeconomics literature indicates that strategic decision making under risk and uncertainty, anticipation of gains and losses, and evaluating payoffs are governed by different neural mechanisms. In other words, when processing information regarding these issues, different areas are activated in the human brain.

Table 3. Examples of neuroeconomics research areas and literature

Literature	Argument	Reference
Decision Making		
Neuro Basis of Financial Risk Taking	Decision-making under uncertainty shows that people's anticipation towards gains and losses is governed by different neural mechanisms	Kuhnen and Knutson, 2005
Self-Referential Thinking and Equilibrium as States of Mind in Games: fMRI Evidence	Decision-making involves both cognition and emotion. Subjects who showed better cooperation between the prefrontal cortex (thinking) and limbic system (feeling) were the most successful when performing the experimental games	Bhatt and Camerer, 2005
Risk		
Selective Activation of the Nucleus Accumbens during Risk-Taking Decision Making	Correlation exists between activation of nucleus accumbens and individual's risk avoidance	Matthew et al., 2004
Anticipation of Increasing Monetary Reward Selective recruits Nucleus Accumbens	When subjects predict risks, nucleus accumbens will be activated	Knutson et al., 2001
Uncertainty		
Distinct Neural Mechanisms of Risks and Ambiguity: A Meta-Analysis of Decision-Making	When performing decision-making under ambiguity, orbitofrontal and inferior parietal cortices showed brain activity	Krain et al., 2006
Characterization of Decision-Making Deficit of Patients with Ventromedial Prefrontal Cortex Lesions	As level of uncertainty increases, activation in the orbitofrontal and inferior parietal cortices increases as well	Bechara et al., 2000;
Neural Systems Responding to Degrees of Uncertainty in Human Decision-Making	Orbitofrontal cortex plays a key role in distinguishing different levels of uncertainty	Hsu et al., 2005
Loss		
Ventromedial Prefrontal Cortex Activation is Critical for Preference Judgments	In risky game that inclines toward loss, insular cortex is activated, which implies negative feeling toward the expectation for loss	Paulus and Frank, 2003

Neuromarketing

By using theories and techniques of cognitive neuroscience, neuromarketing studies the cognitive and affective response of the human brain that reacts to marketing stimuli (Zaltman, 2003). Neuromarketing aims to construct new models to better understand consumers' behavior and consumers' reactions to different marketing stimuli (Lee et al., 2007).

Besides consumer behavior, consumer preferences and purchase choice, and brand advertising are other main areas of research. For example, Deppe et al. (2005) investigate the relationship between consumer's brand choice (e.g., first brand choice and second brand choice) and the cognitive process of decision-making. The McClure et al. (2004) study analyzes brain areas that are activated when subjects are provided with immediate rewards versus delayed rewards, thereby the tradeoffs between impulse and planned purchases.

Table 4 displays the examples of neuromarking research literature in terms of purchasing choice, brand advertising, and rewards.

Table 4. Examples of neuromarketing research areas and literature

Literatures	Arguments and Methods	Author
	Purchase Choice	
Distributed Neuronal Systems Supporting Choice-Making in Real-Life Situations: Differences between Men and Women when Choosing Groceries Detected using MagnetoEncephaloGraphy	Different brain areas are activated in response to choice predictability. With unpredictable choices, brain activation is associated with silent vocalization as well as perception towards immediate vs. delayed rewards (Magneto-encephalogram brain imaging technique, MEG)	Braeutigam et al., 2004
Neural Correlates of Behavioral Preference for Culturally Familiar Drinks	When performing purchasing task that relates to preference judgments and social decision-making, ventromedial prefrontal cortex and limbic system are activated	McClure et al., 2004; Paulus and Frank, 2003
Nonlinear Responses within the Medial Prefrontal Cortex Reveal when Specific Implicit Information Influences Economic Decision Making	A consumer's first choice of brand depends on social decision-making, rather than rational decision-making	Deppe et al. 2005
Brand Advertising		
Real Time Processing of Affective and Cognitive Stimuli in Human Brain Extracted from MEG Signals	Brand advertising stimuli induce massive brain activations	Ioannides et al., 2000;
Brain Waves, Picture Sorts, and Branding Moments (Finding the Right Moment for Brand Advertising during TV commercials)	Examine impact of advertisements on brand development, awareness, and attention	Young, 2002;
Brain-Imaging Detection of Visual Scene Encoding in Long-Term Memory for TV Commercials	Visual images create faster activation in prefrontal cortex, which creates stronger brand recognition	Rossiter et al., 2001
	Rewards	
Neural Correlates of Behavioral Preference for Culturally Familiar Drinks	Immediate rewards primarily activate limbic system; delayed rewards activate lateral prefrontal and inferior parietal cortices	McClure et al., 2004
An fMRI Study of Reward-Related Probability Learning	Increased activation in caudate nucleus indicates that people tend to over-react towards losses than rewards	Delgado et al., 2005

Table 5 summarizes the key brain areas that are activated in response to activities from the decision model which is developed by Herbert A. Simon. The original Decision Model includes three phases: Intelligence, Design and Choice. Later, the model is extended to include Implementation and Review. In this case, we assign mental activities to each phase and localize brain activations towards each process.

Studies With the Use of Eye-Tracking Technology

Fixation

Fixation refers to the state wherein the eyes remain relatively stationary for a certain period of time. However, the interpretation of fixation depends on the contexts. For example, when browsing a web page, higher fixation frequency on specified areas may indicate interest. In a search task, a large number of single fixations and cluster fixations may indicate uncertainty in recognizing the target.

Table 5. Key brain areas activated in response to mental activities from decision model

Key Brain Areas		Prefrontal Cortex					Limbic system		Nucleus Accunmbens	Basal ganglia	Cerebral cortex	Inferior Parietal Cortices
		Dorsolateral			Orbifrontal, ventromedial, basal, orbital	Frontopolar, anterior rostral	Amygdala	Hippocampus		Caudate Nucleus	Insular Cortex	
		Dorsolateral		Ventrolateral								
Decision Model		Porterior dorsolateral	Mid-dorsolateral									
Intelligence	Decision Making	√			√			√				
	Risk								√			
	Uncertainty				√							√
	Ambiguity	√									√	√
Design	Loss										√	
	Rewards		√		√			√	√	√		
Choice	Trust									√		
	Distrust						√				√	
Implementation	Consumer Behavior	√		√				√				√
	Theory of Mind		√									
	Cooperation				√							
	Competition		√									√

Saccades

Saccades refer to the eye movement in between fixations. Examining saccades alone probably cannot reveal much about processing complexity. However, regressive saccades can be regarded as indicating difficulty in processing (Rayner & Pollatsek, 1989).

Scanpaths

A scanpath describes a complete saccade-fixate-saccade sequence. Goldberg (1999) proposed that an optimal scanpath when performing a search task is a straight line that links to the target with short fixation duration around the target.

Blink Rate and Pupil Size

Blink rate and pupil size can measure the degree of cognitive workload. Higher workload may produce lower blink rate whilst a higher blink rate may be caused by fatigue (Bruneau et al., 2002; Brokkings et al., 1996). Pupil size and blink rate can also be affected by many other factors such as ambient light levels (Goldberg & Wichansky, 2003). For these reasons, pupil size and blink rate are less often used in eye tracking research.

Table 6 summarizes the eye-tracking literature and metrics used.

Technical Issues in Eye-Tracking Research

When conducting experiments with an eye tracker, researchers should be aware of the limits of this technology and how these limits will impact the collected data. For example, experimenters should decide which measurements will be measured in advance. If the intention is to analyze fixations, the eye tracker needs to be optimized to detect fixations (Karn et al., 2000). Some eye tracking machines may require subjects to wear devices or take off their prescription lenses while others will not work well when

Table 6. Examples of eye-tracking literature and metrics

Literature	Metric	Measurements	Reference
Fixation-Derived Metrics			
Computer interface evaluation using eye movements: Methods and constructs	Number of fixations overall	Search task: More overall fixations indicate less search efficiency or poor interface layout	Goldberg & Kotval, 1999
In search of salience: A response time and eye movement analysis of bookmark recognition	Fixations per area of interest	More fixations on a particular area indicate the object is more noticeable or more important	Poole et at., 2004
Eye fixations and cognitive processes	Fixation duration	A longer fixation duration indicates that the information is difficult to extract or more attractive	Just & Carpenter, 1976
What attracts the eye to the location of missed and reported breast cancers? Measuring team situation awareness by means of eye movement data	Eye Gaze (sum of all fixations within a specific area)	Usage: measure attention distribution among targets or measure anticipation of situation awareness	Mello-thoms et al., 2004; Hauland, 2003
An eye-movement analysis of web-page usability	Fixation spatial density	Search Task: - Fixations concentrated on a small area indicate focused and efficient searching - Evenly spread fixations reflect widespread and inefficient search	Cowen et al., 2002
Computer interface evaluation using eye movements: Methods and constructs	Repeat fixations	Higher repeat fixations indicate that the information lacks meaningfulness or visibility	Goldberg & Kotval, 1999
Saccade-Derived Metrics			
Computer interface evaluation using eye movements: Methods and constructs	Number of saccades	More saccades indicate more searching	Goldberg & Kotval, 1999
Eye tracking in web search tasks: Design implications	Saccade amplitude	Larger saccades indicate more meaningful cues	Goldberg et al., 2002
The Reading Assistant: Eye gaze triggered auditory prompting for reading remediation.	Regressive saccades (regressions)	Regressions indicate less meaningful cues	Sibert et al., 2000
Scanpath Derived Metrics			
Computer interface evaluation using eye movements: Methods and constructs	Scanpath duration	A longer-lasting scanpath indicates less efficient scanning	Goldberg & Kotval, 1999
Eye tracking in web search tasks: Design implications	Scanpath length	A longer scanpath indicates less efficient searching or poor interface layout	Goldberg et al., 2002
Computer interface evaluation using eye movements: Methods and constructs	Spatial density	Smaller spatial density indicates more direct search	Goldberg & Kotval, 1999
Computer interface evaluation using eye movements: Methods and constructs	Transition matrix (Search order from one area to another)	Efficiency: direct scanpath to target Uncertainty: back and forth between areas	Goldberg & Kotval, 1999; Hendrickson 1989
Computer interface evaluation using eye movements: Methods and constructs	Scanpath regularity	Deviation from a "normal" scanpath indicates search problems or poor interface layout	Goldberg & Kotval, 1999
101 Spots or how do users read menus?	Scanpath direction	Scannpath indicates search strategy (e.g., top-down and bottom-up)	Altonen et al., 1998

subjects display huge movements. Another issue is the set minimum duration to define a fixation. Given different time sets in the eye tracking system, interpretations of the records can be varied dramatically (Inhoff & Radach, 1998).

FUTURE RESEARCH QUESTIONS AND DIRECTIONS IN COGNITIVE NEUROSCIENCE IN INFORMATION SYSTEMS

Dimoka et al. (2007) propose a term "Neuro-IS" that refers to the idea of applying cognitive neuroscience theories, methods, and tools in Information System (IS) research. IS researchers typically collect data by conducting surveys, lab experiments interviews, case studies, or using secondary data and analytical models. With the availability of neurophysiological tools, cognitive neuroscience research presents another dimension for IS researchers to pursue.

Cognitive neuroscience creates opportunities for IS research in several areas: (1) identifying IS phenomena that can be studied using cognitive neuroscience; (2) identifying existing IS theories and constructs that can be explained or expanded using cognitive neuroscience; and (3) using cognitive neuroscience as a vehicle to collect data for empirical studies. IS research can also help shed light on the workings of the brain and contribute to the cognitive neuroscience literature.

In the following subsections, we will present some future research directions for cognitive neuroscience research in IS.

Studying Constructs in Information Systems Development

Many constructs used in information systems development such as data flow, entities, and objects are designed and created by researchers with little or no theoretical foundation (Siau 2004; Siau and Loo 2006; Siau and Rossi 2011; Siau and Wang 2007). Some of these constructs have been argued and presented as natural and corresponding to human's information processing (Siau and Tan 2005, 2006, 2008, 2009). Cognitive neuroscience provides a means to study such claims and assertions. Neurophysiological tools can be used to identify those IS constructs that do not correspond to functionalities and processes of human brain. By excluding "unnatural" constructs, researchers will be able to develop better constructs that are easier for human comprehension and that will help enhance the developmental success of information systems.

Encouraging Technology Adoption and Use by Individuals

With the aid of neurophysiological tools such as fMRI, researchers can better understand the nature of constructs related to the adoption and use of systems. Some of the neural activities that correlate to constructs have been identified, such as perceived usefulness and ease of use (Dimoka and Davis, 2008). Other constructs such as enjoyment and anxiety and the neural activities related to these constructs are still not well understood and can be studied using cognitive neuroscience.

There are opportunities to identify new determinants of system adoption and use. By designing systems that help enhance system utility and user friendliness, and by establishing direct usability criteria based on neurophysiological data, systems can be developed with enhanced adoption by individual users.

Assessing Information Overload and Cognitive Overload

IS researchers often design systems to avoid information and cognitive overload. In order to do so, they try to reduce information complexity and provide decision-making support so that users are able to simplify their decision-making process. In other aspects, researchers also attempt to enhance users' information processing capabilities by representing information in a more comprehensive way (Galbraith, 1974) with the use of visualization tools. In spite of these endeavors, measurements of information and cognitive overload are still elusive in the literature (Payne et al., 1992). Consequently, a direct measurement is needed in order to measure information and cognitive overload. Neurophysiological tools provide the means for such a direct measurement. For example, eye-tracking tools can capture whether a user finds it difficult to identify information by observing how her or his eyes wander on a computer screen. By measuring brain activity when users undertake cognitive tasks, researchers can test whether the IT artifacts reduce information and cognitive overload.

Improving E-Commerce Transactions

In addition to encouraging individual's adoption of system, there are opportunities to enhance economy by enhancing online consumers' adoption of e-commerce (Venkatesh and Davis 2000). Among the literature that promotes the adoption and use of commercial websites (e.g., Cenfetelli, 2004; Pavlou and Fygenson, 2006), several criteria have been evaluated such as usefulness, ease of use (e.g. navigation), trust (e.g., willingness to input bank information), privacy (e.g., personal address) and security (e.g., deception). What is still unclear is whether these criteria are distinctive from each other.

Cognitive neuroscience can help in identifying or developing an e-commerce model that will result in higher adoption. Researchers can use cognitive neuroscience techniques to detect hidden predictors of e-commerce adoption. For example, in order to define whether information deception is one of the concerns for online consumers, researchers can track how users detect such deceptions and thereby developing a pattern or model. Besides enhancing credibility of e-commerce, there are also opportunities to take advantage of consumers' cooperative behavior and social communities. Cognitive neuroscience can also help in interactive e-commerce website design to enhance consumer engagement and increase the likelihood of future purchases.

In summary, identifying the neural activities that correlate to antecedents of websites and consumer adoptions could help better understand the nature and dimensionality of these antecedents, and thereby resulting in e-commerce model with more validity and adoption, and one that corresponds to the functionality of human brains and behaviors.

Developing IS Strategy

The literature of IS strategy has been focused on achieving requirement capabilities such as operational capabilities (e.g., performing daily activities), dynamic capabilities (e.g., reconfiguration), and improvisational capabilities (e.g., response to changes in a real time manner) (e.g., Pavlou and El Sawy 2006, 2010). Although these capabilities have clear distinctions in academic theories, it is still unclear whether they will correspond to organizational functionality in real business (Eisenhardt and Martin 2000).

Recently, IS strategy has shifted its focus to the intensive decision making under turbulent environment. According to Ernst and Paulus (2005), people will undertake calculation which is an emotional process before they can reach a decision. As discussed earlier, there is a rich neurophysiological literature regarding decision making under uncertainties. These studies will contribute to IS strategy that requires quick decision-making under uncertainty and constantly changing environments (Pavlou and El Sawy, 2006).

Enhancing Development and Implementation of Organizational Systems

When organizations implement ERP systems, one of the common issues is the gap between standard functionalities provided by the system package and the organization's requirements. This is also a factor that decides whether the system is successfully implemented (Rolland and Prakash, 2000). However, this gap is difficult to define because most ERP systems are very complex and there is a lack of appropriate measurements to define users' needs, especially from users' perspectives. Neurophysiological tools can help minimize the misfit. For example, the cognitive fit theory (Vessey, 1991) can be applied to examine the interaction between problem representation and problem-solvings, and create metrics to measure misfit in ex ante analysis. Cognitive neuroscience can help in this aspect.

Integrating Various Neurophysiological Tools

Researchers are also attempting to seamlessly integrate various neurophysiological tools. For example, some research studies are attempting to collect data from both the EEG and the eye-tracker. This will help to enhance the reliability and validity of results, and provide a means to cross-validate various neurophysiological data. The success of cognitive neuroscience will depend on the researchers' ability to integrate various neurophysiological data. This is not an easy feat because the timing of data collection needs to be precisely synchronized wherein a split second delay may distort the data.

Complementing Existing Data Collection Techniques

The value of cognitive neuroscience lies in combining neurophysiological data with other data collected from traditional sources. Neurophysiological tools should not be considered as a replacement, but a complement to the existing data collection tool and how these tools work together will define the benefits achieved from the new tools. One future direction for cognitive neuroscience research is to investigate how neurophysiological tools can be combined and integrated with other data collection techniques.

Validating Results From Cognitive Neuroscience Studies

Despite the fact that existing literature has indicated high correlations between the brain and psychometric measures (Vul et al., 2009), there are still concerns regarding whether this correlation is artificially inflated (e.g., Lieberman et al., 2009). In order to assess the true extent of the correlations between the brain and different IS constructs which may vary from cognitive to emotional ones, future research is needed to address this issue.

In addition to re-examining existing studies, neurophysiological tools may enrich IS theories by uncovering new constructs that have been ignored or that have not been able to be investigated due to the lack of appropriate neurophysiological tools.

CONCLUSION

Cognitive neuroscience is still in its infancy. Nevertheless, cognitive neuroscience has a lot of potential and the effective application of its theories and neurophysiological tools will benefit IS research. It should, however, be recognized that neurophysiological tools are not silver bullets. It is important for IS researchers to assess when the existing tools are insufficient and when to use neurophysiological tools. A rule-of-thumb is that when existing tools can adequately measure a research question, neurophysiological tools may not be necessary unless neurophysiological data can provide additional insights.

As new kids on the block, neurophysiological tools have generated much interest and excitement. Despite the abundant advantages of each research tool, usually no single tool is sufficient and neurophysiological tools are no exception. For IS research, it is necessary to use data from various sources with different measurements and neurophysiological measurement is simply one type of it.

REFERENCES

Altonen, A. et al.. (1998). 101 Spots, or how do users read menus? *Proceedings of CHI'98 Human Factors in Computing Systems* (pp. 132- 139). NY: ACM Press. doi:10.1145/274644.274664

Bechara, A., Tranel, D., and Damasio, H. (2000). Characterization of the Decision-Making Deficit of Patients with Ventromedial Prefrontal Cortex Lesions. *Brain*, 123(11), 2189-2202.

Bhatt, M., & Camerer, C. F. (2005). Self-Referential Thinking and Equilibrium as States of Mind in Games: fMRI Evidence. *Games and Economic Behavior*, *52*(2), 424–459. doi:10.1016/j.geb.2005.03.007

Braeutigam, S. et al., (2004). The Distributed Neuronal Systems Supporting Choice-Making in Real-Life Situations: Differences between Men and Women when Choosing Groceries Detected using Magnetoencephalograhy. *European Journal of Neuroscience*, 20(1), 293-302.

Braver, T. S., Cohen, J. D., Jonides, J., Smith, E. E., & Noll, D. C. (1997). A parametric study of prefrontal cortex involvement in human working memory. *NeuroImage*, *5*(1), 49–62. doi:10.1006/nimg.1996.0247 PMID:9038284

Brookings, J. B., Wilson, G. F., & Swain, G. R. (1996). Psychophysiological responses to changes in workload during simulated air traffic control. *Biological Psychology*, *42*(3), 361–377. doi:10.1016/0301-0511(95)05167-8 PMID:8652753

Bruneau, D., Sasse, M. A., & McCarthy, J. D. (2002). The eyes never lie: The use of eye tracking data in HCI research. *Proceedings of the CHI'02 Workshop on Physiological Computing*. NY: ACM Press.

Camerer, C., Loewenstein, G., & Prelec, D. (2005). Neuroeconomics: How Neuroscience Can Inform Economics. *Journal of Economic Literature*, *43(1)*, 9–64. doi:10.1257/0022051053737843

Cenfetelli, R. (2004). Inhibitors and Enablers as Dual Factor Concepts in Technology Usage. *Journal of the AIS*, 5(11-12), 472-492.

Cohen, J. D., Perlstein, W. M., Braver, T. S., Nystrom, L. E., Noll, D. C., Jonides, J., & Smith, E. E. (1997). Temporal dynamics of brain activation during a working memory task. *Nature*, *386*(6625), 604–608. doi:10.1038/386604a0 PMID:9121583

Cowen, L., Ball, L. J., & Delin, J. (2002). An eye-movement analysis of web-page usability. In X. Faulkner, J. Finlay, & F. Détienne (Eds.), *People and Computers XVI— Memorable yet Invisible: Proceedings of HCI 2002* (pp. 317-335). London: Springer- Verlag Ltd.

CurlikShors, D. & Shors, T.J. (2012). Training your brain: Do mental and physical (map) training enhance cognition through the process of neurogenesis in the hippocampus? *Neuropharmacology*, *64*(1), 506–514. PMID:22898496

Debener, S., Ullsperger, M., Siegeland, M., & Engel, A. K. (2006). Single-trial EEG-fMRI reveals the dynamics of cognitive function. *Trends in Cognitive Sciences*, *10*(12), 558–563. doi:10.1016/j.tics.2006.09.010 PMID:17074530

Delgado, M. R., Miller, M. M., Inati, S., & Phelps, E. A. (2005). An fMRI Study of Reward-Related Probability Learning. *NeuroImage*, *2005*(24), 862–873. doi:10.1016/j.neuroimage.2004.10.002 PMID:15652321

Deppe, M., et al., (2005). Nonlinear Responses within the Medial Prefrontal Cortex Reveal when Specific Implicit Information Influences Economic Decision Making. *Journal of Neuroimaging*, 15(2), 171-182.

Dimoka, A., & Davis, F. D. (2008). Where Does TAM Reside in the Brain. Proceedings of IGS '08 (p. 169).

Dimoka, A., Pavlou, P. A., & Davis, F. (2010, December 11-14). Neuro-IS: The Potential of Cognitive Neuroscience for Information Systems Research. *Proceedings of the 28th International Conference on Information Systems*, Montreal, Canada (pp. 687-702).

Dimoka, A., Pavlou, P. A., & Davis, F. (2010). Research Complementary: Neuro-IS: The Potential of Cognitive Neuroscience for Information Systems Research. *Information Systems Research*, *22*(4), 687–702. doi:10.1287/isre.1100.0284

Eisenhardt, K., & Martin, J. (2000). Dynamic Capabilities: What Are They? *Strategic Management Journal*, 21(6), 1105-1121.

Ernst, M., & Paulus, M. P. (2005). Neurobiology of Decision Making: A Selective Review from a Neurocognitive and Clinical Perspective. *Biological Psychiatry*, *58*(8), 597–604. doi:10.1016/j.biopsych.2005.06.004 PMID:16095567

Galbraith, J. R. (1974). Organization Design: An Information Processing View. *Interfaces*, *4*(3), 28–36. doi:10.1287/inte.4.3.28

Galbraith, J. R. (1974). Organization Design. *An Information Processing View*, *4*(3), 28–36.

Goel, V., & Dolan, R. J. (2001). The functional anatomy of humor: Segregating cognitive and affective components. *Nature Neuroscience*, *4*(3), 237–238. doi:10.1038/85076 PMID:11224538

Goldberg, H. J., & Kotval, X. P. (1999). Computer interface evaluation using eye movements: Methods and constructs. *International Journal of Industrial Ergonomics*, *24*(6), 631–645. doi:10.1016/S0169-8141(98)00068-7

Goldberg, H. J., & Wichansky, A. M. (2003). Eye tracking in usability evaluation: A practitioner's guide. In J. Hyönä, R. Radach, & H. Deubel (Eds.), *The mind's eye: Cognitive and applied aspects of eye movement research* (pp. 493–516). Amsterdam: Elsevier. doi:10.1016/B978-044451020-4/50027-X

Goldberg, J. H., Stimson, M. J., Lewenstein, M., Scott, N., & Wichansky, A. M. (2002). Eye tracking in web search tasks: Design implications. *Proceedings of the Eye Tracking Research and Applications Symposium 2002* (pp. 51-58). NY: ACM Press. doi:10.1145/507072.507082

Goldman-Rakic, P. S. (1988). Topography of cognition: Parallel distributed networks in primate association cortex. *Annual Review of Neuroscience*, *11*(1), 137–156. doi:10.1146/annurev.ne.11.030188.001033 PMID:3284439

Hauland, G. (2003). Measuring team situation awareness by means of eye movement data. In *Proceedings of HCI International '03* (Vol 3, pp. 230-234). Mahwah, NJ: Lawrence Erlbaum Associates.

Hendrickson, J. J. (1989). Performance, preference, and visual scan patterns on a menu-based system: Implications for interface design. *Proceedings of CHI'89 Human Factors in Computing Systems Conference* (pp. 217-222). NY: ACM Press. doi:10.1145/67450.67492

Hsu, M., Bhatt, M., Adolphs, R., Tranel, D., and Camerer, C.F. (2005). Neural Systems Responding to Degrees of Uncertainty in Human Decision-Making. *Science*, 310(5754), 1680-1683.

Inhoff, A. W., & Radach, R. (1998). Definition and computation of oculomotor measures in the study of cognitive processes. In G. Underwood (Ed.), *Eye guidance in reading, driving and scene perception* (pp. 29–53). New York: Elsevier. doi:10.1016/B978-008043361-5/50003-1

Ioannides, A. A., Liu, L., Theofilou, D., Dammers, J., Burne, T., Ambler, T., & Rose, S. (2000). Real Time Processing of Affective and Cognitive Stimuli in the Human Brain Extracted from MEG Signals. *Brain Topography*, *2000*(13), 11–19. doi:10.1023/A:1007878001388 PMID:11073090

Just, M. A., & Carpenter, P. A. (1976). Eye fixations and cognitive processes. *Cognitive Psychology*, *8*(4), 441–480. doi:10.1016/0010-0285(76)90015-3

Karn, K., Goldberg, J., McConkie, G., Rojna, W., Salvucci, D., Senders, J., & Wooding, D. et al. (2000). "Saccade pickers" vs. "fixation pickers": The effect of eye tracking instrumentation on research. *Proceedings of the Eye Tracking Research and Applications Symposium 2000* (pp. 87-88). NY: ACM Press.

Kheirbeck, M. A., & Hen, R. (2011). Dorsal vs ventral hippocampal neurogenensis: Implications for cognition and mood. *Neuropsychopharmacology*, *36*(1), 373–374. doi:10.1038/npp.2010.148 PMID:21116266

Knutson, B., Adams, C.M., Fong, G.W., & Hommer, D. (2001). Anticipation of Increasing Monetary Reward Selectively Recruits Nucleus Accumbens. *Neuroscience,* 21(16), RC 159.

Krain, A., Wilson, A.M., Arbuckle, R., Castellanos, F.X., & Milham, M.P. (2006). Distinct Neural Mechanisms of Risk and Ambiguity: A Meta-Analysis of Decision-Making. *Neuroimage*, 32(1), 477-484.

Kuhnen, C., &Knutson, B. (2005). The Neural Basis of Financial Risk Taking. *Neuron*, 47(5), 763-770.

Lee, N., Broderick, A. J., & Chamberlain, L. (2007). What Is "Neuromarketing"? A Discussion and Agenda for Future Research. *International Journal of Psychophysiology*, 63(2), 199-204.

Lieberman, M. D. (2009). Social Cognitive Neuroscience. In S. T. Fiske, D. T. Gilbert, & G. Lindzey (Eds.), *Handbook of Social Psychology* (5th ed.). New York: McGraw-Hill.

Mander, B. A., Rao, V., Lu, B., Saletin, J. M., Lindquist, J. R., Ancoli-Israel, S., & Walker, M. P. et al. (2013). Prefrontal atrophy, disrupted NREM slow waves and impaired hippocampal-dependent memory in aging. *Nature Neuroscience*, *16*(3), 357–364. doi:10.1038/nn.3324 PMID:23354332

Markowitsch, H. J., & Staniloiu, A. (2011). Amygdala in action: Relaying biological and social significance to autobiographical memory. *Neuropsychologia*, *49*(4), 718–733. doi:10.1016/j.neuropsychologia.2010.10.007 PMID:20933525

Matthews, S.C., Simmons, A.N., Lane, S.D., & Paulus, M.P. (2004). Selective Activation of the Nucleus Accumbens during Risk-Taking Decision Making. *Neuroreport*, 15(13), 2123-2127.

Mcclure, S.M., Li, J., Tomlin, D., Cypert, K.S., Montague, L.M., & Montague, P.R. (2004). Neural Correlates of Behavioral Preference for Culturally Familiar Drinks. *Neuron*, 44(2), 379-387.

Mello-Thoms, C., Nodine, C. F., & Kundel, H. L. (2004). What attracts the eye to the location of missed and reported breast cancers? *Proceedings of the Eye Tracking Research and Applications Symposium 2002* (pp. 111-117). NY: ACM Press.

Owen, A. M., McMillan, K. M., Laird, A. R., & Bullmore, E. (2005). The n-back working memory paradigm: A meta-analysis of normative studies. *Human Brain Mapping*, *25*, 46–59. doi:10.1002/hbm.20131 PMID:15846822

Paulus, M.P. & Frank, L.R. (2003). Ventromedial Prefrontal Cortex Activation is Critical for Preference Judgments. Neuroreport, 14(10), 1311-1315.

Pavlou, P. A., & El Sawy, O.A. (2006). From IT Leveraging Competence to Competitive Advantage in Turbulent Environments. Information Systems Research, 17(3), 198-227.

Pavlou, P. A., &El Sawy, O. A. (2010). The "Third Hand": IT-Enabled Competitive Advantage in Turbulence through Improvisational Capabilities. Information Systems Research, 21(3), 443-471.

Pavlou, P. A., & Fygenson, M. (2006). Understanding and Predicting Electronic Commerce Adoption: An Extension of the Theory of Planned Behavior. MIS Quarterly, 30(1), 115-143.

Payne, J. W., Bettman, J. R., & Johnson, E. J. (1992). Behavioral Decision Research: A Constructive Processing Perspective. *Annual Review of Psychology*, *43*(1), 87–131. doi:10.1146/annurev.ps.43.020192.000511

Poole, A., Ball, L. J., & Phillips, P. (2004). In search of salience: A response time and eye movement analysis of bookmark recognition. In S. Fincher, P. Markopolous, D. Moore, & R. Ruddle (Eds.), *People and Computers XVIII-Design for Life: Proceedings of HCI 2004*. London: Springer-Verlag Ltd.

Price, J. L. (1999). Prefrontal cortical networks related to visceral function and mood. *Annals of the New York Academy of Sciences*, *877*(1), 383–396. doi:10.1111/j.1749-6632.1999.tb09278.x PMID:10415660

Rayner, K. (1998). Eye movements in reading and information processing: 20 years research. *Psychol. Bull.*, 124(3), pp. 372-422.

Rayner, K., & Pollatsek, A. (1989). The psychology of reading. Englewood Cliffs, NJ: Prentice Hall, 21(4), pp. 531-532

Rolland, C., & Prakash, N. (2000). Bridging the Gap Between Organizational Needs and ERP Functionality. *Requirements Engineering*, 5(3), 180-193.

Rossiter, J.R., Silberstein, R.B., Nield, G., & Harris, P.G. (2001). Brain-Imaging Detection of Visual Scene Encoding in Long-Term Memory for TV Commercials. *Advertising Research*, 41(2), 13-21.

Rustichini, A. (2005). Neuroeconomics: Present and Future. *Games and Economic Behavior*, *52*(2), 201–212. doi:10.1016/j.geb.2005.05.004

Siau, K. (2004). Informational and Computational Equivalence in Comparing Information Modeling Methods. *Journal of Database Management*, *15*(1), 73–86. doi:10.4018/jdm.2004010103

Siau, K., & Loo, P. (2006). Identifying Difficulties in Learning UML. *Information Systems Management*, *23*(3), 43–51. doi:10.1201/1078.10580530/46108.23.3.20060601/93706.5

Siau, K., & Rossi, M. (2011). Evaluation Techniques for Systems Analysis and Design Modelling Methods: A Review and Comparative Analysis. *Information Systems Journal*, *21*(3), 249–268. doi:10.1111/j.1365-2575.2007.00255.x

Siau, K., & Tan, X. (2005). Improving the Quality of Conceptual Modeling Using Cognitive Mapping Techniques. *Data & Knowledge Engineering*, *55*(3), 343–365. doi:10.1016/j.datak.2004.12.006

Siau, K., & Tan, X. (2006). Cognitive Mapping Techniques for User-Database Interaction. *IEEE Transactions on Professional Communication*, *49*(2), 96–108. doi:10.1109/TPC.2006.875074

Siau, K., & Tan, X. (2006). Using Cognitive Mapping Techniques to Supplement UML and UP in Information Requirements Determination. *Journal of Computer Information Systems*, *46*(5), 59–66.

Siau, K., & Tan, X. (2008). Use of Cognitive Mapping Techniques in Information Systems Development. *Journal of Computer Information Systems*, *48*(4), 49–57.

Siau, K., & Tian, Y. (2009). A Semiotics Analysis of UML Graphical Notations. *Requirements Engineering*, *14*(1), 15–26. doi:10.1007/s00766-008-0071-7

Siau, K., & Wang, Y. (2007). Cognitive Evaluation of Information Modeling Methods. *Information and Software Technology*, *49*(5), 455–474. doi:10.1016/j.infsof.2006.07.001

Sibert, J. L., Gokturk, M., & Lavine, R. A. (2000). The Reading Assistant: Eye gaze triggered auditory prompting for reading remediation. *Proceedings of the Thirteenth Annual ACM Symposium on User Interface Software and Technology* (pp. 101-107). NY: ACM Press.

Van der Linden, M., Juillerat, A., & Adam, S. (2003). Cognitive intervention. In R. Mulligan, M. Van der Linden, & A. C. Juillerat (Eds.), *The Clinical Management of Early Alzheimer's Disease* (pp. 169–233). Mahwah, NJ: Erlbaum.

Venkatesh, V., & Davis, F.D. (2000). A Theoretical Extension of the Technology Acceptance Model: Four Longitudinal Field Studies. *Management Science,* 46(2), 186-204.

Vessey, I. (1991). Cognitive Fit: A Theory-Based Analysis of the Graphs Versus Tables Literature. *Decision Sciences*, 22(2), 219-240.

vom Brocke, J., & Liang, T.-P. (2014). Guidelines for Neuroscience Studies in Information Systems Research. *Journal of Management Information Systems*, April, 19.

Vul, E., Harris, C., Winkielman, P., & Pashler, E. (2009). Puzzlingly High Correlations in fMRI Studies of Emotion, Personality, and Social Cognition. *Perspectives on Psychological Science*, *4*(4), 274–290. doi:10.1111/j.1745-6924.2009.01125.x PMID:26158964

Williams, R. W., & Herrup, K. (1988). The control of neuron number. *Annual Review of Neuroscience*, *11*(1), 423–453. doi:10.1146/annurev.ne.11.030188.002231 PMID:3284447

Young, C. (2002). Brain Waves, Picture Sorts, and Branding Moments. (Finding the Right Moment for Brand Advertising during Television Commercials). *MPA*.

Zak, P. (2004). Neuroeconomics. *Philosophical Transactions of the Royal Society of London. Series B, Biological Sciences*, (359): 1737–1748. PMID:15590614

Zaltman, G. (2003). *How Consumers Think.* Boston, Mass.: Harvard Business School Press.

This research was previously published in the Journal of Database Management (JDM), 27(1); edited by Keng Siau, pages 58-73, copyright year 2016 by IGI Publishing (an imprint of IGI Global).

Chapter 10
Exploring Perception, Cognition, and Neural Pathways of Stereo Vision and the Split-Brain Human Computer Interface

Gregory P. Garvey
Quinnipiac University, USA

ABSTRACT

This chapter examines research from psychology of perception and cognition as well as select developments in the visual arts that inspired the design of the split-brain user interface developed for the interactive documentary Anita und Clarence in der Hölle: An Opera for Split-Brains in Modular Parts (Garvey, 2002). This experimental interface aims at 'enhanced' interaction while creating a new aesthetic experience. This emergent aesthetic might also be described as induced artificial cognitive dissonance and recalls select innovations in the rise of modernism notably the experiments of the Surrealists. The split-brain interface project offers a model for further investigations of human perception, neural processing and cognition through experimentation with the basic principles of stereo and binocular vision. It is conceivable that such an interface could be a design strategy for augmented or virtual reality or even wearable computing. The chapter concludes with a short discussion of potential avenues for further experimentation and development.

INTRODUCTION

Form is henceforth divorced from matter. In fact, matter as a visible object is of no great use any longer, except as the mould on which form is shaped. Give us a few negatives of a thing worth seeing, taken from different points of view, and that is all we want of it. (Oliver Wendell Holmes, 1859)

DOI: 10.4018/978-1-5225-5478-3.ch010

The split-brain user interface was first developed for the interactive installation *Anita und Clarence in der Hölle* (trans. "Anita and Clarence in Hell"): *An Opera for Split-Brains in Modular Parts*. This project used documentary video (C-SPAN 1991) from the 1991 Senate Judiciary Committee hearings on the nomination of Judge Clarence Thomas to be Associate Justice of the United States Supreme Court (United States Congress Senate Committee on the Judiciary 1993). During these televised hearings lawyer Anita Hill came forward to testify under oath and accused nominee Clarence Thomas of sexual harassment when she worked under him at the United States Equal Opportunity Employment Commission in Washington, D.C. Thomas vehemently denied these charges as "a high tech lynching." On October 15, 1991 the full Senate voted 52 to 48 to confirm Thomas as a Justice on the United States Supreme Court.

The split-brain interface presents the testimony of Anita Hill simultaneously with that of Clarence Thomas before the Senate Judiciary Committee. The principle behind a split-brain interface is to deliver two independent video and audio streams separately to each eye and ear so that the content of those streams is experienced or perceived independently and simultaneously by each hemisphere of the brain.

These two diametrically opposed versions of the same events delivered in parallel to each hemisphere of the brain induce a kind of artificial cognitive dissonance. The right and left hemisphere of the viewer hears a different version of the events. The 'whole' brain is left to sort out the truth.

The title of the original installation (Garvey 2002), "refers literally to the ordeal of the two main participants. It is a story with tragic characters, a clear conflict and a well-defined beginning, middle and end. In the tradition of grand opera, it is an epic narrative that emerged from the realm of the personal and private to play out on the national televised stage. Against a backdrop of gender, race and class, this drama pits the political forces of the right and left, and a man and a woman, against one another in a lurid spectacle of 'she said, he said.'" This work was originally developed as a co-production at the Banff New Media Institute in Alberta, Canada (Banff Centre 1999) during a residency and was later part of the exhibition *Sleuthing the Mind* in the fall of 2014 at Pratt Manhattan Gallery in New York City curated by Ellen K. Levy (2014).

The original prototype developed at the Banff Centre required the viewer to wear the Virtual Research Systems Head Mounted Display (1998–2000) to view the digital video. The 2014 installation at the Pratt Manhattan Gallery used the ScreenScope (Stereo Aids, n.d.) handheld mirror stereoscope for viewing the digital video displayed side-by-side on a large flat screen display (Figure 1 shows the Screenscope and monitor at the Pratt Manhattan Gallery). Normally a stereoscope creates the perception of depth by projecting two slightly different images to each eye. The brain fuses these two images into a single coherent percept (perception) of three dimensions.

When the two eyes receive independent images this is technically known "dichoptic" stimulation (Blake 2005). The split-brain or dichoptic interface uses the ScreenScope viewer so the left eye sees only video of Anita Hill and simultaneously the right eye only sees the video of Clarence. Therefore this video installation included in the Sleuthing the Mind exhibit was titled *The Split-Brain (Dichoptic) Interface: Thomas v. Hill* (1999/2014). Through headphones the left ear hears only the testimony of Anita Hill while the right ear hears that of Clarence Thomas (Figures 2 and 3 show a user viewing the installation).

Figure 1. The ScreenScope Stereoscopic Viewer for the split-brain (dichoptic) interface: Thomas v. Hill (1999/2014) at Pratt Manhattan Gallery in New York City curated by Ellen K. Levy
(© 2015, Gregory P. Garvey. Used with permission)

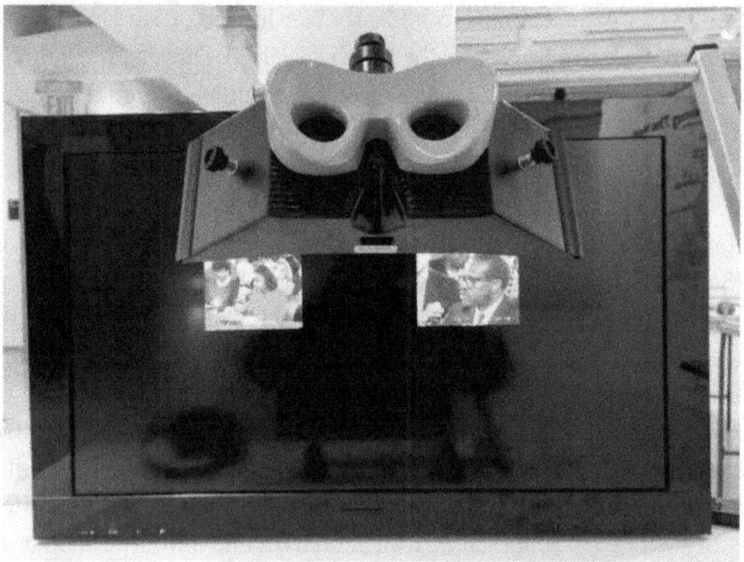

Figure 2. The split-brain (dichoptic) interface: Thomas v. Hill (1999/2014) at Pratt Manhattan Gallery in New York City curated by Ellen K. Levy
(© 2015, Gregory P. Garvey. Used with permission)

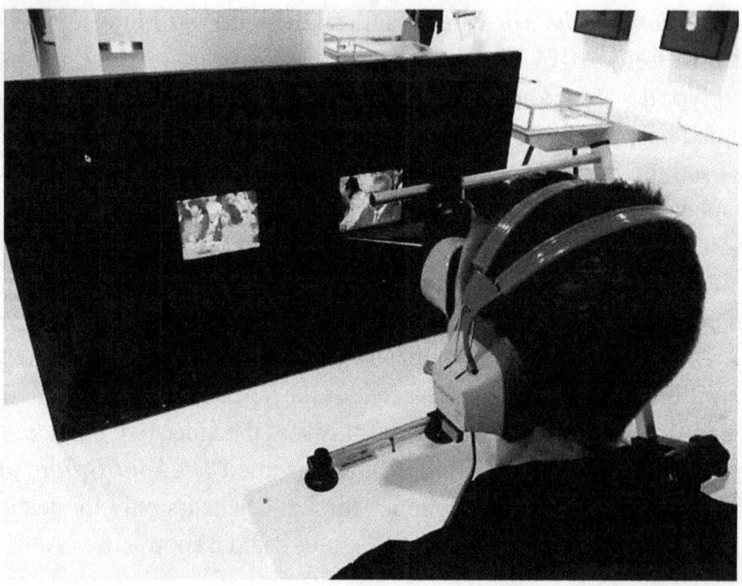

Figure 3. The split-brain (dichoptic) interface: Thomas v. Hill (1999/2014) at Pratt Manhattan Gallery in New York City
(© 2015, Gregory P. Garvey. Used with permission)

BACKGROUND

The split-brain interface uses a technique similar to the Divided Visual Field technique (Banich 2003), which exploits the lateralization of the visual system. The visual pathways as shown in Figure 4 exit each eye and continue to the same side (temporal) hemisphere and also split at the optic chiasma and cross to the opposite side hemisphere. In the scientific literature the terms lateralization, hemispheric specialization or localization are used interchangeably. When a visual stimulus is presented in the left visual field (LVF) it is initially processed by the right cerebral hemisphere. Similarly, when a visual stimulus is presented in the right visual field (RVF) it is initially processed by the left cerebral hemisphere.

Fibers from nasal (medial) and temporal (lateral) sides of the retina of each eye exit at the back of the eyeball. The temporal fibers continue to the Lateral Geniculate body (nucleus) on the same side hemisphere. The fibers from the nasal retina cross the Optic Chiasma and continue to the Pulvinar nuclei making up 40% of the thalamus. Fibers continue to the cortex of the Occipital Lobes (Figure 4 shows the visual pathways).

With normal stereovision, what is seen in the left visual field is viewed by the nasal retina of the left eye and the temporal retina of the right eye. Conversely, what is seen in the right visual field is viewed by the nasal retina of the right eye and the temporal retina of the left eye (Figure 5 shows these relationships).

The lens of the eye functions as a biconvex lens so the image on the retina is inverted and left-right reversed (Figure 6 shows how the image is reversed and inverted).

The split-brain interface presents digital video on only one half of each visual field. The digital videos registered to the lateral (temporal) edge to the center of the right and left visual fields and in principle, are seen only by the nasal retinas. Conversely the digital videos registered to the medial (nasal) edge to the center of the right and left visual fields are seen only by the temporal retinas.

Figure 4. The visual pathways based on Gray's Anatomy depiction of the optic nerves & nuclei, optic chiasma and the optic lobes in a human brain. Plate 722
(Adopted from Gray (1918) Public Domain)

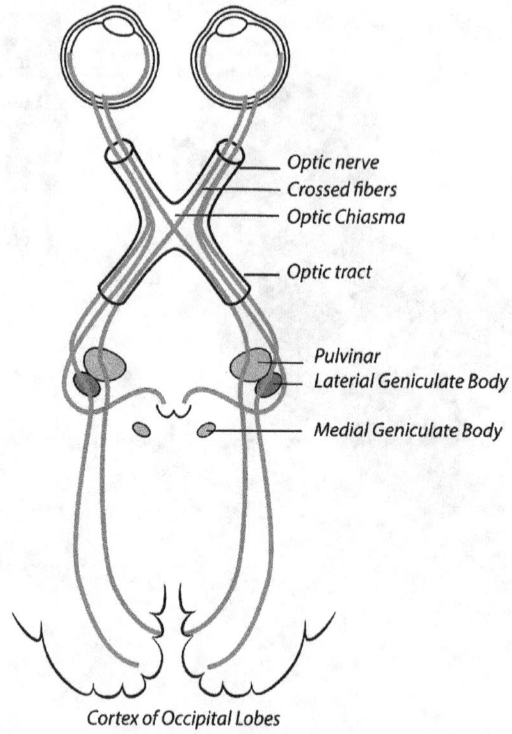

Figure 5. Each eye has a biconvex lens so the image viewed in the left visual field is viewed by the nasal retina of the left eye and also by the temporal retina of the right eye. The image is inverted, left-right reversed
(Adopted from Byrne (1997–Present) © 2015, Gregory P. Garvey. Used with permission)

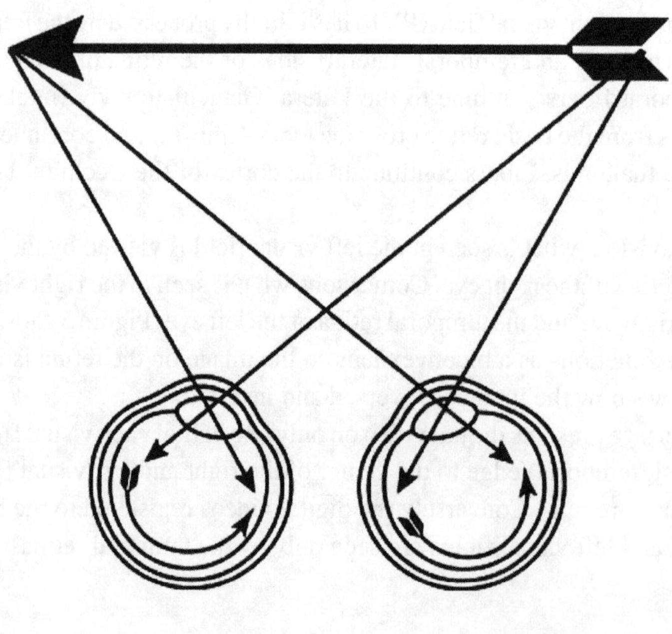

Figure 6. The lens of the eye functions as a biconvex lens so the image on the retina is inverted and left-right reversed
(Adopted from Uzwiak (n.d.) © 2015, Gregory P. Garvey. Used with permission)

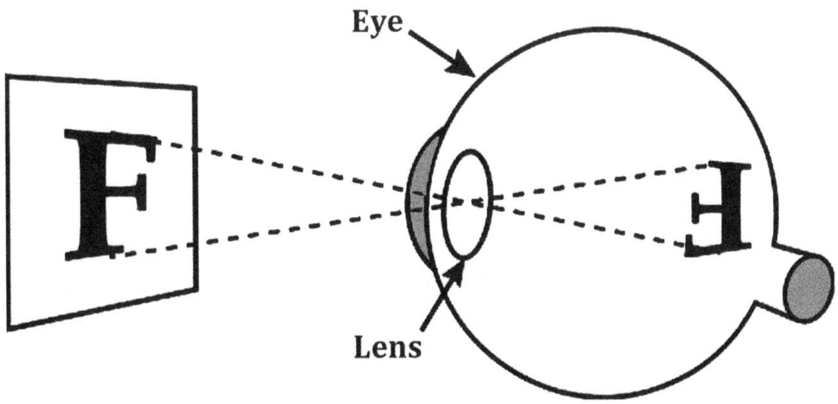

When properly holding the ScreenScope, the viewer looks straight ahead and fixates both eyes on a point or target that is effectively 'in the middle.' The viewer "sees" each speaker separately with the right and left eye. Therefore Clarence Thomas seen only by the right eye is perceived by left brain, while simultaneously Anita Hill seen only by the left eye is perceived by the right brain. While binocular and retinal rivalry, hemispheric dominance and suppression are factors (discussed below) most viewers can perceive a fusion of both separate inputs where the heads and faces merge into a composite perceived image.

Split-Brain Research

The split-brain interface was inspired by accounts of split-brain research, which led to important discoveries regarding specialization of brain functions in the left and right hemispheres in both animals and humans (Gazzaniga 2005). Neuropsychologist Roger Sperry (1968), who won the Nobel Prize in 1981, is generally credited with launching split-brain research conducting the first split-brain studies with Joseph E. Bogen at California Institute of Technology (Caltech). Sperry's collaborators also included his graduate student at the time, Michael S. Gazzaniga who separately made remarkable findings based on studies performed with split-brain patients (Gazzaniga, Bogen, & Sperry, 1962). These patients had undergone a surgical procedure that severs the bundle of connections between the two hemispheres known as corpus callosum. Sperry (1966) observed of these patients: "Everything we have seen indicates that the surgery has left these people with two separate minds, that is, two separate spheres of consciousness. What is experienced in the right hemisphere seems to lie entirely outside the realm of awareness of the left hemisphere. This mental division has been demonstrated in regard to perception, cognition, volition, learning and memory."

This procedure was first performed by Van Wagenen and Herren, in 1939 in order to limit the spread of seizures from one hemisphere to the other on a group of 26 people in Rochester, New York with intractable cases of epilepsy (Van Wagenen and Herren 1940). With a callosotomy the hippocampal commissure along with the corpus callosum are removed. A more radical procedure known as a commissurotomy, sections the corpus callosum, hippocampal commissure and the anterior commissure (Hirstein

2005). These procedures are considered a treatment of last resort and have been largely discontinued with the advent of more powerful drugs that could control epileptic seizures. Roughly thirty years after the initial experiments (Gazanniga 2005), Vogel and Bogel performed a complete commissurotomy (c) on a World War II veteran plagued with severe seizures due to a war time head injury (Bogen & Vogel, 1962). Gazanniga (1982, 2005) was able to study this patient known as "WJ" for the next five years. Later researchers were able to locate select others who had undergone the same procedure and study them.

These studies were designed to take advantage of the visual pathways where a visual stimulus presented in the left visual field is seen by the right hemisphere and a visual stimulus presented in the right visual field is seen by the left hemisphere. A typical experimental procedure involved having a right handed split-brain subject to sit in front of a screen and focus on a central fixation point (Kitterle, Christman, & Hellige, 1990; Rapaczynski & Ehrlichman, 1979; Van Kleek, 1989; Weismann & Banich, 1999, 2000).

If an image is flashed for 100–200 milliseconds in the left visual field it is perceived only by the right hemisphere. The brevity of the stimulus does not leave time for the subject to shift his or her eyes away from the central fixation point (Hirstein 2005). The left hemisphere, which has the capacity for language has not seen the visual stimulus and cannot say what it was. Miller (Wolman 2012) notes in regard to these experiments that "the right hemisphere is experiencing its own aspect of the world that it can no longer express, except through gestures and control of the left hand".

Other experiments used goggles or lenses to maintain image separation to guarantee lateralization of the visual stimulus. Bourne (2006) recommends using for stimulus presentation either eye-tracking equipment or electrooculography. Sperry (1968) used a procedure that permitted the split-brain subject to use one hand to feel unseen objects. An image is flashed for 100 milliseconds (to prevent the eyes from moving and engaging the other visual field).

While Sperry died in 1994 Gazzaniga continued his split-brain tests with these patients. Using a similar procedure Gazzaniga reports another experiment where the right brain sees a wintery scene of a snow man in front of a house covered in snow and the left brain sees a chicken claw. Eight cards with images are placed before the subject. Four of the images are related to the snow scenes and the other four images are related to the image of the chicken claw. Both the left and right hemisphere can see all eight cards. When told to select cards the split-brain subject chose with his right hand (left hemisphere) an image of chicken. With his left hand (right hemisphere) the subject chose an image of a shovel. Why asked why he made these choices the split brain subject said: "Oh, that's simple. The chicken claw goes with the chicken, and you need a shovel to clean out the chicken shed." Gazzaniga (1998) explains that because the split-brain subject's left brain did not know what the right brain saw (the snowy scene) the left brain rationalizes or confabulates the choice of a shovel as necessary to "clean out the chicken shed."

From numerous experiments he concluded that for right handed split-brain subjects the left hemisphere of the brain is dominant in language related skills such as writing, speaking, mathematical calculation and reading. Gazzaniga (2005) further argues that the right hemisphere is specialized for visuo-spatial processing. "Studies with split-brain patients have revealed right hemisphere superiority for various tasks involving such components as part–whole relations (Nebes 1972), spatial relationships (Nebers 1973), apparent motion detection (Forster, Corballis, & Corballis, 2000), mental rotation (Corballis & Sergent, 1988), spatial matching (Corballis, Funnell & Gazzaniga, 1999) and mirror image discrimination (Funnell, Corballis, & Gazzaniga, 1999)". However these capabilities may be distributed differently with left-handed or ambidextrous patients (Geschwind & Crabtree 2003). In a majority of left-handed split-brain subjects speech functions are in both hemispheres. Few left-handers have speech localized in the right hemisphere (McCarthy & Warrington, 1990).

The lateralization of skills in the two hemispheres is thought to be a result of the evolutionary history of bilaterally symmetric animals inherited from a common ancestor that appear over 570 million years ago. Left right symmetry apparent in humans in utero confers obvious survival benefits: having symmetric limbs make locomotion quick and energy efficient. As a basic body plan this is left right symmetric led to two eyes, ears, antennae and two hemispheres emerges in the neocortex – the outer layers of the brain. However there are asymmetries such as the heart or the appendix and differences between the hemispheres. Damage to Broca's area in the left hemisphere impairs speech but damage to the corresponding area in the right hemisphere does not. The right half of the facial fusiform area on the underside of the brain is critical for recognizing faces. This can be tested by viewing a face with the left eye (hence the right hemisphere) and subjects do a better job at recognizing faces than if they look only with their right eye. It is speculated that the optic nerve crosses to the opposite side hemisphere so the appropriate hemisphere controls the side of the body that correspondences to what is seen in retina.

Issues, Controversies, Problems

Since the original pioneering work by Sperry, Gazzaniga and others, split-brain research has continued at leading institutions primarily in the United States and in Italy. These studies have largely confirmed the same results (Gazzaniga 2005): "Severing the entire callosum blocks the interhemispheric transfer of perceptual, sensory, motor, gnostic and other forms of information in a dramatic way, allowing us to gain insights into hemispheric differences as well as the mechanisms through which the two hemispheres interact." Questions have been raised on whether some connections persisted at the cortical level (the cerebral cortex) in split-brain patients or that there were still subcortical (a part of the brain below the cerebral cortex) connections that remain untouched by surgical procedures. Lambert (1991) has shown there is evidence for interhemispheric integration persisting after surgeries. Franz, Waldie, & Smith, (2000) designed experiments that required split-brain subject to perform bimanual skills and concluded that such skills in split-brain subjects are still coordinated at the subcortical level. Other research (Corballis 1994) indicates that the two hemispheres share processing resources.

In the normal, 'intact' brain, the two hemispheres work together. Bourne (2006) describes four experimental methods used to measure interhemispheric cooperation using the Divided Field methodology. These include the Poffenberger paradigm (Poffenberger, 1912), still in today (e.g., used by Sperry and Gazzaniga as described above), where visual targets are presented to both the left and right visual field and the subject responds with either the opposite or same side hand as the stimulus: "the crossed uncrossed difference can be used as an estimate of the amount of time it takes for information to be transferred from one hemisphere to another." This is known as interhemispheric transfer time or IHTT. A second method is the redundant target paradigm (Todd, 1912) where the reaction times of presenting just one stimulus to one eye (one hemisphere) is compared with presenting two copies of the same stimulus to both eyes (and hemispheres). A faster reaction in the later instance indicates likely interhemispheric cooperation. Similarly a third method (Dimond and Beaumont 1972) compares the time to process pairs of stimuli presented unilaterally versus pairs presented bilaterally. A fourth method (Banich & Shenker, 1994) involves presenting three stimuli using alphabetic (Weissman & Banich, 2000), geometric (Weissman & Banich, 1999) and face examples (Compton, 2002). Bourne (2006) describes that stimuli are presented: "two in the top half of the display, one in each visual field, and the third in the bottom half of the display. The participant's task is to decide whether the bottom stimulus matches either of the stimuli presented above it."

The validity of these four experimental methods have been tested with both normal and split-brain subjects. For example in bilateral trials split-brain patients have no advantage (Mohr, Pulvermuller, Rayman, & Zaidel, 1994). In addition to work using functional Magnetic Resonance Imaging (fMRI) new brain imaging techniques combined with behavioural methodologies such as ITHH (Colvin, Funnell, Hahn, & Gazzaniga, 2005) promise further advances in understanding functional connectivity and recruitment through the corpus callosum between the two hemispheres in healthy individuals (Gazzaniga 2005). Gazzaniga (2008) describes the advantage of interhemispheric cooperation as follows: "The right hemisphere maintains an accurate record of events, leaving the left hemisphere free to elaborate and make inferences about the material presented. In an intact brain, the two systems complement each other, allowing elaborative processing without sacrificing veracity." Much has been made of the differences in capacities between the left and right brain hemispheres especially in the popular press spawning a mini industry of self-help and having a huge influence on educational theory.

Nobuyuki Kayahara's Spinning Dancer illusion (see Other Online Resources in the Appendix) has been promoted as a test for left or right brain dominance. However it is simply an example of an ambiguous or bi-stable image (Parker-Pope 2008). Several websites offer 'scientifically validated' self-administered tests to determine "which side of your brain do you wake up on in the morning? Find out with this test!" (psychtests n.d.). After taking the results are reported on a simple one dimensional scale from 0% to 100%. If a score is closer to 0% the personality is "more characteristically left-brained." A score closer to 100% reflects a right brain personality. A score in the middle represents a balance between the left and right brain "a level of perfect harmony – rather than trying to dominant each other, they work together to create a unique and well-balanced "you." Your spontaneous, impulsive, and free-flowing right brain creates an exciting and adventurous world, while your left brain helps you make sense of it and keep track of everything." The website includes the following disclaimer that "This report is intended for personal growth purposes only." However such psychological assessments are utilized by human resource professionals, therapists, athletic coaches and consultants and can lead to real impacts in hiring and retention.

In a paper that was greatly influential outside of neuroscience, Joseph E. Bogen (1975) noted that, "the brain is double, in the sense that each cerebral hemisphere is capable of functioning independently, each in manner different from the other." His argument is essentially that the research of the time suggested that a dichotomous theory of intelligence could be reflected, as Arthur Jensen suggests in a more diversified "curricula, instructional methods and educational goals and values that will make it possible to for children ranging over a wider spectrum of abilities and proclivities genuinely benefit from their years in school" (Jensen 1972). Many readers and later authors seized upon Bogan's discussion of "two kinds of intelligence" or parallel "ways of knowing" and his linkage to hemispheric specialization. Bogan (1975) quotes cognitive psychologist Neisser: "Historically, psychology has long recognized the existence of two different forms of mental organization. The distinction has been given many names: 'rational' vs. 'intuitive,' 'constrained' vs. 'creative,' 'logical' vs. 'prelogical,'…".

Neisser himself makes a distinction between "sequential processing" vs. "multiple processing" which involves simultaneous or independent thinking (Bogan, 1975). Bogan compiled a list of authors associated with dichotomous pairs of terms that represented two "types of intelligence" or "cognitive styles." In the 1980s a series of academic publications addressed the potential impacts of split-brain theory on education and the arts (e.g. Brooks, 1980; LeCompte & Rush, 1981; Henry, 1981; Hopkins, 1984; McLuhan, 1979). Murr and Williams, observe that spatial reasoning, symbolic processing and pictorial interpretation associated with the right brain are ignored in education. They call for "whole brain" learning where both hearing and seeing a story forces "sensory connections" to both the left and right hemispheres. The authors conclude with this celebratory invocation of the personal computer that should "become the facilitator of whole-brain learning, the essence of the corpus callosum, bridling the left and right hemispheres and connecting the book and the video cassette tape, the newspaper and television, visual statics and dynamics, science and the liberal arts, the West and the East, creating a world convergence toward common perceptions and singular realities" (Murr and Williams, 1988).

In many ways Betty Edwards' *Drawing with the Right Side of the Brain* epitomizes this faith in the under utilized powers of the right brain. Edwards introduces practical techniques for achieving accurate representations from observational drawing and its success can be measured in the number of editions published since the first edition. Much of the popularized versions of left-brained or right-brained tests originate in the demonstrated lateralization of brain function. This research has shown that the left hemisphere controls features of language and logical while the right hemisphere is specialized for processing visual comprehension and spatial information. However, research using neuroimaging techniques (Nielsen et al. 2013) challenges these earlier results: "*It has been conjectured that individuals may be left-brain dominant or right-brain dominant based on personality and cognitive style, but neuroimaging data has not provided clear evidence whether such phenotypic differences in the strength of left-dominant or right-dominant networks exist.*" The exaggerations and over generalization of pop-psychology about right-brain/left-brain theories have largely been debunked (Rogers 2013). Later research has shown that the brain is not nearly as dichotomous as once thought. For example, recent research has shown that abilities in subjects such as math are strongest when both halves of the brain work together. Today, neuroscientists know that the two sides of the brain collaborate to perform a broad variety of tasks and that the two hemispheres communicate through the corpus callosum.

The exhibition of the Split-Brain interface does not use the rigorous controls of a laboratory experiment. In the gallery setting users sit and lean slightly forward to peer through ScreenScope which is attached to an armature. There can still be a great deal of variation in how each individual sees the display. While the ideal is to achieve fusion of the two separate images of Clarence Thomas and Anita Hills this is not always possible for all viewers. As noted earlier binocular rivalry can lead to the suppression of one of the 'incongruent' or dissimilar images–and the viewer sees only one view. If the viewer has uncorrected amblyopia (lazy eye), which occurs in up to 5% of the population (Webber & Wood, 2005) it is likely the individual is viewing with one eye. It is generally thought the cause resides in the brain where one eye is 'turned off' to compensate for double vision (Levi 2013). This can also occur with strabismus (misaligned eyes or heterotropia) often referred colloquially as being cross–eyed or wall-eyed. Other factors can inhibit has severe refractive problems (anisotropia) colloguially known as extreme near or far sightedness or astigmatism. Should any of these conditions be present the viewer may be unable to see equally each image. The ability to view straight ahead with both through the ScreenScope with both eyes replaces a central fixation point or a chin and forehead rest used experimentally to stabilize the viewer's gaze. If the viewer is unable to fixate straight ahead then it is unlikely that an image is seen only by the opposite side cerebral hemisphere. Rather the eye fixates in such a way that the image falls on both the temporal (lateral) and nasal (medial) retina.

The Interpreter Mechanism

Yet with right handed split-brain patients Gazzaniga theorized that the left hemisphere served as a kind of "interpreter mechanism" with a tendency to fabricate or confabulate explanations. Gazzaniga describes an experiment, which displayed two stories separately to each hemisphere of a split-brain subject.

While this individual's left hemisphere was dominant verbally, the right hemisphere had some verbal capabilities. "Two stories are presented, one to each half-brain. The left hemisphere quickly reports its story, followed by the right hemisphere offering its story in bits and pieces. After the left brain hears these semantic items, it combines both stories into yet a new one" (Gazzaniga 1985). This experiment in particular led to the conception of a split-brain interface.

The source story would normally be read from left to right as follows:

Story 1: *Mary Ann, May Come, Visit Into, The Town, Ship Today.*

The source story is decomposed into two stories made by selecting every other word from the original resulting in Story 2 and 3.

Story 2: *Ann Come Into Town Today*
Story 3: *Mary May Visit The Ship*

Using a specially constructed screen following the principles described above, word pairs are displayed so that the right and left hemisphere of the split-brain subject each see a single word. For example:

Left Side of Screen	Right Side of Screen
Right Hemisphere (Story 3)	Left Hemisphere (Story 2)
Mary	Ann

Word pairs from Story 2 and 3 continued to be shown one word at a time as follows:

Left Side of Screen	Right Side of Screen/
Right Hemisphere (Story 3)	Left Hemisphere (Story 2)
May	Come
Visit	Into
The	Town
Ship	Today

Gazzaniga reports that when the split-brain subject was asked to recall the story, the subject responded "Ann come into town today." This is the story seen by the verbally dominant left hemisphere. The subject was then prompted to respond if that was the whole story. According to Gazzanniga the subject's right hemisphere "blurted out" an invented or confabulated story based on what it saw: "on a ship . . . to visit . . . to visit Ma!'" The subject was asked again to repeat the entire story again. Now that the left hemisphere has heard what the right hemisphere said the subject synthesized the two source stories into a new story: "Ann came into town today to visit Ma on the boat!" (Gazzaniga, 1985).

Gazzaniga speculates that this "interpreter mechanism" of the left hemisphere has a "creative narrative talent" which "is constantly looking for order and reason, even when there is none – which leads it continually to make mistakes. It tends to over generalize, frequently constructing a potential past as opposed to a true one." Gazzaniga suggests that the left hemisphere is prone to invent an explanation based on partial information it already knows. By contrast the right hemisphere is thought to be literal and truthful.

These intriguing results led to consideration of what if two related stories were displayed separately to the left and right hemispheres of a "normal subject" with an intact corpus callosum? How would two conflicting stories heard or read separately by each hemisphere affect interpretation and understanding? A split-brain interface was designed to test this conjecture–to independently target each hemisphere of a normal brain by delivering video and audio directly to the separate visual and auditory pathways.

Binocular Vision, Depth Perception and Retinal Disparity

With binocular vision (Latin bini for double combined with oculus for eye) a creature uses two eyes together creating overlapping field of views, which allows for the perception of depth. In contrast with monocular vision a creature may have two eyes, but each eye sees a different image. Fish, many birds, reptiles and lizards have monocular vision where the two eyes are located on the sides of the head (lateralized) and see two separate fields of view. The eyes of the chameleon are controlled by a rotating turret like, anatomical structure. This allows each eye to see independently. The chameleon can also move the eyes to converge the fields of view to focus the eyes together with overlapping visual fields. Some birds of prey are able to converge their eyes forward to create overlapping fields of view and thereby more accurate perception of depth.

Depth perception can be further classified into monocular (single eye) and binocular (two eyes) cues. Examples of monocular depth include: motion parallax when moving objects nearby quickly pass but distant object move slowly or even appear stationary; single point, two point and three point perspective–one point perspective is seen with the recession of a railroad track from a vantage point in the middle between the two rails. Three point perspective can be seen with tall buildings; aerial or atmospheric perspective occurs as objects get further and further in the distance they lose sharpness and the color becomes muted losing saturation (purity of the hue) and contrast (difference between black and white) due to thickness of the atmosphere; overlap or interposition–when one object overlaps another or blocks the view it is perceived to be closer; texture gradient discrimination–textures close by appear sharp and distinct whereas textures in the distant appear blurry and lack detail.

In human binocular vision two eyes are front facing and the visual fields overlap (Figure 7 shows the overlap of the two visual fields forming the binocular visual field). Typically in an adult with normal binocular vision the two eyes are, on average, 6.5 centimeters apart. Due to the location of the eyes, the retina of each eye sees near objects from a slightly different angle. This difference between the two retinal images is called retinal disparity. Ocular convergence occurs when coordinated movements of each eye look at the same object (Herring 1868). When the object is closer the degree of turning the eyes inward is greater. The region of binocular overlap is 120 degrees wide by 135 degrees in height (Wandell, n.d.). The brain fuses the two similar images from each eye into the perception of a single composite visual image known as binocular fusion. This single mental image is sometimes called the Cyclopean eye because the fused image is perceived to be between the two eyes (Julesz, 1971).

This area of overlap provides visual cues for the perception of depth, which is known as stereopsis. Depth perception makes it possible to view the world in three dimensions. Stereopsis can be readily demonstrated by holding a finger in front of the nose while focusing on a point in the far distance. Focusing on the finger causes convergence where both eyes turn inward toward each other. By closing and opening one eye after the other it is possible to observe the difference in angle of view of each eye and the resulting retinal disparity as the left and right eye see different sides of the finger.

Figure 7. Binocular visual field
(Adopted from Figure 7 in Aprile, Ferrarin, Padua, Di Sipio, Simbolotti, Petroni, Tredici & Dickmann (2014) © 2015, Gregory P. Garvey. Used with permission).

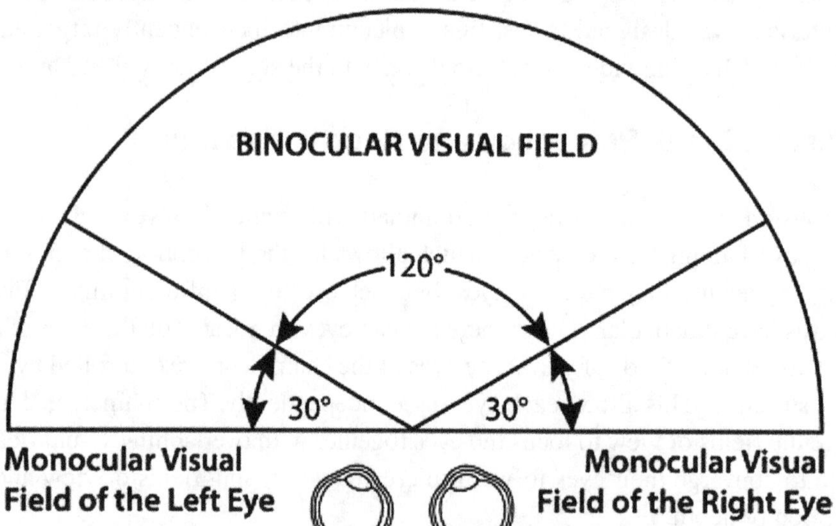

Stereopsis and retinal disparity were described by Sir Charles Wheatstone (1838) in his seminal monograph published by the Royal Society, Wheatstone initially suggests these principles can be deduced by simple perspective: "The appearances, which are by this simple experiment rendered so obvious, may be easily inferred from the established laws of perspective; for the same object in relief is, when viewed by a different eye, seen from two points of sight at a distance from each other equal to the line joining the two eyes." He illustrates this with the following figure, which shows a 'wireframe' of a cube as if seen from the left (a) and right (b) eyes (Figure 8 shows the cube as stereo image pair):

Without the aid of a stereoscope, the reader can fuse the wireframe images of the cube by using the cross-eyed viewing method (Simanek n.d.). This is accomplished by viewing the two cubes at a distance of approximately 6 to 7 inches and then gradually cross the eyes so the two images of the cubes begin to overlap. When three images appear focus on the center image and try to bring it into sharp focus and it will appear to be a 3 dimensional image. Normally the cross-eyed viewing method requires that the image to be seen by the right eye is on the left and the image to be seen by the left eye is on the right. This is particularly true with photographic pairs. However the cross-eyed method works with the above images because the cube is rendered as a wireframe and all 6 sides are visible. The illusion of 3D is possible because of corresponding or congruent points that fall on the retina (Figure 9 shows how these points fall on the retina).

In his original monograph Wheatstone (1838) describes retinal disparity as: "the projection of two obviously dissimilar pictures on the two retinæ when a single object is viewed, while the optic axes converge, must therefore be regarded as a new fact in the theory of vision." He restates this principle as follows and then poses a question which led to the design of the stereoscope: "It being thus established that the mind perceives an object of three dimensions by means of the two dissimilar pictures projected by it on the two retinæ, the following question occurs: What would be the visual effect of simultaneously presenting to each eye, instead of the object itself, its projection on a plane surface as it appears to

Figure 8. A recreation of original figure 13 from Wheatstone illustrating Wireframe Cube Image Pair.
(Adapted from Wheatstone (1838). Public Domain)

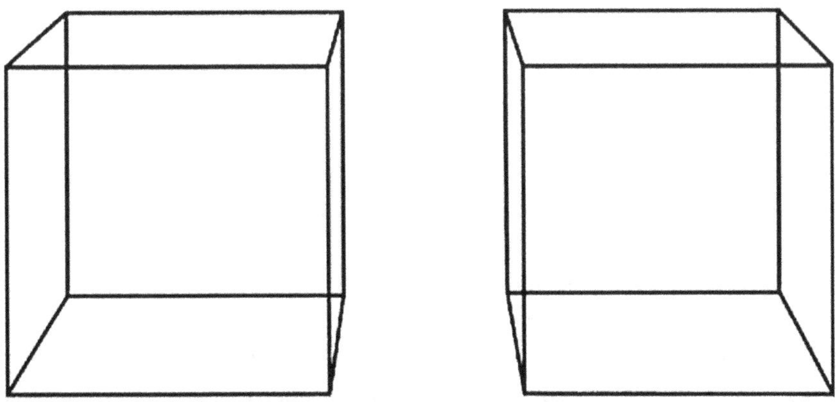

Figure 9. Corresponding points on retina
(Adopted from Encyclopedia Britannica. (1994) © 2015, Gregory P. Garvey. Used with permission)

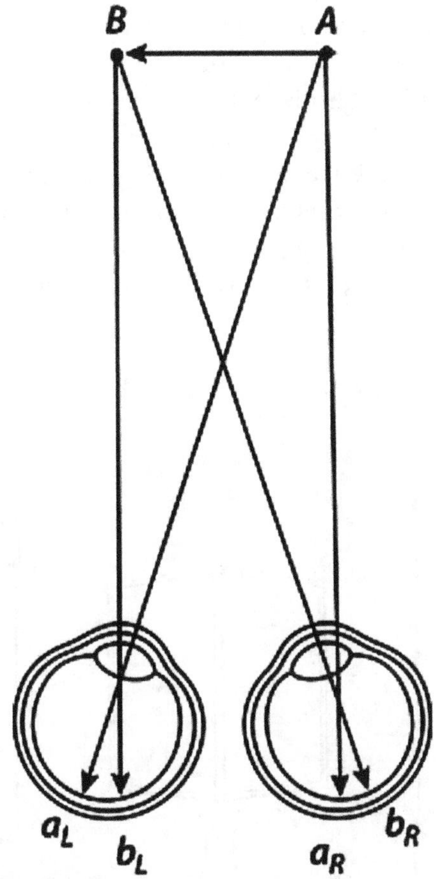

that eye?" Wheatstone answered this question with his design of the first stereoscope (Figure 10 shows the original design of the Wheatstone Stereoscope). Wheatstone's design is seen from the front, which depicts the functional representation of how the stereoscope works.

Instead of the Object Itself

Writing about the stereoscope twenty-one years after Wheatstone's invention, Oliver Wendell Holmes (1859) inspired by Wheatstone's conjecture ("instead of the object itself, its projection on a plane surface as it appears to that eye?") designed a low cost stereoscope that quickly outsold more expensive European designs. In 1861 Holmes introduced a low cost hand held stereoscope (Figure 11 shows an advertisement for Holmes' stereoscope). He deliberately chose not to file a patent on his design – a kind of 'open source" invention which arguably was the first Virtual Reality viewer!

It soon became one of the most popular designs of its time. Commenting on its success Holmes wrote: "There was not any wholly new principle involved in its construction, but, it proved so much more convenient than any hand-instrument in use, that it gradually drove them all out of the field, in great measure, at least so far as the Boston market was concerned."

Wendell proclaimed a new epoch "in the history of human progress" where "Form is henceforth divorced from matter. In fact, matter as a visible object is of no great use any longer, except as the mould on which form is shaped. Give us a few negatives of a thing worth seeing, taken from different points of view, and that is all we want of it." What is remarkable about this statement is that Holmes could be describing Virtual Reality. The 'mould' is the polygonal model. The virtual camera can simulate 'different points of view' and all we need is the motion tracking head mounted display.

The Stereoscopes of the nineteenth century were further enabled by the still new technology of photography. Holmes recounts the necessary inventions of his time that led him to design an affordable hand held stereoscope. He acknowledges Wheatstone's invention and then mentions the development of

Figure 10. The Wheatstone Stereoscope
Wheatstone (1838). (Ⓟ Public Domain).

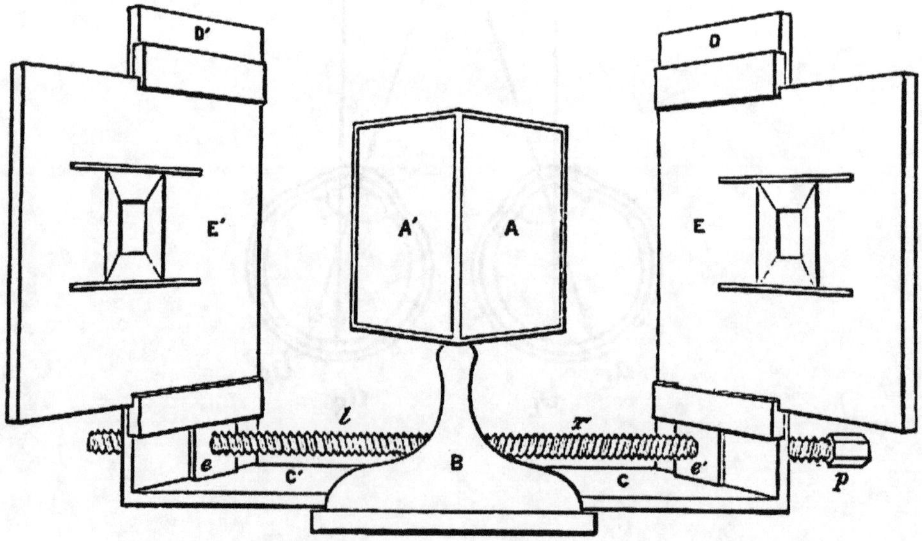

Figure 11. The Holmes stereoscope, with the inventions and improvements added by Joseph L. Bates, circa 1869)
(Public Domain, Unknown, circa 1869)

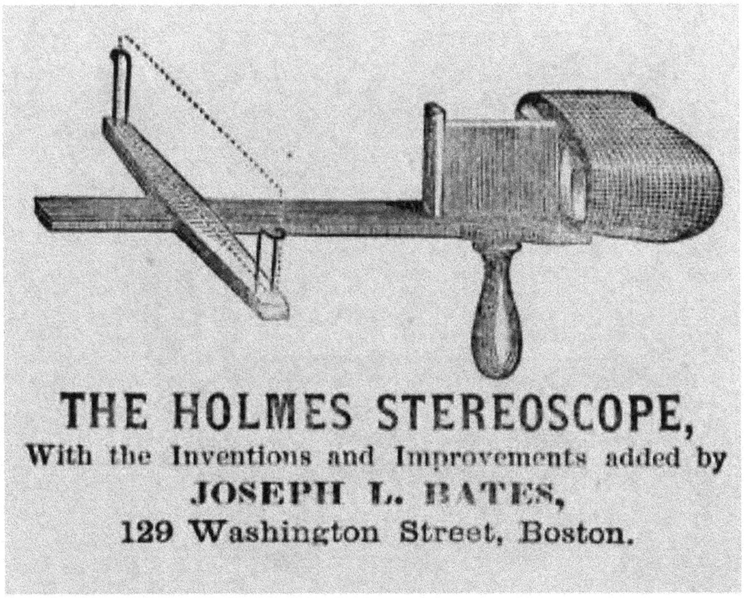

the Daguerrotype in 1837 which "has fixed the most fleeting of our illusions, that which the apostle and the philosopher and the poet have alike used as the type of instability and unreality." Next is the light fast and permanent photograph affixed to a paper substrate first invented by William Henry Fox Talbot (Chisholm 1911) which "has completed the triumph, by making a sheet of paper reflect images like a mirror and hold them as a picture." Holmes celebrates the invention of photography as "This triumph of human ingenuity is the most audacious, remote, improbable, incredible" but he laments "It has become such an everyday matter with us, that we forget its miraculous nature, as we forget that of the sun itself, to which we owe the creations of our new art." Holmes goes one to describe the principles of binocular vision and stereopsis: "These exceptions illustrate the every-day truth, that, when we are in right condition, our two eyes see two somewhat different pictures, which our perception combines to form one picture, representing objects in all their dimensions, and not merely as surfaces." Two photographs of the same subject taken from slightly different angles (determined by the disparity of interpupillary distance) create the illusion of depth.

Holmes next describes stereo-photography: "A first picture of an object is taken, then the instrument is moved a couple of inches or a little more, the distance between the human eyes, and a second picture is taken. Better than this, two pictures are taken at once in a double camera." He mentions two ways to view a 'stereograph'. One is by squinting but this method as with the cross eyed method described above is difficult and fatiguing. The other method is to cut a convex lens in two, grind down one side to be flat and to join the two resulting lens together. Declaring this a 'squinting magnifier' it affords the ability to "with its right half we see the right picture on the slide, and with its left half the left picture, it squints them both inward so that they run together and form a single picture." David Brewster is credited with the introduction of lenses. Brewster's stereoscope from 1849 (Figure 12 shows Brewster's stereoscope) is a lenticular design (having lenses) permitted a more compact housing (Brewster 1856).

Figure 12. The Brewster Lenticular Stereoscope circa 1849
(© Public Domain).

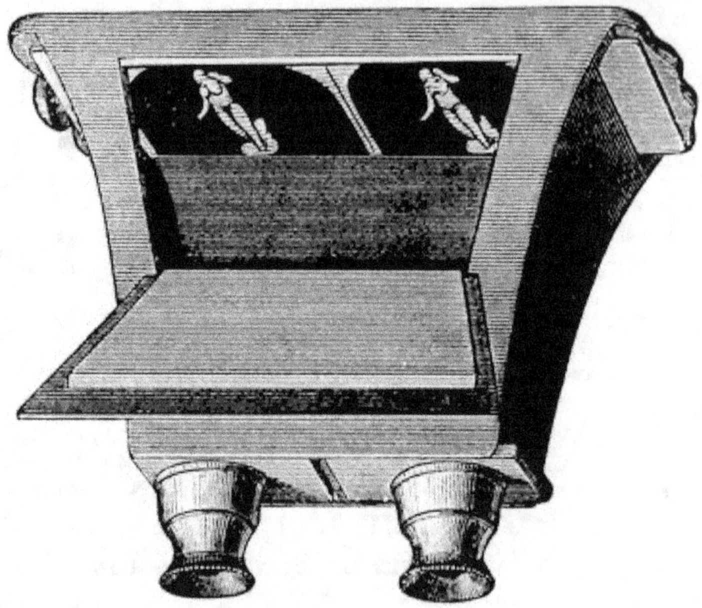

Holmes declares the stereograph of Ann Hathaway's Cottage in Shottery, England (the birthplace of Shakespeare's wife) to be "the most perfect, perhaps, of all the paper stereographs we have seen." (Figure 13 shows the Hathaway Cottage Stereograph). A copy of this stereograph is in the collection of the Boston Public Library and is reproduced here under a creative commons license (Keystone View Company1879–1930).

Figure 13. The Ann Hathaway Cottage, Shottery, England
(© Creative Commons BY-NC-ND).

He repeats his argument that "Matter in large masses must always be fixed and dear; form is cheap and transportable. We have got the fruit of creation now, and need not trouble ourselves with the core. Every conceivable object of Nature and Art will soon scale off its surface for us." Holmes envisions a future having billions of pictures requiring vast libraries "where all men can find the special forms they particularly desire to see as artists, or as scholars, or as mechanics, or in any other capacity. Already a workman has been travelling about the country with stereographic views of furniture, showing his employer's patterns in this way, and taking orders for them. This is a mere hint of what is coming before long." Holmes' predictions were in part dead on. Eventually mail order catalogs like Sears, Roebuck and Co. featured photography to promote hundreds of different goods including furniture and entire homes. However stereoscopes were still too cumbersome to be of practical use to salesmen. The Holmes stereoscope remained in the home and parlor and was relegated to the status of a curiosity and diversion. Interestingly one hundred twenty years later Jaron Lanier, Virtual Reality pioneer and proselytizer did predict that one day you could design your kitchen in Virtual Reality while inhabiting the body of a octopus (Huffman n.d.).

Binocular Rivalry

When discordant views are seen separately by the left and right eyes there are alternating moments of dominance and suppression. Wheatstone (1838) accurately described what happens when each eye views a dissimilar image having corresponding and non-corresponding, discordant or incongruent elements. (Figure 14 shows a recreation of Wheatstone's figure 25 depicting the letter "S" and "A") "each [the letter "S" and "A"] presented at the same time to a different eye, the common border will remain constant, while the letter within it will change alternately from that which would be perceived by the right eye alone to that which would be perceived by the left eye alone. At the moment of change the letter which has just been seen breaks into fragments, while fragments of the letter which is about to appear mingle with them, and are immediately after replaced by the entire letter."

Wade and Tgo (2013) credits Porta (1593) with first describing what happens when each eye sees two different images. Porta used a partition so each eye sees a different page of a different book: "To separate the two eyes, let us place a book before the right eye and read it; then someone shows another book to the left eye, it is impossible to read it or even see the pages, unless for a short moment of time the power of seeing is taken from the right eye and borrowed by the left." (Porta, 1593, pp. 142–143)

Figure 14. A recreation of Wheatstone's original figure 25 illustrating dissimilar image pair *(Adopted from Wheatstone (1838) Public Domain).*

Blake (2001) notes that Porta observed an essential feature of binocular rivalry where the vision in his right eye was dominant and his left eye suppressed. Later Dutour (1760) described an alternation between the perceived color when each eye viewed a different color (O'Shea 1999), Charles Wheatstone did a more systematic study using outline drawings. Randolf Blake (2001) points out that Wheatstone correctly observed "key aspects of rivalry, including the complete suppression of one of two discordant stimuli, the alternations in dominance between the eyes, the spatial fragmentation of the two images during times of transition, and the dependence of predominance on the physical characteristics of the rival stimuli." When different images are presented simultaneously and independently to the right and left eye "observers are presented with an ambiguous stimulus which supports two distinct interpretations, perception alternates between these interpretations in a random manner." This alternation in perceptual dominance is part of a more general phenomenon, known as perceptual bi-stability (Blake 2001), which can also be described as perceptual rivalry.

In addition to Wheatstone's original work Blake recounts other important work on perceptual instability including binocular rivalry (Levelt, 1968; Blake, 1989; Logothetis, 1998; Blake, 2001; Tong, 2001), ambiguous motion displays (Wallach, 1935, 1996; Hupé and Rubin, 2003), ambiguous depth perception (Necker, 1832; Rubin, 1921, 2001), and random dot kinematograms (Julesz, 1971). Bitstable images are another example of perceptual instability. The Schroeder Stairs, Necker Cube, and the Vase/Faces (each shown by Figure 15) are well-known examples that can be perceived with monocular viewing (Breese, 1899; Campbell and Howell, 1972). (For *The Young Lady Versus Old Lady Optical Illusion* see Other Online Resources in the Appendix). For some researchers these bistable images are hypothesized to result from underlying common mechanisms (Andrews and Purves, 1997). O'Shea (2004) suggests that the processing of binocular rivalry occurs at a lower level: "Processing of the other studied bistable phenomena must involve neurons sensitive to the orientation, colour and location of stimuli on the retina. Such neurons are located at low levels of the visual system." The alternation between figure and ground (Rubin, N. 2001) and the underlying neuronal mechanisms also likely play a role.

Elsewhere researchers (Meenes, 1930) also describe subjects reporting "patchwork" rivalry similar to Wheatstone's account "fragmentation of the two images" in place an alternation between the dominance of one eye's view versus the other eye's view. Longer viewing times of so-called rival (discordant) targets or images lead to an increase in reporting the perception of this "patchwork" rivalry (Hollins & Hudnell, 1980).

Measurable features of binocular rivalry include that of predominance of the total view time for a rival pattern, the stimulus strength as compared with the rival pattern. Stimulus variables also include contour density, spatial frequency, motion, and pattern contrast (Levelt, 1965). Blake (2001) notes that "a 'stronger' rival target (e.g., one that is higher contrast than the other) enjoys enhanced predominance, defined as the total percentage of time that a given stimulus is visible during an extended viewing period."

An important finding is the Eye Exchange Experiment (see Online Demonstrations in the Appendix). Normally we are not aware of which eye is dominant–we simply see the stimulus pattern regardless of which eye is doing "the seeing." This experiment allows a subject to view two competing patterns and one is dominant. However the subject can press a button, which causes the two rival patterns to switch from one eye to the other. Under controlled conditions Blake (1979) drew the following conclusion: "the dominant pattern abruptly becomes invisible and the previously suppressed pattern becomes dominant, implying that it was the region of an eye that was dominant, not a particular stimulus." This outcome holds for meaningful objects such house or face patterns (Blake, Westendorf, & Overton 1979; Walker & Powell, 1979).

Figure 15. Common bi-stable images: Schroeder Stairs; Necker Cube; Face/Vase (Public Domain).

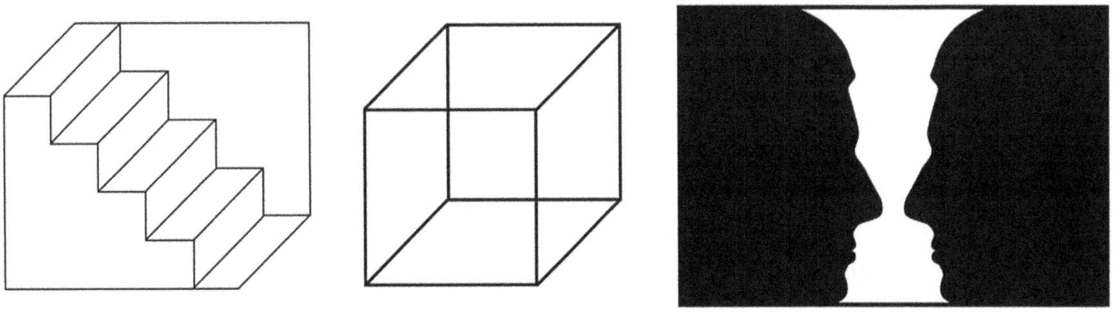

The Cheshire Cat

The phenomenon of binocular rivalry was considered to be something of a "laboratory curiosity" but has been of greater interest in neuroscience for not only understanding perception but as a way to study the neural basis of visual awareness at the level of individual neurons (Logothetis, 1998; Engel et al, 1999) and studies using brain imaging (Tong et al, 1998; Polansky et al, 2000). Blake and Logothetis (2002) argue that "Instead, multiple neural operations are implicated in rivalry, including: registration of incompatible visual messages arising from the two eyes; promotion of dominance of one coherent percept; suppression of incoherent image elements; and alternations in dominance over time."

To further the understanding of the links between visual awareness and consciousness Crick and Koch (1992, 2003) underscored the importance of the need for new theories and research into the neural basis of attention and short term memory supported by studies at the molecular, neurobiological level in conjunction with clinical imaging studies. They speculate that binocular rivalry may involve sets of neurons firing that correspond directly to visual awareness. The Cheshire Cat effect is vivid demonstration of binocular rivalry, attention and the relationship to awareness. The viewer holds a mirror up to the nose so the field of vision is divided. The mirror reflects a blank wall on the right seen only by the right eye. The left eye sees only a cat seated on stool. If the viewer waves his/her right hand in the visual field of the right eye (which sees the blank wall reflected in the mirror) and then into the visual field of the left eye (which sees the cat), "The result is that the cat may disappear. Or if the viewer was attentive to a specific feature before the hand was waved, those parts —— the eyes or even a mocking smile —— may remain (Crick and Koch 1992)." The phenomenon of the lingering facial feature, led to this effect being called the Cheshire Cat Effect referring to the depiction of Cheshire Cat's smile in Lewis Carroll's (the pseudonym of Charles Lutwigde Dodgson) Alice's Adventures in Wonderland (1865). Important research continues (Blake et al. 2014) in how binocular rivalry offers a powerful framework for the study of the neural correlates of conscious and visual perception.

A comprehensive list of research papers on binocular rivalry compiled by is provided in the Appendix in the Other Online Resources section.

The Auditory Pathways and the Split-Brain Interface

The split-brain interface requires the use of stereo headphones to properly experience the work. The proper orientation of the headphones should be where the left and right stereo channels match what is seen in the left and right visual fields. Specifically for *The Split-Brain (Dichoptic) Interface: Thomas v. Hill* Anita Hill is seen by the left eye and heard by the left ear. Similarly Clarence Thomas is seen by the right eye and heard with the right ear. However where the ScreenScope viewer exploits the visual pathways to deliver the digital video signal to the opposite side hemisphere, the auditory pathways while initially lateralized (starting with sounds heard in each ear) project to both sides the brain in a complex of neural connections still poorly understood.

The ear has three major anatomical parts: the outer, middle and inner ear. The sense organs of the inner govern both hearing, posture equilibrium, head and eye movements (Hawkins n.d.) When sound waves reach the ear the outer ear or pinna reflects and attenuates the sound. These changes contain information about direction. The auditory canal is approximately 2 centimeter in length serves to amplify the sound. As sound pressure waves impinge on the ear drum or the tympanic membrane the wave energy is converted to mechanical energy in the air filled middle ear. The mechanical energy is conducted by ossicles–the malleus (hammer), the incus (anvil) and stapes (the stirrup), which articulates (connects) to the oval window. Working together the ossicles mechanically convert the lower pressure waves to higher pressure waves required for the fluid filled inner ear. The cochlea in the inner converts the mechanical wave information to nerve impulses (Figure 16 shows the major anatomical features of hearing). Within the cochlea the organ of Corti has hair cells that directly transform the liquid wave forms into nerve impulses. It is beyond the scope of this chapter to describe further anatomical and bio chemical details of the inner ear or to cover the vestibular system that governs balance.

Nerve fibers from the auditory nerve innervate (connect to) these hair cells, continue and join the vestibular nerve forming the vestibule cochlear nerve also know as the eighth cranial nerve. The auditory

Figure 16. The anatomy of the human ear
(Adopted from Chittka and Brockmann (2005) Public Domain)

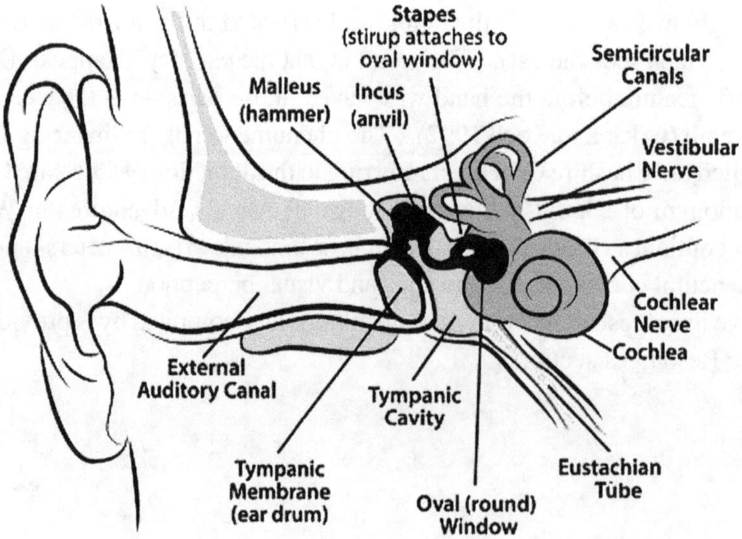

pathways from each ear are bilateral meaning they go to both sides of the brain. The cochlear nerve connects to cochlear nuclei, which in turn sends auditory information to olivary nuclei on both sides of the brain stem. The medial olivary nuclei respond to arrival time differences and the lateral olivary nuclei respond to amplitude differences. The cochlear nuclei also send information up to the mid brain to the inferior colliculus in both sides of the brain. Pathways continue from inferior colliculus on each side to the medial geniculate nucleus and then to the primary cortex (Gray n.d.) located in the temporal lobe (Figure 17 shows the auditory pathways and the location of the primary auditory cortex). Because the complexity of projections to both sides of the brain the auditory pathways have been difficult to study.

As previously noted (Garvey, 2002), hearing simultaneously the opening statements of the two protagonists (Thomas and Hill) is a kind of multi-tasking polyphony (literally many voices). Testimony is given slowly by actors in this drama. Each word of their sworn testimony is carefully articulated. Rather than hearing each word at the same moment, the spoken words alternate. For many viewers (listeners) it is possible to follow both. The viewer/listener multi-tasks by shifting attention and focus rapidly back and forth from one speaker to the other.

Another factor at play may be what is called selective auditory attention or selective listening. This phenomenon was first described by cognitive scientist Colin Cherry (1953) became known as the cocktail problem where one is able to follow one conversation against a background of other noisy conversations. Cherry's experimental method played two different messages to the left and right ear. Subjects were instructed to pay attention to just one message in one ear while ignoring the other message in the unattended ear. This is known as shadowing.

Dichotic listening tests were introduced by Broadbent (1954) and further developed by Kimura (1961a, 1961b) and are used to test different models of attention. With the dichotic method two different auditory stimuli are presented to each ear and the subject is tested on the ability to accurately identify these sounds. Dichotic listening tests offer another experimental method for the study of laterality or hemispheric differentiation. Broadbent concluded from his early work (Broadbent, 1954), that subjects attended to a message one ear at a time. Kimura (1961a, 1961b) used a dichotic auditory technique where

Figure 17. The auditory pathways and the primary auditory cortex
(Public Domain, Recreation under CC BY-SA 3.0 and GFDL)

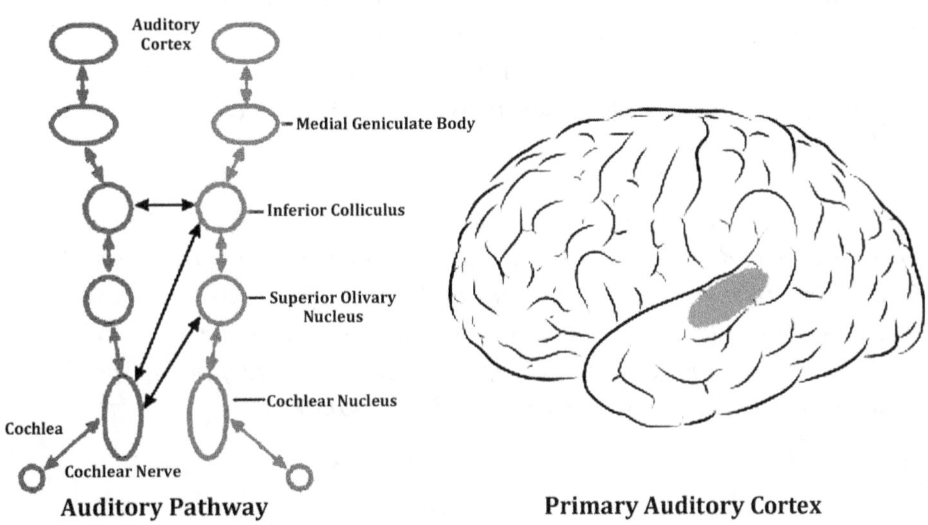

competing verbal stimuli are presented to the two ears simultaneously. Kimura describes a "right-ear effect." Also known as "right-ear advantage" it is thought this is due to left hemispheric dominance for language processing and the crossing of auditory pathways from the ear to the brain.

Moray (1959) presented subjects' own names to their unattended ear and they responded with recognition. Treisman (1964) contributed to an attenuation model of attention based on work that showed that subjects understood words heard by the unattended ear if these words were somehow related to the message that the subject was attending to. Perhaps this is the better explanation for ability of most listeners to follow and grasp the testimony of both Hill and Thomas.

An interesting historical footnote to a discussion of the neural pathways of hearing is the work of Julian Jaynes (1976, 2000). Jaynes proposed an intriguing theory of the rise of the conscious mind. In Jaynes account prior to 1000 BCE humans were preconscious or possessed what Jaynes called the bicameral mind. Auditory hallucinations in the right hemisphere were interpreted as the voices of the gods. Commands heard in the right hemisphere were communicated via the corpus callosum to the left hemisphere and immediately put into action "to hear is to obey." The development of the technology of writing contributed to the "breakdown" this "bicameralism" of the left and right hemispheres. This in turn led to the development of consciousness, as we understand it today. While criticized as being an untestable theory Jaynes' account offers compelling ideas that are fertile ground for artistic explorations.

The Arts and Ideas

While the concept of a "split-brain interface" is directly inspired by experiments with split-brain patients there are many other discoveries and developments in the arts as well as in neuroscience and psychology. Binocular disparity and dissimilar visual inputs have long be observed and commented on. In 1664 Rènè Descartes published *Traité de l'homme* (Treatise on Man). According to Howard and Rogers (1995) Descartes based his account of vision on Kepler's ideas of image formation but retained "Galen's notion of animal spirits and ventricular projection." An illustration (Figure 18 shows this image) from this treatise accurately depicts binocular vision, depicting how the optic nerve of each eye projects back to the same side of the brain so that fibers originating in the retina project to the lining of the ipsilateral cerebal ventricle. According to Descartes "particles" from each point on the retina are then mapped to the pineal gland (the purportive center of the soul) to form a single image. In Descartes account the pineal gland, in turn manipulates "fluids" to control muscles.

A century and half before Wheatstone designed the first stereoscopic Sébastian Le Clerc accurately described retinal disparity and took Descartes to task. In 1679 Le Clerc wrote the following criticism of Descartes' theories of vision:

Monsieur Descartes having considered that according to his principles external objects should make an impression on both eyes, and that the soul neverless had only one perception believed that the images of the same object found in the two eyes are reunited in the brain, but if this great genius had reflected a little more on the demonstrations which he gave in his Treatise on Man, he would have recognized that the images in the two eyes although produced by the same object, are different, and because of these differences their reunion is impossible. (Le Clerc, 1679, 44-46).

Figure 18. Illustration showing binocular vision from Traité de l'homme by Rènè Descartes (1644) (Public Domain).

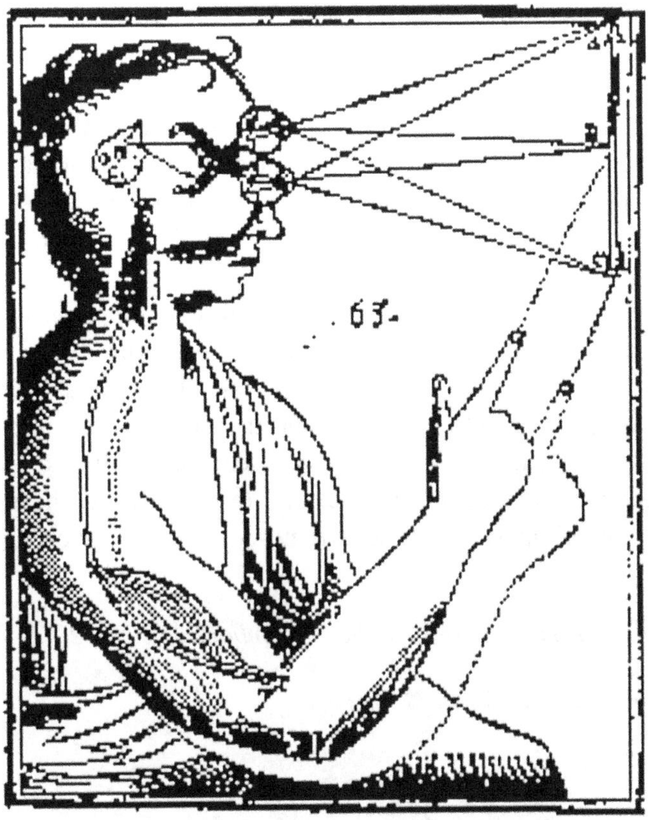

It turns out that Descartes was right about the projection of fibers from the retina of the eye into the brain where a unitary image is formed. However his claim that this took place in the pineal gland was wrong.

It was noted above that Dutour (1760) first described an alternation between the perceived color when each eye viewed a different color (O'Shea 1999). The 19th century French scientist Michel Eugène Chevreul (1860) noticed similar properties of colors and termed this the principle of simultaneous contrast. Chevreul developed dyes and in describing the perceptual effects of juxtapositions of pure patches of color. Chevreul's work inspired in part, the experiments of the French Impressionists (e.g. Monet) with broken brushwork. At a certain distance individual brushstrokes of pure color would perceptually fuse and be perceived as a third color. A patch of yellow next to blue is perceived as green. Subsequently the post-impressionists (notably George Seurat) used small dots or points of color to created vibrant effects. This became known pointillism or divisionism. Such techniques suggest that the artifacts of the artwork are perceived as a subjective experience seen only in the mind's eye. Moreover these techniques opened the door to the exploration of viewing simultaneously dissimilar elements in images.

An important development in parallel was the shift in the subject matter of art: paintings of gods, goddesses, heroes, kings, generals and aristocrats gave way to the depiction of the working class, the common people, the everyday and the pedestrian. As a genre till life painting allowed artists the luxury of exploring colors, textures and simulations, which set the stage for abstraction. In the late Renaissance Giuseppe Arcimboldo (1527–1650) did a series of paintings on the subject of the four seasons. Winter, spring, summer and fall depicted faces in profile made up of plants and vegetables commonly available at that time of year. Contemporary American artist Philip Haas (Figure 19 shows Haas's outdoor sculpture) created scale models of these paintings in fiberglass as monumental sculptures.

The Spanish surrealist Salvador Dalí was fascinated by such work. Many of Dalí's paintings incorporate bi-stable images. For Dalî these images opened the door to a kind of reasoning madness essential to his paranoiac-critical method. The surrealists explored "abnormal" mental states pursuing techniques such as automatic writing in an attempt to unlock the unconscious (inspired by the writings of Freud) to stimulate creativity and challenge conventional notions of reality.

Dalí's pursuit of "optical insecurity" led to an interest in stereoscopic images. Dalí painted pairs of photorealistic stereoscopic images, which were to be viewed by means of two angled mirrors placed at 45 degrees, similar to the first stereoscopes invented by David Brewster and Charles Wheatstone. One painting "Fire in the Borgo; Is Athens Buring?" presents the viewer with copies of two paintings by Raphael: "The School of Athens" and "Fire in the Borgo." (see Other Online Resources in the Appendix). While each painting depicts similar architectural features (e.g. an arch) each painting is dissimilar. Dalí unified the paintings by superimposing a spiral of multi-colored squares on each of the two paintings.

Figure 19. Philip Haas. The Four Seasons (After Arcimboldo), 2010–2011. Painted and pigmented fiberglass
(Photograph © 2015, Gregory P. Garvey. Used with Permission).

These squares are offset on each painting to create a vivid stereoscopic three-dimensional effect. Today the reader can easily experiment with similar setup by placing two mirrors at 90 degrees to one another and setting up either printed images or even display monitors to be reflected by the mirrors (see figure 10). To view the effect position one self so you see the reflected images from each mirror as indicated in Wheatstone's original depiction.

The French surrealist writer, Raymond Roussel (1935, 1995) describes how he used "a tiny pendant opera-glass, of which each tube, two millimeters across and made to go tight against the eye, contains a photograph on glass: one of Cairo bazaars, the other of a quay at Luxor". He found inspiration for his writing in such unusual perceptual effects. Viewing two dissimilar images in this fashion is a direct historical precursor to the Split-Brain Interface.

It is part of a tradition of experimentation for visual artists to explore different mental, psychological and perceptual effects as part of the artistic process. The reader can easily experiment with such aesthetic perceptual effects in which the viewer has to consciously struggle to merge two dissimilar images. The reader could create a similar viewing experience by the use of two mobile phones and a piece of cardboard. Hold the cardboard in front of ones nose so that each eye sees only the image or video on the phone immediately in front of the eye. The reader can use Google Cardboard (https://www.google.com/get/cardboard/) along with a compatible Android phone to explore related optical effects that deliver dissimilar images to each eye.

Today artists, computer graphic modelers, animators, game designers and virtual world builders use these techniques in combination to create compelling immersive and engaging 3D experiences. For the computer graphics artist and game designer software and game engines render 3D views where many of these effects are by product of the rendering engine. Yet many digital games explore alternative worlds consisting of two and half dimensional spaces for side-scrolling platformer games. These games restrict movement along a 2D plane yet have limited depth that provides the illusion of dynamic movement through motion parallax.

In conventional drawing, painting and animation a student learns how to create each effect separately and then in combination. In basic drawing typically one point perspective is introduced first, followed by two point and three point techniques. So called 'scientific perspective' based on projective geometry can then be combined with other visual depth cues such as relative object size, overlap, motion parallax and atmospheric perspective to create a vivid visual experience of depth and space. While it can be tedious to analyze individual objects it heightens observational skills. In combination with learning about art history students learn that there are alternative systems of visual representation such as the canon of representation of ancient Egypt, the paraline, isometric or non-converging elevated viewpoints of seen in Chinese and Japanese screen paintings or of course the ground breaking experiments of the cubists in combining multiple angles of viewpoints in a single 2D framed painting. The fragmentation of viewpoints into individual facets or planes became an aesthetic end in itself.

Inspiration

Wheatstone's original monograph (1838) includes an illustration of a stereoscopic image pair depicting a single line. (Figure 20 shows Wheatstone's original figure 23). He describes this illustration as follows: "Present, in the stereoscope, to the right eye a vertical line, and to the left eye a line inclined some degrees from the perpendicular (Figure 23); the observer will then perceive, as formerly explained, a line, the extremities of which appear at different distances before the eyes."

Figure 20. Wheatstone's original figure 23 demonstrating stereoscopy
(Adopted from Wheatstone (1838), © Public Domain).

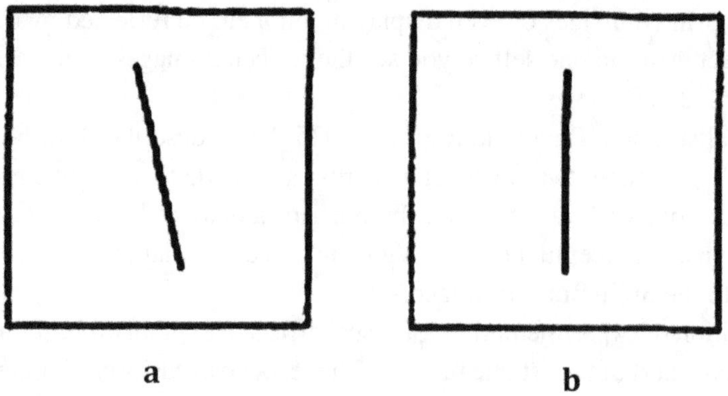

This simple illustration from Wheatstone's monograph inspired the design of an interactive installation entitled *The Wheatstone Stereoscopic Random Line Pair Generator with Fitness Function for Non-objective Art* (Garvey 2011). Wheatstone's stereoscopic pair shown above by Figure 20 inspired the initial random pair generation (Figure 21 shows that initial seed pair initial pair). Viewers in the gallery wear stereoscopic glasses in order to see the stereoscopic effect. The motion of each viewer will trigger the projection of randomly positioned stereoscopic pairs of lines projected on each wall of the gallery forming a dynamic, evolving, emergent, stochastic, 'non-objective' composition. The goal is the creation of 3D generated stereoscopic wire-frame pairs both as stills and animations. The brain attempts to fuse the two inputs into one coherent but virtual image seen only in the "mind's eye."

The Generative Composition

In the actual installation motion by the viewer wearing a stereoscopic headset will trigger the generation of randomly positioned pairs of lines that are displayed within demarcated rectangular areas. These designated rectangular areas are like picture frames in which an abstract or non-objective composition emergences. In the original proposal, audio is created by using line length to calculate audible frequencies. Z-depth will be mapped to generate loudness or softness. The 2D Fitness Function is a simple

Figure 21. The Wheatstone Seed based on figure 19.
(© 2011, Gregory P. Garvey. Used with permission).

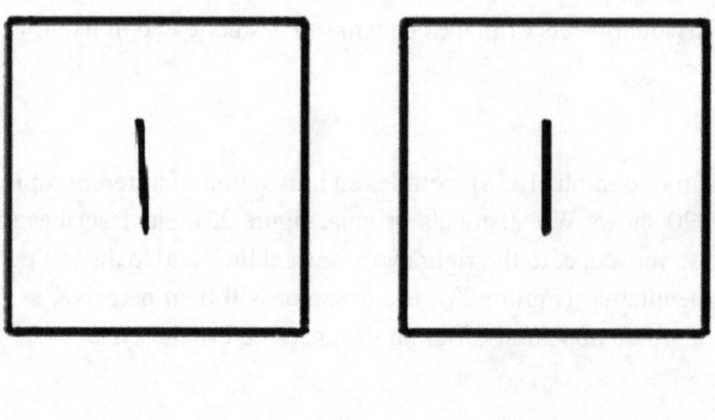

evolutionary algorithm described as follows: "Overlapping lines will be discarded but the position of these lines will be used to generate a new pair that is constrained to a best fit without overlap. Surviving pairs will be added to the database. The attributes (position, length) of 'parents' are inherited by next generation with random perturbations (Garvey, 2011)." As viewers move through the gallery new stereoscopic pairs are generated. If the newly generated line pairs overlap (Figure 22 shows a sample overlapping stereoscopic pair), those overlapping lines are discarded. The elimination of overlapping lines is done to preserve the illusion of 3D depth.

New lines are generated that inherit properties of the initial pair. If these lines are a best fit and don't overlap they 'survive' into the next generation (Figure 23 shows non-overlapping pair).

As the system responds to the viewer's movement a complex image emerges (Figure 24 shows an sample emergent image after 10 generations).

The displayed images continue to add stereo image pairs building complexity for multiple generations until the viewer leaves the gallery space (Figure 25 shows 50 generations).

The original proposal for the installation of *The Wheatstone Stereoscopic Random Line Pair Generator with Fitness Function for Non-objective Art* (Garvey 2011) planned for multiple displays updating continuously triggered by viewers' movements in the gallery space. (Figure 26 is a computer graphics visualization of the proposed installation).

Figure 22. Random pair generation
(© 2011, Gregory P. Garvey. Used with permission).

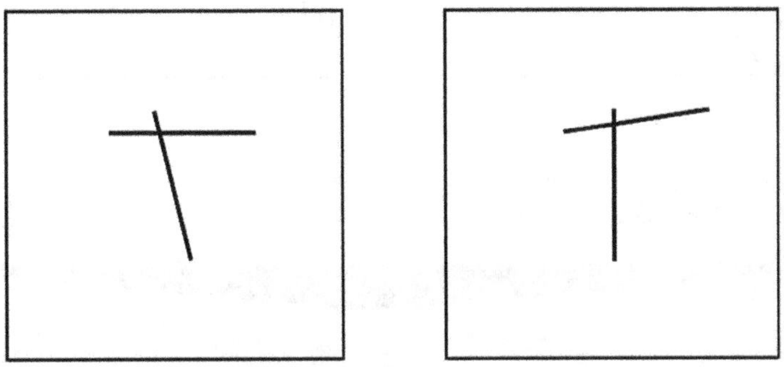

Figure 23. After discarding overlapping lines, replacement is with non-overlapping best fit lines
(© 2011, Gregory P. Garvey. Used with permission).

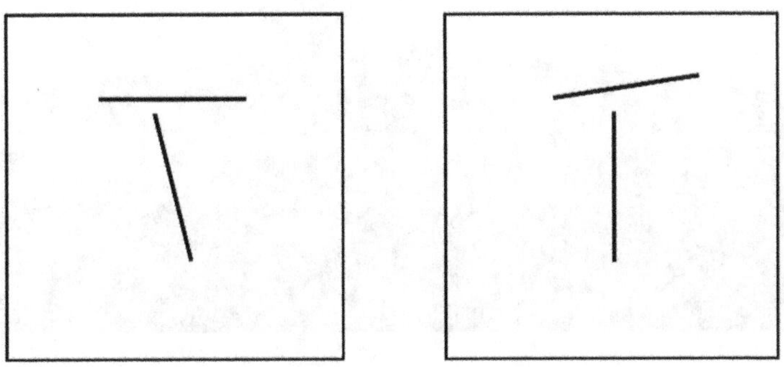

Figure 24. After 10 generations of random stereoscopic pairs
(© 2011, Gregory P. Garvey. Used with permission).

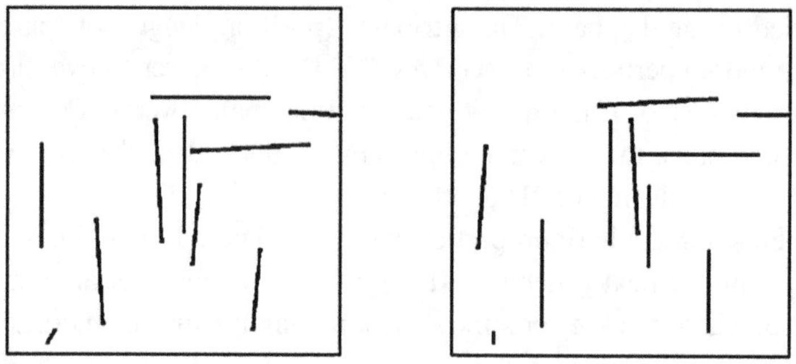

Figure 25. After 50 generations of random stereoscopic pairs
(© 2011, Gregory P. Garvey. Used with permission).

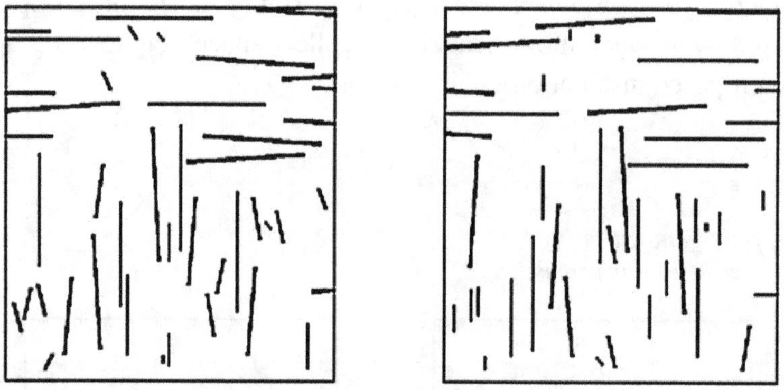

Figure 26. 3D Computer graphics visualization of installation
(© 2011, Gregory P. Garvey. Used with permission).

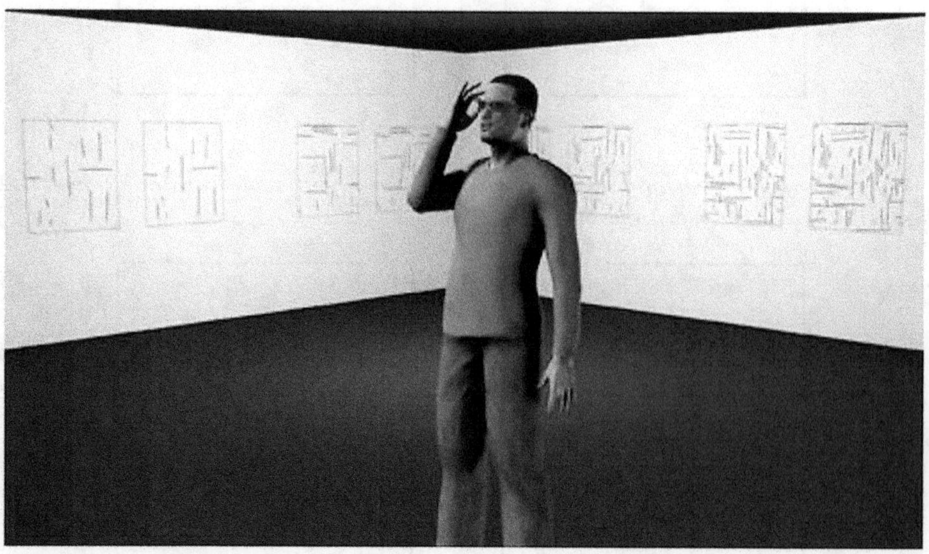

Homage to the Square

Joseph Albers (1975) explored the "interaction of color" first as a series of paintings and then in book form. This publication and the documentation of the painting series were hugely influential especially for artists studying color theory. Albers built upon the discoveries of Chevreul (discussed above). By reducing or abstracting to a simple depiction of a flat square within a square Albers systematically explored hundreds of color relationships: analogous, complementary and value contrasts. Using Albers Homage to The Square series as a point of departure a 2D computer generated animation [Garvey 2005] was created entitled "Homage to Square: Suprematist Composition." This series of work is given a name that also references the early 20th century Suprematist paintings of the Russian constructivist Kasimir Malevich. Malevich sought to reduce art to pure sensation by the use of pure form. This series explores color relationships much like the work of Albers while keeping a strict restriction of the simple form of the square.

The installation shows an animation of two similarly size squares presented side by side. In each square another smaller square is inscribed. Colors pairs continuously vary from warm to neutral to cool (Figures 27, 28, and 29 show sample inputs of different warm, cool and neutral values).

When viewed the ScreenScope Viewer most people can fuse the two dissimilar inputs into a coherent overlay of one square on top of the other (Figure 30 shows a simulation of what the viewer's perception of the overlapping images).

SOLUTIONS AND RECOMMENDATIONS

Recent developments in Virtual Reality Head mounted displays offer accessible technology to explore stereovision and possibly effect of binocular rivalry. With the introduction of low cost headsets like the Oculus Rift, virtual reality is now an accessible and robust technology for playing 3D games and even

Figure 27. Warm pairs
(© 2005, Gregory P. Garvey. Used with permission).

**For a more accurate representation see the electronic version.*

Figure 28. Cool pairs
(© 2005, Gregory P. Garvey. Used with permission).

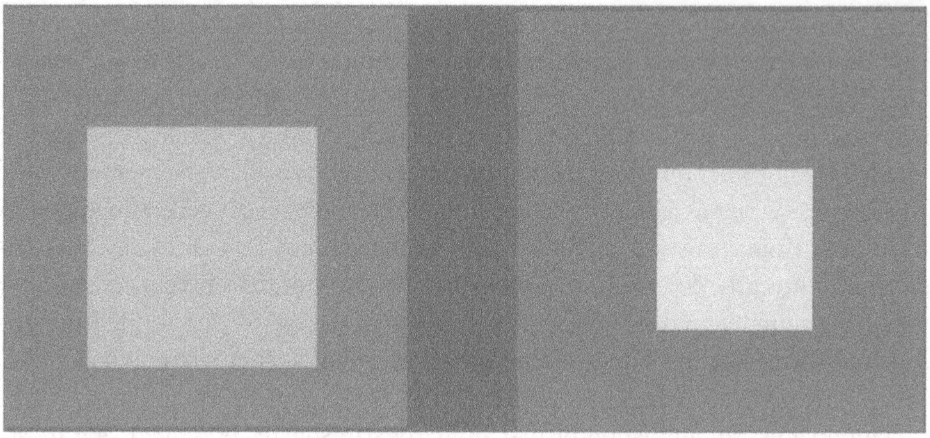

**For a more accurate representation see the electronic version.*

Figure 29. Neutral pairs
(© 2005, Gregory P. Garvey. Used with permission).

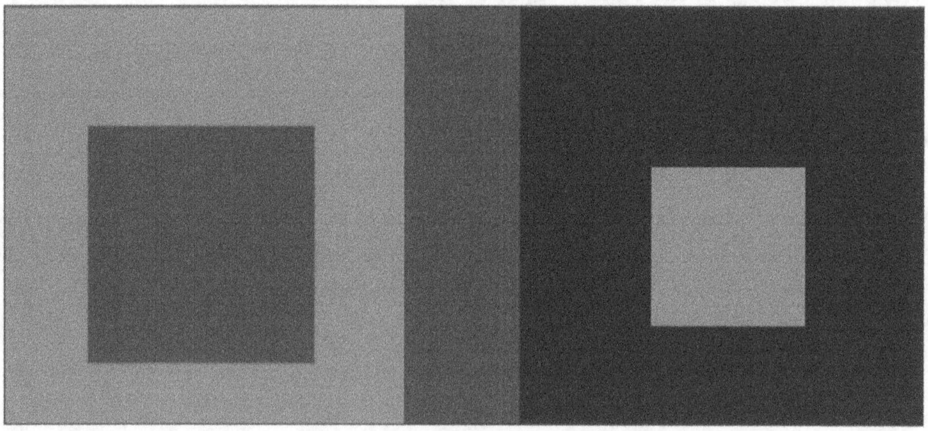

**For a more accurate representation see the electronic version.*

viewing movies. The Oculus Rift combines stereo-optics and head-tracking to offer a fully immersive 3 dimensional viewing experience. In 2014 Facebook paid 2 billion dollars to acquire Oculus spurring other manufacturers to rush to market competing headsets. Other leading systems include Sony's Project Morpheus, HTC Vive (Steam/Valve), Samsung's Gear VR, Microsoft HoloLens and Zeiss VR One. The FOVE VR includes eye tracking and simulated depth-of-field. The Avegant Glyph is unique in its lightweight minimalist design using an array of micro-mirrors to reflect images directly to the lens onto the retina of the eye. Google's Cardboard offers an inexpensive do-it-yourself solution that is compatible with Android Phones.

These modern day systems not only have the virtue of affordability, but have largely solved latency issues that triggered motion sickness and headaches after prolonged use that plagued the VR systems of the late eighties and early nineties. All of these systems seek to reproduce a 3 dimensional viewing experience by using principles of stereopsis.

Figure 30. Simulation of the merger into one image of two dissimilar inputs for Homage to the Square
(© 2005, Gregory P. Garvey. Used with permission).

FUTURE RESEARCH DIRECTIONS

Visual illusions offer a rich resource for further exploration and for controlled experiments in perception. For example the Müller-Lyer or Zöllner illusions (Figure 31 shows the Müller-Lyer illusion. Figure 32 shows the Zöllner illusion) could be used as a basis for a new fitness function for an interactive generative work. At the end of this chapter a list of online resources for optical illusions is provided. Blind spot, random dot stereograms and spreading waves of dominance demonstrations are rich in possibilities for further artistic experimentation. The simple demonstration of retinal disparity using the finger held

Figure 31. Recreation of Müller-Lyer Illusion
(Public Domain).

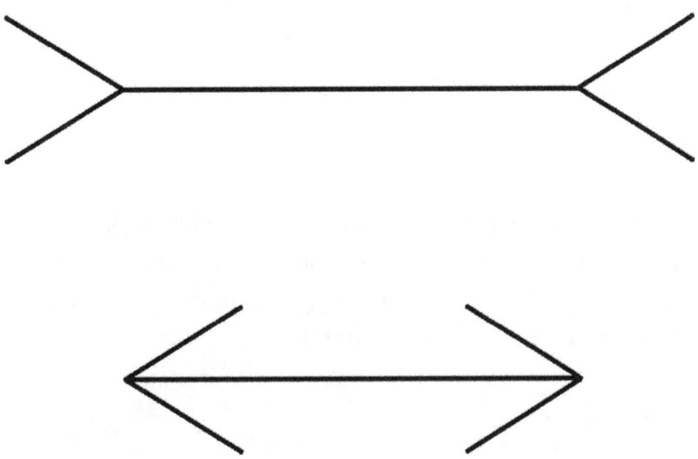

Figure 32. Recreation of Zöllner Illusion
(© Public Domain)

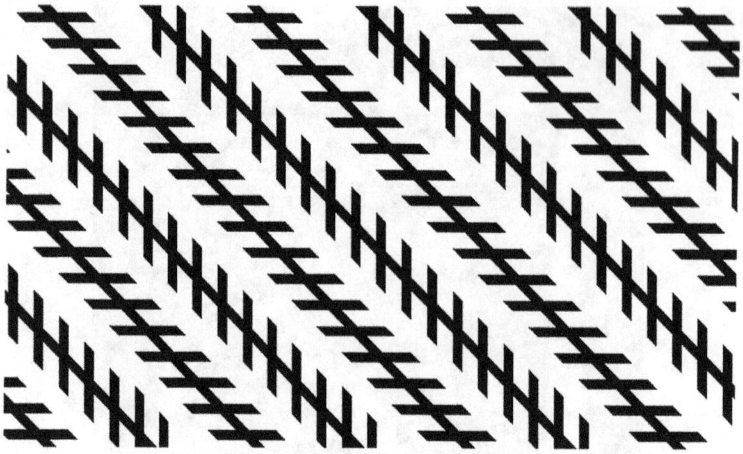

close to the center of the face described above is another area underexplored by artists. The Cheshire Cat Effect also discussed, suggests intriguing possibilities for additional experiments that explore binocular rivalry combining dissimilar inputs with motion erasure effects.

Game engines such as UNITY or UNREAL offer the prospect of real time generation of 3D stereoscopic images that create an immersive environment when viewed with low cost headsets like the Oculus Rift. Motion in this case is generated by head movement. Point of view (POV) is a natural way to determine what is rendered and visible since game engine rendering automatically discards hidden line and occluded polygons. The artistic works describe above that employ dissimilar inputs using recorded digital video, simple line 'drawings' or flat fields of color point the way towards the expanded possibilities of 3D generated stereoscopic pairs of incongruent images. How will the brain attempt to fuse such 3D dissimilar inputs into the singular perception of the 'cyclopean' eye? Other possible ways to monitor user interaction includes low cost eye tracking for using gaze and focus of attention as triggers for algorithmic generation of image pairs. The S2 Eye Tracker along with Eyeworks software is a powerful and reliable system. Other systems include the EYE-TRAC Head Mounted Display and the Mobile Eye from Applied Science Laboratories. (see Other Online Resources in the Appendix). While not technically not an eye or motion tracking affordance Google Glass is a fairly reliable technology that enables augmented reality approaches.

CONCLUSION

There are striking examples of Roman mosaic floors patterns, which depict a kind of faux 3 dimensional space. In particular a pattern of cubes are depicted where three shades of tesserae are used to create a repeating 3-D box pattern made from polygonal lozenge shapes (Figure 33 shows a copy of one such pattern from the Piazza della Vittoria in Palermo, Italy).

Figure 33. First century 3D box pattern mosaic from the Terme di Diocleziano, Palermo Italy
(Adopted from Ace of the Fungal Kingdom (2006) (© Creative Commons Noncommercial license. http://creativecommons.org/licenses/by-nc-sa/2.5/ BY-NC-SA 2.5)

Such patterns have long engaged and delighted the eye. The mind's eye is literally captivated by the illusion of a 3D shape that emerges from a pattern on a flat form. This is a kind of play that visual artists have long explored. It is pursued also in a spirit of investigation and exploration leading to the creation and discovery of striking visual illusions. Exploiting the human visual system opens doors to new experiences that can result from leveraging the power of 3D computer graphics, binocular vision and stereoscopy, coupled to new 3D viewing devices supported by an understanding the visual pathways. Today we can open the doors of perception through a determined study and understanding of the "eye" and "brain."

REFERENCES

10 cool optical illusions. (n.d.). Retrieved September 17, 2015 from http://psychology.about.com/od/sensationandperception/tp/cool-optical-illusions.htm

Ace of the Fungal Kingdom. (2006). *Terme di Diocleziano.* Retrieved December 10, 2015, from http://www.tiedyedfreaks.org/ace/diocleziano/DSCN5314crop.jpg

Aids, S. (n.d.). *ScreenScope - Mirror Stereoscope.* Retrieved August 3, 2015, from http://www.stereo-aids.com.au

Albers, J. (1975). *Interaction of Color* (Revised Edition). New Haven, CT: Yale University Press.

Andrew, T. J., & Purves, D. (1997). Similarities in normal and binocular rivalrous viewing. *Proceedings of the National Academy of Sciences of the United States of America, 94*(18), 9905–9908. doi:10.1073/pnas.94.18.9905 PMID:9275224

Animation, P. (n.d.). Retrieved September 17, 2015 from http://www.psy.vanderbilt.edu/faculty/blake/rivalry/BR.html

Aprile, I., Ferrarin, M., Padua, L., Di Sipio, E., Simbolotti, C., Petroni, S., & Dickmann, A. et al. (2014, July). Walking strategies in subjects with congenital or early onset strabismus. *Frontiers in Human Neuroscience*, *8*, 484. doi:10.3389/fnhum.2014.00484 PMID:25071514

Banff Centre New Media Institute. (1999). BNMI Co-Production Archives 'G'. *The Split-Brain Human Computer User Interface*. Retrieved July 10, 2015, from http://www.banffcentre.ca/bnmi/coproduction/archives/s.asp#thesplit

Banich, M. T. (2003). The divided visual field technique in laterality and interhemispheric integration. In K. Hughdahl (Ed.), *Experimental Methods in Neuropsychology* (pp. 47–63). New York: Kluwer. doi:10.1007/978-1-4615-1163-2_3

Banich, M. T., & Shenker, J. I. (1994). Investigations of interhemispheric processing: Methodological considerations. *Neuropsychology*, *8*(2), 263–277. doi:10.1037/0894-4105.8.2.263

Binocular rivalry bibliography. (n.d.). Retrieved October 2, 2015, from https://sites.google.com/site/oshearobertp/publications/binocular-rivalry-bibliography

Blake, R. (1989). A neural theory of binocular rivalry. *Psychological Review*, *96*(1), 145–167. doi:10.1037/0033-295X.96.1.145 PMID:2648445

Blake, R. (2001). A Primer on Binocular Rivalry, Including Current Controversies. *Brain and Mind*, 2(1), 5–38. doi:10.1023/A:1017925416289

Blake, R. (2005). Landmarks in the History of Binocular Vision. In D. Alais & R. Blake (Eds.), *Binocular Rivalry* (pp. 1–27). Cambridge, MA: MIT Press.

Blake, R., Brascamp, J., & Heeger, D. J. (2014). Can binocular rivalry reveal neural correlates of consciousness? *Philosophical Transactions of the Royal Society of London. Series B, Biological Sciences*, *369*(1641). doi:10.1098/rstb.2013.0211 PMID:24639582

Blake, R., & Logothetis, N. K. (2002, January). Visual Competition. *Nature Reviews. Neuroscience*, *I*(1), 13–21. doi:10.1038/nrn701 PMID:11823801

Blake, R., Westendorf, D., & Overton, R. (1979). What is suppressed during binocular rivalry? *Perception*, *9*(2), 223–231. doi:10.1068/p090223 PMID:7375329

Blind Spot. (n.d.). Retrieved September 17, 2015 from http://www.exploratorium.edu/snacks/blind_spot/index.html

Bogen, J. E. (1975, Spring). Some Educational Aspects of Hemispheric Specialization. *UCLA Educator*, *17*(2), 24–32.

Bogen, J. E., & Vogel, P. J. (1962). Cerebral commissurotomy in man. *Bulletin of the Los Angeles Neurological Society*, *27*, 169–172.

Bourne, V. J. (2006). The divided visual field paradigm: Methodological considerations. *Laterality*, *11*(4), 373–393. doi:10.1080/13576500600633982 PMID:16754238

Breese, B. B. (1899). On inhibition. *Psychological Monographs*, *3*(1), 1–65. doi:10.1037/h0092990

Brewster, D. (1856). *The Stereoscope; its History, Theory, and Construction, with its Application to the fine and useful Arts and to Education: With fifty wood Engravings*. London: John Murray.

Broadbent, D. E. (1954). The role of auditory localization in attention and memory span. *Journal of Experimental Psychology*, *44*, 51–55. doi:10.1037/h0056491 PMID:13152294

Brooks, R. (1980, January). Hemispheric Differences in Memory: Implications for Education. *The Clearing House: A Journal of Educational Strategies, Issues and Ideas*, *53*(5), 248–250. doi:10.1080/00098655.1980.9959221

Byrne, J. H. (Ed.). (1997–Present). *Chapter 14. Visual Processing: Eye and Retina: Figure 14.7*. Neuroscience Online. Department of Neurobiology and Anatomy. The University of Texas Medical School at Houston. Retrieved November 21, 2015, from http://neuroscience.uth.tmc.edu/s2/chapter14.html

C-SPAN. (1991). *Thomas Confirmation Hearings*. Retrieved September 6, 2015, from http://www.c-span.org/search/?searchtype=Videos&sort=Newest&seriesid[]=24

Campbell, F. W., & Howell, E. R. (1972). Monocular alternation; a method for the investigation of pattern vision. *The Journal of Physiology*, *225*, 19–21. PMID:5074381

Carroll, L. (1865). *Alice's Adventures in Wonderland*. London: MacMillan and Company.

Cherry, C. (1953). Some experiments on the recognition of speech, with one and with two ears. *The Journal of the Acoustical Society of America*, *25*(5), 975–979. doi:10.1121/1.1907229

Cheshire Cat Experiment. (n.d.). Retrieved September 17, 2015 from http://www.exploratorium.edu/snacks/cheshire_cat/

Chittka, L., & Brockmann, A. (2005). Perception Space–The Final Frontier. *PLoS Biology*, *3*(4), e137. doi:10.1371/journal.pbio.0030137 PMID:15819608

Colvin, M. K., Funnell, M. G., Hahn, B., & Gazzaniga, M. S. (2005). Identifying functional channels in the corpus callosum: Correlating interhemispheric transfer time with white matter organization. *Journal of Cognitive Neuroscience*, *139*(suppl. 5), 2409–2419.

Compton, R. J. (2002). Interhemispheric interaction facilitates face processing. *Neuropsychologia*, *40*(13), 2409–2419. doi:10.1016/S0028-3932(02)00078-7 PMID:12417469

Contrast Influences Predominance. (n.d.). Retrieved September 17, 2015 from http://www.psy.vanderbilt.edu/faculty/blake/rivalry/BR.html

Corballis, M. C. (1994). Split decisions: Problems in the interpretation of results from commissurotomized subjects. *Behavioural Brain Research*, *64*(1-2), 163–172. doi:10.1016/0166-4328(94)90128-7 PMID:7840883

Corballis, M. C., & Sergent, J. (1988). Imagery in a commissurotomized patient. *Neuropsychologia*, *26*(1), 13–26. doi:10.1016/0028-3932(88)90027-9 PMID:3362338

Corballis, P. M., Funnell, M. G., & Gazzaniga, M. S. (1999). A dissociation between spatial and identity matching in callosotomy patients. *Neuroreport*, *10*(10), 2183–2187. doi:10.1097/00001756-199907130-00033 PMID:10424695

Crick, F. C., & Koch, C. (1992, September). The Problem of Consciousness. *Scientific American, 267*(3), 152–159. doi:10.1038/scientificamerican0992-152 PMID:1502517

Crick, F. C. & Koch. (2003 February). A Framework for Consciousness. *Nature Neuroscience, 6*(2), 119-126. Retrieved November 2, 2015, from http://www.klab.caltech.edu/koch/crick-koch-03.pdf

Descartes, R. (1644). *Traité de l'homme*. Paris: Angot.

Dimond, S., & Beaumont, G. (1972). Processing in perceptual integration between the cerebral hemispheres. *British Journal of Psychology, 63*(4), 509–514. doi:10.1111/j.2044-8295.1972.tb01300.x PMID:4661080

Dutour, É. F. (1760). Discussion d'une question d'optique. Mémoires de Mathématique et de Physique Présentés par Divers Savants. *L'Académie des Sciences, 3*, 514-530.

Edwards, B. (2012). *Drawing on the Right Side of the Brain: The Definitive* (4th ed.). New York: Tarcher/Penguin.

Encyclopaedia Britannica. (1994). Depth Perception: Correspondence of Points. *Encyclopaedia Britannica*. Retrieved December 20, 2015, from http://www.britannica.com/topic/depth-perception

Engel, A. A. K., Fries, P., Konig, P., Brecht, M., & Singer, W. (1999). Temporal binding, binocular rivalry and consciousness. *Consciousness and Cognition, 8*(2), 128–151. doi:10.1006/ccog.1999.0389 PMID:10447995

Exchange, E. (n.d.). Retrieved September 17, 2015 from http://www.psy.vanderbilt.edu/faculty/blake/rivalry/BR.html

Experiments. (n.d.). Retrieved September 17, 2015 from http://www.richardgregory.org/experiments/

Forster, B. A., Corballis, P. M., & Corballis, M. C. (2000). Effect of luminance on successiveness discrimination in the absence of the corpus callosum. *Neuropsychologia, 38*(4), 441–450. doi:10.1016/S0028-3932(99)00087-1 PMID:10683394

Framing Game. (n.d.). Retrieved December 26, 2015 from http://www.vision3d.com/frame.html

Franz, E. A., Waldie, K. E., & Smith, M. J. (2000). The effect of callosotomy on novel versus familiar manual actions: A neural dissociation between controlled and automatic processes. *Psychological Science, 11*(1), 82–85. doi:10.1111/1467-9280.00220 PMID:11228850

Funnell, M. G., Corballis, P. M., & Gazzaniga, M. S. (1999). A deficit in perceptual matching in the left hemisphere of a callosotomy patient. *Neuropsychologia, 38*, 441–450. PMID:10509836

Garvey, G. P. (2002). The Split Brain Human Computer User Interface. *Leonardo, 35*(3), 319–325. doi:10.1162/002409402760105352

Garvey, G. P. (2005). *Homage to Square: Suprematist Composition*. Flash Animation. Artist Collection.

Garvey, G. P. (2011). The Wheatstone Stereoscopic Random Line Pair Generator with Fitness Function for Non-objective Art. *Proceedings of the 8th Conference on Creativity & Cognition*. Atlanta, GA: The High Museum of Art. doi:10.1145/2069618.2069728

Gazzaniga, M. S. (1982). Split brain research: A personal history. *Cornell University Alumni Quarterly.*, *45*, 2–12.

Gazzaniga, M. S. (1985). *The Social Brain. Discovering the Networks of the Mind.* New York, NY: Basic Books, Inc.

Gazzaniga, M. S. (1998, June). The Split Brain Revisited. *Scientific American.* Retrieved July 20, 2015, from http://www.scientificamerican.com/article/the-split-brain-revisited

Gazzaniga, M. S. (2005). Forty-five years of split-brain research and still going strong. *Nature Reviews. Neuroscience*, *6*(8), 653–659. doi:10.1038/nrn1723 PMID:16062172

Gazzaniga, M. S. (2008). Spheres of Influence. *Scientific American*, *19*(3), 32–39. Retrieved from http://issuu.com/samuelsantos52/docs/scientific_american_mind_vol_19__3_ PMID:18783046

Gazzaniga, M. S., Bogen, J. E., & Sperry, R. W. (1962). Some functional effects of sectioning the cerebral commissures in man. *Proceedings of the National Academy of Sciences of the United States of America*, *48*(10), 1765–1769. doi:10.1073/pnas.48.10.1765 PMID:13946939

Geschwind, D. H., & Crabtree, E. (2003). Handedness and Cerebral Laterality. In H. Whitaker (Ed.), *Concise Encyclopedia of Brain and Language* (pp. 221–230). Oxford, UK: Elsevier Science.

Gray, H. (1918). *Anatomy of the Human Body.* Philadelphia, PA: Lea and Febiger. Retrieved December 2, 2015, from https://commons.wikimedia.org/wiki/File:Gray722.png

Gray, L. (n.d.). Auditory System: Pathways and Reflexes. In *Neuroscience Online, the Open-Access Neuroscience Electronic Textbook.* The University of Texas Health Science Center at Houston (UTHealth). Retrieved September 6, 2015, from http://neuroscience.uth.tmc.edu/

Hawkins, J. E. (n.d.). The Human Ear Anatomy. *Encyclopeadia Britannica.* Retrieved September 6, 2015, from http://www.britannica.com/science/ear

Henry, L. H. (1981). The Economic Benefits of the Arts: A Neuropsychological Comment. *Journal of Cultural Economics*, *5*(1), 52–60. doi:10.1007/BF00189206

Herring, E. (1868). The Theory of Binocular Vision. Plenum.

Hirstein, W. (2005). *Brain Fiction: Self-deception and the Riddle of Confabulation.* Cambridge, MA: MIT Press.

Hollins, M., & Hudnell, K. (1980). Adaptation of the binocular rivalry mechanism. *Investigative Ophthalmology & Visual Science*, *19*, 1117–1120. PMID:7410003

Holmes, O. W. (1859, June). The Stereoscope and the Stereograph. Boston: *The Atlantic Monthly.* Retrieved September 12, 2015, from http://www.theatlantic.com/magazine/archive/1859/06/the-stereoscope-and-the-stereograph/303361/

Hopkins, R. L. (1984, Summer). Education and the Right Hemisphere of the Brain: Thinking in Patterns for a Computer World. *Journal of Thought*, *19*(2), 104–114.

Howard, I. P., & Rogers, B. J. (1995). *Binocular Vision and Stereopsis*. Oxford, UK: Oxford University Press.

Huffman, K. R. (n.d.). *The Electronic Arts Community Goes Online: A Personal View*. Retrieved September 30, 2015, from http://www.lehman.cuny.edu/vpadvance/artgallery/gallery/talkback/issue3/centerpieces/huffman.html

Hupé, J., & Rubin, N. (2003, March). The Dynamics of Bi-stable Alternation in Ambiguous Motion Displays: A Fresh Look at Plaids. *Vision Research*, *43*(5), 531–548. doi:10.1016/S0042-6989(02)00593-X PMID:12594999

Illusions, V. (n.d.). Retrieved November 2, 2015, from http://faculty.washington.edu/chudler/chvision.html

Jaynes, J. (1976, 2000). The Origin of Consciousness in the Breakdown of the Bicameral Mind. New York: Houghton Mifflin/Mariner Books.

Jensen, A. R. (1972, October 12). Genetics and Education. A second look. *New Scientist*, 96–99.

Julesz, B. (1971). *Foundations of Cyclopean Perception*. Cambridge, MA: MIT Press.

Keystone View Company. (1879-1930). *Keystone View Company*. Retrieved September 18, 2015, from http://ark.digitalcommonwealth.org/ark:/50959/sq87dd56f

Kimura, D. (1961a). Cerebral dominance and the perception of verbal stimuli. *Canadian Journal of Psychology*, *15*(3), 166–171. doi:10.1037/h0083219

Kimura, D. (1961b). Some effects of temporal-lobe damage on auditory perception. *Canadian Journal of Psychology*, *15*(3), 156–165. doi:10.1037/h0083218 PMID:13756014

Kitterle, F. L., Christman, S., & Hellige, J. B. (1990). Hemispheric differences are found in the identification, but not the detection, of low versus high spatial frequencies. *Perception & Psychophysics*, *48*(4), 297–306. doi:10.3758/BF03206680 PMID:2243753

Lambert, A. J. (1991). Interhemispheric interaction in the splitbrain. *Neuropsychologia*, *29*(10), 941–948. doi:10.1016/0028-3932(91)90058-G PMID:1762673

Le Clerc, S. (1679). Discours Touchant de Point de Veue, dans lequel il es prouvé que les chose qu'on voit distinctement, ne sont veues que d'un oeil. In Destined for Distinguished Oblivion: The Scientific Vision of William Charles Wells (1757-1817). New York: Springer-Science+Business Media.

LeCompte, N., & Rush, J. C. (1981). The Jack Sprat Syndrome: Can Split-Brain Theory Improve Education by Including the Arts? *The Journal of Education*, *163*(4), 335–343.

Levelt, W. (1965). *On Binocular Rivalry*. Soesterberg, The Netherlands: Institute for Perception RVO-TNO.

Levi, D. (2013). Linking assumptions in amblyopia. *Visual Neuroscience*, *30*(5-6), 277–287. doi:10.1017/S0952523813000023 PMID:23879956

Levy, E. K. (2014). Sleuthing the Mind: Curator's Introduction. *Leonardo*, *47*(5), 427–428. doi:10.1162/LEON_a_00864

Logothetis, N. K. (1998). Single units and conscious vision. *Philosophical Transactions of the Royal Society of London. Series B, Biological Sciences*, *353*(1377), 1801–1818. doi:10.1098/rstb.1998.0333 PMID:9854253

Magic Eye Random Dot Stereograms. (n.d.). Retrieved September 17, 2015 from http://www.magiceye.com/

McCarthy, R. A., & Warrington, E. K. (1990). *Cognitive Neuropsychology: A Clinical Introduction*. New York: Academic Press.

McLuhan, M. H. (1979). Figure and Grounds in Linguistic Criticism. Interpretation of Narrative by Mario J. Valdés, Owen J. Miller. *A Review of General Semantics, 36*(3), 289-294.

Meenes, M. (1930). A phenomenological description of retinal rivalry. *The American Journal of Psychology*, *42*(2), 260–269. doi:10.2307/1415275

Mohr, B., Pulvermuller, F., Rayman, J., & Zaidel, E. (1994). Interhemispheric cooperation during lexical processing is mediated by the corpus-callosum: Evidence from the split-brain. *Neuroscience Letters*, *181*(1-2), 17–21. doi:10.1016/0304-3940(94)90550-9 PMID:7898762

Moray, N. (1959). Attention in dichotic listening: Affective cues and the influence of instructions. *The Quarterly Journal of Experimental Psychology*, *11*(1), 56–60. doi:10.1080/17470215908416289

Murr, L. E., & Williams, J. B. (1988). Half-Brained Ideas about Education: Thinking and Learning with Both the Left and Right Brain in a Visual Culture. *Leonardo*, *21*(4), 413–419. doi:10.2307/1578704

Nebes, R. (1972). Superiority of the minor hemisphere in commissurotomized man on a test of figural unification. *Brain*, *95*(3), 633–638. doi:10.1093/brain/95.3.633 PMID:4655286

Nebes, R. (1973). Perception of spatial relationships by the right and left hemispheres of a commissurotomized man. *Neuropsychologia*, *7*, 333–349. PMID:4792179

Necker, L. A. (1832). Observations on some remarkable optical phaenomena seen in Switzerland; and on an optical phaenomenon which occurs on viewing a figure of a crystal or geometrical solid. *London and Edinburgh Philosophical Magazine and Journal of Science*, *1*(5), 329–337.

Neisser, U. (1966). Cognitive Psychology. New York: Appleton.

Nielsen, J. A., Zielinski, B. A., Ferguson, M. A., Lainhart, J. E., & Anderson, J. S. (2013). An evaluation of the left-brain vs. right brain hypothesis with resting state functional connectivity magnetic resonance imaging. *PLOS One*. Retrieved September 17, 2015, from http://www.plosone.org/article/info%3Adoi%2F10.1371%2Fjournal.pone.0071275

Nobuyuki Kayahara's Spinning Dancer Illusion. (n.d.). Retrieved September 19, 2015, from http://www.procreo.jp/labo/silhouette.swf

O'Shea, R. P. (1999). *Translation of Dutour (1760)*. Retrieved October 1, 2015, from https://sites.google.com/site/oshearobertp/publications/translations/dutour-1760

O'Shea, R. P. (2004, June). Psychophysics: Catching the Old Codger's Eye. *Current Biology*, *14*(12), R478–R479. doi:10.1016/j.cub.2004.06.014 PMID:15203021

Optical Illusion Pictures. (n.d.). Retrieved September 17, 2015 from http://brainden.com/optical-illusions.htm

Optical Illusions & Visual Phenomena 123 of them. (n.d.). Retrieved September 17, 2015 from http://michaelbach.de/ot/

Parker-Pope, T. (2008). *The Truth About the Spinning Dancer*. Retrieved October 1, 2015, from http://well.blogs.nytimes.com/2008/04/28/the-truth-about-the-spinning-dancer/?_r=0

Poffenberger, A. (1912). Reaction time to retinal stimulation with special reference to the time lost in conduction through nervous centers. *Archives de Psychologie*, *23*, 1–73.

Polansky, A., Blake, R., Braun, J., & Heeger, D. (2000). Neuronal activity in human primary visual cortex correlates with perception during binocular rivalry. *Nature Neuroscience*, *3*(11), 1153–1159. doi:10.1038/80676 PMID:11036274

Porta, J. B. (1593). *De Refractione. Optices Parte. Libri Novem*. Carlinum and Pacem, Naples Retrieved October 1, 2015, from http://testyourself.psychtests.com/testid/3178

Psychology Concepts. (n.d.). Retrieved October 1, 2015, from http://www.psychologyconcepts.com/depth-perception/

Rapaczynski, W., & Ehrlichman, H. (1979). Opposite visual hemifield superiorities in face recognition as a function of cognitive style. *Neuropsychologia*, *17*(6), 645–652. doi:10.1016/0028-3932(79)90039-3 PMID:522978

Rivalry, F. (n.d.). Retrieved September 17, 2015 from http://www.psy.vanderbilt.edu/faculty/blake/rivalry/BR.html

Rogers, M. (2013). *Researchers debunk myth of "right brain" and "left-brain" personality traits*. University of Utah, Office of Public Affairs. Retrieved September 17, 2015, from http://www.plosone.org/article/info%3Adoi%2F10.1371%2Fjournal.pone.0071275

Roussel, R. (1995). *How I Wrote Certain of My Books* (T. Winkfield, Trans.). Boston, MA: Exact Change. (Original work published 1935)

Rubin, E. (2001). Readings in perception. In S. Yantis (Ed.), *Visual perception: Essentials readings*. Philadelphia: Psychology Press. (Original work published 1921)

Rubin, N. (2001). Figure and ground in the brain. *Nature Neuroscience*, *4*(9), 857–858. doi:10.1038/nn0901-857 PMID:11528408

Severed Corpus Callosum. (2008, June 25). Scientific American Frontiers. Retrieved September 17, 2015, from https://www.youtube.com/watch?v=82tlVcq6E7A

Simanek, D. (n.d.). *How to View 3D Without Glasses*. Retrieved September 28, 2015, from http://www.lhup.edu/~dsimanek/3d/view3d.htm

Smith, E. E. (2015, July 27). One Head, Two Brains. *The Atlantic*. Retrieved August 24, 2015, from http://www.theatlantic.com/health/archive/2015/07/split-brain-research-sperry-gazzaniga/399290/

Sperry, R. (1966). Brain bisection and the neurology of consciousness. In J. C. Eccles (Ed.), Brain and conscious experience, (pp. 298-313). New York: Springer-Verlag.

Sperry, R. (1968). Hemisphere Deconnection and Unity in Conscious Awareness. *The American Psychologist, 23*(10), 723–733. doi:10.1037/h0026839 PMID:5682831

Spreading Waves of Dominance. (n.d.). Retrieved September 17, 2015 from http://www.psy.vanderbilt.edu/faculty/blake/rivalry/BR.html

The S2 Eye Tracker. Eyeworks software. (n.d.). Retrieved September 25, 2015, from (http://www.mirametrix.com/products/?gclid=CN3Jl8Xz5sgCFZWRHwodaBEBdA)

The blind spot. (n.d.). Retrieved September 17, 2015 from http://faculty.washington.edu/chudler/chvision.html

The Young Lady Versus Old Lady Optical Illusion. (n.d.). Retrieved September 22, 2015, from http://brainden.com/face-illusions.htm

Todd, J. W. (1912). *Reaction to multiple stimuli.* New York: The Science Press. doi:10.1037/13053-000

Tong, F., Nakayama, K., Vaughan, J. T., & Kanwisher, N. (1998). Binocular rivalry and visual awareness in human extrastriate cortex. *Neuron, 21*(4), 753–759. doi:10.1016/S0896-6273(00)80592-9 PMID:9808462

Treisman, A. (1964). Verbal cues, language and meaning in selective attention. *The American Journal of Psychology, 77*(2), 206–209. doi:10.2307/1420127 PMID:14141474

Unitary vs. Piecemeal Rivalry. (n.d.). Retrieved September 17, 2015 from http://www.psy.vanderbilt.edu/faculty/blake/rivalry/BR.html

United States. Congress. Senate. Committee on the Judiciary. (1993). Nomination of Judge Clarence Thomas to be Associate Justice of the Supreme Court of the United States: hearings before the Committee on the Judiciary, United States Senate, first session ... Washington: U.S. G.P.O. For sale by the U.S. G.P.O., Supt. of Docs. Unknown. *The Holmes stereoscope, with the inventions and improvements added by Joseph L. Bates.* Center for the History of Medicine: OnView. Retrieved October 4, 2015, from http://collections.countway.harvard.edu/onview/items/show/6277

Uzwiak, A. (n.d.). *Vision.* Retrieved December 2, 2015, from http://www.rci.rutgers.edu/~uzwiak/AnatPhys/Sensory_Systems.html

Van Kleek, M. H. (1989). Hemispheric differences in global versus local processing of hierarchical visual stimuli by normal subjects: New data and a meta-analysis of previous studies. *Neuropsychologia, 27*(9), 1165–1178. doi:10.1016/0028-3932(89)90099-7 PMID:2812299

Van Wagenen, W. P., & Herren, R. Y. (1940). Surgical division of commissural pathways in the corpus callosum relation to spread of an epileptic attack. *Archives of Neurology and Psychiatry, 44*(4), 740–759. doi:10.1001/archneurpsyc.1940.02280100042004

Virtual Research Systems. (1998–2000). *Company Profile: Virtual Research Systems, Inc.* Retrieved September 4, 2015, from http://www.virtualresearch.com/company.html

Wade, N. J. (2000). *A Natural History of Vision.* Cambridge, MA: MIT Press.

Wade, N. J., & Tgo, T. T. (2013). Early views on binocular rivalry. In S. M. Miller (Ed.), *The constitution of visual consciousness: Lessons from binocular rivalry, Edition: Advances in Consciousness Research* (Vol. 90, pp. 77–108). Philadelphia: John Benjamins Publishing Company. doi:10.1075/aicr.90.04wad

Walker, P., & Powell, D. J. (1979). The sensitivity of binocular rivalry to changes in the nondominant stimulus. *Vision Research, 19*(3), 247–249. doi:10.1016/0042-6989(79)90169-X PMID:442549

Wallach, H. (1935). Uber visuell wahrgenommene Bewegungsrichtung. *Perception, 25*, 1319–1368.

Wandell, B. A. (n.d.). *Useful Numbers in Vision Science*. Retrieved September 4, 2015, from http://web.stanford.edu/group/vista/cgi-bin/wandell/useful-numbers-in-vision-science/

Webber, A. L., & Wood, J. (2005). Amblyopia: Prevalence, Natural History, Functional Effects and Treatment. *Clinical & Experimental Optometry, 88*(6), 365–375. doi:10.1111/j.1444-0938.2005.tb05102.x PMID:16329744

Weissmann, D. H., & Banich, M. T. (1999). Global local inference modulated by communication between the hemispheres. *Journal of Experimental Psychology. General, 128*(3), 283–308. doi:10.1037/0096-3445.128.3.283 PMID:10513397

Weissmann, D. H., & Banich, M. T. (2000). The cerebral hemispheres cooperate to perform complex but not simple tasks. *Neuropsychology, 14*(1), 41–59. doi:10.1037/0894-4105.14.1.41 PMID:10674797

Wheatstone, C. (1838). Contributions to the physiology of vision – Part the first. On some remarkable, and hitherto unobserved, phenomena of binocular vision. *Philosophical Transactions of the Royal Society of London. Series B, Biological Sciences, 128*, 371–394. Retrieved from https://www.stereoscopy.com/library/wheatstone-paper1838.html

Wolman, D. (2012, March 14). The split brain: A tale of two halves. *Nature, 483*(7389), 260–263. doi:10.1038/483260a PMID:22422242

Zimmer, C. (2009, April 15). The Big Similarities & Quirky Differences Between Our Left and Right Brains. *Discover*. Retrieved October 1, 2015, from http://discovermagazine.com/2009/may/15-big-similarities-and-quirky-differences-between-our-left-and-right-brains

ADDITIONAL READING

Alasis, D., & Blake, R. (2005). *Binocular Rivalry*. Cambridge, MA: MIT Press.

American Psychological Association. *Retrieved* September 2, 2015, from http://apa.org/

Annett, M. (2002). *Handedness and Brain Asymmetry: The Right Shift Theory*. New York: Psychology Press.

Banich, M. T. (1998). The missing link: The role of interhemispheric interaction in attentional processing. *Brain and Cognition, 36*(2), 128–157. doi:10.1006/brcg.1997.0950 PMID:9520311

Benson, D. F., & Zaidel, E. (Eds.). (1985). *The dual brain: Hemispheric specialization in humans*. New York: Guilford Press.

Blake, R., & Sekuler, R. (2006). *Perception* (5th ed.). New York: McGraw Hill.

Bogen, J. E. (1985). The dual brain: some historical and methodological aspects. In D. F. Benson & E. Zaidel (Eds.), *The Dual Brain*. New York: Guilford Press.

Breese, B. B. (1909). Binocular rivalry. *Psychological Review*, *16*(6), 410–415. doi:10.1037/h0075805

Brown, L. (1993). *The New shorter Oxford English Dictionary on Historical Principles*. Oxford, England: Clarendon.

Concepts, P. http://www.psychologyconcepts.com/ *Retrieved September 8, 2015* Psychology Concepts. Retrieved September 6, 2015, from http://www.psychologyconcepts.com/depth-perception/

Consmelli, D., & Thompson, E. (2007). Mountains and valleys: Binocular rivalry and the flow of experience. Subjectivity and the Body. *Consciousness and Cognition*, *16*(3), 623–641. doi:10.1016/j.concog.2007.06.013 PMID:17804257

Crick, F. (1996). Visual perception: Rivalry and consciousness. *Nature*, *379*(6565), 485–486. doi:10.1038/379485a0 PMID:8596623

Department of Neurobiology and Anatomy. Neuroscience Online. Retrieved October 30, *2015*, from http://neuroscience.uth.tmc.edu/index.htm

Deutsch, D. (1974, September). An auditory illusion. *Nature*, *25*(5473), 307–309. doi:10.1038/251307a0 PMID:4427654

DIY Hacks and How Tos (n.d.). *3D Stereoscopic Photography*. Retrieved September 28, 2015, from http://www.instructables.com/id/3D-Stereoscopic-Photography/step3/How-to-View-Cross-eyed-3D-Images/

Doherty, P. (1995). *The Cheshire Cat and Other Interactive Experiments in Perception*. New York: Wiley.

Duane's Clinical Ophthalmology. (2004, October 25). New York: Lippincott Williams & Wilkins.

Eye, M. (n.d.) Retrieved September 20, 2015 from http://www.magiceye.com/

Gardner, H. (1978). What We Know (and Don't Know) about the Two Halves of the Brain. *Harvard Magazine*, *80*, 24–27.

Gazzaniga, M. S. (2005). Forty-five years of split-brain research and still going strong. *Nature Reviews. Neuroscience*, *6*(8), 653–659. doi:10.1038/nrn1723 PMID:16062172

Greenstein, B., & Greenstein, A. (2000). *Color Atlas of Neuroscience: Neuroanatomy and Neurophysiology*. New York: Thieme.

Gregory, R. L. (1974). *Concepts and Mechanisms of Perception*. London: Duckworth.

Gregory, R. L. (1992). Evolution of the Eye and Visual System. J. R. Cronly-Dillon & R. L. Gregory (Eds.). Vol. 2 of Vision and Visual Dysfuction. London: Macmillan.

Gregory, R. L. (1995). *The Artful Eye* (R. L. Gregory, J. Harris, P. Heard, & D. Rose, Eds.). Oxford: Oxford University Press.

Gregory, R. L. (1998). *Eye and Brain: The Psychology of Seeing* (5th ed.). Princeton: Princeton University Press.

Gregory, R. L. (1978). Illusions and Hallucinations. E. C. Carterette, & M. P. Freidman (eds.) Handbook of Perception 9 (Chapter 9).

Gregory, R. L., & Gombrich, E. (Eds.). (1973). *Illusion in Nature and Art*. London: Duckworth.

Helige, J. B. (2001). *Hemispheric Asymmetry: What's Right and What's Left*. Cambridge, MA: Harvard University Press.

Hugdahl, K., & Davidson, R. J. (Eds.). (2003). *The Asymmetrical Brain*. Cambridge: MIT Press.

Human, B. F. T. Online Psychology Laboratory. Retrieved September 10, 2015, from http://opl.apa.org/contributions/EC/BrainFly.htm

Ivry, R. B., & Robertson, L. C. (1998). *The Two Sides of Perception*. Cambridge: MIT Press.

Joseph, R. (1990). *Neuropsychology, Neurospyschiatry, and Behavioral Neurology*. New York: Plenum Press. doi:10.1007/978-1-4757-5969-3

Kellogg, R. T. (2016). *Fundamentals of Cognitive Psychology*. Thousand Oaks, CA: Sage Publishing.

Kosslyn, S. M. (1987). Seeing and imagining in the cerebral hemispheres: A computational approach. *Psychological Review*, *94*(2), 148–175. doi:10.1037/0033-295X.94.2.148 PMID:3575583

Levy, J., Trevarthen, C., & Sperry, R. A. (1972). Perception of bilateral chimeric figures following hemispheric deconnexion. *Brain*, *95*(1), 61–78. doi:10.1093/brain/95.1.61 PMID:5023091

Lipton, L. (1982). *Foundations of the Stereo-Scopic Cinema, A Study in Depth. NewYork*. Van Nostrand Reinhold.

MarekP. (Ed.). (n.d.) Online Psychology Laboratory. Retrieved September 8, 2015, from http://opl.apa.org/Main.aspx

Miller, M. B., Sinnot-Armstrong, W., Young, L., King, D., Paggi, A., Fabri, M., & Gazzaniga, M. S. et al. (2010). Abnormal Moral Reasoning in complete and partial callosotomy patients. *Neuropsychologia*, *48*(7), 2215–2220. doi:10.1016/j.neuropsychologia.2010.02.021 PMID:20188113

Monaghan, P., & Pollmann, S. (2003). Division of labor between the hemispheres for complex but not simple tasks: An implemented connectionist model. *Journal of Experimental Psychology. General*, *132*(3), 379–399. doi:10.1037/0096-3445.132.3.379 PMID:13678374

Motz, B. A., James, K. H., & Busey, T. A. (2012). The Lateralizer: A tool for students to explore the divided brain. *Advances in Physiology Education*, *36*(3), 220–225. doi:10.1152/advan.00060.2012 PMID:22952261

Ornstein, R. (1977). *The Psychology of Consciousness*. New York: Harcourt.

Sensation and Perception. Online Psychology Laboratory. *Retrieved* September 8, *2015*, from http://opl.apa.org/Resources.aspx#Sensation

Sherwood, L. (2004). *Human Physiology: From Cells to Systems* (7th ed., pp. 197–208). Belmont, CA: Brooks/Cole Cengage Learning.

The Society for the Teaching of Psychology. Division 2 of the American Psychological Association. *Retrieved* September 10, *2015*, from http://teachpsych.org/page-1588384

Unknown, "The Holmes stereoscope, with the inventions and improvements added by Joseph L. Bates, " *Center for the History of Medicine: OnView*, Retrieved November 29, 2015, http://collections.countway.harvard.edu/onview/items/show/6277

Valyus, N. A. (1962). *Stereoscopy*. New York: The Focal Press.

Wade, N. J. (1996). Descriptions of visual phenomena from Aristotle to Wheatstone. *Perception*, *25*(10), 1137–1175. doi:10.1068/p251137 PMID:9027920

Wagemans, J. (Ed.). (2015). *The Oxford Handbook of Perceptual Organization*. Oxford, UK: Oxford University Press. doi:10.1093/oxfordhb/9780199686858.001.0001

Wandell, B. A. (1995). *Foundations of Vision*. Sunderland, MA: Sinauer Associates.

Whitaker, H. A. (2010). *Concise Encyclopedia of Brain and Language*. Oxford, UK: Elsevier Ltd.

Wikipedia, (n.d.). The Auditory System. Retrieved September 5, 2015, from https://en.wikipedia.org/wiki/Auditory_system

Yeh, Y. Y., & Silverstein, L. D. (1990). Limits of Fusion and Depth Judgment in Stereoscopic Color Displays. *Human Factors*, *32*, 45–60.

Zaidel, E., & Iacoboni, M. (Eds.). (2003). *The Parallel Brain: The Cognitive Neuroscience of the Corpus Callosum*. Cambridge: MIT Press.

KEY TERMS AND DEFINITIONS

Binocular Vision: Is in which creatures having two use them together. The word binocular comes from two roots, *bini* for double, and *oculus* for eye. https://en.wikipedia.org/wiki/Binocular_vision.

Cerebral Cortex: Is the cerebrum's (brain) outer layer of neural tissue in humans and other mammals. It is divided into two cortices, along the sagittal plane: the left and right cerebral hemispheres divided by the medial longitudinal fissure. The cerebral cortex plays a key role in memory, attention, perception, awareness, thought, language, and consciousness. The human cerebral cortex is 2 to 4 millimetres (0.079 to 0.157 in) thick. https://en.wikipedia.org/wiki/Cerebral_cortex#Areas.

Cerebral Hemispheres: The vertebrate cerebrum (brain) is formed by two cerebral hemispheres that are separated by a groove, the medial longitudinal fissure. The brain can thus be described as being divided into left and right cerebral hemispheres. Each of these hemispheres has an outer layer of grey matter, the cerebral cortex, that is supported by an inner layer of white matter. In eutherian (placental see:

https://en.wikipedia.org/wiki/Eutheria) mammals, the hemispheres are linked by the corpus callosum, a very large bundle of nerve fibers. Smaller commissures, including the anterior commissure, the posterior commissure and the hippocampal commissure also join the hemispheres and these are also present in other vertebrates. These commissures transfer information between the two hemispheres to coordinate localized functions. https://en.wikipedia.org/wiki/Cerebral_hemisphere.

Commissures or Commissural Fibers: The fibers or transverse fibers are coherent white-matter structures that connect the two hemispheres of the brain. https://en.wikipedia.org/wiki/Commissural_fiber.

Commissurotomy: In, as a treatment for severe epilepsy, the corpus callosum, or the area of the brain that connects the two hemispheres, would be completely bisected. By eliminating the connection between the two hemispheres of a patient's brain, electrical communication would be cut off greatly diminishing the amount and severity of the epileptic seizures. For some, seizures would be completely eliminated. https://en.wikipedia.org/wiki/Commissurotomy.

Corpus Callosotomy: Corpus callosotomy is a palliative surgical procedure for the treatment of seizures. As the corpus callosum is critical to the interhemispheric spread of epileptic activity, the procedure seeks to eliminate this pathway. The corpus callosum is usually severed in order to stop epileptic seizures. Once the corpus callosum is cut, the brain has much more difficulty sending messages between the hemispheres. Although the corpus callosum is the largest white matter tract connecting the hemispheres, some limited interhemispheric communication is still possible via the anterior commissure and posterior commissure. When tested in particular situations, it is obvious that information transfer between the hemispheres is reduced.

Corpus Callosum: (/ˈkɔrpəs kəˈloʊsəm/; Latin for "tough body"), also known as the callosal commissure, is a wide, flat bundle of neural fibers about 10 cm long beneath the cortex in the eutherian brain at the longitudinal fissure. It connects the left and right cerebral hemispheres and facilitates interhemispheric communication. It is the largest white matter structure in the brain, consisting of 200–250 million contralateral axonal projections. https://en.wikipedia.org/wiki/Corpus_callosum.

Cortical: 1. of, relating to, or consisting of cortex. 2. involving or resulting from the action or condition of the cerebral cortex. http://www.merriam-webster.com/dictionary/subcortical.

Cyclopean Image: Is a single mental image of a scene created by the brain by combining two images received from the two eyes. Cyclopean image is named after the mythical Cyclops with a single eye. Literally it refers to the way stereo sighted viewers perceive the centre of their fused visual field as lying between the two physical eyes, as if seen by a cyclopean eye. Alternative terms for cyclopean eye include third central imaginary eye and binoculus. https://en.wikipedia.org/wiki/Cyclopean_image.

Dichoptic: (From the Greek words δίχα *dicha*, meaning "in two," and ὀπτικός *optikos*, "relating to sight") is viewing a separate and independent field by each eye. In dichoptic presentation, stimulus A is presented to the left eye and then stimulus B is presented to the right eye. https://en.wikipedia.org/wiki/Dichoptic_presentation.

Electrooculography: Is a technique for measuring the corneo-retinal standing potential that exists between the front and the back of the human eye. The resulting signal is called the electrooculogram. Primary applications are in and in recording.

Monocular Vision: Is in which both are used separately. By using the eyes in this way, as opposed by, the is increased, while is limited. The eyes of an animal with monocular vision are usually positioned on opposite sides of the animal's head, giving it the ability to see two objects at once. The word monocular comes from the root, *mono* for one, and the root, *oculus* for eye. https://en.wikipedia.org/wiki/Monocular_vision.

Split-Brain: Split-brain is a lay term to describe the result when the corpus callosum connecting the two hemispheres of the brain is severed to some degree. It is an association of symptoms produced by disruption of or interference with the connection between the hemispheres of the brain. The surgical operation to produce this condition results from transection of the corpus callosum, and is usually a last resort to treat refractory epilepsy. https://en.wikipedia.org/wiki/Split-brain.

Stereopsis: (From the Greek στερεο- *stereo-* meaning "solid", and ὄψις *opsis*, "appearance, sight") is a term that is most often used to refer to the perception of depth and 3-dimensional structure obtained on the basis of visual information deriving from two eyes by individuals with normally developed binocular vision. https://en.wikipedia.org/wiki/Stereopsis.

Subcortical: Of, relating to, involving, or being a part of the brain below the cerebral cortex. http://www.merriam-webster.com/dictionary/subcortical.

This research was previously published in Knowledge Visualization and Visual Literacy in Science Education edited by Anna Ursyn, pages 28-76, copyright year 2016 by Information Science Reference (an imprint of IGI Global).

APPENDIX

Online Demonstrations

10 cool optical illusions. Retrieved September 17, 2015 from http://psychology.about.com/od/sensation-andperception/tp/cool-optical-illusions.htm

The blind spot. Retrieved September 17, 2015 from http://faculty.washington.edu/chudler/chvision.html

Blind Spot. Retrieved September 17, 2015 from http://www.exploratorium.edu/snacks/blind_spot/index.html

Cheshire Cat Experiment. Retrieved September 17, 2015 from http://www.exploratorium.edu/snacks/cheshire_cat/

Contrast Influences Predominance. Retrieved September 17, 2015 from http://www.psy.vanderbilt.edu/faculty/blake/rivalry/BR.html

Experiments. Retrieved September 17, 2015 from http://www.richardgregory.org/experiments/

Eye Exchange. Retrieved September 17, 2015 from http://www.psy.vanderbilt.edu/faculty/blake/rivalry/BR.html

(The) Framing Game. Retrieved December 26, 2015 from http://www.vision3d.com/frame.html

Fusion Rivalry. Retrieved September 17, 2015 from http://www.psy.vanderbilt.edu/faculty/blake/rivalry/BR.html

Magic Eye Random Dot Stereograms. Retrieved September 17, 2015 from http://www.magiceye.com/

Optical Illusion Pictures. Retrieved September 17, 2015 from http://brainden.com/optical-illusions.htm

Optical Illusions & Visual Phenomena 123 of them. Retrieved September 17, 2015 from http://michaelbach.de/ot/

Probe Animation. Retrieved September 17, 2015 from http://www.psy.vanderbilt.edu/faculty/blake/rivalry/BR.html

Spreading Waves of Dominance. Retrieved September 17, 2015 from http://www.psy.vanderbilt.edu/faculty/blake/rivalry/BR.html

Stereo Pictures for Cross-Eyed Viewing. Retrieved September 17, 2015 from http://www.lhup.edu/~dsimanek/3d/stereo/3dgallery.htm

Unitary vs. Piecemeal Rivalry. Retrieved September 17, 2015 from http://www.psy.vanderbilt.edu/faculty/blake/rivalry/BR.html

Visual Illusions. Retrieved November 2, 2015, from http://faculty.washington.edu/chudler/chvision.html

Other Online Resources

Binocular rivalry bibliography. Retrieved October 2, 2015, from https://sites.google.com/site/oshearobertp/publications/binocular-rivalry-bibliography

Fire in the Borgo; Is Athens Buring? Retrieved September 10, 2015, from http://www.wikiart.org/en/salvador-dali/athens-is-burning-the-school-of-athens-and-the-fire-in-the-borgo-1980

Nobuyuki Kayahara's Spinning Dancer Illusion. Retrieved September 19, 2015, from http://www.procreo.jp/labo/silhouette.swf

The S2 Eye Tracker, Eyeworks software. Retrieved September 25, 2015, from (http://www.mirametrix.com/products/?gclid=CN3Jl8Xz5sgCFZWRHwodaBEBdA).

EYE-TRAC Head Mounted Display, the Mobile Eye from Applied Science Laboratories. Retrieved September 26, 2015, from (http://www.asleyetracking.com/Site/).

Severed Corpus Callosum (2008, June 25), Scientific American Frontiers. Retrieved September 17, 2015, from https://www.youtube.com/watch?v=82tlVcq6E7A

The Young Lady Versus Old Lady Optical Illusion. Retrieved September 22, 2015, from http://brain-den.com/face-illusions.htm

Chapter 11
Development and Evaluation of Neuroscience Computer-Based Modules for Medical Students:
Instructional Design Principles and Effectiveness

Kathryn L. Lovell
Michigan State University, USA

ABSTRACT

Interactive neuropathology computer-based teaching modules and other neuroscience computer-based resources were developed to provide individualized self-paced content information accompanied by images and self-assessment questions with feedback, along with problem-solving cases to facilitate application of neuroanatomy, neurology, and neuropathology concepts to patient cases. Initial implementation occurred in three curricula for second-year medical students. Evaluation of the modules was conducted using quantitative and qualitative methods to determine features of the modules that were important for students. This chapter will describe the instructional design principles that evaluation results identified as important and effective for student learning, and compare those to current principles for effective multimedia instructional design identified in a variety of research. Especially important principles applied in the neuroscience modules included cognitive load theory, retrieval practice and self-assessment, feedback, and learner control.

INTRODUCTION

Interactive computer-based teaching modules covering neuropathology content for second-year medical students were first developed by the author when interactive computer programs (e.g. Hypercard) and videodiscs with randomly accessible images became available (images from the second edition of Slice of Life videodisc, coordinated by Suzanne Stensaas, Ph.D., were utilized; most of the neuropathology images were provided by Margaret Z. Jones, M.D.). These factors were ideal for pathology content where

DOI: 10.4018/978-1-5225-5478-3.ch011

gross and microscopic images are essential to the learning process. The goal of the modules was to provide individualized self-paced content information accompanied by images and self-testing questions with feedback, along with problem-solving cases to facilitate application of neuroanatomy, neurology, and neuropathology concepts to patient cases.

The development process included consideration of instructional design principles, production of initial topics (neoplastic disorders and cerebrovascular disorders), use-testing to determine student perceptions of technical and instructional effectiveness, and revisions based on student input. All images and diagrams were open source (Creative Commons or similar licensing). The interactive modules have been used continuously in the Colleges of Human Medicine and Osteopathic Medicine at Michigan State University since 1988, and repeated evaluations have demonstrated student support for the effectiveness of the instructional design principles. Periodic changes have been made based on changes in software and hardware, and access to additional images, and additional neuroscience units have been created for specific teaching goals.

The objectives of this chapter are to describe the development and structure of the neuropathology teaching modules and other units, including the instructional design principles utilized; to summarize results of evaluation of the modules by students; and to compare features of the modules with instructional design principles based on theories of learning and experimental evidence from other research.

BACKGROUND

Initial Structure and Implementation of Neuropathology Modules

Two neuropathology topics (neoplastic disorders and cerebrovascular disorders) were selected for initial module development and use-testing by students before additional topic modules were generated. The initial computer-based modules consisted of:

- Pre-test.
- Lesson utilizing text, diagrams and images (both normal and abnormal) with descriptions.
- Post-test (with immediate feedback for correct and incorrect answers).
- Clinical simulation emphasizing reinforcement of concepts in the lesson.
- Glossary.

A randomly accessible table of contents was available to permit learner control over lesson sequencing. The content of the units were based on paper modules with images provided in 35mm slide carousels developed previously by Dr. Margaret Z. Jones. Images in the modules were from the second edition of Slice of Life videodisc, coordinated by Suzanne Stensaas, Ph.D., University of Utah); most of the neuropathology images used were provided by Margaret Z. Jones, M.D., Michigan State University).

Hodgins & Lovell (1988) described the implementation of the initial two units in three curricular formats in the Colleges of Human Medicine and Osteopathic Medicine at Michigan State University (systems curriculum, discipline-based courses, problem-based independent study curriculum). The modules were designed to be usable by students in any order at any time, and could be used individually or in groups, and were required or optional for different sets of students. A log-in system and data tracking system recorded data during student use of the modules to determine the time spent in each component of each module.

The first formal evaluation of the two initial modules was described by Lovell et al. (1991). It should be noted that some of the evaluation related to using the new computer technology that was just beginning to be available in medical education. Students were given an evaluation form when the videodisc was checked out in the computer laboratory, and most students (over 80 for each module) completed the form. The evaluation form had three parts: Likert scale rating of effectiveness of different parts of the units; student preference for specific features of the units; and written comments. In addition, focus groups were organized to get feedback from students about hardware, utility of the modules to help students achieve levels of mastery, appropriateness of module content, organizational factors, motivational factors, and self-testing components within the module. Results indicated that students rated all aspects of the modules as very valuable in helping master the material, including content, images, diagrams, and review questions.

In addition, many of the open-ended comments emphasized the value and effectiveness of the interactive videodisc instructional modules in helping them master course content. About half of the students used the modules on a voluntary basis, even though the content was also covered in lectures and written course materials. Focus group session comment summaries included the following points: the modules were best suited for acquiring basic concepts and principles and practicing with case studies; the glossary and visual ability to compare normal and abnormal images were helpful; the organization was excellent and fit into the overall course; the reference points in each lesson were helpful. The self-evaluation components (post-test questions) were especially effective for students, and they suggested increasing the number of test questions.

After getting preliminary feedback on the two initial modules, additional modules were developed to encompass all of the neuropathology topics taught for second-year medical students, and further evaluation was conducted on effectiveness of specific features, to be discussed further in a different section of this chapter.

Cognitive Load Theory

Cognitive load theory describes learning effectiveness in terms of the capacity of working memory, which is limited. Working memory is divided into intrinsic, germane and extraneous loads and has a total that cannot be exceeded during the process of learning new material (Lau, 2014, Paas et al., 2010, Mayer, 2010, Mayer and Moreno 2010, Sweller 1994, and van Merrienboer and Sweller, 2010). Intrinsic load refers to the inherent difficulty of the material and typically cannot be altered by instructional interventions without altering the content to be learned (e.g. simplification). Extraneous load is imposed by instructional procedures, i.e. the attention paid towards the presentation of the material as compared to the intrinsic difficulty of the material itself. Germane load refers to the understanding of information through which to integrate the new knowledge into one's permanent memory, i.e. meaningful learning. Part of germane load involves connecting new information elements to existing frameworks in long-term memory. In cognitive load theory, intrinsic and extraneous cognitive loads are additive; if working memory is exceeded in a learning task, minimizing extraneous load can increase working memory capacity for germane load. Thus, reduction of extraneous cognitive load is a major goal of designing multimedia instructional materials in order to increase learning effectiveness.

Retrieval Practice (Testing Effect) and Self-Assessment

Karpicke and Blunt (2011) and Karpicke and Roediger (2008) demonstrated that a combination of study and tests is more effective than spending the same amount of time reviewing the material in some other way, such as rereading it. Rohrer and Pashler (2010) pointed out that surveys of college students show that most of them study almost entirely by rereading, with self-testing relatively rarely employed. In the author's experience this is also true for medical students, either because self-testing materials are not available or there is lack of awareness of the value of this approach. For many medical school pre-clinical courses, there is great variability in the availability of self-testing questions that are consistent with the objectives of the material to be tested over in a high-stakes course examination. Cognitive psychology research has demonstrated that repeated testing of information in any format produces superior retention compared to repeated study, a finding that is found across a wide variety of materials and experimental conditions (Larsen et al., 2008, Karpicke and Roediger, 2008, Karpicke and Blunt, 2011, Schmidmeier et al., 2011, Wood, 2009, and Rohrer and Pashler, 2010).

There are different types of tests that can be used to practice retrieval, including vocabulary tests, true-false statements, multiple-choice questions, short-answer questions, and essay or free recall questions. All are effective, but tests that require effortful retrieval of information, such as short-answer tests or free recall after reading a passage, or concept mapping, promote better retention and meaningful learning than tests based on recognition only such as multiple-choice tests. These results support the theory that retrieval practice enhances learning by retrieval-specific mechanisms that are more effective than episodes of repeated study.

According to Karpicke and Blunt (2011):

Not only does retrieval produce learning, but a retrieval event may actually represent a more powerful learning activity than an encoding event. This research suggests a conceptualization of mind and learning that is different from one in which encoding places knowledge in memory and retrieval simply accesses that stored knowledge. Because each act of retrieval changes memory, the act of reconstructing knowledge must be considered essential to the process of learning.

In the studies on retrieval practice, students also demonstrated inability to judge their own learning proficiency when they were asked to predict performance on an upcoming test. Karpicke and Blunt (2011) examined students metacognitive knowledge of the effectiveness of different types of learning activities (i.e. students predicted the percentage of information they would remember in 1 week). Students predicted that repeated studying would produce the best long-term retention on both factual and inference questions, and that practicing retrieval would produce the worst retention, even though the opposite was true. Thus students are not able to assess their level of competence or effectiveness of study strategies prior to an examination. This is also true for many medical students in the author's experience. An opportunity and encouragement for testing (or self-testing or self-assessment) is critical for students to assess their own level of knowledge and to most effectively learn and retain material.

Open Access and Creative Commons

One of the challenges in development of multimedia materials that will be open for all (open access) to use (through CD-ROM distribution or freely available on the Internet) is finding open access images or diagrams that illustrate the content, especially in the fields based on visual identification, such as anatomy, pathology and radiology. Carroll (2013) states that openness requires making the literature freely accessible under liberal terms that permit nearly all reuses so long as the author receives credit for the work when it's republished or adapted. Copyright owners seeking to grant permission to everyone have issued public licenses broadening the range of permitted uses, subject to certain conditions, e.g. attribution, use only for educational purposes, use only for non-profit endeavors. Creative Commons licenses (http://creativecommons.org/) are the most widely used for most educational materials, according to Carroll (2013). The Wikipedia community, for example, has adopted the Creative Commons Attribution ShareAlike license.

From 1985 to 2007, the Slice of Life Project (http://slice.utah.edu/sol/), based at the University of Utah, served as a nonprofit, cooperative venture creating and sharing educational materials using computers, multimedia, and new media in health care, health sciences, and medical education. Many people were instrumental in the Slice of Life Project, notably Dr. Suzanne S. Stensaas, who conceived of the videodisc and marshaled the cooperation and contributions of her colleagues, educators and research professionals in the health sciences and medical education. All of the images on the videodiscs could be used and adapted for non-profit educational purposes under a Creative Commons license.

The Health Education Assets Library (HEAL; http://library.med.utah.edu/heal/) was formed as a multi-institutional repository of multimedia learning objects for health sciences education. HEAL's vision was to create a premier national digital library for all levels of health sciences education by providing free and easy access to a large number of high-quality educational materials. This collection, which is no longer funded, still provides online access to many types of open source images, diagrams and videos.

More recently other initiatives have been developed with similar goals, at a variety of levels. For example, Open.Michigan (http://open.umich.edu/index.html) is a University of Michigan initiative that enables faculty, students, and others to share their educational resources and research with the global learning community. The Open Education Consortium (http://www.oeconsortium.org/) is a global network for open education. MedEdPORTAL (www.mededportal.org/), provided by the Association of American Medical Colleges, makes available stand-alone modules following a rigorous peer review process.

MAIN FOCUS OF THE CHAPTER

Evaluation Results for Neuropathology Modules

A series of evaluations by Lovell et al. (1991), Parkhurst et al. (1991), Lovell et al. (1993) and Lovell and Hodgins (2001) were conducted involving the neuropathology modules using videodisc images, and the CD-ROM version that included additional neuroscience practice applications and case studies. Results indicated that students found the units to be very effective in enhancing learning. The greatest assets of the units were in their visual impact, their ability to engage the learner as a participant rather than a passive listener (due to the interactive design), and self-assessment options. Several students noted that the units helped lengthen their attention span, and fostered problem solving rather than simple memorizing. When asked about specific features, students consistently mentioned the post-tests were especially useful, and the feedback explaining incorrect responses as well as reinforcing the rationale for the correct response, were relevant and extremely helpful.

Features of Neuropathology Modules that students considered very important included:

- High quality visual images that helped reinforce concepts.
- Flexibility in controlling location (easy navigation).
- Simultaneous display of both text and complete image or diagram.
- Indication of place in program (e.g. how many screens in remainder of section).
- Self-testing opportunities with explanations for right and wrong answers.
- Compatibility between the content and review questions.
- Flexibility in time allocation for completing the modules.
- Case studies to facilitate application of concepts to patient cases and attain higher level course objectives.

One objective in creating the modules was to accommodate different learner styles or characteristics. Students reported different patterns of use of the learner control options: pacing control, sequence control, content control, and whether to view the materials in groups or individually. The modules satisfied a variety of objectives including: as an advance organizer to acquire basic concepts; as a supplement to the lecture; for review; and for some students as a replacement for lecture.

Development of Neuroscience Resources for Specific Teaching Goals

After changes in the curricular structure for the Colleges of Human Medicine and Osteopathic Medicine, and based on changes in technology involving personal computers, further development occurred to build on the neuropathology teaching units, and to help students master areas that have been traditionally difficult and in which they have been observed to perform lower than expected on performance-based evaluations (Mavis et al., 1998).

Conversion of Neuropathology Modules to CD-ROM Format and Addition of Neuroanatomy Review Unit

The neuropathology modules were converted to a CD-ROM format, and a basic clinical neuroanatomy unit was added. Other units were developed to enhance problem-solving and integration of clinical and pathological concepts. Technical support and instructional design input was provided by Mark Hodgins. The CD-ROM was produced in the college facilities and distributed free of charge to students. Use was optional, but students were informed that the CD-ROM content covered the same material as the written neuropathology exams for the course, and that additional material for neuroanatomy review and problem-solving practice related to localization of lesions was available.

As part of the evaluation process after the CD-ROM was provided to students, Lovell and Hodgins (2001) showed that most students providing input (79% response rate for voluntary survey) used the CD-ROM. Response to the statement "The CD was effective in enhancing my understanding of the neuropathology objectives" was: Agree 13%, Strongly agree 85%. Response to the statement "The CD was effective in my review of neuroanatomy and localization of lesions" was: Agree 21%, Strongly agree 76%. Representative comments included the following: "The CD-ROM made studying less painful and learning easier;" "Questions had explanations and facilitated my learning greatly;" "I enjoyed having the pictures along with the readings. I also liked the fact that on the quiz questions answers were given as to why something was right or wrong;" "It helped my time management and assessment of learning." Since localization of lesions covered in neuroanatomy content is critical to diagnosis of a neurological disorder based on clinical presentation for a patient, the combination of units enhanced student understanding of the combined content.

Integration of Neuropathology and Patient Case Presentations

Another set of practice questions covering all neuropathology content with neuroanatomy reinforcement was developed after students performed more poorly than expected on an exam that included basic questions related to neuropathology images. This unit was designed to integrate a clinical scenario with the expected gross or microscopic pathology (clinic-pathological correlation), using 26 cases (Case A to Case Z). It was designed with short-answer questions with explanations rather than multiple-choice questions for several reasons: students would need to exert more effortful recall over a wider range of content, multiple concepts could be addressed more efficiently, and multiple-choice questions can be cumbersome to write.

An example of one scenario is "A 56-year-old man had been in good health. During an evening at home with his wife, he reported the sudden onset of a severe headache, and then lost consciousness. The patient died after several hours in the hospital. What is the diagnosis indicated by the history and brain section shown?" The image shows a coronal brain section with blood in the subarachnoid space and the lateral and third ventricles. Students can click a button "Show answer and explanation." The answer and explanation provided when a student clicks the button is: "The coronal section shows blood in the subarachnoid space, between the arachnoid and pia. Thus the diagnosis is subarachnoid hemorrhage. In this patient, a lumbar puncture would show bloody CSF. The most common cause of non-traumatic subarachnoid hemorrhage is ruptured saccular (berry) aneurysm. The history is also typical for ruptured saccular aneurysm, including sudden onset of headache ("worst headache of my life") and loss of consciousness ("Case History", n.d.).

This set of questions, designed to help students integrate pathology with features of a patient case, also included questions from multiple types of neuropathology disorders. The teaching modules were organized around topics (e.g. cerebrovascular disorders, neoplastic disorders, etc) and the post-test for each module included questions only on that topic. In contrast, the Neuropathology Case Examples (Cases A-Z) required students to think across topics, the process of interleaving. Rohrer & Pashler (2010) discuss interleaving as a process of combining testing or practice of different skills in a way that improves memory as compared to testing or practice blocked by type of skill. E.g. when baseball players practiced hitting three types of pitches that were either blocked or interleaved, interleaving improved performance on a subsequent test in which the batters did not know the type of pitch in advance. Blocked practice is important for initial learning, but provides students with a crutch that is unavailable during a cumulative exam, and in the real-world situations for which they are being trained. It is not surprising if they often struggle when asked to demonstrate a skill they have not previously practiced. More sets of practice problems utilizing interleaving across a number of topics can help students master the integration needed for comprehensive problem-solving related to patient cases.

Guided Problem-Solving Cases for Localization of Lesions

A set of 15 neuroanatomy problem-solving cases was developed to help students work through localization of lesions. These started with a patient scenario, e.g. "A 60-year-old man collapsed while at work. After he regained consciousness in the emergency room, neurological exam was performed with the following abnormalities noted: paralysis of the right arm and leg, dysarthria, deviation of the tongue to the left when protruded, loss of vibration, proprioception and discriminative touch sensation on the right side of the body (the face has normal sensation)." Relevant open-ended questions are asked and the student receives the answer and explanation when a button is clicked. These cases give students the opportunity to determine their answer before seeing the correct answer, and help with meaningful learning as well as self-assessment of their level of understanding ("Case History, n.d.).

Elements of Neuropathology Self-Instructional Modules and Neuroscience Units and Relation to Principles of Learning

The sections below focus on several features of the neuroscience self-instructional modules and discuss how they relate to instructional design principles to promote meaningful learning. These features can be viewed at http://learn.chm.msu.edu/neuropath/.

Self-Assessment and Feedback

As discussed in the Background section, retrieval practice markedly enhances learning and subsequent performance on a test. The Neuropathology Modules were designed with multiple-choice practice questions to facilitate student understanding of the content, and to facilitate their self-assessment of their ability to understand and apply the content. Some of the questions were at the recall level, and some required integration and application of concepts to a patient scenario. Some were intended to cover only chunks of material and some covered larger sets of material (interleaving). All were designed to enhance understanding through retrieval practice. Students consistently rated the practice questions in these modules as one of the most effective features and requested additional practice questions. Schmidmaier et al. (2011) found that in medical students learning clinical nephrology, repetitive testing promoted better recall than repetitive studying after 1 week; however after 6 months general recall was poor and no difference between the restudy and retest groups was observed.

Sets of cases for additional problem-solving were designed with open-ended questions, expecting students to retrieve an answer before clicking to get the answer and explanation. Rohrer and Pashler (2010) discuss the findings that tests that require effortful retrieval of information, such as short-answer questions, promote better retention than tests based on recognition only such as multiple-choice tests. Larsen et al. (2008) emphasize that production tests (e.g. short-answer, fill-in-the-blank) which require the test-taker to construct a response lead to better retention than recognition tests, presumably because production tests require more effortful retrieval of information from memory. The instructional design principles are consistent with students high rating of these in promoting understanding of the material.

The self-testing (multiple-choice questions and open-ended questions) in the modules all included feedback and explanations. Larsen et al. (2008) discuss the importance of feedback. Although testing improves retention in the absence of feedback, providing feedback enhances the benefits of testing by correcting errors and confirming correct responses. Also, the content of the feedback message is important, and should include the correct answer as well as other important information. The feedback for the modules was constructed to give the correct answer as well as give key information that students should focus on in answering similar questions.

The practice questions used as self-assessment also provide an opportunity for students to judge their own mastery of the content. Schumacher et al. (2013) states that most learners possess a limited ability to identify their knowledge gaps, and this was demonstrated in medical students by Schmidmaier et al. (2011). The limited ability of learners to assess and fill knowledge gaps plays out in study habits as well. Re-reading material is a common learning strategy but familiarity with information is often mistaken with knowing and understanding; thus individuals tend not to study what they perceive they already know, even if that perception is incorrect. Studies by Karpicke and Blunt (2011) also indicate that students are not able to accurately predict their performance on an exam. Thus it is important to provide students multiple opportunities to assess their own knowledge, both with respect to basic facts and to higher-level application of concepts.

Design of Content Presentation to Minimize Extraneous Processing in Cognitive Load

The Neuropathology Modules were designed to be straightforward concise descriptions of the key concepts in neuroscience and neuropathology content important for second-year medical students. Mayer (2008) discussed findings that students performed better on a problem-solving transfer test after receiving a concise lesson rather than an expanded lesson. According to the cognitive theory of multimedia learning, inserting extraneous material may cause learners to engage in extraneous processing – by using their processing capacity to attend to and process material that is not essential to building a mental model of the to-be-learned system. Learners given an expanded lesson may have less cognitive capacity for processing the essential material and therefore may be less likely to build a learning outcome that can be used to generate useful answers on a transfer test.

Efforts were made to write learning objectives and put key concepts in bold font to draw attention to these points. Also key features of images were represented in a caption of the image, with arrows, to emphasize the important terminology and put it in perspective. Mayer (2008) discussed the instructional design concept of signaling - the process of highlighting the essential material in the lesson. People learn better from a multimedia lesson when essential words are highlighted. Signaling can help guide the learner's attention toward the essential material.

When images were added to screen pages, the images were placed next to the text describing the concept displayed in the image, with the caption also visible on the same page. Mayer (2008) discussed the principle of spatial contiguity, i.e. that people learn better when corresponding words and pictures are presented near rather than far from each other on the page or screen.

Recommendations to manage cognitive load by Van Merrienboer and Sweller (2010) included replacing multiple sources of information, distributed either in space or time, with one integrated source of information. The Neuropathology Modules contained all of the neuropathology content taught for second-year medical students in our curriculum, and thus provided an integrated source of information to manage cognitive load and enhance learning.

Learner Control

The navigation of Neuropathology Modules in all forms (videodisc units, CD-ROM and Internet) was designed to be as flexible and straightforward as possible, and provide learner control over all elements, including the questions (which could be skipped if desired). Users could complete one or more units at a sitting, or could complete part of a unit, and the units were self-paced. This not only facilitated student variation in type of use, but also facilitated implementation of the module in various types of curricula, and for use as both initial learning modules, and for review in clerkships or residency. Learner control and flexible navigation were highly rated by students and fulfilled several goals of instructional design.

Clark and Mayer (2011) discuss learner control and navigation from several perspectives. For example, they recommend giving experienced learners more control. There are more significant differences between effectiveness of courses with learner or program control where learners have lower levels of prior knowledge and some evidence to suggest that some agency at higher levels has a positive effect. Plus there are the benefits in terms of reducing dropout and increasing engagement through agency. Pacing control generally means that learners spend more time on tasks but that more is learned and transferred better to real life situations. It is important that users can easily keep track of where they are and find other elements of the program. Clark and Mayer (2011) indicate that clear navigation leads to better retention of information. In the neuropathology and neuroscience modules, the clear flexible navigation was commended by students and the learner control of pacing in the modules enhanced engagement through agency and made it possible for students to effectively utilize the material at any level – initial learning, review within the course, review for board exams, review in clerkships or residency.

Learner control also can accommodate different learning approaches and styles of individual students. Friedlander et al. (2011) discuss individual learning and point out that individuals have various types of intelligence and show differences in the types of learning that they employ best. The flexible navigation with learning control also permits adaptability in curriculum implementation to meet variable curriculum objectives. I.e., the topics can be used in any order and at any time in the medical school curriculum to fit school and user needs in the context of blended and online learning formats. The structure of the current modules, which can be used either online or downloaded to a local device also permits student flexibility to use the modules in any location, even if an internet connection is not available.

SOLUTIONS AND RECOMMENDATIONS

The following are a few selected recommendations for the design of multimedia teaching materials that the author considers important based on student input and on results of cognitive psychology studies of learning principles.

- **Enhance Knowledge of Evidence-Based Learning Strategies by Medical Education Faculty and Content Developers, in Order to Promote the Application of Evidence-Based Instructional Design Principles:** Many of the faculty presentations in medical schools do not follow these principles, although there has been improvement in the last few years. Levinson (2010) wrote: “The majority of medical school lectures or grand rounds presentations run contrary to best practices in instructional design, often consisting of Power-Point slides with redundant text, ‘death by bullet points’ and – where graphics are used at all – trivial, seductive augmentation that probably reduces learning.” One of the reasons is that, according to Levinson (2010), “institutions under-fund quality curriculum content development, evaluation and improvement.” Increasing the use of strategies that have been demonstrated to be effective in medical education, as developed by educators that know something about the topic, the audience and how people learn, would result in more efficient and effective learning by medical students and promote problem-solving and retention for application in clinical settings (Friedlander et al., 2011, Rohrer and Pashler, 2010, Mayer, 2008, Levinson, 2010, and Schumacher et al 2013).

- **Intentionally Utilize the Testing Effect (Retrieval Practice) to Help Students Learn:** Multiple practice tests of various formats (e.g. multiple-choice and short-answer questions) with feedback should be integrated into teaching materials or made easily available to students. This enhancement would help students learn and retain the content. While not as effective as short-answer or recall questions, the use of representative multiple-choice questions would also help students manage the multiple-choice question format of high-stakes testing, including many medical school course exams and licensing exams. Larsen et al. (2008) discussed studies indicating that feedback is crucial to learning from tests, and whenever possible feedback should include both answers and explanations to promote student understanding of the reason for errors. This was emphasized by Levinson (2010) who stated: "At the macro level of curriculum and programme design, one would like to see the integration of more opportunities for deliberate and mixed practice with expert feedback."
- **Enhance Availability of Multimedia Materials Using High Quality Open Source Images, Diagrams and Animations That Could Be Used Without Copyright Issues:** One challenge the author and colleagues have encountered while developing multimedia teaching modules was finding enough appropriate high-quality digital materials. There are currently several sources of these images, but a developer has to search in many venues to find what is available and has limited options if no source is found. The availability of multiple diagrams, animations, images, videos to provide enhancement of concepts, multidisciplinary integration, and application to patient cases would enhance the effectiveness of medical education materials.

FUTURE RESEARCH DIRECTIONS

Cook et al. (2008) reported a meta-analysis of internet-based learning interventions in the health professions, with the conclusion that Internet-based learning is associated with large positive effects compared with no intervention, and recommended that future research should directly compare different Internet-based interventions. There are a variety of directions for such research and the discussion below focuses on only a few of these.

It has been well established that applying the principles of multimedia instructional design, retrieval practice, and cognitive load theory result in impressive effects on retention in non-health professions learners. It is expected that these effects will also apply to medical students. However, there has been little replication or extension of the findings in medical education settings. Future research should investigate which elements of instructional design are most effective for different types of content in medical education. This is important in order to plan development of multimedia materials that are highly effective but also consider the cost of design and production. For example, Ruiz et al. (2009) conducted a critical literature review related to animations and state: "Medical educators have used animations in a variety of computer-assisted learning applications, but few comparative studies have been published and the evidence is inconclusive. Research outside medical education shows conflicting results for studies comparing animations with static images. This may reflect differences in cognitive load induced by animation, or differences in the type of motion being illustrated. The benefits of animations may also vary according to learner characteristics such as prior knowledge and spatial ability. Features of animation that appear to facilitate learning include permitting learner control over the animation's pace, allowing learners to interact with animations and splitting the animation activity into small chunks."

More recent studies by Daly et al. (2016), Berney et al. (2015), Hoyek et al. (2014), and Keedy et al. (2011) have compared effectiveness of static versus animated images, and found no significant differences in student performance. Further research is needed to examine when and how more complex animated images or high-fidelity simulations can substantially enhance meaningful learning, and when static images, or other low-fidelity techniques, are just as, or more, effective, since production of multimedia instruction using more straightforward low-fidelity techniques is less costly and less time-intensive. Challenges in development of multimedia self-instructional materials include faculty and staff time, and it is important to distinguish the relative effectiveness of each type of instructional design for the intended purpose.

Research on effective support for different learning characteristics or learning styles is also important. Medical students need different degrees of assistance in integrating challenging concepts from different disciplines and in application of content to patient cases or interpretation of experimental findings. Interactive computer-based materials can provide options for those students who need specific types of assistance, e.g. a longer amount of time to cover the material, more opportunity to practice with explanations as part of the feedback, easily available resources to review previously learned concepts, and examples of the type of integration and application expected.

Another important direction is to enhance self-assessment for students to assist both with understanding of concepts and application of concepts to patient cases or interpretation of data. Students are not good at judging their own competence, and both low-fidelity computer instructional programs with self-testing options and high-fidelity simulations are needed to a greater extent. Research should focus on when simpler low-fidelity approaches are sufficient to enhance student identification of their own learning needs or practice, and when more expensive high-fidelity approaches are necessary.

CONCLUSION

Medical education will benefit from integration of current understanding of learning theory and evidence with educational strategies and curricular design. Important instructional design principles include cognitive load theory and importance of retrieval practice as part of the learning process. Medical students are adult learners with different backgrounds and abilities and learner characteristics. Self-paced interactive multimedia materials can substantially promote retention and understanding for students at a variety of levels of mastery. Self-instructional materials using appropriate instructional design principles should be available for students learning basic science material at the pre-clinical level, for students studying for board examinations, and for students in clerkships that are learning to apply material to diagnosis and treatment of patient problems. These can be available anyplace/anytime for the specific needs of each student and improve the efficiency and effectiveness of meaningful learning.

REFERENCES

Berney, S., Betrancourt, M., Molinari, G., & Hoyek, N. (2015). How spatial abilities and dynamic visualizations interplay when learning functional anatomy with 3D anatomical models. *Anatomical Sciences Education*, *8*(5), 452–462. doi:10.1002/ase.1524 PMID:25689057

Carroll, M. W. (2013). Creative Commons and the Openness of Open Access. *The New England Journal of Medicine*, *368*(9), 789–791. doi:10.1056/NEJMp1300040 PMID:23445090

Case History: Case B. (n.d.). Retrieved from: http://learn.chm.msu.edu/neuropath/content/neuropath_cases/neuropathology_cases/CaseB.html

Clark, R. C., & Mayer, R. E. (2011). *E-Learning and the science of instruction: Proven guidelines for consumers and designers of multimedia learning*. Chichester, UK: John Wiley and Sons Ltd. doi:10.1002/9781118255971

Cook, D. A., Levinson, A. J., Garside, S., Dupras, D. M., Erwin, P. J., & Montori, V. M. (2008). Internet-based learning in the health professions: A meta-analysis. *Journal of the American Medical Association*, *300*(10), 1181–1196. doi:10.1001/jama.300.10.1181 PMID:18780847

Daly, C. J., Bullock, J. M., Ma, M., & Aidulis, D. (2016). A comparison of animated versus static images in an instructional multimedia presentation. *Advances in Physiology Education*, *40*(2), 201–205. doi:10.1152/advan.00053.2015 PMID:27105738

Friedlander, M. J., Andrews, L., Armstrong, E. G., Aschenbrenner, C., Kass, J. S., Ogden, P., & Viggiano, T. R. et al. (2011). What can medical education learn from the neurobiology of learning? *Academic Medicine*, *86*(4), 415–420. doi:10.1097/ACM.0b013e31820dc197 PMID:21346504

Hodgins, M. W., & Lovell, K. L. (1988). CNS Neoplasms - Interactive videodisc instructional unit. *Bulletin of Pathology Education*, *13*, 89–90.

Hoyek, N., Collet, C., Di Rienzo, F., De Almeida, M., & Guillot, A. (2014). Effectiveness of three-dimensional digital animation in teaching human anatomy in an authentic classroom context. *Anatomical Sciences Education*, *7*(6), 430–437. doi:10.1002/ase.1446 PMID:24678034

Karpicke, J. D., & Blunt, J. R. (2011). Retrieval practice produces more learning than elaborative studying with concept mapping. *Science*, *331*(6018), 772–775. doi:10.1126/science.1199327 PMID:21252317

Karpicke, J. D., & Roediger, H. L. (2008). The critical importance of retrieval for learning. *Science*, *319*(5865), 966–968. doi:10.1126/science.1152408 PMID:18276894

Keedy, A. W., Durack, J. C., Sandhu, P., Chen, E. M., OSullivan, P. S., & Breiman, R. S. (2011). Comparison of traditional methods with 3D computer models in the instruction of hepatobiliary anatomy. *Anatomical Sciences Education*, *4*(2), 84–91. doi:10.1002/ase.212 PMID:21412990

Larsen, D. P., Butler, A. C., & Roediger, H. L. III. (2008). Test-enhanced learning in medical education. *Medical Education, 42*(10), 959–966. doi:10.1111/j.1365-2923.2008.03124.x PMID:18823514

Lau, K. H. V. (2014). Computer-based teaching module design: Principles derived from learning theories. *Medical Education, 48*(3), 247–254. doi:10.1111/medu.12357 PMID:24528459

Levinson, A. J. (2010). Where is evidence-based instructional design in medical education curriculum development? *Medical Education, 44*(6), 536–537. doi:10.1111/j.1365-2923.2010.03715.x PMID:20604847

Lovell, K. L., Haf, J., & Hodgins, M. (1991). Development of neuropathology interactive videodisc instructional units. *Teaching and Learning in Medicine, 3*(3), 156–158. doi:10.1080/10401339109539501

Lovell, K. L., & Hodgins, M. W. (2001). Implementation and evaluation of neuroscience technology resources for second-year medical students. In *Proceedings of Slice of Life/Computers in Health Care Annual Meeting*. Salt Lake City, UT: University of Utah.

Lovell, K. L., Parkhurst, P. E., Sprafka, S. A., Hodgins, M. W., & Bean, P. L. (1993). Quantitative and qualitative evaluation of interactive videodisc instructional modules in preclinical neuropathology education. *Teaching and Learning in Medicine, 5*(1), 3–9. doi:10.1080/10401339309539579

Mavis, B. E., Lovell, K. L., & Ogle, K. S. (1998). Why Johnnie cant apply neuroscience: Testing alternative hypotheses using performance-based assessment. *Advances in Health Science Education, 3*(3), 165–175. doi:10.1023/A:1009794026466 PMID:12386438

Mayer, R. E. (2008). Applying the science of learning: Evidence-based principles for the design of multimedia instruction. *The American Psychologist, 63*(8), 760–769. doi:10.1037/0003-066X.63.8.760 PMID:19014238

Mayer, R. E. (2010). Applying the science of learning to medical education. *Medical Education, 44*(6), 543–549. doi:10.1111/j.1365-2923.2010.03624.x PMID:20604850

Mayer, R. E., & Moreno, R. (2010). Nine ways to reduce cognitive load in multimedia learning. *Educational Psychologist, 38*(1), 43–52. doi:10.1207/S15326985EP3801_6

Paas, F., Renkl, A., & Sweller, J. (2010). Cognitive load theory and instructional design: Recent developments. *Educational Psychologist, 38*(1), 1–4. doi:10.1207/S15326985EP3801_1

Parkhurst, P. E., Lovell, K. L., Sprafka, S. A., Hodgins, M. W., & Bean, P. L. (1991). Evaluation of interactive videodisc instructional modules in preclinical neuropathology education. *Journal of Medical Education Technology, 2*, 17–21.

Rohrer, D., & Pashler, H. (2010). Recent research on human learning challenges conventional instructional strategies. *Educational Researcher, 39*(5), 406–412. doi:10.3102/0013189X10374770

Ruiz, J. G., Cook, D. A., & Levinson, A. J. (2009). Computer animations in medical education: A critical literature review. *Medical Education*, *43*(9), 838–846. doi:10.1111/j.1365-2923.2009.03429.x PMID:19709008

Schmidmaier, R., Ebersbach, R., Schiller, M., Hege, I., Holzer, M., & Fischer, M. R. (2011). Using electronic flashcards to promote learning in medical students: Retesting versus restudying. *Medical Education*, *45*(11), 1101–1110. doi:10.1111/j.1365-2923.2011.04043.x PMID:21988625

Schumacher, D. J., Englander, R., & Carraccio, C. (2013). Developing the master learner: Applying learning theory to the learner, the teacher, and the learning environment. *Academic Medicine*, *88*(11), 1–11. doi:10.1097/ACM.0b013e3182a6e8f8 PMID:24072107

Sweller, J. (1994). Cognitive load theory, learning difficulty, and instructional design. *Learning and Instruction*, *4*(4), 295–312. doi:10.1016/0959-4752(94)90003-5

Van Merrienboer, J. J. G., & Sweller, J. (2010). Cognitive load theory in health professional education: Design principles and strategies. *Medical Education*, *44*, 85–93. doi:10.1111/j.1365-2923.2009.03498.x PMID:20078759

Wood, T. (2009). Assessment not only drives learning, it may also help learning. *Medical Education*, *43*(1), 5–6. doi:10.1111/j.1365-2923.2008.03237.x PMID:19140992

KEY TERMS AND DEFINITIONS

Cognitive Load: Total amount of mental effort being used in working memory. Intrinsic cognitive load is the effort associated with a specific topic; extraneous cognitive load refers to the way information or tasks are presented to a learner; germane cognitive load refers to the work put into creating a permanent store of knowledge.

Computer-Based Teaching Module: Teaching format in which a multimedia program serves as a single source for learning.

Creative Commons: Organization promoting open access to Internet resources, and defining public licensing use for all types of copyrighted works.

Interleaving: Process of combining testing or practice of different skills in a way that improves memory as compared to testing or practice blocked by type of skill.

Learner Control: The process by which a user determines how to utilize different elements of a computer-based teaching module.

Meaningful Learning: A deep understanding of the material, which includes organization, integration and ability to apply the knowledge to new situations.

Multimedia Instruction: Presenting words and pictures (including images and diagrams) that are intended to promote learning.

Retrieval Practice: A process used as part of learning in which information must be recalled from memory; retrieval practice plays a critical role in consolidating learning.

Self-Assessment: The process of testing oneself to determine how much knowledge has been acquired or how well the knowledge can be applied.

Testing Effect: Evidence that students remember material better when they are tested over material than when they use repeated study but are not tested.

This research was previously published in Advancing Medical Education Through Strategic Instructional Design edited by Jill Stefaniak, pages 262-276, copyright year 2017 by Medical Information Science Reference (an imprint of IGI Global).

Chapter 12
A Strategic Perspective on Using Symbolic Transformation in STEM Education:
Robotics and Automation

Jack M. Rappaport
Brilliance Consulting, USA

Stephen B. Richter
West Chester University, USA

Dennis T. Kennedy
La Salle University, USA

ABSTRACT

This paper describes and implements an innovative model for teaching Science, Technology, Engineering and Mathematics (STEM) that enhances the decision making process of students considering a major or a career in STEM fields. The model can also be used as a decision making tool for educators interested in stressing the importance of STEM for career enhancement and for society as a whole. The model creates analogies and metaphors for various STEM topics using the contents of popular music videos. Theories of neuroscience, the interdisciplinary study of the nervous system, are used to describe and validate our decision making model. Concepts such as, embodied cognition, mirror neurons and the connection between emotion and cognition, are used to explain how the brain processes the information and multi-modal stimuli generated by our model. The model was implemented using the topic of automated decision processes in robotics and automation with a group of university and high school students and teachers. The impact of the model was evaluated using the National Science Foundation (NSF) frameworks for evaluating informal science projects. The results indicate that the model using symbolic transformation to teach STEM can have a significant impact on students' attitude towards STEM and the decision making process about their careers.

DOI: 10.4018/978-1-5225-5478-3.ch012

INTRODUCTION

In this paper, we propose an innovative educational model to enhance STEM learning. STEM is the acronym that refers to the disciplines of science, technology, engineering and mathematics. The model creates analogies for various STEM topics using the content of popular music videos including the lyrics of popular songs, still images, dance, other movements of humans and inanimate objects, facial expressions and the general motion pictures. Because the music can instill an emotional aspect to a presentation, our model is particularly innovative in its application to STEM topics that are not normally associated with emotions. The model can enhance the role of the imagination in the learning process and engage students with diverse academic and cultural backgrounds. We implement our model for the study of robotics and automation and measure its effectiveness using frameworks proposed for the evaluation of informal STEM projects. Most of the analogies presented in this paper as examples of our model are based upon robotics and automation topics, many with themes of fashion and dance.

As depicted in Figure 1, theories from the fields of education, psychology, philosophy, linguistics are cited to support our model. The theory of metaphor plays a central role in the theory of cognitive linguists. For example, George Lakoff (1989,1999) believed that all or nearly all thought is essentially metaphorical. Various writings dealing with the theory of art and communication theory study the cognitive significance of visual metaphors in works of art and other visual media.

In addition, we cite theories from the field of neuroscience and its allied disciplines, philosophy and psychology, to support our model. Researchers in these fields have studied embodied cognition and the embodied mind. They have argued that all aspects of cognition, including decision making, are shaped by aspects of the body. These physical aspects include the motor system and the perceptual system that are built into the brain and effect the body's interaction with the environment (Rosch, Thompson & Varela, 1991). In social and cognitive psychology, research on embodied cognition encompasses issues such as social interaction and decision-making (Borghi & Cimatti, 2010). This research supports the embodied cognition view that the motor system influences cognition, just as the mind influences bodily actions. Furthermore, Edelman (2004) and Damasio (1999) have outlined the connection between the body, individual structures in the brain and aspects of the mind such as consciousness, emotion, self-awareness and will.

Consequently, the study of neuroscience provides the basis for understanding how cognitive information is created by superimposing stimuli from various modes, semantic, visual, sensorimotor, and auditory. Lyrics from songs can be used to represent various aspects of robotic and manufacturing processes. Visual metaphors are also developed by aligning images from music videos with various aspects of robotic processes and automated systems. Motion from the music videos, like dance, can represent either real or animated motions of robotic or manufacturing processes. Some of the more technical aspects of robotics, such as coded commands and numerical/geometric models, are also represented by various forms of dance and visual metaphors.

Building on the concept of embodied cognition, neuroscience tells us that effective decision making is not possible without the motivation and meaning provided by emotional input. For example, Antonio Damasio had a patient who underwent neurosurgery for a tumor and lost the part of his brain connecting the frontal lobes, that control reason, with the emotions. Rather than this making the patient more rational, he became paralyzed by every decision in life. In other words, feelings provide an essential component of human decisions (Jarrett, 2014; Seth, 2014).

Figure 1. Conceptual framework

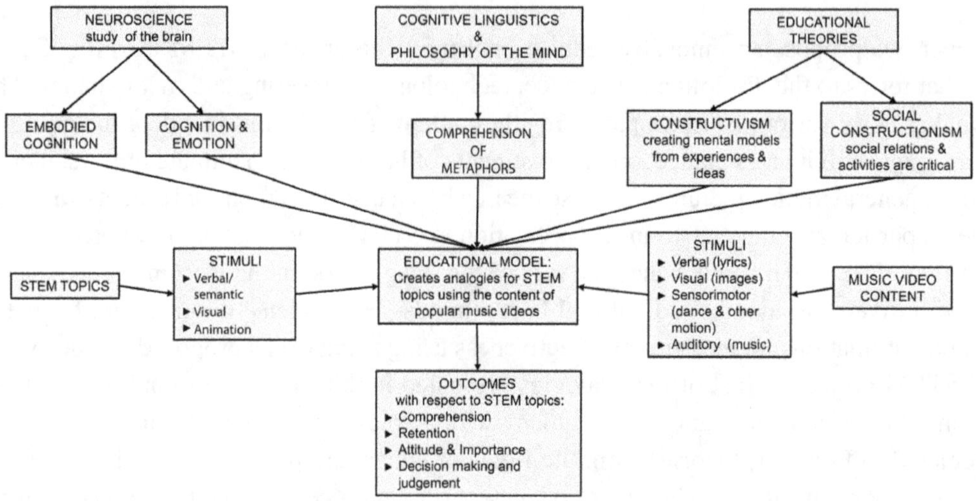

An implementation and outcome assessment of the model employs the frameworks for evaluating informal science projects outlined as part of an NSF report (Friedman, 2008) and by the Afterschool Alliance (2011, 2013). The results indicate that the model can have a significant impact on the motivation to learn more about STEM topics and on the attitudes towards and the perception of the importance of STEM topics in our society.

THE CHALLENGE FOR STEM EDUCATION: CONSTRUCTIVISM AND CONSTRUCTIONISM

As universities and secondary schools continue a national effort to invest in STEM education, teachers and faculty are faced with challenges to develop innovative approaches to teaching STEM. At the U.S. News STEM Solutions Conference in Washington D.C., academic leaders agreed that traditional approaches to teaching STEM, in large lecture halls with little attention to hands on work, has not proved successful in recruiting and training students for a future in STEM fields. Similarly, Hunter R. Rawlings, the president of the Association of American Universities, acknowledged that universities have not done a good job of engaging students in the classroom with STEM education (Brody, 2014).

Many innovative educational ideas can be viewed within the context of constructivism. This theory of knowledge, proposed by Jean Piaget (1950), argues that humans generate knowledge and meaning from an interaction between their experiences and their ideas to create mental models for understanding the world around them. Based in part on the ideas of constructivists like Vygostsky (1978), contemporary psychologist Jerome Bruner (1961) has asserted that all reality is a narrative construction in the imagination of individuals. Dewey (1938) had a strong impact on the constructivism movement and argued that education and learning are social and interactive processes. From this perspective, the school itself is a social institution through which social reform can and should take place. In addition, he believed that students thrive in an environment where they are allowed to experience and interact with the curriculum, and all students should have the opportunity to take part in their own learning.

Dewey believed that disequilibrium, a feeling of having a problem or being upset, was potentially a productive condition for the growth of the mind. It was a temporary sense of disequilibrium that motivated people to restore an also temporary state of equilibrium. This shift or oscillation between equilibrium and disequilibrium was at the center of the educational process. For growth to occur, the increase of a student's ability to frame and pursue his own purposes, the student needs a stake in the learning process. It is the process of restoring equilibrium through the process of inquiry that provides the satisfaction of the journey (Eisner, 2002).

The juxtaposition of stimuli from opposing environments, for instance the STEM topics and the music videos, can be seen as a symbol of the imbalance or disequilibrium between the educational system and the real world. Our music video model engages students to understand various technologies using a process of inquiry that involves comparison and analogy. The comprehension of the metaphors can create a balance between the rigors of a formal educational system and the social life of the students. For inner city youth, this can be the dichotomy between School and the Street.

Other forms of disequilibrium can occur as a result of social barriers, feelings of disenchantment with the educational process or the lack of appreciation of the importance of STEM skills. The process of inquiry initiated by our music video model can be a convenient way for the student to dispel these imbalances and help to facilitate a more meaningful approach to learning. Innovative education involves encouraging and incentivizing teachers to take the risk of a new approach without fear of reprisals. The results from the implementation of our model should increase the risk threshold of STEM teachers and lessen the desire of administrators to impose restraints or sanctions for taking these risks.

Dance metaphors can enhance the learning process through the alignment of a person's gender, age, career, cultural, socioeconomic, regional and national identities (Wilce, 2009) with the process of creative constructions and reconstruction of self-identities (Hanna, 2015). Music videos are generally targeted to the youth and contain many social and cultural manifestations of their real life experiences. Integrating these into the educational process allows students to reflect on their own experiences and direct their own learning based upon these experiences. Our model also has the potential to engage students with diverse cultural backgrounds and particular demographic groups, including African Americans, Latinos or females, that are underrepresented in the scientific and engineering professions. In addition, our model can be used to integrate students with diverse cultural backgrounds within the same learning environment. For example, hip-hop/rap music is generally enjoyed by youth from all socio-economic and ethnic classes.

Ultimately, we want the students to develop shared constructs that are formed by the metaphors from the common experiences of music and dance. Constructionism, a theory related to constructivism, places a critical emphasis on particular constructions of the subject that are external and shared (Papert, 1991). In social constructionism, an extension of constructionism, Shaw (as cited in Kafai & Resnick, 1996, ch. 3) observes that social relations and social activities are critical components of the shared outcomes and artifacts at work in the learning cycle. Popular music videos, including the possibility of engaging the student in active dance, form a convenient platform to develop these social constructions. Students can also learn to develop their own metaphors that can play an important role in the constructionist model by fostering a recursive process of internalization and externalization .

The metaphors can enhance students' awareness of themselves an owner-protagonist in the learning process as they identify with one of the protagonists in the video. The music videos can provide stimuli that become symbols of important aspects of a their cultural landscape, thus enhancing a sense of personal esteem and value.

ANALOGIES IN SCIENCE

Analogy is a cognitive process of transferring information or meaning from a particular subject, the analogue or source, to another particular subject, the target, with a linguistic expression corresponding to such a process. Analogy has been studied and discussed since classical antiquity by philosophers, scientists and lawyers. The last few decades has shown an increased interest in analogy, most notably in cognitive science. The Greek philosophers Plato and Aristotle saw analogy as a shared abstraction. Analogous objects did not share necessarily a relation, but also an idea, a pattern, a regularity, an attribute, an effect or a function. In our project, we create analogies between an object or concept of technology as the target and the title or lyrics of a particular song as the analogue or source.

Throughout history, analogies have played an important role in scientific discoveries, not as proof, but as inspiration. Analogies have also played an important role in explaining those discoveries (Kaiser, 1989). For example, Johannes Kepler, the famous seventeenth-century astronomer, wrote: "I especially love analogies, my most faithful masters, acquainted with all the secrets of nature" (Vickers, 1984, p. 149). Kepler, who discovered laws of planetary motion, used analogies to help explain his discoveries: "I am much occupied with the investigation of the physical causes. My aim in this is to show that the celestial machine is to be likened not to a divine organism but rather to a clockwork" (Holton, 1973, p. 72). Steven Chu, who received the 1997 Nobel Prize in physics, used analogies to discover and explain his method of slowing down atoms with laser light. Chu calls his laser-light method "optical molasses" (Glynn, 2007).

The use of analogies in teaching of scientific studies is not new. For example, The Teaching With Analogies Model (Glynn, 2007, 2008) uses analogies to build conceptual bridges for students between what is familiar, an analog concept, and what is new, a target concept. The Teaching With Analogies Model is used extensively in science education to introduce students to concepts that represent complex, hard-to-visualize systems with interacting parts, for example an atom, a cell, photosynthesis, an electric circuit, an ecosystem. The model guides teachers, textbook authors, and website designers in their use of analogies. Research has shown that using the model increases students' learning and interest (Glynn, 2007, 2008).

METAPHORS USING MUSIC VIDEOS

Most of our analogies are metaphors. A metaphor is a type of analogy that describes a subject by asserting that it is, on some point of comparison, the same as another otherwise unrelated object. The Philosophy of Rhetoric (1936) by I. A. Richards describes a metaphor as having two parts: the tenor and the vehicle. The tenor is the subject to which attributes are ascribed. The vehicle is the object whose attributes are borrowed. In our model, the subject or tenor of the metaphor consists of the particular scientific or technological concept that is to be described. The vehicle is the content from the music video whose attributes are borrowed to describe the subject or tenor. For example, the manipulator arms of an automobile manufacturing robot, the subject, can be described using either still image or dance motion from Chris Brown's video "I Can Transform Ya" that is based upon a robotic theme.

The metaphors can be semantic, visual, sensorimotor, auditory or based upon a combination of modes. A purely verbal metaphor can align a STEM topic with the lyrics of a particular song. Visual metaphors can juxtapose or superimpose two different visual images into a single metaphorical unit. Metaphors can also be based upon motion and animation. Therefore, dance choreography can be aligned to the motions of various manufacturing processes, for example casting or welding.

The music can also be the source of metaphors as the style or moods of particular songs can be aligned to various manufacturing processes. The music can add an emotional element which can enhance the learning process and retention of information. Humor can also play a role in reducing stress, thus enhancing the ability to learn and retain information. Creating and understanding metaphor can be particularly effective when used in conjunction with humor (Hanna, 2015).

Our music video model is designed for high school and university students and is implemented on PowerPoint 2013. The images or moving pictures can be juxtaposed using split screens or superimposed using a transparency process. The music, obtained from iTunes, and the video clips are embedded within the PowerPoint presentation. The EasiestSoft Video Converter is used to convert the videos that are downloaded from YouTube and the iTunes into a form that is compatible with the PowerPoint program.

THE PHILOSOPHY OF METAPHORS

Originally, many philosophers did not consider metaphors a legitimate form of rational discourse. For example both Hobbes (1651) and Locke (1690) professed that metaphors can insinuate wrong ideas and mislead judgment. In the 1930's, logical positivists like Ayer and Carnap assumed that metaphors have no real meaning or verification conditions (Camp, 2006).

However, in more recent times many philosophers and linguists have taken a more friendly position with respect to the metaphor arguing that they have cognitive significance (Black, 1962; Beardsley, 1962; Weinrich, 1966; Levin, 1977). Some theorists (Grice, 1975; Searle, 1979; Martinich, 1984) used pragmatic theories to claim that the speaker of a metaphor can communicate a distinct propositional content. Contextualists have probably been the most friendly to the notion of metaphor (Sperber & Wilson, 1955; Bezuidenhout, 2001; Carston, 2002; Recanti, 2004) emphasizing that metaphor can be a pervasive and natural aspect of ordinary language use. Guttenplan (2005) puts forward a novel philosophical account of metaphor and justifies the use of metaphor as fundamental to language and the study of language. Robinson (2011) believes that using metaphor to describe causation increases its explanatory power while diminishing many of the problems associated with the standard analysis of it.

In our model, we use metaphors based upon dance as well as on ordinary language. In *Languages of the Brain*, Galaburda, Kosslyn and Christen (2002) argue that there are multiple possible "languages of thought" that play different roles in the brain and anthropologist Margaret Clegg would classify dance as a language. Hanna (2015) considers dance as a form of nonverbal language and like verbal language has vocabulary (locomotion and gestures), grammar or syntax (dance traditions) and semantics (meaning). Furthermore, she states that the most common way of encoding meaning in dance is through metaphor.

Cognitive linguists, like George Lakoff, probably grant the most central role of the metaphor in our conceptual system. Essentially, they believe that nearly all thought is essentially metaphorical: "our ordinary conceptual system, in terms of which we both think and act, is fundamentally metaphorical in nature" (Lakoff & Johnson, 1980). Lakoff and his colleagues are interested in establishing the metaphorical nature of ordinary thought about familiar topics, claiming that we cannot think about these topics on their own terms because they are too abstract and must therefore filter them with more concrete domains. This can be relevant for the conceptualization of scientific topics that can seem particularly abstract and foreign to the average student.

In our model, we use metaphors as anchors for comprehension. There are several models of metaphor comprehension (Camp, 2006) and we consider two of them, category transfer and feature matching. These models can be useful to understand how metaphors can function as a means of learning in our educational model.

The category transfer model focuses on the entire complex schema associated with the source or vehicle, and thus captures the global organizational effect that is an important part of metaphorical comprehension. Goodman (1968, p.73) says that metaphor is not an isolated term or concept but a "whole apparatus of organization." Glucksberg and Keysar (1993, p. 421) make a similar point by saying that metaphors present "a patterned complex of properties in one chunk". Using the category transfer model we abstract away from the specific, concrete feature of the subjects to produce a general schema for the source or vehicle. For example, the music video dance routines and robotics process can be perceived and comprehended in "one whole" blending together the abstract properties of agility, flexibility, dexterity and continuity to form a holistic unit that forms the basis of the comparison. This process is similar to the idea of Gestalt psychology or gestaltism and is based upon the principle that the human eyes sees objects in their entirety before looking at the individual parts (Kofka, 1935; Wertheimer, 1945; Kohler 1929).

However, this approach focuses exclusively on the source or vehicle and has a hard time explaining the very different effects that can be produced by applying the same schema to different topics. If we think of an automotive robot in terms of a dance routine, the result might be very different if we think of an automated bottling plant in terms of the same dance routine. Rather than locating all the action exclusively in the source, feature-matching models (Ortony, 1979, 1988; Fogelin, 1988) operate through a direct comparison between the source and the topic. As a result, they can explain why the same source can have different effects when applied to different topics, and how metaphors can lead us to notice very specific features of the topic. In this way, different manufacturing processes can be aligned to specific choreographies used by the dancers in music videos.

On the other hand, the feature matching model has a hard time accounting for the metaphor's holistic effects. Therefore, some highly salient features of the source may go unmatched, for example the ability of the dancer to communicate with an audience, move to the beat of the music and presumably learn new motions and techniques more naturally than a robot. Similarly, we may notice some properties of the robot that cannot be matched directly to the dancer, such as the ability of the robot to go underground, or to be programmed by a computer. The category-transfer model mitigates many of these issues by focusing on the metaphor's holistic effects.

In general, we use dance motions to represent robotic motions, and not the other way around. The dance represents the source and the robot the target in the metaphor. However, in pure artificial intelligence, the robot is intended to simulate all of the qualities of a human being, and perhaps have even more. In this sense, the dancer can function as an internal representation of itself as defined by Dilworth (2004). The dancer is both a human being and a machine at the same time, so the target and the source of the metaphor are embodied in the same entity.

THEORIES OF NEUROSCIENCE: SEMANTIC EMBODIMENT

One aspect of embodied cognition is that semantic and conceptual information is, at least in part, based upon on information in action and perception systems of the brain. This concept, often referred to as semantic grounding, is the basis for embodied theories that the meaning of at least some words, concepts and constructions are based upon objects and action. Pulvermuller (1999) proposed that neuronal circuits, cell assemblies distributed over sensory, motor and multimodal association areas of the brain, are the neurobiological correlates of meaningful words and constructions.

Semantic circuits consist of cell assemblies that bind modality specific semantic information into a more abstract multimodal, and in a sense ''amodal'' and ''modality-unspecific', representation (Fuster, 1995; Pulvermuller, 2012). These semantic circuits are widely distributed and can reach into modality-specific and multimodal areas of the cortex. Crucially, semantic circuit topographies, the distribution of semantic circuits over the cortex, can reflect aspects of the category-specific meanings they carry. This theoretical perspective covers all aspects of cognition sometimes claimed to be missing from some versions of embodiment theory, including mechanisms for abstraction, generalization and symbol combination (Pulvermuller, 2012). Specifically, correlated activity in sensory and motor brain systems, especially in the cortex, and pre-existing neuroanatomical connections drive the formation of major building blocks of cognition, language and learning (Pulvermuller, 2013).

The process of semantic embodiment forms an important basis of our metaphorically based education model. The lyrics of the songs are linked to the action of the videos, but they are also linked to the technological processes via the metaphors. In an analogous way, semantic or conceptual information of STEM topics can be aligned to the images or dance motions of the music videos. Scientist and dancer Blake Martin observes "dance creation is the necessary use of 'embodied metaphor', in which the creation is all at once signified, signifier, and communicator" (Hanna, 2015, pp. 56-57).

The metaphors can be constructed using purely perceptual or action based modalities, perhaps linking the dance motion from Madonna's "Vogue" with images or animated motion of a CAD/CAM process. This alignment process can also be facilitated by the use of language and meaning. For our example, the metaphor works in part because the term "Vogue" means casting in the sense of design and casting in the world of fashion that can be linked to a similar concept in the manufacturing sector.

Based on his study of the neurobiological basis of metaphor, Seitz (2005) observes that people may be prewired to recognize or create metaphoric analogies across disparate sensory, perceptual, and affective domains of experience (Seitz, 2005). He proposes four basic kinds of metaphors that are uniquely mapped onto specific brain networks. These are mainly nonconceptual and include: (a) perceptual-perceptual (b) cross-modal (c) movement-movement and (d) perceptual-affective mappings.

Seitz (2005) refers to many examples where children can make metaphorical associations using moving objects. A preschool child can note the metaphorical similarity between a spinning top and a dancing ballerina in a match-to-sample or similar nonverbal, for example filmed, format (Dent, 1984). Likewise, infants (18–24 months) can note the enactive similarity between a paper crayon cover that has slipped off and putting on an article of clothing, "I am putting your clothes on crayon" (Winner, McCarthy & Gardner, 1980).

Metaphors can enhance learning as an ongoing process of building the brain's cerebral pathways. Learning and creating memory can be compared to the process of chiseling and shaping as our individual brain wiring diagrams are being molded (Damasio, 2010).

METAPHORS USING LYRICS

We can align the lyrics of popular songs to create metaphors. Table 1 gives examples of metaphors using the lyrics of "I Can Transform Ya" linked to various robotic or manufacturing processes. Many of

Table 1. Metaphors using lyrics

Lyrics from "I Can Transform Ya"	STEM Topic
I Can Transform Ya	Flowchart of production system
Make it so new you never want to go back to the old you	Image of a CAD/CAM diagram
No I can't dance but I can dance on ya	Welding robots in automobile production line
Anything you want I can get it for ya	Materials handling system
I can transform her to a Ducati	Machine tending
Iced out everything break it like an Eskimo	Mechanical grinding and cutting
Need a ride I can range you up	Circular conveyor belt
I can have you swag surfing	Continuous flow processing
I'll take you to where its warmer	Robots operating in high temperatures
Swiss on the beat Chris move your feet	Honda research robot Asimo walking
See potential in ya	Developing artificial intelligence for Asimo
I turn you from a human to a Carter	Developing human skills for Asimo
See me in the video	Asimo greeting guests at the research lab
And she gon transform like Optimus Prime	A humanoid robot research project
Cause her form puts me in a trance	Image of a robot with emotion
Swag low ill build you up	Boston dynamics research robot
Weak knees Ill stand you up	Boston dynamics research robot
No longer be the passenger	Robot helping another robot
Wanna fly we can go Anywhere you wanna go	Amazon drone project
I can change your life	Robots used in micro surgery
My black card they don't deny that	Diploma in engineering
Jimmy Choos in Italy Louie V in Tokyo	Robot taking your handbag and coat at party
Money I can change you up	Automated bitcoin process

the lyrics are action based as they are directly linked to many of the dance and motion in the videos. As such, the semantic content is grounded in visual, sensorimotor and auditory, music, stimuli.

VISUAL METAPHORS

V.C. Aldrich (1968) defines visual metaphor as a fusion or interanimation of two visual images, A and B, whose colors, forms or positions cause us to link them visually into a single, though complex, metaphorical unit C.

The idea of visual metaphors (see Table 2) can be understood within the tradition of the Gestalt psychology or gestaltism. In German, gestalt means "essence or shape of an entity's complete form." Gestaltism is based on the operational principle that the brain is holistic, parallel, and analog, with self-organizing tendencies. The principle maintains that the human eye sees objects in their entirety before perceiving their individual parts, suggesting the whole is greater than the sum of its parts. Early 20th century theorists, Kurt Koffka (1935), Max Wertheimer (1945), and Wolfgang Köhler (1929), saw objects as perceived within an environment according to all of their elements taken together as a global construct. This gestalt or whole form approach sought to define principles of perception, seemingly innate mental laws that determined the way objects were perceived (Smith, 1988). Gestalt psychologists stipulate that perception is the product of complex interactions among various stimuli contrary to the behaviorist approach to understanding the elements of cognitive processes (Carlson & Heth, 2010).

Table 2. Visual metaphors

Video Image	STEM Image	Properties of Metaphorical Unit
Guns N' Roses drummer	Welding operator	Energy; productivity; pride of work
Gun N' Roses singer in shouting/excitable state	Sparks from welding process	Energy release from motivated worker or controlled process
Gun N' Roses guitar player	Automated control machinery	Worker control of complex technology
Dance choreography from "I Can Transform Ya"	Manipulator arms of welding robots	Dexterity, agility, flexibility, efficiency of motion
Image of Ace Hood and his community in "Hustle Hard"	Flowchart symbolizing lean manufacturing	Ingenuity involved in improving productivity and minimizing waste
Poses of Justin Timberlake that generate various angles for body parts from "Cry Me a River"	Geometrical/numerical analysis for robot motion control; parameter encoding	Precision, motion analysis
G Unit rap star with hat and chain at a "cool" angle	Geometrical/numerical analysis for robot motion control; parameter encoding	Precision, motion analysis
Britney Spears kicking a chair in "Stronger" video	Google one-legged kicking robot	Strength, durability
Choreography in Britney Spear's show using ropes and masks	Robot with wires connected to a circuit board	Control, action within control limits
Image of a man controlled in Meghan Trainor's "My Future Husband"	Robotic software commands	Control, predictability
Image of Nicki Minaj's eye in video "Superbass"	Image of robot vision	Importance of vision for security, analysis and control
Aerial shot of Superbowl halftime stage (N' Sync, Aerosmith, Britney Spears show)	Image of a circuit board	Complex connections; rapid connections
Image of Ace Hood with a Bugatti from "Bugatti"	Images of various applications of CAD/CAM	Quality of design
Various virtual reality poses in Taylor Swift's "Style"	Images of various applications of CAD/CAM	Creativity of design
Choreography in Usher's "Hey Daddy"	Image of a robotic "motherboard"	Integration of parts of motherboard system
Various poses of Lil Wayne in his video "I Am Not a Human Being"	Image of neural network system; simulation of brain functions	The development of artificial intelligence
Riding on a yacht in "Ride" with Ace Hood and Trey Songz	Image and motion of an automated conveyor belt	Motivation for automated assembly lines and continuous flow processes
Scary pose of Chris Brown in Nicki Minaj's "Only"	Futuristic image of a robot dangerous to society	Moral and societal issues concerning the dangers of robots

VISUAL METAPHORS IN ART AND OTHER MEDIA

Visual metaphors have played an important role in the interpretations of art. For example, complex metaphorical structures in art, including metalepses, metaphors for metaphors, existed as far back as the Renaissance (Vicari, 1993). Many art analysts have studied Picasso's use of metaphor. Because Picasso's Cubism style threatened to create super abstract ideas that were too abstract and far from reality, Picasso invented collage in which he pasted everyday objects unto the paintings. It was this interplay of elements that advanced his use of metaphor. Proweller (1971, p. 245) in his paper on Picasso's Guernica can be quoted:

One can say that, in a study of this sort, the painting of 'disparate phenomena' is in an analogous way an instance of 'visual metaphor'; herein, meaning is derived from the mere interplay of elements, the metaphorical situation itself 'selecting, emphasizing, suppressing and organizing features of the principal objects,' that is, accommodating a measure of insight that would otherwise remain incommunicable.

The use of visual metaphors is also widespread in many media forms, magazines, cartoons, TV and visual displays. They can be used for advertising purposes or more general purposes to communication information in creative ways and to engage the viewer. A variety of researchers have studied ways in which visual metaphors have been used in business magazines, cartoons, advertising and visual displays (Koller, 2005; El Refaie, 2003; Jeong, 2008; Kaplan, 1990; Dent-Read, Klein & Eggleston, 1994). Overall, the research shows how widespread the use of visual metaphors is in our society.

VISUAL METAPHORS AND NEUROSCIENCE

Many theories of embodied cognition are concerned with the perceptual system and the way the brain interprets data supplied by vision. Half the sensory data going to the brain is visual. The visual system "creates representations in the brain (in the form of neural codes) that require… more information than the modest amount the brain receives from the eyes. That additional information is created within the brain," explains Nobel laureate molecular biologist and neuroscientist Eric Kandel (2012, pp. 233, 219). The image on the retina is deconstructed into electrical signals that portray an object. These signals moving through the brain are recoded based upon prior knowledge and reconstructed and elaborated into the image that we perceive. Vision, as other sensory systems, is a creation of the brain, the supreme enabler (Fischer & Whitney, 2014).

Francis Crick (1994) the co-discover of the structure of DNA spent the last several decades of his career studying the marvels of conscious visual perception. Crick argued that while we see to have a picture of the world in our brain, we actually have a symbolic representation, a hypothesis, of that world. Machines, like computers and TV sets, present us with pictures, and inside these machines we would find arrangement of parts and circuits that process the encoded data. Similarly what we would expect to find in the brain is a representation of the visual scene in some symbolic form. Thus, what we see in the mind's eye goes dramatically beyond what is present in the image cast on the retina. Luckily for us, the brain generates a hypothesis from the raw data that is remarkably accurate. It is in the construction of these representations of the visual world that we see the brain's creative processes at work.

Using our visual images, students are challenged to create rich and meaningful images of the scientific subject based upon what they see. A recoding process develops as students perceive the STEM and music video images, either juxtaposed or superimposed on each other. This process is best described by neuroscientist Antonio Damasio (2010, pp. 18, 70):

Minds emerge when the activity of small circuits (of neurons) is organized across large networks so as to compose momentary patterns that represent things and events situated outside of the brain, either in the body or the external world, as well as in the brain as it processes other patterns… A spectacular consequence of the brain's incessant and dynamic mapping is the mind.

The brain processes visual information in two ways. One process is on a fine scale, by the foveal cones, for a parts analysis. The other is on a coarse scale, processed by the peripheral cones, for a holistic analysis. Livingstone (2002) uses the example of the Mona Lisa to show that we can perceive things in peripheral vision, like Mona Lisa's smile, that we miss in central vision. She supplies evidence for this by simulating the blurring effects of periphael vision on Mona Lisa's face showing that the edge of Mona's Lisa mouth seems much less turned upward when examined with central rather than peripheral vision. She uses this finding to illustrate that we can perceive things in peripheral vision that we can miss in central vision. This lends further support for our model because peripheral vision would play a key role in perceiving images that are juxtaposed together or superimposed upon each other.

Many of our visual metaphors could be seen as examples of nonconscious stimuli (Baars, 1989). A nonconscious stimulus is subliminal when the information is so reduced as to make it undetectable as in the case of superimposing two images. A nonconscious stimulus is called a preconscious stimulus when it is potentially visible in a given trial but is not consciously perceived because of temporary distraction or inattention. In our model, this may happen with one of the images when we juxtapose two images together. Dehaene and Changeux (2011) focus on the issue of how an external or internal piece of information goes beyond nonconscious processing and gains access to conscious processing using various neuroimaging and neurophysiological data. A visual stimulus that is masked and remains invisible can nevertheless affect behavior and brain activity at multiple levels (Kouider & Dehaene, 2007; Van den Bussche, Van den Noortgate & Reynvoet, 2009).

METAPHORS USING DANCE

Dance plays a significant role in our model because it represents a core element in many if not most popular music videos. Many students like to dance primarily in social situations and as a means of self-expression. Many of them often identify with their favorite pop dancing stars. The dance of popular music videos can be used to create a personal bond or sense of empathy between students and the STEM topics when used in a metaphorical context.

The ability to create metaphors between dance and scientific subjects should not be surprising. Science and dance are not alien disciplines. Sylvie Leotin (2012), as both a ballerina and a scientists noticed many commonalities. Scientists attending a dance performance will undoubtedly relate to the physicality and geometry of dance. The movement through time, the geometry of interactions, the symmetry of the lines, the balance of the bodies. All these have parallels in the physical interactions that occur in every science, from astronomy and physics, to chemistry and biology.

Scientists need to empathize with their materials, and immerse themselves into the problems they seek to illuminate. Einstein visualized travelling astride a speeding light beam, and pondered what the world would look like if he traveled at the velocity of light. Nobel Laureate Barbara McClintock imagined being the genes of the corn plants she studied, even claiming to become their "friend" (Leotin, 2012).

Watching dance is a good way to create empathy. Since antiquity, dancers have been great translators and purveyors of emotions and meaning. They inhabit music, characters, objects, and give life to them before our mesmerized eyes. We can learn from them. MacArthur Fellow John Cairns generated valuable insights about bacterial processes by dancing his experiments (Leotin, 2012).

The use of dance as an educational tool is justified by studying the history of scientific discoveries. For example many breakthrough discoveries resulted from scientists seeing links between their profession and other fields. Leonardo da Vinci and other Renaissance polymaths are prime illustrations. C. H. Waddington, a celebrated embryologist, was a dancer. He linked his work to the unfolding of a set of dance instructions, causing him to rethink embryology as a process rather than a mechanism, and resulting in a novel approach (Leotin, 2012). Albert Einstein thought in pictures and embodied the development of his theory of relativity through muscular feeling and body sensation (Rothenberg, 1979).

who studies the creative process across disciplines might agree that it is virtually identical, even with real differences in materials and goals. Studies also show that scientists with creative avocations are often more successful than those without (Root-Bernstein & Root-Bernstein, 1999). Perhaps this is because they are able to comprehend problems with greater breadth, simultaneously linking intuitive Anyone and subjective ways of feeling with objective and communicable ways of knowing.

MEANING IN DANCE

Symbolization is a key concept in dance just as in verbal language. People are prewired to recognize or create metaphoric analogies across disparate sensory, perceptual and affective domains of experience (Seitz, 2005). The potential to create metaphors using dance is almost limitless. The meaning of a message a dancer intends may not be always the same as the one the viewer receives. Just as in our model, we filter the meaning of the dance to develop a STEM learning experience.

The idea of using dance as a symbolic representation of science is not new. The goal of the international competition Dance Your Ph.D., sponsored by the journal Science, is to see which scientists can best explain their graduate work through interpretive dance. Australians have dominated in recent years with a physicist winning in 2011 for his stop-motion dance about titanium hips and a chemist in 2012 for his old-time burlesque about aluminum crystals.

Hanna (2015) has developed a grid, a semantic modification of Charles Pierce's semiotic forms of verbal meaning, as a tool to discover different dimensions of meaning in dance. The grid has devices and spheres representing various ways in which dancers embody the imagination. The grid consists of six symbolic devices to encode meaning including metaphor, the primary source of symbolization used in our model. The devices for encapsulating meaning in dance are assumed to operate in one of nine spheres. We use several of these devices: "total human body in action", "specific movements", the "intermesh of movements with another medium", "dance as a vehicle for another medium" and "presence". Hanna (2015) defines "presence" as the "emotional impact of projected sensuality, raw animality, transcendence or charisma of dance." This characterization applies to many popular videos that are based upon a romantic or sensual theme.

EMBODIED COGNITION, MIRROR NEURONS, EMPATHY AND LEARNING THROUGH DANCE

The theory of embodied cognition points to a strong bond between the body and the mind, providing ample support for the use of dance metaphors in our model. The brains makes explicit maps of the structures that compose the body and also maps the functional states naturally assumed by those body components. Thanks to the brain, the body becomes a natural topic of the mind.

The brain's representation of the body has another major implication. Because we can depict our own body states, we can more easily simulate the equivalent body states of others. The range of phenomena denoted by the word empathy owes a lot to this arrangement. The connections to others occurs not just by visual images, language and logical inference but also by the actions with which we can portray the movements of others. These actions can include somatosensory representations of movements and visual representations of movements.

The concept of dance as a form of nonverbal language is also strongly supported by neuroscience. For example the process of making a dance engages some of the same components in the brain for conceptualization, creativity and memory as do verbal poetry or prose (Cross, Hamilton & Grafton, 2006). Broca's area of the brain, which is activated in various linguistic tasks, is also active in encoding complex human movements (Clerget et al. 2009). Fadiga et al. (2006) shows that observation of meaningful hand shadows resembling moving animals and grasping and manipulation, activates the frontal language area, demonstrating that Broca's area plays a role in interpreting the actions of others. Broca's area assembles and decodes speech sounds in the same way it interprets body language (Hanna, 2015).

We use images of static body posture in many of our visual metaphors. However visual perception research has demonstrated a difference between the processing of dynamic body movement and the processing of static body posture. Vicary et al. (2014) showed that the recognition of body movement based only upon posture is possible but is significantly poorer than recognition based on the entire movement stimulus. These findings provide a persuasive argument to incorporate the full dance motions into a presentation.

The observation of dance is a special case of observing body movements. Kinesthetic empathy and the phenomena of mirror neurons occur when we see in a human body movement that we experience vicariously in our nerves and muscles. The observed movement evokes associations we would have had if the original movement had been ours (Rizzolatti & Destro, 2008; Calvo-Merino et al., 2005, 2006; Grove, Stevens & McKechnie, 2005). Iacoboni (2005, 2009) has argued that mirror neurons are the neural basis of the human capacity for emotions (Blakeslee, 2006). Preston and Frans de Waal (2002), Decety (2002, 2004), Gallese (1998, 2001) and Keysers (2011) have independently argued that the mirror neuron system is involved in empathy.

As noted in the mirror neuron research, true motor experience in performing observed dance movements leads to greater involvement of the cognitive-neuron system than only observing movement. In addition, Blasing, Puttke and Schack (2009) shows that activation of the cortical motor system was based only on mental stimulation of the movement, and not based upon observing lists of verbal items. Research has also demonstrated that a person's mental representation of movement may depend on what is being danced and what the observer can do (Hanna, 2015).

The cerebellum is located behind the brain stem and the rear of the cerebrum. It coordinates multiple sets of voluntary muscle movements and their timing, maintains posture, balance and equilibrium. Despite its small size it has more than half the neurons in the entire brain, which reflects it importance to learning. Brain scans show that a dancer who visualizes a dance choreography without actually moving stimulates the cerebellum (Olsson, Jonsson, Larsson & Nyberg, 2008; Jones & Viamontes, 2010).

The connection of mirror neurons to empathy and learning gives strong support to our use of dance metaphors as a way to engage students in the learning process. students who either can dance or would like to imitate the dance in the videos should have a greater response. While research shows that the mental representation of dance depends on what the observer can do, nonetheless many students, regardless of their ability or desire to dance, have an affinity for watching popular dance videos and identify with many of the performers. Table 3 describes some of the dance metaphors that we have developed for our educational model.

SOCIALIZATION AND DANCE AS A MEDIUM FOR LEARNING

Constructivism, and its outgrowth, social constructionism, play an important role in our educational model. Cognitive research has shown the importance of community for learning. Specifically, research has considered the conception of "computational participation" for learning computer coding and computational thinking that moves beyond the individual to focus on wider social networks (Kafai & Burke, 2014). In "Youth Video Productions of Dance Performances", Peppler and Kafai (as cited in Kafai, Peppler & Chapman, 2009, ch. 6) discuss the role of youth video productions of dance performances that are popular among youth in the Computer Clubhouses. The Clubhouses are social networks funded by the Intel Foundation to spur the learning of digital technologies in afterschool community centers. The Clubhouse dance video activity presents multiple opportunities for youth to draw upon their individual areas of expertise and take on strong leadership roles.

Dance is known to be a form of nonverbal communication for children at play (Hanna, 1986, 1987). We also consider the possibility of students participating in active dance as part of our educational model. Uniting their aspirations to dance with a STEM learning process can provide interesting synergies for learning. Dance can also help to reduce the stress that often impedes learning and can be effective in sparking the brains of all students, particularly the at risk, low income youngsters.

Psychiatrist John Ratey (2008) explains how each person's capacity to master new and remember old information is improved by biological changes in the brain brought on by physical activity. Thus, dance as a form of exercise can trigger the chemical brain-derived neurotropic factor that supports the health of young neurons, encourages the growth of new ones and fortifies connections among the neurons (Hanna, 2015).

OTHER SENSORIMOTOR BASED METAPHORS

We also consider metaphors based on non-dance body motions, such as the gestures of a rapper, and use motions of inanimate objects, for example cars or planes to represent many of the STEM topics. In particular, the streaming police cars chasing the rap star 50 Cent in the video "Stunt 101" is used as a metaphor for the automobile assembly line. The motion of a yacht in Ace Hood's "Ride" can be a symbol for a continuous flow conveyor belt.

In many cases, the dance is part of the background of the main scene and therefore does not serve a central role in the metaphor. This is the concept of the sphere of dance defined by Hanna (2015, p. 61) as

Table 3. Dance metaphors

Music Video Dance	STEM Image or Animation	Properties of Metaphorical Unit
Fast, fluid dance choreography from "Livi' la Vida Loca"	Teaching a robot to replicate a path based upon numerical/geometric path control	Precision; flow; replication
Fast stop and start motion of dancer in the video "Ignition"	Teaching a robot to replicate a path	Precision based upon discrete changes in path direction
Humanoid robot-like body motions by Chris Brown in "I Can Transform Ya"	Walking and running motions of Asimo the Honda research humanoid robot	The essential qualities of the human mind and body: either real or simulated
Choreography of Chris Brown that elevates the body from the ground level	Motion of various military research robots developed by Boston, since bought by Google	Durability, flexibility, ruggedness
Bruce Springsteen inviting a dance partner from the audience in "Dancing in the Dark"	Robot "sees" in 3D in order to develop the optimum grip	Intelligence, ability to predict and conform, clairvoyance
Britney Spears and dance troupe performing in "3"	Process of layering used in 3-D printing	Technological innovation in the development of manufacturing using 3-D printing
Choreography from Madonna's video "Vogue"	CAD/CAM: computer aided design and manufacturing	The importance of shape, proportion, and measurement in design
Choreography from Britney Spears Las Vegas show using ropes and masks	Robots with wires connected to a circuit board	Control, action within control limits
Choreography from "I Can Transform Ya"	Manipulator arms of welding robots	Dexterity, agility, flexibility, efficiency
Choreography from Usher's DJ's Got us Fallin in Love using stop and start photography of motion	Robot control and sensors	Position, velocity, force and distance
Choreography from Justin Bieber's "Baby"	Parameter encoding in robotic control	Precision in distance and geometric measurement
Choreography with "Freak Show" in Britney Spears Las Vegas Show	Robot navigation, path planning and trajectory following	Command and control for path trajectories
Choreography of Britney Spears for "Toxic" in Las Vegas Show	Robots going into dangerous and hazardous places	Ability to survive in and adapt to dangerous places
Complex choreography in Paula Abdul's "Cold Hearted"	Flowchart representing reactive control process	Robotic control based upon reactive sensors
Fast paced dancing in Mark Ronson's "Uptown Funk" featuring Bruno Mars	Flowchart of various hardware and software components for robotic control	Speed and precision in robotic control

"dance as a vehicle for another medium". Nonetheless, in such situations the dance can be considered part of the entire landscape of sensorimotor stimuli that are enacted in the comprehension of the metaphor.

MUSIC METAPHORS

Neuroscience theory shows a strong relationship between music and cognition as well as emotion and cognition. Thus music plays an important role in enhancing the comprehension of the STEM topics in our educational model.

The concept of music as a communication channel of meaning was explored by Edward Meyer's *Emotion and Meaning in Music* (1956). This is perhaps the first major treatise on music, written by a

Western music theorist, that relies heavily on psychological insights and psychologically-based arguments in describing music. In the Preface, Meyer explicitly acknowledges his debt to Koffka, one of the main proponents of Gestalt psychology. Note that we have already considered the Gestalt psychology as a way to explain the conceptualization of visual metaphors.

MUSIC AND NEUROSCIENCE

From the standpoint of modern cognitive science the relationship between music and language has barely begun to be explored but the appeal of such research is easy to understand. The two systems of music and language perform remarkably similar interpretative feats, converting complex acoustic sequences into perceptually discrete elements, words or chords that convey rich meanings (Patel, 2008).

The cognitive neuroscience of music has become increasingly concerned with the brain basis for musical aesthetics and musical emotion. It is distinguished from related fields, such as cognitive musicology, by its reliance on direct observations of the brain and use of such techniques as functional magnetic resonance imaging (fMRI) and transcranial magnetic stimulation (TMS).

In our model, music is typically connected with some form of dance. Not only is music a frequent component of dance, but it uses the same skills. For example the ability to tap to a beat requires cognitive, linguistic, motor and perceptual skills (Patel, 2008; Brown et al., 2006). Tapping to a beat is a specialized, complex process that calls upon a wide-ranging network of auditory, motor and pre-frontal brain areas. The time-keeping tapping performance appears to be associated with attention and reading (Tierney & Kraus, 2013).

Timing is one of the three elementary motor sensory functions of the brain that are involved with a musical performance. The other two are sequencing and spatial organization. Although neural mechanisms involved in timing movement have been studied rigorously over the past 20 years, much remains controversial. The ability to phrase movements in precise time has been accredited to a neural metronome or clock mechanism where time is represented through oscillations or pulses (Buhusi & Meck, 2005; Ivry & Spencer, 2004; Spencer, Zelaznik, Diedrichson & Ivry, 2003; Wing 2002). An opposing view to this metronome mechanism has also been hypothesized stating that it is an emergent property of the kinematics of movement itself (Spencer et al., 2003; Wing, 2002; Mauk & Buonomano, 2004).

The entwinement of music and dance in the music videos parallels the similar functions they span in the neurological process particularly as it relates to timing and rhythm. In our educational model we can capitalize on these synergies by engaging the students in the rhythmic beat of the songs as a way of enhancing the domain of neurological stimuli and gaining the most from the cross sensory modalities of vision, action and music. Table 4 describes some of the music metaphors that we developed for our educational model.

COGNITION AND EMOTIONS

Emotions play an important role in our educational model. Emotions and feelings can be triggered by the images, dance and music of the videos and can be powerful tools to increase student engagement and retention as well as create personal empathy with information in the presentation. Positive emotions and feeling can also reduce stress which can be a barrier to learning.

Table 4. Metaphors based upon music

Song Title or Concept	Music Style Mood	Video Content	STEM Topic
"Stunt 101", G Unit	Fast paced rap music	Car detail location, streaming police cars	Automotive manufacturing technology
"I Can Transform Ya", Chris Brown	Rhythmic R&B song with a robotic theme	Choreography with robotic theme	Motions of various types of robots; production flow charts
"23", Mike Will Made-It featuring Miley Cyrus	Upbeat rap music appealing to high school students	Students carry out their "free expression" in high school	Manufacturing of the future: innovation as a calling for youth
"Vogue", Madonna	Popular dance piece with creative theme	Design and casting in the fashion industry	CAD/CAM
"Livin' la Vida Loca", Ricky Martin	Popular Latin dance number with carefree mood	Fast paced dancing to Latin music	Tracking robotic gestures using human motion
"Shake It Off", Taylor Swift	Big hit pop number with upbeat mood catering to youth	High quality dance number with rich choreography; finger and hand motions	Programming ASIMO to make a drink; hand and finger motion
"Stronger", Britney Spears	Assertive, bold mood	Assertive choreography	One legged Google robot
"Dancing in the Dark", Bruce Springsteen	Rock song with upbeat dancing spirit	Bruce picks a girl from the audience to dance with	"seeing" in 3D with optimum grip
"Beauty and the Beat", Justin Bieber	Upbeat carefree party style music	Justin swims underwater in at a pool party	Marine robots
"Toy Soldier", Britney Spears	Pop song with sensual overtones	Military style choreography	Military drones
"I've Got You Under My Skin", Katherine McFee	Old time romantic classic	Tactile functions: instruments and micrphone	Haptic technology
"Mirrors", Lil Wayne	Soulful, introspective hip-hop/rap song	Blood stained images that reflect looking at inner self	Robotic surgery
"Perfume", Britney Spears	Soulful pop song	Britney in a colorful outfit at her Las Vegas show	Electronic nose
"Who Gon Stop Me", Kanye West	Hip-hop/rap song with determination	Animated sharks in tanks at rap concert	Navy shark drone
"Out Here Grindin", DJ Khaled	Hip-hop/rap song with bold, arrogant mood	Rappers seeming to take over the world	The "singularity", when robots become more powerful than humans

Dance as well as other forms of bodily movement either observed in the videos or performed by the participants can be a strong force in eliciting emotions. Emotion is semantically linked to movement, as the Latin word *emovere* means to "move out". To quote Hanna (1983, p. 5): "Through movement, emotions announce themselves inside our bodies; hearts pound, backs stiffen, stomachs churn, hands tremble, and faces blush".

Researchers have long studied and debated the relationship between emotions and reasoning, particularly as it pertains to learning and development. The emotional brain, the limbic system has the power to open or close access to learning and memory, so emotions can be viewed as an on/off switch for learning. Furthermore, emotions play a crucial role in the evolutionary development of the brain and functioning of the neurological system (Vail, 1994):

In the quest to understand human behavior, many have tried to overlook emotion, but to no avail. Behavior and mind, conscious or not, and the brain that generates them, refuses to reveal their secrets unless emotion (and the many factors that hide under its name) is factored in and given its due. Damasio (2010, p.108)

Emotions have traditionally been thought of as disrupting cognitive activity. However neuroscientists have found that they can play an important role in the cognitive process. Historian of science Ruth Leys (2011) notes that a message consciously received may be overridden by the unconscious affect that resonances with the source of the message. Positive emotion can increase our concentration power and emotions play a role in virtually all brain functions (Davidson, 2012). Neuroscientist Joseph LeDoux (1998) informs us that the connections from the emotional systems to the cognitive systems are stronger than the connections from the cognitive systems to the emotional systems. He concludes that the struggle between thought and emotion may be ultimately resolved by a harmonious integration of reason and passions in the brain that will allow future humans to use their emotions more effectively in their daily lives.

Neuroscience theory provides ample evidence of the connection between the emotions induced by music and brain activity. For example, emotions induced by music activate similar frontal brain regions compared to emotions elicited by other stimuli (Croom 2012). Schmidt and Trainor (2001) discovered that the positive vs. negative valence of musical segments was distinguished by patterns of frontal EEG activity. Blood and Zatorre (2001) found that music is able to create an incredibly pleasurable experience that can be described as "chills". They used PET to measure changes in cerebral blood flow while participants listened to music that they knew to give them the "chills" or any sort of intensely pleasant emotional response.

By providing scientific evidence of the connection between emotions and learning as well emotions and music, neuroscience research provides strong support for our approach that integrates popular music and dance into the learning process.

CREATING FUTURISTIC IDEAS

Perhaps the most significant part of the model is its potential for enhancing student imagination to create futuristic ideas for technology that go beyond the information that is presented. For Langer (1942) the human mind "is constantly carrying on a process of symbolic transformation of the experiential data that come to it," causing it to be "a veritable fountain of more or less spontaneous ideas". This process of creativity can be initiated by our metaphorical model and developed in a way to create ideas for innovative technological inventions, such as autonomous vehicles and E-textiles. One could also conceive the possibility of robotic software coding using direct body motions that would correspond to the various robotic commands, move, grip, bend, restore, etc., thus bypassing the traditional tedious and time consuming coding process.

As these examples show, the imagination required to comprehend the metaphors for existing technologies can become a platform to create futuristic ideas.

IMPLEMENTATION AND PROGRAM EVALUATION

In this section, we describe the implementation of our model and present an assessment based upon certain outcome measures. We base our analysis on research and studies in the area of assessing informal science education and in particular afterschool STEM programs. As noted in several reports, assessing the full impact of informal science education (ISE) projects is complex (NRC 2007, 2009, 2012; Friedman, 2008; Noam, Shah, Gomer, Wylie, Buehler & Papain, 2013).

DELIVERABLE PRODUCT

Our educational model was formally implemented in a series of basic computer science classes at West Chester University. The setting was designed to test and evaluate the effectiveness of the model on STEM learning. A less formal implementation of the program was also performed with the Robo Lancers from Central High School in Philadelphia, one of the best public high schools in the nation.

The implementation of the model at West Chester University involved of series of presentations in several sections of an introductory computer science class over a period of several semesters. PowerPoint presentations were based on metaphors that aligned various robotic topics with popular music videos using many of the examples cited in Tables 1-4. The metaphors were developed across several modalities including semantic, linguistic, conceptual, visual, sensorimotor including dance, music and emotion.

A wide range of topics were included from (a) technical topics concerning robotic software, parameter encoding, robotic software coding, robotic vision, and circuit boards to (b) broader topics including applications in business and society and potential benefits and dangers of robots in the future. The presentation is amenable to different environments from formal classrooms to afterschool activities or lab settings and can be structured to include various demonstrations of real robots and other artifacts including active learning experiences.

FORMATIVE EVALUATION

We found formative evaluation, the iterative testing of learning strategies to improve them as they are being developed, to be particularly useful in the development and implementation of our model. Because the presentations occurred over two semesters of classes, we were able to gain feedback from the students and the team members allowing us to try different strategies within the parameters of the project. Following are some examples of either formal suggestions or indirect feedback that were successfully adapted to improve the learning model:

- Limit the number of split screens to two;
- Make maximum use of superimposing videos and images using the PowerPoint transparency function;
- Ascertain when the music was appropriate and when it was distracting;
- Determining which songs or videos were the most popular;
- Finding the appropriate style of presentation based upon trial and error.

There was a strong consensus that superimposing the STEM and music images and videos on each other, as opposed to juxtaposing them, was the most effective way to present the metaphors. Some students thought the music was sometimes distracting if played for too long a period or when it seemed to interfere with the verbal part of the presentation. It also became clear that opinions of the students on each part of the presentation varied depending upon the popularity of the song or the personal preferences for particular songs or performers. For this reason, it seemed better to choose songs or videos with the greatest overall popularity and that were more recent.

It also became clear that an optimal strategy was for each presenter to develop his own style of presentation. However, it was considered useful for each presenter to understand some of the psychological, philosophical and neuroscience underpinnings of the model. For example, an understanding of the relationship between mirror neurons, empathy and learning and the concept of semantic embodiment can help the presenter develop effective ways to link the various STEM topics to visual images, dance and other sensorimotor stimuli of the music videos. The style of presentation would most likely depend upon the type of audience, inner city versus suburban youth, public versus private school students or high school versus university students, science majors versus liberal arts majors.

SUMMATIVE EVALUATION: INFORMAL ASSESSMENT

In this section, we present the results of an informal assessment based on observation of the project using the impacts framework for informal science projects developed by Dierking in the National Science Foundation's (NSF) Framework for Evaluating Informal Science Projects (Friedman, 2008 ch. 3). The impact categories suggested in this report are: Knowledge, Engagement, Attitude, Behavior, Skills and Other specific to the topic. These categories are theoretically grounded in the informal science education professional literature as well as the more general education literature.

Knowledge

This impact category emphasizes what a participant consciously knows including facts, awareness or understanding that can be stated by the participants in their own words. It can defined as the measurable demonstration of, assessment of, change in, or exercise of awareness, knowledge, understanding of a particular scientific topic, concept, phenomena, theory or careers central to the project (Friedman, 2008).

Potential Indicators: Students self-reporting statements; comments by faculty observers.

Evidence of Impact

Students were randomly questioned after each of the sessions to assess the impact of the presentation on their acquisition of knowledge. The general reaction from the students indicated that the videos increased their level of concentration and attention. There was a strong consensus that the music videos enhanced their interest in the topics. Their statements provided strong evidence that the music videos created a positive impact on their ability to learn. Several university faculty were invited to attend some of the presentations as impartial observers and give feedback as to the impact on knowledge as well as other impact categories. They confirmed that an appropriate amount of information content was imparted to the students, even with the time and attention paid to the music videos.

Engagement

This impact category captures the excitement and involvement of the learners in a topic or area of STEM. This impact is often a focus of projects that aim to engage under-represented participants in STEM.

Potential Indicators: Spontaneous reaction and comments from students; willingness of students to volunteer time to advance the project.

Evidence of Impact

The use of popular music captured the imagination of many of the students. For example, there was a spontaneous reaction to the popular Taylor Swift video "Shake It Off." The style of presentation seemed to play a significant role in translating the enthusiasm generated by the music videos to interest in robotic topics, for instance Asimo, the Honda humanoid robot; Atlas, a military robot; and the Amazon drones. The occasional clapping of the hands and body motions of the students in rhythm to the music provided concrete evidence of the increased engagement by the students. There was also a positive response by many students to volunteer their time to help advance the project by providing feedback on the music videos, technical support in downloading music videos and STEM learning videos, and generating ideas to develop the metaphors.

Attitude

Although this category is similar to engagement, it goes beyond engagement to encompass longer term stances that students take toward the STEM topics.

Indictors: Linkage of popular music videos and pop stars to the promotion of STEM topics.

Evidence of Impact

The illusion created by the videos that the pop stars are engaged in teaching the topics, besides being a critical component of the metaphor, is also somewhat grounded in reality. One of the project goals is to enlist the support of the pop stars to donate time or money to the educational project. The students showed an unqualified level of support for such a plan. Thus, they were asked to suggest the most appropriate pop star that could potentially be enlisted to support such a program taking into account the technical, business, demographic, and strategic importance of each topic, the contents of the relevant music videos and the personalities and persona of the various pop stars.

Behavior

One would not expect to be able to measure the impact of the model on the long term behavior of the students from such an informal assessment process. However, in the next section, we present the results of a randomized post-only experimental design that uses questionnaire data to assess the impact of the project on the motivation of students to pursue careers and studies in STEM fields.

Skills

This impact category is concerned with the procedural aspects of knowing on the part of the students.

Potential Indicators: The use of metaphors that involve constructions based upon skills, for example computer coding and making homemade robots.

Evidence of Impact

Metaphors were developed that aligned various robotic based skills to the music videos. For example, the information implied by the codes of robotic software program were aligned to the dance motions of several videos. Code phrases, like "Nodehandle," "GetParam," and "read_joint_to" were linked to the sensory motor aspects of the dance. Similarly, the artifacts and procedures used to build a homemade robot were linked to the images and motions of various videos like the dance routine from Britney Spears Las Vegas Show involving the use of ropes as a metaphor for the wiring a homemade robot. Other music videos were used to represent various aspects of a robotic motherboard.

Although most of the presentations consisted of lecture, there were some lab based learning sessions involved programming and robotic constructions. As in the lecture portion of the presentation, the use of the music videos appeared to increase the interest and engagement in the acquisition of skills during these lab sessions.

SUMMATIVE EVALUATION: EXPERIMENTAL DESIGN AND STATISTICAL RESULTS

Experimental designs are not always more appropriate than naturalistic methods, but they provide a structured framework that can make the results more plausible. We use a randomized post-only design in which participants are randomly assigned to a group that experiences a presentation including metaphorical content based on our music video model or to a group that sees the same STEM content without the metaphors. This design can be summarized as follows:

R XO

R YO

where R represents random assignment, X represents the presentation that includes the music video metaphors, and Y represents the presentation that includes the same STEM content as presentation X but without the metaphors. In some cases, relevant You Tube educational videos were used in presentation Y.

We developed a series of questions that attempts to measure the effect of the model on three broad learning outcomes identified by the Defining Youth Outcomes for Stem Learning in the Afterschool (2013) study. This study yielded consensus about three major outcomes for children and youth participating in afterschool STEM programs and a set of indicators and sub-indicators that support these outcomes. The three outcomes are: (a) interest in STEM and STEM learning activities; (b) capacities to engage in STEM learning activities; (c) appreciating the value of STEM goals and STEM learning activities. These outcomes are reflected in the phrases: (a) "I like to do this"; (b) "I can do this"; (c) "I think this is important." These broad developmental outcomes and indicators of learning reflect constructs found in evaluation reports of afterschool STEM programs (Afterschool Alliance, 2011) as well as the research literature pertaining to human development (Hidi & Renninger, 2006; Holland, Lachicotte, Skinner & Cain, 1998; Lave & Wenger, 1991), youth development (Barber, Stone, Hunt & Eccles, 2005; Eccles, 2005) and science learning (NRC, 2007 & 2009). To measure the effect of our model on these broad outcomes, we developed a questionnaire format that offered six statements or factoids about each of eleven robotic topics. Of these six statements, two represent each of the three outcome categories we designated as: (a) Interest/Curiosity; (b) Skills/Properties, where the term properties includes reference to the detailed, technical aspects; (c) Importance/Attitude. Each statement was designed to reflect an aspect of each outcome categories. The factoid statements are provided in the Appendix. The participants were asked to rank order the six statements from 1 to 6 (1= highest rank) for each topic in terms of their perceived importance. Table 5 presents the section of the instrument that relates to Topic 1: Manufacturing Technology. The other ten topics in the appendix are surveyed with the same format. The survey was administered to classes of computer science students at West Chester University during the Spring semester 2015.

Each of the 11 topics was assigned to a separate page with a visual metaphor related from the music video for the topic displayed at the top of the page. The 6 statements or factoids related to the topic were written below the image. Although the categories were not identified to the students who participated in the survey, the first two statements corresponded to Interest/Curiosity, the third and fourth statements corresponded to Skills/Properties and the last two statements corresponded to Importance/Attitude.

We tabulated the number of statements that were ranked either 1st or 2nd in each of the three categories, for each of the eleven topics, for Presentation X and Presentation Y. The average rank of the statements in each category is also computed for each presentation. The results are presented in Table 6.

There were 32 respondents for Presentation X and 22 respondents for Presentation Y. The total frequency for the factoids ranked either 1st or 2nd for Presentation X is equal to 704, representing the first

Table 5. Outcome questionnaire

Topic 1: Manufacturing Technology	
Ranking 1 - 6	Based on your impressions from the PowerPoint presentation that you have seen, rank the following statements form 1 (highest) – 6 (lowest)
	Manufacturing technology is a creative process.
	It can be both personally and financially rewarding to work in todays advanced manufacturing technology environment.
	Knowledge and skills about manufacturing technology are very marketable in today's economic environment.
	The tools and technologies that can assist manufacturing are constantly changing to keep up with the competitive nature of today's business and consumer needs.
	Manufacturing technology is important because it helps to make better quality products at a lower cost.
	Manufacturing technology affects our everyday lives because of all the products that it helps to produce.

Table 6. Summary of questionnaire results

Outcome Category	Frequency (%) of Statements Ranked 1st or 2nd for Presentation		Average Rank for Presentation	
	X	Y	X	Y
Interest / Curiosity	213 (30.2%)	237 (48.9%)	3.40	2.65
Skills / Properties	132 (18.8%)	124 (25.7%)	4.85	4.00
Importance / Attitude	359 (51.0%)	123 (25.4%)	2.25	3.85
Total (Frequency)	704 (100%)	484 (100%)	10.5	10.5

two ranks for each of the eleven topics for 32 respondents (704= 2×11×32). Similarly there are a total of 484 factoids ranked either 1st or 2nd for Presentation Y (484= 2×11×22) which has 22 respondents. Note that the sum of the average ranks across all three outcomes for both presentations is 10.5 = (1 + 2 +…+ 6) / 2, because there are 6 questions per topic that are ranked.

The results show a significantly larger influence of Presentation X on the Importance/Attitude category compared with Presentation Y, using a Wilcoxon rank-sum test at the .01 level of significance. This provides some validation that the music video presentation increased student awareness of the importance of robots in our society and their appreciation of the topic beyond that normally obtained from a traditional learning experience. These result are consistent with the findings of the Afterschool Alliance study (2013) that the participants were most confident about the impact of traditional stem learning on: (a) the curiosity or interest about STEM topic, outcome category 1; (b) the indicators of doing STEM learning activities, outcome category 2. However, this study also found that the participants were less confident about the contribution of traditional stem learning programs to student appreciation of the value of STEM in society and their ability to exercise STEM relevant life and career skills, outcome category 3. It is in the domain of outcome category 3 that our music video model has potential to supplement traditional stem learning models.

Another survey consisting of 10 questions was administered to the students who had participated in the previous survey. Of the 10 questions, 9 were chosen to reflect salient aspects of the three categories studied in the previous survey. Questions 1 and 5 relate to Skills/Properties, while questions 2, 5, 6, 7, 8 and 9 relate to Importance/Attitude. Question 10 assesses how enjoyable and engaging was the presentation, relating to Interest/Curiosity. The questions were posed to measure the extent to which the presentations had an immediate impact on the students' attitude and perception of learning about robots and general engineering and science topics. The results are presented in Tables 7 and 8.

While both presentations received overall favorable ratings regarding their impact on knowledge and interest, the results show Y is viewed more favorably than X on the question 1 concerning knowledge about robotics. This is not surprising because Presentation Y consisted entirely of educational YouTube videos on robotic processes that contained more technical details than the music video presentation. For questions 3, 4 and 7, the two presentations received similar patterns of response. However, for six of the questions, Presentation X received either slightly or moderately more favorable results. Three of these, questions 5, 9 and 10, involve the personal attitude of the student to the presentation related to outcome category 3. Questions 2 and 6 involve the importance of engineering and science to society while 8 indicates the consequence to encourage others to pursue the field. These three are related to the importance aspect of outcome category 3.

Table 7. Ratings for presentation X (n=32)

Questions	STA	MA	SLA	N	D
1. The presentation significantly expanded my knowledge about robotic processes	3	5	12	7	4
2. The presentation showed the value of robots in our society	6	10	10	4	2
3. The presentation spurred my interest in robots	4	6	8	8	6
4. The presentation motivated me to learn more about engineering	3	6	9	7	7
5. The presentation reduced my anxiety about learning engineering or science	6	8	8	6	4
6. The presentation increased my perception of the importance of engineering or science for our society	6	9	11	4	2
7. The presentation increased the likelihood that I might pursue a career in engineering or science	3	3	6	10	10
8. I am more likely to encourage a friend to pursue a career related to science or engineering	4	8	12	4	4
9. The presentation affected me in a personal way	5	7	10	6	4
10. The presentation was enjoyable and engaging	7	10	10	4	1

STA = Strongly agree, MA = Moderately agree, SLA = Slightly Agree, N = Neutral, D = Disagree

Table 8. Ratings for presentation Y (n=22)

Questions	STA	MA	SLA	N	D
1. The presentation significantly expanded my knowledge about robotic processes	6	8	5	2	1
2. The presentation showed the value of robots in our society	3	4	7	6	2
3. The presentation spurred my interest in robots	3	3	5	6	5
4. The presentation motivated me to learn more about engineering	3	4	6	5	4
5. The presentation reduced my anxiety about learning engineering or science	2	3	7	6	4
6. The presentation increased my perception of the importance of engineering or science for our society	3	4	5	7	3
7. The presentation increased the likelihood that I might pursue a career in engineering or science	2	3	5	6	6
8. I am more likely to encourage a friend to pursue a career related to science or engineering	3	4	5	6	4
9. The presentation affected me in a personal way	1	4	5	8	4
10. The presentation was enjoyable and engaging	1	5	3	3	4

STA = Strongly agree, MA = Moderately agree, SLA = Slightly Agree, N = Neutral, D = Disagree

Overall, the results of the questionnaires seems to indicate that the music video presentation improved the attitude of the students toward the STEM topics and increased their perception of the importance of the topics. Thus, the presentation seems to add some value not normally achieved by traditional approaches. As mentioned above, this seems consistent with the results of the study of the Afterschool Alliance (2013) indicating a weakness in the ability of afterschool STEM programs to have a significant impact on the attitudes toward STEM topics.

The results of the formative evaluation, the informal summative evaluation and the statistical results of the experimental design, coupled with the theories of neuroscience, indicate that the model enhanced the achievement of the outcomes proposed in our conceptual framework listed in Figure 1, namely comprehension, retention, attitude and importance, and decision making and judgement.

CONCLUSION AND FUTURE RESEARCH DIRECTIONS

Our model can be viewed within the context of the STEAM initiative. The "A" for arts is added to the STEM acronym. In July 2012, PBS published a case study about the benefits of STEAM education. The PBS study cited a report from the National Endowment of the Arts that students from lower socioeconomic backgrounds are more likely to aspire to college and score better in writing and science if they studied the arts (Chen, 2015). In November of 2014, EdWeek published an article debating if and how the arts can be integrated into a STEM curriculum (Jolley, 2015).

Popular music and dance, including rap and hip-hop, may not represent the kind of art that the founders of STEAM originally intended, but our study shows that this genre of art can be integrated with STEM education. Although the general popularity of a work of art is not usually considered an adequate measure of quality, popularity can be a useful way to measure the extent to which a work of art can be used to influence the greatest number of people in the shortest amount of time.

Postulating a neuroscience basis for our model is interesting in itself, but also has a practical value by motivating the presenters to develop creative styles of presentation. The ability of the presenter to coordinate the different stimuli, semantic, perception, sensory motor, affective and auditory, is undoubtedly challenging but feasible. Various neuroscience studies suggest that these signals are not so distinct as far as the brain is concerned, and thus can be coordinated into a cognitive unit. Future research could consider the possibility of using functional resonance imaging (fMRI) to investigate differences in brain activity using our approach versus traditional approaches.

Expert practitioners make choices based on a wide variety of factors: who the children are, what the potential of the learning setting is, etc. Seasoned practitioners combine these situational factors with the understanding, theories, and beliefs they have developed through experience and reflecting on the views of trusted experts (Smylie, 1989). In the end, practical knowledge often has more impact than research-based knowledge on what actually happens in the field (Coburn, Honig & Stein, 2009; Coburn & Stein, 2010; Nelson, Leffler, & Hansen, 2009). Ultimately, the effectiveness of the model would depend upon the practical application of the model given the type of student, their educational level and predisposition towards different educational approaches.

Future research could investigate the effects of the model on students already predisposed to learning STEM topics. It would be insightful to compare the effects on students who have higher science and mathematics grades or wh are participating in science based clubs with the average students. One could also adapt the study to measure the effects of the model on the motivation of STEM professionals to continue or advance their careers.

The possibility of gaining the support of celebrated pop stars adds an exciting dimension to the project. In general, harnessing the energy of youth normally devoted to popular music and redirecting it to the study of science and technology could add a positive dimension to the challenge of STEM education and awareness.

REFERENCES

Aldrich, V. C. (1968). Visual metaphor. *Journal of Aesthetic Education*, *2*(1), 73–86. doi:10.2307/3331241

Alliance, A. (2011). STEM learning in afterschool: An analysis of impacts and outcomes. Retrieved from http://www.afterschoolalliance.org/STEM-Afterschool-Outcomes.pdf

Alliance, A. (2013). Defining youth outcomes for stem learning in afterschool. Retrieved from http://www.afterschoolalliance.org/STEM-Outcomes-2013.pdf

Baars, B. J. (1989). *A cognitive theory of consciousness*. Cambridge, Mass: Cambridge University Press.

Barber, B. L., Stone, M. R., Hunt, J. E., & Eccles, J. S. (2005). Benefits of activity participation: The roles of identity affirmation and peer group norm sharing. In J. L. Mahoney, R. W. Larson, & J. S. Eccles (Eds.), *Organized activities as contexts of development: extracurricular activities, after-school and community programs* (pp. 185–210). Mahwah, NJ: Lawrence Erlbaum Associates.

Beardsley, M. (1962). The metaphorical twist. *Philosophy and Phenomenological Research*, *22*(3), 293–307. doi:10.2307/2104415

Bezuidenhout, A. (2001). Metaphor and what is said: A defense of a direct expression view of metaphor. *Midwest Studies in Philosophy*, *25*(1), 156–186. doi:10.1111/1475-4975.00044

Black, M. (1962). *Metaphor. Models and Metaphors* (pp. 25–47). Ithaca, NY: Cornell UP.

Blakeslee, S. (2006, January 10). Cells that read minds. *New York Times*.

Blasing, B., Puttke, M., & Schack, T. (2009). The cognitive structure of movements in classical dance. *Psychology of Sport and Exercise*, *10*(3), 350–360. doi:10.1016/j.psychsport.2008.10.001

Blood, A. J., & Zatorre, R. J. (2001). Intensely pleasurable responses to music correlate with activity in brain regions implicated in reward and emotion. *Proceedings of the National Academy of Sciences of the United States of America*, *98*(20), 11. doi:10.1073/pnas.191355898 PMID:11573015

Borghi, A. M., & Cimatti, F. (2010). Embodied cognition and beyond: Acting and sensing the body. *Neuropsychologia*, *48*(3), 763–773. doi:10.1016/j.neuropsychologia.2009.10.029 PMID:19913041

Brody, R. (2014). Developing new stem teaching methods. *U.S. News & World Report*, (April): 25.

Brown, S., Martinez, M. J., & Parsons, L. M. (2006). The neural basis of human dance. *Cerebral Cortex*, *16*(8), 1157–1167. doi:10.1093/cercor/bhj057 PMID:16221923

Bruner, J. S. (1961). The art of discovery. *Harvard Educational Review*, *31*(1), 21–32.

Buhusi, C. V., & Meck, W. H. (2005). What makes us tick? functional and neural mechanisms of interval timing. *Nature Reviews. Neuroscience*, *6*(10), 755–765. doi:10.1038/nrn1764 PMID:16163383

Calvo-Merino, B., Glaser, D. E., Grezes, J., Passingham, R. E., & Haggard, P. (2005). Action observation and acquired motor skills: An fMRI study with expert dancers. *Cerebral Cortex*, *15*(8), 1243–1249. doi:10.1093/cercor/bhi007 PMID:15616133

Calvo-Merino, B., Glaser, D. E., Grezes, J., Passingham, R. E., & Haggard, P. (2006). Seeing or doing? influence of visual or motor familiarity in action observation. *Current Biology*, *16*(19), 1905–1910. doi:10.1016/j.cub.2006.07.065 PMID:17027486

Camp, E. (2006). Metaphor in the mind: The cognition of metaphor philosophy. *Compass*, *1*(2), 154–170. doi:10.1111/j.1747-9991.2006.00013.x

Carlson, N. R., & Heth, C. D. (2010). *Psychology the science of behaviour* (pp. 20–22). Ontario, CA: Pearson Education Canada.

Carston, R. (2002). *Thoughts and utterances: The pragmatics of explicit communication*. Oxford: Blackwell. doi:10.1002/9780470754603

Chen, Kelly, & Cheers, Imani (Retrieved 7 March 2015). STEAM Ahead: Merging Arts and Science Education. PBS News Hour (television broadcast). PBS. Retrieved 7 March 2015.

Clerget, E., Winderickx, A., Fadiga, L., & Olivier, E. (2009). Role of broca's area in encoding sequential human actions: A virtual lesion study. *Neuroreport*, *20*(16), 1496–1499. doi:10.1097/WNR.0b013e3283329be8 PMID:19809371

Coburn, C. E., Honig, M. I., & Stein, M. K. (2009). What's the evidence on district's use of evidence? In J. D. Bransford, D. J. Stipek, N. J. Vye, L. Gomez, & D. Lam (Eds.), *The role of research in educational improvement* (pp. 67–87). Cambridge: Harvard Education Press.

Coburn, C. E., & Stein, M. K. (Eds.). (2010). *Research and practice in education: Building alliances, bridging the divide*. Lanham, Maryland: Rowman & Littlefield.

Crick, F. (1994). *The astonishing hypothesis: The scientific search for the soul*. New York: Charles Scribner's Sons.

Croom, A. M. (2012). Music, neuroscience, and the psychology of well-being: A précis. *Frontiers in Psychology*, *2*, 1–15. doi:10.3389/fpsyg.2011.00393 PMID:22232614

Cross, E. S., Hamilton, A. F. C., & Grafton, S. T. (2006). Building a motor simulation de noto: Observation of dance by dancers. *NeuroImage*, *31*(3), 1257–1267. doi:10.1016/j.neuroimage.2006.01.033 PMID:16530429

Damasio, A. (1999). *The feeling of what happens: Body and emotions in the making of consciousness*. New York: Harcourt Brace.

Damasio, A. (2010). *Self comes to mind: Constructing the conscious brain*. New York: Random House.

Davidson, R. J. with Begley, S. (2012). The emotional life of your brain: How its unique patterns affect the way you think feel, and live- and how you can change them. New York: Hudson Street Press.

Decety, J. (2002). Naturaliser l'empathie [Empathy naturalized]. *L'Encéphale*, *28*, 9–20. PMID:11963348

Decety, J., & Jackson, P. L. (2004). The functional architecture of human empathy. *Behavioral and Cognitive Neuroscience Reviews*, *3*(2), 71–100. doi:10.1177/1534582304267187 PMID:15537986

Deheane, S., & Changeux, J. P. (2011). Experimental and theoretical approaches to conscious processing. *Neuron, 70*(2), 200–219. doi:10.1016/j.neuron.2011.03.018 PMID:21521609

Dent, C. H. (1984). The developmental importance of motion information in perceiving and describing metaphoric similarity. *Child Development, 55*(4), 1607–1613. doi:10.2307/1130029 PMID:6207996

Dent-Read, C. H., Klein, G., & Eggleston, R. (1994). Metaphor in visual displays designed to guide action. *Metaphor and Symbolic Activity, 9*(3), 211–232. doi:10.1207/s15327868ms0903_4

Dewey, J. (1938). *Experience and education*. New York: Simon and Schuster.

Dilworth, J. (2004). Internal versus external representation. *The Journal of Aesthetics and Art Criticism, 62*(1), 23–36. doi:10.1111/j.1540-594X.2004.00132.x

Eccles, J. S. (2005). The present and future of research on activity settings as developmental contexts. In J. L. Mahoney, R. W. Larson, & J. S. Eccles (Eds.), *Organized activities as contexts of development: Extracurricular activities, after-school and community programs* (pp. 353–371). Mahwah, NJ: Lawrence Erlbaum Associates.

Edelman, G. (2004). *Wider than the sky: The phenomenal gift of consciousness*. New Haven: Yale University Press.

Eisner, E. W. (2002). *The educational imagination: The design and evaluation of school programs* (3rd ed.). Upper Saddle River, NJ: Merrill Prentice Hall.

El Refaie, E. (2003). Understanding visual metaphor: The example of newspaper cartoons. *Visual Communication, 2*(1), 75–95. doi:10.1177/1470357203002001755

Fadiga, L., Craighero, L., Destro, M. F., Finos, L., Cotilon-Williams, N., Smith, A. T., & Castiell, U. (2006). Language in shadow. *Social Neuroscience, 1*(2), 77–89. doi:10.1080/17470910600976430 PMID:18633777

Fischer, J., & Whitney, D. (2014). Serial dependence in visual perception. *Nature Neuroscience, 17*(5), 738–743. doi:10.1038/nn.3689 PMID:24686785

Fogelin, R. (1988). *Figuratively speaking*. New Haven: Yale University Press.

Friedman, A. (Ed.). (2008). Framework for evaluating impacts of informal science education projects. Retrieved from http://insci.org/resources/Eval_Framework.pdf

Fuster, J. M. (1995). *Memory in the cerebral cortex–An empirical approach to neural networks in the human and nonhuman primate*. Cambridge, MA: MIT Press.

Galaburda, A. M., Kosslyn, S. M., & Christen, Y. (Eds.). (2002) The languages of the brain. Cambridge, M.A: Harvard University Press.

Gallese, V. (2001). The "shared manifold" hypothesis: From mirror neurons to empathy. *Journal of Consciousness Studies, 8*, 33–50.

Gallese, V., & Goldman, A. I. (1998). Mirror neurons and the simulation theory. *Trends in Cognitive Sciences*, *2*(12), 493–501. doi:10.1016/S1364-6613(98)01262-5 PMID:21227300

Glucksberg, S., & Keysar, B. (1993). How metaphors work. In A. Ortony (Ed.), *Metaphor and thought* (2nd ed., pp. 401–424). Cambridge: Cambridge University Press. doi:10.1017/CBO9781139173865.020

Glynn, S. M. (2007). Methods and strategies: The teaching-with-analogies model. *Science and Children*, *44*(8), 52–55.

Glynn, S. M. (2008). Making science concepts meaningful to students: Teaching with analogies. In S. Mikelskis-Seifert, U. Ringelband, & M. Brückmann (Eds.), *Four decades of research in science education: From curriculum development to quality improvement* (pp. 113–125). Münster, Germany: Waxmann.

Goodman, N. (1968). *Languages of art: An approach to a theory of symbols*. New York: Bobbs-Merrill Company.

Grice, H. Paul (1975). Logic and conversation. Syntax and Semantics (Ed. P. Cole & J. L. Morgan) (Vol. 3, pp. 41–58). . New York: Academic Press.

Grove, R., Stevens, C., & McKechnie, S. (Eds.). (2005). *Thinking in four dimensions: Creativity and cognition in contemporary dance*. Carlton, Australia: Melbourne University Press.

Guttenplan, S. P. (2005). *Objects of metaphor*. Oxford University Press. doi:10.1093/0199280894.001.0001

Hanna, J. L. (1983). *The performer-audience connection*. Austin, Texas: University of Austin Press.

Hanna, J. L. (1986). Interethnic communication in children's own dance, play, and protest. In Y. Y. Kim (Ed.), *Interethnic communication, 10. International and intercultural communication annual* (pp. 176–198). Newbury Park, CA: Sage.

Hanna, J. L. (1987). To dance is human: A theory of nonverbal communication (Rev. 1979 ed.). Chicago: University of Chicago Press.

Hanna, J. L. (2015). *Dancing to learn: The brain's cognition, emotion and movement*. Lanham, Maryland: Rowan and Littlefield.

Hidi, S., & Renninger, K. A. (2006). The four-phase model of interest development. *Educational Psychologist*, *41*(2), 111–127. doi:10.1207/s15326985ep4102_4

Hobbes, T. (1651). *Leviathan; or, the matter, form, and power of a commonwealth ecclesiastical and civil*. London: Andrew Crooke.

Holland, D., Lachicotte, W. Jr, Skinner, D., & Cain, C. (1998). *Identity and agency in cultural worlds*. Cambridge: Harvard University Press.

Holton, G. (1973). *Thematic origins of scientific thought*. Cambridge, MA: Harvard University Press.

Iacoboni, M. (2005). Grasping the intentions of others with one's own mirror neuron system. *PLOS Biology*, *3*(3), 0529-0535.

Iacoboni, M. (2009). Imitation, empathy and motor neurons. *Annual Review of Psychology*, *60*(1), 653–670. doi:10.1146/annurev.psych.60.110707.163604 PMID:18793090

Ivry, R. B., & Spencer, R. M. (2004). The neural representation of time. *Current Opinion in Neurobiology*, *14*(2), 225–232. doi:10.1016/j.conb.2004.03.013 PMID:15082329

Jarrett, C. (2014, March 10). The neuroscience of decision making explained in 30 seconds. *Wired.com.* Retrieved from http://www.wired.com

Jeong, S. (2008). Visual metaphor in advertising: Is the persuasive effect attributable to visual argumentation or metaphorical rhetoric? *Journal of Marketing Communications*, *14*(1), 59–73. doi:10.1080/14697010701717488

Jolly, A. (2015). Stem vs. steam: do the arts belong? *Education Week Teacher.* Retrieved from http://www.EdWeek.org

Jones, L., & Viamontes, G. I. (2010). Practical applications of mind-body phenomena: The modulation of motor performance by action visualization. *Psychiatric Annals*, *40*(8), 381–387. doi:10.3928/00485713-20100804-04

Kafai, Y. B., & Burke, Q. (2014). *Connected code: Why children need to learn programming*. Cambridge, MA: The MIT Press.

Kafai, Y. B., Peppler, K. A., & Chapman, R. N. (Eds.). (2009). *The computer clubhouse: Constructionism and creativity in youth communities.* New York, London: Teachers College Press.

Kafai, Y. B., & Resnick, M. (Eds.). (1996). *Constructionism in practice: Designing, thinking and learning in a digital world.* Mahway, New Jersey: Lawrence Erlbaum Associates.

Kaiser, W. (1989). Analogien in Physik und i'm 19. Und 20. Jahrhundert. (Analogies in physics and technology in the 19th and 20th centuries). *Berichte zur Wissenschaftsgeschichte*, *12*(1), 19–34. doi:10.1002/bewi.19890120104

Kandel, E. R. (2012). *The art of insight: The quest to understand the unconscious in art, mind and brain, from Vienna 1900 to the present.* New York: Random House.

Kaplan, S. J. (1990). Visual metaphors in the representation of communication technology. *Critical Studies in Mass Communication*, *7*(1), 37–47. doi:10.1080/15295039009360162

Keysers, C. (2011). *The empathetic brain.* Kindle Books.

Koffka, K. (1935). *Principles of gestalt psychology.* London: Routledge.

Kohler, W. (1929). *Gestalt psychology.* New York: Liveright.

Koller, V. (2005). Designing cognition: Visual metaphor as a design feature in business magazines. *Information Design Journal +. Document Design (Amsterdam)*, *13*(2), 136–150.

Kouider, S., & Dehaene, S. (2007). Levels of processing during nonconscious perception: A critical preview of visual masking. *Philosophical Transactions of the Royal Society of London. Series B, Biological Sciences, 362*(1481), 857–875. doi:10.1098/rstb.2007.2093 PMID:17403642

Lakoff, G., & Johnson, M. (1980). *Metaphors we live by*. Chicago: U of Chicago Press.

Lakoff, G., & Johnson, M. (1999). *Philosophy in the flesh: The embodied mind and its challenge to western thought*. New York: Basic Books.

Lakoff, G., & Turner, M. (1989). *More than cool reason: A field guide to poetic metaphor*. Chicago: University of Chicago Press. doi:10.7208/chicago/9780226470986.001.0001

Langer, S. (1942). *Philosophy in a new key: A study in the symbolism of reason, rite and art*. Cambridge, MA: Harvard University Press.

Lave, J., & Wenger, E. (1991). *Situated learning: Legitimate peripheral participation*. Cambridge: Cambridge University Press. doi:10.1017/CBO9780511815355

LeDoux, J. (1998). *The emotional brain*. New York: Simon and Schuster.

Leotin, S. (2012). What scientists can learn from ballet. *Huffington Post.* Retrieved from http://www.huffingtonpost.com

Levin, S. (1977). *The semantics of metaphor*. Baltimore: Johns Hopkins University Press.

Leys, R. (2011). The turn to affect: A critique. *Critical Inquiry, 37*(3), 434–472. doi:10.1086/659353

Livingstone, M. (2002). *Vision and art: The biology of seeing*. New York: Abrams.

Locke, J. (1690). *An essay concerning human understanding*. London: T. Basset. doi:10.1093/oseo/instance.00018020

Martinich, A. P. (1984). A theory for metaphor. *Journal of Literary Semantics, 13*(1), 35–56. doi:10.1515/jlse.1984.13.1.35

Mauk, M. D., & Buonomano, D. V. (2004). The neural basis of temporal processing. *Annual Review of Neuroscience, 27*(1), 307–340. doi:10.1146/annurev.neuro.27.070203.144247 PMID:15217335

Meyer, L. B. (1956). *Emotion and meaning in music*. Lambertville, NJ: Phoenix Books.

National Research Council. (2007). *Taking science to school: Learning and teaching science in grades K-8*. Washington, D.C.: National Academies Press.

National Research Council. (2009). *Learning science in informal environments: People, places, and pursuits*. Washington, D.C.: The National Academies Press.

National Research Council. (2012). Assessment of informal and afterschool science learning. Retrieved from http://sites.nationalacademies.org/DBASSE/BOSE/DBASSE_071087#.ULkF72eJun4

Nelson, S. R., Leffler, J. C., & Hansen, B. A. (2009). *Toward a research agenda for understanding and improving the use of research evidence*. Portland, OR: Northwest Regional Educational Laboratory.

Noam, G., Shah, A., Gitomer, D., Wylie, C., Buehler, J., & Papazian, A. (2013). Dimensions of success: An observation tool for stem programming in out-of-school time. *Program in Education, Afterschool, and Resiliency (PEAR), Harvard University and McLean Hospital.*

Olsson, C.-J., Jonsson, B., Larsson, A., & Nyberg, L. (2008). Motor representations and practice affect brain systems underlying imagery. *The Open Neuroimaging Journal*, *2*(1), 5–13. doi:10.2174/1874440000802010005 PMID:19018312

Ortony, A. (1979). Beyond literal similarity. *Psychological Review*, *86*(3), 161–180. doi:10.1037/0033-295X.86.3.161

Ortony, A. (1988). Are emotion metaphors conceptual or lexical? *Cognition and Emotion*, *2*(2), 95–103. doi:10.1080/02699938808408066

Papert, S. A. (1991). Situating Constructionism. In I. Harel & S.A. Papert (Eds.), Constructionism. Norwood, NJ: Ablex.

Patel, A. D. (2008). *Music, language and the brain*. New York: Oxford University Press.

Piaget, J. (1950). *The psychology of intelligence*. New York: Routledge.

Preston, S. D., & de Waal, F. B. M. (2002). Empathy: Its ultimate and proximate bases. *Behavioral and Brain Sciences*, *25*, 1–72. PMID:12625087

Proweller, W. (1971). Picasso's "Guernica": A study in visual metaphor. *Art Journal*, *30*(1), 240–248.

Pulvermuller, F. (1999). Words in the brain's language. *Behavioral and Brain Sciences*, *22*(2), 253–336. doi:10.1017/S0140525X9900182X PMID:11301524

Pulvermuller, F. (2012). Meaning and the brain, the neurosemantics of referential, interactive, and combinatorial knowledge. *Journal of Neurolinguistics*, *25*(5), 423–459. doi:10.1016/j.jneuroling.2011.03.004

Pulvermuller, F. (2013). Semantic embodiment, disembodiment or misemdodiment? In search of meaning in modules and neuron circuits. *Brain and Language*, *127*(1), 86–103. doi:10.1016/j.bandl.2013.05.015 PMID:23932167

Ratey, J., & Hagerman, E. (2008). Spark: The revolutionary new science of exercise and the brain. New York: Little, Brown.

Recanti, F. (2004). *Literal meaning*. Cambridge: Cambridge University Press.

Richards, I. A. (1936). *The philosophy of rhetoric*. Oxford: Oxford University Press.

Rizzolatti, G., & Destro, M. F. (2008). Mirror neurons. *Scholarpedia*, *3*(1), 2055. doi:10.4249/scholarpedia.2055

Robinson, R. C. (2011). Causation as metaphor. *Rupkatha Journal On Intermediate Studies in Humanities*, *3*(1), 181–190.

Root-Bernstein, R., & Root-Bernstein, M. (1999). *Spark of genius: The thirteen thinking tools of the world's most creative people*. New York: Houghton Mifflin.

Rosch, E., Thompson, E., & Varela, F. J. (1991). *The embodied mind: Cognitive science and human experience* (1992 ed.). MIT Press.

Rothenberg, A. (1979). Einstein's creative thinking and the general theory of relativity. *The American Journal of Psychiatry*, *136*(1), 38–43. doi:10.1176/ajp.136.1.38 PMID:365000

Schmidt, L. A., & Trainor, L. J. (2001). Frontal brain electrical activity (EEG) distinguishes valence and intensity of musical emotions. *Cognition and Emotion*, *15*(4), 487–500. doi:10.1080/02699930126048

Searle, J. (1979). Metaphor. In *Expression and meaning: Studies in the theory of speech acts*. Cambridge: Cambridge University Press. doi:10.1017/CBO9780511609213.006

Seitz, J. A. (2005). The neural, evolutionary, developmental, and bodily basis of metaphor. *New Ideas in Psychology*, *23*(2), 74–95. doi:10.1016/j.newideapsych.2005.11.001

Seth, E. (Ed.), (2014). *30- second brain*. London: Icon Books Ltd.

Smith, B. (Ed.), (1988). *Foundations of gestalt theory*. Munich, Vienna: Philosophia Verlag.

Smylie, M. A. (1989). Teachers' views of the effectiveness of sources of learning to teach. *The Elementary School Journal*, *89*(5), 543–558. doi:10.1086/461591

Spencer, R. M., Zelaznik, H. N., Diedrichson, J., & Ivry, R. B. (2003). Disrupted timing of discontinuous but not continuous movements by cerebellar lesions. *Science*, *300*(5624), 1437–1439. doi:10.1126/science.1083661 PMID:12775842

Sperber, D., & Wilson, D. (1955). *Relevance: communication and cognition* (2nd ed.). Oxford: Blackwell.

Tierney, A., & Kraus, N. (2013). The ability to tap to a beat related to cognitive, linguistic and perceptual skills. *Brain and Language*, *124*(3), 225–231. doi:10.1016/j.bandl.2012.12.014 PMID:23400117

Vail, P. (1994). *Emotion: The on/off switch for learning*. Rosemont, NJ: Modern Learning Press.

Van den Buscsche, E., Van den Noortgate, W., & Reynvoet, B. (2009). Mechansims of masked priming: A meta-analysis. *Psychological Bulletin*, *135*(3), 452–477. doi:10.1037/a0015329 PMID:19379025

Vicari, P. (1993). Renaissance emblematic. *Metaphor and Symbolic Activity*, *8*(3), 153–168. doi:10.1207/s15327868ms0803_2

Vicary, S. A., Robbins, R. A., Calvo-Merino, B., & Stevens, C. J. (2014). Recognition of dance-like actions: Memory for static posture. *Journal of Experimental Psychology. General*, *138*, 64–87. PMID:24554278

Vickers, B. (1984). Analogy versus identity: The rejection of occult symbolism, 1580-1680. In B. Vickers (Ed.), *Occult and scientific mentalities in the renaissance* (pp. 95–163). Cambridge, England: Cambridge University Press. doi:10.1017/CBO9780511572999.005

Vygotskii, L. S. (1978). *Mind in society: The development of higher mental processes*. Cambridge, MA: Harvard University Press.

Weinreich, U. (1966). Explorations in semantic theory. In T. A. Sebeok (Ed.), *Current trends in linguistics, 3: Theoretical foundations* (pp. 395–477). The Hague: Mouton Press.

Wertheimer, M. (1945). *Productive thinking*. New York, NY: Harper.

Wilce, J. Jr. (2009). *Language and emotions. New York*. Cambridge: Cambridge University Press. doi:10.1017/CBO9780511626692

Wing, A. M. (2002). Voluntary timing and brain function: An information processing approach. *Brain and Cognition*, *48*(1), 7–30. doi:10.1006/brcg.2001.1301 PMID:11812030

Winner, E., McCarthy, M., & Gardner, H. (1980). The ontogenesis of metaphor. In R. P. Honeck & R. R. Hoffman (Eds.), *Cognition and figurative language* (pp. 341–361). Hillsdale, NJ: Erlbaum.

This research was previously published in the International Journal of Strategic Decision Sciences (IJSDS), 7(1); edited by Madjid Tavana, pages 39-75, copyright year 2016 by IGI Publishing (an imprint of IGI Global).

APPENDIX: SURVEY FACTOIDS FOR ELEVEN ROBOTIC TOPICS

Topic 1: Manufacturing Technology

Interest/Curiosity

- Manufacturing technology is a creative process;
- It can be both personally and financially rewarding to work in todays advanced manufacturing technology environment.

Skills/Properties

- Knowledge and skills about manufacturing technology are very marketable in today's economic environment;
- The tools and technologies that can assist manufacturing are constantly changing to keep up with the competitive nature of today's business and consumer needs.

Importance/Attitude

- Manufacturing technology is important because it helps to make better quality products at a lower cost;
- Manufacturing technology affects our everyday lives because of all the products that it helps to produce.

Topic 2: Computer Aided Design (CAD)

Interest/Curiosity

- CAD (computer aided design) is a creative process;
- Computer aided design can be of interest to students with graphic design, engineering or communication skills.

Skills/Properties

- Computer aided design output is often in the form of electronic files which can be used for printing, machining or other manufacturing operations;
- CAD can be used to design curves and figures in two dimensional (2-D) space or curves, surfaces and solids in three dimensional (3-D).

Importance/Attitude

- CAD was a revolutionary stage in the engineering world where the roles of draftsmen, designers and engineers began to merge;
- CAD is used in many industries including automotive, shipbuilding and aerospace, is used in the design of medical prosthetic devices and is widely used to create computer animation for special effects in movies and advertising.

Topic 3: Robotic Manipulations

Interest/Curiosity

- Manufacturing robots have a controller which operates like a “brain”;
- Robots can assume properties of humans like perception, assembly, autonomy and artificial intelligence.

Skills/Properties

- Robotics engineers need to use mathematics and science, work with things, work with people and perceive and visualize;
- Robotic engineers need to be able to apply principles of engineering, technology, electronics and computer systems.

Importance/Attitude

- In 2005 there were over 120,000 in use in North American industry, half of them for welding;
- Manufacturing robots increase accuracy, repeat ability, and throughput.

Topic 4: Honda - The Humanoid Robot

Interest/Curiosity

- Asimo the humanoid robot is frequently used in demonstrations around the world to encourage the study of science and mathematics;
- The robot has made public appearances around the world including the New York Stock Exchange and Disneyland.

Skills/Properties

- The robot was the first to incorporate predicted body movements;
- Honda introduced a new ASIMO in 2005 with increased walking speed and in 2007 added intelligence capabilities enabling several ASIMOs to work in collaboration with each other.

Importance/Attitude

- Humanoid robots have great potential to improve society by carrying out many tasks more efficiently than humans;
- Humanoid robots have the potential to assist the elderly and disabled in their daily activities.

Topic 5: Military Robots

Interest/Curiosity

- Boston Dynamics (recently purchased by Google) has produced a variety of robots mainly for military purposes including Big Dog, Cheetah, Littledog and Petman;
- Called "the world's most ambitious legged robot", Big Dog is designed to carry 340 pounds alongside a soldier at 4 miles per hour, traversing rough terrains at up to 35 degrees incline.

Skills/Properties

- Big Dog uses sensors for locomotion include joint position, joint force, ground contact, ground load, a gyroscope (a device for measuring orientation) and a stereo vision system;
- Big Dog also uses LIDAR, a remote sensing technology that for distance that measures distance by illuminating a target with a laser and analyzing the reflected light.

Importance/Attitude

- Various robots of Boston Dynamics can be very useful in military operations, for example hauling heavy materials over rough terrains or to test protective clothing used by soldiers for protection against chemical warfare agents;
- Eventually robots may have the capability of fighting wars on their own, although there may be some ethical implications of this.

Topic 6: Marine Robots

Interest/Curiosity

- Marine robotic systems are some of the most eye catching tools a scientist can use;
- Marine robots can go to some of the most exotic areas like Greenland and the Antarctica where it investigated the mysterious ice caverns and the deep Caribbean.

Skills/Properties

- One of the challenges is to be able to interpret the mass of data and imagery collected by the robots;

- Skills in various areas like engineering science, ecology, biology and geoscience can be useful when working with marine robots.

Importance/Properties

- Marine robots promise to open up the ocean to humans in new ways and allow scientists, industry and the military address problems in new ways;
- Underwater autonomous vehicles (UAV) can collect high resolution imagery and data that can be used in the fields of engineering science, ecology, biology and geoscience.

Topic 7: Amazon Drones

Interest/Curiosity

- Amazon has asked the Federal Aviation Administration (FAA) for an exemption from rules prohibiting from using drones for commercial purposes;
- Amazon wants to innovate and it knows that it can't under the FAA's burdensome regulatory regime.

Skills/Properties

- Amazon has quickly developed highly-automated aerial vehicles with advanced capabilities;
- They claimed to have developed sense and avoid sensors that allow the drones to sense obstacles and avoid collisions.

Importance/Attitude

- The drones will be able to carry 86% of the products in the company's inventory;
- The drones should be capable of reducing the time and potentially the cost associated with customer orders.

Topic 8: 3D Printing

Interest/Curiosity

- The cost of 3D printers has decreased dramatically since 2010 with printers that used to cost $20,000 now costing less than $1,000;
- 3D printing has spread into the world of clothing with fashion designers experimenting with 3D printed bikinis, shoes and dresses.

Skills/Properties

- 3D printable model can be created with a computer aided design package or via a 3D scanner;
- Before printing a 3D model from an STL file, it must first be processed by a piece of software called a "slicer" which converts the model into the model into a series of thin layers.

Importance/Attitude

- 3D printing may have a profound impact on the world as the coming of the factory did; it is likely to disrupt every field that it touches;
- 3D printing has a wide range of applications, not just manufacturing, but in science, medicine, art, and communication.

Topic 9: Haptic Feedback

Interest/Curiosity

- Haptic feedback is commonly used in arcade games. In 1976, Sega's motorbike game Moto-Cross was the first game to use haptic feedback which caused the handlebars to vibrate during a Collison with another vehicle;
- In 2008, Apple's MacBook and MacBook Pro started incorporating a "Tactile Touchpad" design with button functionality and haptic feedback incorporated into the tracking surface.

Skills/Properties

- Haptic feedback uses various technologies, for example electromagnetic motors, electroactive polymers, electrostatics and subsonic audio waves;
- More advanced haptic techniques can create customizable effects using low latency chips.

Importance/Attitude

- Haptic interfaces for medical simulation can be useful for performing remote surgery;
- Future applications of haptic technology cover a wide spectrum of human interaction with technology which may result in advancements in gaming, movies, manufacturing, medical and other industries.

Topic 10: Robotic Surgery

Interest/Curiosity

- The first surgical robot was the Heartthrob developed and used for the first time in Vancouver in 1983;

- The very first surgical operation using a robot for an orthopedic procedure was performed in Vancouver in 1984; over 60 arthroscopic operations were performed using this procedure in the first 12 months and the device was featured in a 1985 National Geographic video on industrial robots.

Skills/Properties

- In the case of minimally invasive surgery the surgeon uses one of five methods to control the instruments, either using a telemanipulator or using computer control;
- In the case of enhanced open surgery, autonomous instruments replace traditional steel tools performing certain actions with much smoother and feedback controlled motions than could be performed by the human hand.

Importance/Attitude

- Robotic surgery offers many advantages to patients, for example shorter hospitalization, reduced pain and discomfort and faster recovery times;
- Robotic surgery offers many advantages to surgeons, for instance greater visualization, enhanced dexterity and greater precision.

Topic 11: Robotic Software

Interest/Curiosity

- Some robot software aim at developing intelligent mechanical devices. Although common on science fiction stories, such programs have yet to become common-place in reality and much research is still required in the field of artificial intelligence;
- Due to the highly proprietary nature of robot software, most manufacturers of robot hardware provide their own software. Fortunately, there are lots of similarities between the robots and their respective programming languages.

Skills/Properties

- Robotic engineering is a well-paid field, but requires specific skills like robotic programming skills. For example software for industrial robots consists of data objects and lists of instructions, known as program flow;
- Other programming languages for robotic engineers are visual programming, scripting languages and parallel languages which provide the robotic engineer with the skills necessary to control the robots for a variety of applications.

Importance/Attitude

- Regardless which language is used, the purpose of robotic software is to develop applications for robots that either entertain or help people. For example, scheduling software for mobile robots in factories helps manufacturing firms reduce the cost of production and improve the quality of the final product;
- Tasking software which includes simple drag-n-drop interfaces are used in a variety applications, including setting up delivery routes, security patrols and visitors. These kinds of applications illustrate how robots can be used to improve customer service in many service industries by reducing the waiting time and the cost of delivery.

Section 3
Marketing and Management

Chapter 13
Neuromarketing Perspective of Consumer Choice

Salim Lahmiri
ESCA, Morocco

ABSTRACT

Behavioural research attempts to study how individuals make decisions and interact and influence other individuals, organizations, markets and society. In this regard, applied neuroscience in human decision-making has gained an increasing attention in recent decades with emergence of two disciplines; namely neuroeconomics and neuromarketing. Indeed, neuroeconomics has emerged as a multidisciplinary research area that integrates knowledge from neuroscience, psychology, and economics to better understand economic decision making and to specify more accurate models of choice and decision. In particular, neuroeconomics is becoming an attractive area of study and research in financial decision making with particular emphasis on understanding investor sentiment and fear when faced to different investment opportunities characterized by various scenarios. In particular, it aims to understand and explain consumer decision process and influence of marketing key factors on consumer choice. As a result, companies may define appropriate marketing strategies based on neuromarketing studies.

INTRODUCTION

Behavioural research attempts to study how individuals make decisions and interact and influence other individuals, organizations, markets and society (Birnberg & Ganguly, 2012). In this regard, applied neuroscience in human decision-making has gained an increasing attention in recent decades with emergence of two disciplines; namely neuroeconomics and neuromarketing. Indeed, neuroeconomics has emerged as a multidisciplinary research area that integrates knowledge from neuroscience, psychology, and economics to better understand economic decision making and to specify more accurate models of choice and decision. In particular, neuroeconomics is becoming an attractive area of study and research in financial decision making with particular emphasis on understanding investor sentiment and fear when faced to different investment opportunities characterized by various scenarios. Indeed, a rich and abundant a literature has been devoted to such exciting field of study in behavioural financial economics.

DOI: 10.4018/978-1-5225-5478-3.ch013

Besides, neurmarketing combines neuroscience and marketing studies to understand marketing-relevant human behaviour by using particularly the advantages of physiological measurement modalities such as functional magnetic resonance imaging (fMRI), magneto-encephalography (MEG), and electroencephalography (EEG). In particular, these neuroimaging modalities are used in conjunction with relevant variables in marketing theories to study problems in consumer behaviour, and response to brands and advertisements (Stewart, 1984, 1985).

In particular, it aims to understand and explain consumer decision process and influence of marketing key factors on consumer choice. As a result, companies may define appropriate marketing strategies and techniques based on neuromarketing studies major findings. This interest continues to grow with more research works in the subject. For instance, Young (2002) used EEG signals to examine whether specific moments within advertising are primarily responsible for brand development and attention. Rossiter et al. (2001) used EEG signals to investigate memory and information processing in the context of visual scene recognition. Ioannides et al. (2000) and Ambler et al. (2000) used MEG signals to study the effectiveness of cognitive and affective advertisements on cortical centers. Lusk et al, (2015) used fMRI to examine consumer choice and associated brain activation.

With the increase of using marketing research in business strategy, some topics have attracted a large attention in neuromarketing; including consumer choice (Khan et al, 2011; Lusk, 2012; Lusk et al, (2015), advertising (Kenning et al, 2007; Plassmann et al, 2007; Mostafa, 2012), and branding (Ma, Wang, Shu, & Dai, 2008). The purpose of this chapter is to present recent works in neuromarketing with applications in consumer choice and related brain activated areas. The purpose is to shed light on brain activated areas as a response to specific marketing stimuli to better understand consumer choice from a physiological point of view.

The chapter will be organized as follows. In Section 1, we present a general view of consumer choice based on economic theory and its limits. In Section 2, we present recent works in neuromarketing which are related to consumer choice. This section will also shed light on the effect of advertising and branding on consumer choice from a physiological point of view. Then, future directions will be provided in Section 3. Finally, we conclude in Section 4.

CONSUMER DECISION-MAKING: A BACKGROUND

Traditionally, marketing research is essentially about understanding, explaining, and predicting individual, group, and organisational behaviour relevant to markets (Lee et al, 2007). However, with the development of technologies used by neuroscientists to directly study cortical activity, psychological and physiological sciences have adopted such techniques to improve understanding of the brain and cognition (Lee et al, 2007). In this regard, marketing science is starting adopting neuroimaging techniques to understand consumer decision-making process and preferences. Indeed, the benefits of physiological measurement for marketing have been pointed out since 1980s (Petty & Cacioppo, 1983; Weinstein et al., 1984). For instance, physiological responses can be obtained when customers are directly participating in the consumer behaviour and choice studies. In such studies, the interest is always about understanding the consumer choice-making by using advanced neuroimaging-based research methods. For instance, Braeutigam et al. (2001, 2004) investigated the difference between predictable and unpredictable choices and found that different brain regions are activated according to choice predictability. In addition, unpredictable choices eliciting activity in brain regions were associated with silent vocalisation and judgement of rewards.

Gonzalez et al, (2005) concluded that the cognitive effort in the prefrontal cortex is lower when a sure gain is expected. Daw et al, (2006) found evidence of both the orbitofrontal cortex and the ventromedial prefrontal cortex being involved in the processing of different alternatives and potential outcomes. Wallis (2007) concluded that the orbitofrontal cortex is associated with the evaluation of trade-offs and the expected capacity of outcomes to satisfy customer's needs. Rilling and Sanfey (2011) found that dorsolateral prefrontal cortex responds to social norms while the ventrolateral prefrontal cortex could motivates social norm compliance.

Other neuroscience related studies found that some specific brain regions are influenced by reward. From a neuromarketing perspective, studying such particular brain regions is interesting in consumer research domain. For example, Knutson and Wimmer (2007) found that the striatum and its components; including putamen, caudate nucleus and nucleus accumbens; are all involved in evaluation of rewards with respect to expectations. In addition, Fliessbach et al., (2007) concluded that the influence of social factors on the reward-related activity in the striatum.

To summarize, Solnais et al, (2013) stated that neuroscience contributes to marketing research at four levels:

1. Consumer decision-making and the formation of consumer preferences,
2. The engagement of the brain's reward system by marketing,
3. Consumer's motivational and emotional responses, and
4. The neural foundations of consumers' attention and memory.

In particular, by using neuroimaging techniques one may study the effect of marketing factors (product, advertising, branding, pricing) on consumer behaviour; namely decision-making process, rewards, memory, and emotions. At decision-making process level, neuroscience can help understanding how consumer preferences are formed and how they could be predicted based on neural activity. In addition, neuroscience could help understanding how attractive marketing stimuli affect brain's reward system and perceived value. Furthermore, it could explain product and advertising memory and its effect on consumer behaviour. Finally, neuroscience may help exploring effect of marketing stimuli on consumer motivations and emotions (Solnais et al, 2013).

Indeed, in neuroeconomics and behavioural finance it was found that emotions and sentiment greatly affect economic agent decision and also stock market trend (ref). For instance, according to Somatic Marker Hypothesis (Bechara et al., 2000; Bechara et al., 2002), decision-making is a process guided by emotions. Besides, investor sentiment was found to be helpful to predict stock market trend and classify its patterns (Lahmiri, 2011a; 2011b; 2012; 2013; 2014; Lahmiri et al, 2014a; 2014b). In this regard, the reader may consult Lahmiri (forthcoming) for a review on the relationship between economic decision, emotion and related brain activated regions.

ADVERTISING, BRANDING, AND CONSUMER BRAIN

Neuromarketing may be defined as the practical implementation of the knowledge brought by consumer neuroscience for managerial purposes (Hubert & Kenning, 2008). Indeed, it is the study of the neural conditions and processes that underlie consumption, their psychological meaning, and their behavioural consequences (Reimann, Schilke, Weber, Neuhaus, & Zaichkowsky, 2011). In this regard, several studies in neuromarketing were conducted to study from a neuroscience perspective the response of consumer to advertising and branding.

In order to understand relationship between advertising and neuronal activity, Ohm et al, (2010) conducted a study to identify frontal cortex activation in reaction to TV advertisements. For instance, they compared three consecutive creative executions of the world-famous Sony Bravia advertisements (''Balls'', ''Paints'', and ''Play-Doh''). The authors found that dominant reactions in left hemisphere were present only in response to one of the tested ads – ''Balls''. In this regard, target group respondents reacted to both emotional informational part of the advertisement; namely product benefit, product, and brand exposure scenes. However, they have not found similar pattern for the remaining two advertisements. Therefore, the authors concluded that frontal asymmetry measure may be a diagnostic tool in examining the potential of advertisements to generate approach related tendencies. Vecchiato et al, (2010a) estimated the power spectral distribution of the electrical signals of a mannequin gathered during the observation of the video related to the commercial advertisings and of those acquired during the observation of the naturalist documentary. They found that electric data from the mannequin's head presents statistical significant differences in power spectra during the visualization of a commercial advertising when compared to the power spectra gathered during a documentary, when no adjustments were made on the alpha level of the multiple univariate tests performed. Jones et al, (20102) explored metacognitive theory that states consumers track fluency feelings when buying by examining consumer EEG signals. They used event-related potential (ERP) measures as twenty high math anxiety (High MA) and nineteen low-math anxiety (Low MA) consumers made buying decisions for promoted (e.g., 15% discount) and non-promoted products. They found that ERP correlates of higher perceptual and conceptual fluency were associated with buys, however only for High MA females under no promotions. In contrast, High MA females and Low MA males demonstrated greater amplitude, associated with enhanced conceptual processing, to prices of buys relative to non-buys under promotions. In summary, they concluded that perceptual and conceptual processes interact with anxiety and gender to modulate brain responses during consumer choices. In his study, Mostafa (2012) used fMRI data to study the neural mechanisms associated with human and non-human sounds' perception in advertising by employing a block design paradigm in which participants heard human versus non-human sounds in different sets of advertisements. The experimental results showed that, compared to nonhuman sounds, human sounds elicited greater activation in several areas in or around the primary auditory cortex. Thus, he concluded that different types of sounds are processed in different functional brain pathways. In other words, he concluded that visual, affective and linguistic information are processed in different cortical regions in the brain. In their work, Khushaba et al, (2013) used a discrete choice experiment to measure individual's preferences. Participants were required to make a series of choices and indicate their most and their least favourite options. For instance, participants were asked to evaluate three cracker features to come with an overall evaluation of each cracker: favourite and least favourite cracker. In this context, the authors investigated the brain activities taking place during decision-making. In particular, they studied the EEG activities related to the choices made by participants. The purpose was to evaluate the cortical activity of the different brain regions and the interdependencies among the EEG signals from these regions. They found clear phase synchronization between the left and right frontal and occipital regions indicating inter-hemispheric communications during the chosen task. In addition, the authors found clear and significant change in the EEG power spectral activities taking a place mainly in the frontal, temporal, and occipital regions when participants indicated their preferences for their preferred crackers. In addition, analysis of mutual information showed that the various cracker flavours and toppings of the crackers were more important factors affecting the buying decision than the shapes of the crackers. Recently, Yılmaz et al, (2014) analyzed EEG signals to investigate which frequencies and EEG channels could be relatively better indicators of preference (like or dislike decisions) of consumer products, and timing characteristic of "like" decisions during such mental processes. It was found that in low-frequency band, 4 and 5 Hz were the most discriminative frequencies, whilst none of the frequencies provided significant information in high-frequency band.

Besides studying effect of advertising on consumer brain activation, several studies have been focused on the relationship between branding and neuronal activity by examining fMRI data. For instance, Deppe et al, (2005) studied the relationship between brain areas and brand preference and concluded that when choosing between one's favourite brand as compared to a second or lower ranked brand increased activity in the ventral medial prefrontal cortex (vmPFC) and reduced with activity in dorsolateral prefrontal cortex (dlPFC) and in inferior frontal gyrus (IFG); and visual cortex (cuneus/precuneus) are triggered. Schaefer et al. (2006) studies the linkage between brain activation and brand familiarity. They found that activity changes in the MFG correlate with familiar versus unfamiliar brands. Plassmann et al, (2007) studied the effect of brand choice in loyal custmers brain and found that activity in the striatum correlates with brand loyalty to retail brands. Schaefer and Rotte (2007) found that imagining driving a car from one's favourite brand correlates with activity changes in the ventral striatum. In addition, activity in this area also correlates with perceived luxury and sportiness of the car. In an interesting study, Koenigs and Tranel (2008) Examined how brand information biases preference judgments and how vmPFC is affected. They concluded that patients with damage in the vmPFC were not biased by brand information during blind vs. open tasting of Coke and Pepsi. Plassmann et al, (2008) examined the effect of uncertainty on the neural signatures of brand preference. They found that Interaction of brand preference with uncertainty of the decision amplifies the neural correlate of brand preference in the vmPFC. Recently, Esch et al. (2012) examined the relationship between neuronal activity and brand familiarity and brand strength. They found that unfamiliar brand logos versus ''strong'' brands induced activity changes in the IFG "strong" versus unfamiliar brands induced activity changes in the hippocampus and lingual gyrus "strong" versus "weak" brands induced activity changes in the dlPFC/MFG.

Finally, as emotions are closely related to consumer behaviour and decision-making process, there are numerous studies that examined emotional processes by using visual, auditory, and combined stimuli to evoke emotions (Vecchiato et al, 2010b, 2011; Plassmann et al, 2008; Damasio & Carvalho, 2013; Padmala & Pessoa, 2008; Damasio & Carvalho, 2013; Plassmann et al, 2008; Damasio & Carvalho, 2013). Table 1 summarizes the results of some related works.

Table 1. Advertising related emotions and neural activation

Study	Results
Vecchiato et al. (2010b, 2011)	During the observation of the TV commercials cortical activity in the theta and gamma bands was higher and localized in the frontal areas for the commercials judged pleasant when compared with disliked ones
Plassmann et al. (2008), Damasio & Carvalho (2013),	Both subcortical areas like amygdala and cortical areas like prefrontal cortex, cingulate cortex, and temporal cortices are critical in emotion processing
Padmala & Pessoa (2008)	Emotional meaning of a stimulus controls the visual cortex including the visual areas V1 and V2.
Damasio & Carvalho (2013)	The relation between the amygdala and the prefrontal and temporal cortices produces an intentionally experienced sensation of an emotion.
Plassmann et al. (2008), Damasio & Carvalho (2013)	Activity from the amygdala and the remaining parts of the limbic system cannot be sensed directly in surface recordings because they lie deep in the brain.

DIRECTIONS OF FUTURE WOKS

In recent works, several directions for future works were pointed out by researchers. For instance, Ohm et al, (2010) suggested using EEG-based research to investigate which of the tested ads are probably more efficacious in forming mind-based brand equity, to consider integration of EEG-based research with EMG and galvanic skin response (GSR). They also suggested combining EEG-based research with traditional self-report methods, advanced experimental designs, reaction time measures and behavioural indices in the context of neuromarketing. Finally, they pointed out that it is necessary to investigate the effectiveness and assess confidence of various research techniques in consumer research field. In their research, Jones et al, (20102) suggested for future research utilizing their approach along with greater source accuracy techniques to clarify quality of information processing among low math anxiety consumers. The goal is to assess buying or not buying decision-making rather than an altered decision process in this group. In his study, Mostafa (2012) recommended using fMRI-based approach to test whether the activation of auditory cortical areas varies with attentional state in order to demonstrate or refute the hypothesis that auditory cortex activation is independent or just synonymous to conscious speech perception. In addition, he suggested using fMRI methods to test whether the activation in primary and secondary auditory cortex occurs is bilateral or hemisphere dependent. Furthermore, he recommended investigating human brain areas activated by animal sounds versus other environmental sounds, and to focus on testing a range of theoretically-driven marketing research concepts such as signaling theory of the brand or surrealist imagery in advertising. In their work, Khushaba et al, (2013) suggested studying the effect of advertisement stimuli with varying features one-at-a-time to limit the effect of insufficient EEG observations in some choice sets. Indeed, this would reliably estimate the effects of advertisement attribute levels on the EEG measures. In addition, the authors suggested presenting the items in each choice set one-at-a-time, and then observe the choices in each set collectively. More recently, Yılmaz et al, (2014) recommended using large number of subjects to better investigate which frequencies and EEG channels could be relatively better indicators of preference of consumer products, and timing characteristic decisions during such mental processes. In addition, they recommended studying the influence of type or nature of stimulus on variations of EEG channels. Very recently, in an interesting study, Zhao and Zhong (2015) found that consumers may give preference to the carbon labelled products. In other words, consumers are intended to buy the green products. In particular, they found that critical premium, public awareness, university education level and the perceived consumer effectiveness have considerable influences on the purchasing behaviour, whilst the effects of income and attitude towards risk can be negligible. In this regard, we recommend that future works should focus on studying consumer's perception to carbon labelled products from a neuromarketing perspective by examining EEG and fMRI signals.

REFERENCES

Ambler, T., Ioannides, A., & Rose, S. (2000). Brands on the brain: Neuro-images of advertising. *Bus. Strateg. Rev.*, *11*(3), 17–30. doi:10.1111/1467-8616.00144

Bechara, A., Damasio, H., & Damasio, A. R. (2000). Emotion, decision-making, and the orbitofrontal cortex. *Cerebral Cortex*, *10*(3), 295–307. doi:10.1093/cercor/10.3.295 PMID:10731224

Bechara, A., Tranel, D., & Damasio, A. R. (2002). The *somatic marker hypothesis and decision-making*. In F. Boller & J. Grafman (Eds.), *Handbook of neuropsychology: Frontal lobes* (2nd ed.; Vol. 7). Amsterdam: Elsevier.

Birnberg, J. G., & Ganguly, A. R. (2012). Is neuroaccounting waiting in the wings? An essay. *Accounting, Organizations and Society*, *37*(1), 1–13. doi:10.1016/j.aos.2011.11.004

Braeutigam, S., Rose, S. P. R., Swithenby, S. J., & Ambler, T. (2004). The distributed neuronal systems supporting choice-making in real-life situations: Differences between men and women when choosing groceries detected using magnetoencephalography. *The European Journal of Neuroscience*, *20*(1), 293–302. doi:10.1111/j.1460-9568.2004.03467.x PMID:15245501

Braeutigam, S., Stins, J. F., Rose, S. P. R., Swithenby, S. J., & Ambler, T. (2001). Magentoencephalographic signals identify stages in real-life decision processes. *Neural Plasticity*, *8*(4), 241–253. doi:10.1155/NP.2001.241 PMID:12018772

Damasio, A., & Carvalho, G. B. (2013). The nature of feelings:evolutionary and neurobiological origins. *Nature Reviews. Neuroscience*, *14*(2), 143–152. doi:10.1038/nrn3403 PMID:23329161

Daw, N. D., O'Doherty, J. P., Dayan, P., Seymour, B., & Dolan, R. J. (2006). Cortical substrates for exploratory decisions in humans. *Nature*, *441*(7095), 876–879. doi:10.1038/nature04766 PMID:16778890

Deppe, M., Schwindt, W., Kugel, H., Plassmann, H., & Kenning, P. (2005). Nonlinear responses within the medial prefrontal cortex reveal when specific implicit information influences economic decisionmaking. *Journal of Neuroimaging*, *15*(2), 171–182. doi:10.1111/j.1552-6569.2005.tb00303.x PMID:15746230

Esch, F. R., Möll, T., Schmitt, B., Elger, C. E., Neuhaus, C., & Weber, B. (2012). Brands on the brain: What happens neurophysiologically when consumers process and evaluate brands? *Journal of Consumer Psychology*, *22*, 75–85. doi:10.1016/j.jcps.2010.08.004

Fliessbach, K., Weber, B., Trautner, P., Dohmen, T., Sunde, U., Elger, C. E., & Falk, A. (2007). Social comparison affects reward-related brain activity in the human ventral striatum. *Science*, *318*(5854), 1305–1308. doi:10.1126/science.1145876 PMID:18033886

Gonzalez, C., Dana, J., Koshino, H., &Just, M. (2005).The framing effect and risky decisions: Examining cognitive functions with fMRI. *Journal of Economic Psychology*, *26*(1).

Hubert, M., & Kenning, P. (2008). Acurrent overview of consumer neuroscience. *Journal of Consumer Behaviour*, *7*(4-5), 272–292. doi:10.1002/cb.251

Ioannides, A. A., Liu, L., Theofilou, D., Dammers, J., Burne, T., Ambler, T., & Rose, S. (2000). Real time processing of affective and cognitive stimuli in the human brain extracted from MEG signals. *Brain Topography*, *13*(1), 11–19. doi:10.1023/A:1007878001388 PMID:11073090

Jones, W. J., Childers, T. L., & Jiang, Y. (2012). The shopping brain: Math anxiety modulates brain responses to buying decisions. *Biological Psychology*, *89*(1), 201–213. doi:10.1016/j.biopsycho.2011.10.011 PMID:22027087

Kenning, P., Plassmann, H., & Ahlert, D. (2007). Applications of functional magnetic resonance imaging for market research. *Qualitative Market Research*, *10*(2), 135–152. doi:10.1108/13522750710740817

Khan, U., Zhu, M., & Kalra, A. (2011). When trade-offs matter: The effect of choice construal on context effects. *JMR, Journal of Marketing Research*, *48*(1), 62–71. doi:10.1509/jmkr.48.1.62

Khushaba, R. N., Wise, C., Kodagoda, S., Louviere, J., Kahn, B. E., & Townsend, C. (2013). Consumer neuroscience: Assessing the brain response to marketing stimuli using electroencephalogram (EEG) and eye tracking. *Expert Systems with Applications*, *40*(9), 3803–3812. doi:10.1016/j.eswa.2012.12.095

Knutson, B., & Wimmer, G. E. (2007). Splitting the difference: How does the brain code reward episodes? *Annals of the New York Academy of Sciences*, *1104*(1), 54–69. doi:10.1196/annals.1390.020 PMID:17416922

Koenigs, M., & Tranel, D. (2008). Prefrontal cortex damage abolishes brand-cued changes in cola preference. *Social Cognitive and Affective Neuroscience*, *3*(1), 1–6. doi:10.1093/scan/nsm032 PMID:18392113

Lahmiri, S. (2011a). A comparison of PNN and SVM for stock market trend prediction using economic and technical information. *Journal of Computer Applications*, *29*, 24–30.

Lahmiri, S. (2011b). Neural networks and investor sentiment measures for stock market trend prediction. *Journal of Theoretical and Applied Information Technology*, *27*, 1–10.

Lahmiri, S. (2012). Wavelet transform, neural networks and the prediction of S&P price index. *Business Intelligence Journal*, *5*, 235–244.

Lahmiri, S. (2013). Forecasting S&P500 directions using wavelets and support vector machines. *International Journal of Strategic Decision Sciences*, *4*, 78–88. doi:10.4018/jsds.2013010105

Lahmiri, S. (2014). *Information, ondelettes, réseaux de neurones, méthodes numériques, et modélisation et prédiction des séries temporelles boursières: une étude comparative*. (PhD dissertation). University of Quebec at Montreal.

Lahmiri, S. (In press). Economic decision making, emotion, and prefrontal cortex. In *Handbook of Research on Neuroeconomics and the Decision-Making Process*. IGI Global. Accepted doi:10.4018/978-1-4666-9989-2.ch007

Lahmiri, S., Boukadoum, M., & Chartier, S. (2014a). Exploring information categories and artificial neural networks numerical algorithms in S&P500 trend prediction: A comparative study. *International Journal of Strategic Decision Sciences*, *5*(1), 76–94. doi:10.4018/IJSDS.2014010105

Lahmiri, S., Boukadoum, M., & Chartier, S. (2014b). A supervised classification system of financial data based on wavelet packet and neural networks. *International Journal of Strategic Decision Sciences*, *4*(4), 72–84. doi:10.4018/ijsds.2013100105

Lee, N., Broderick, A. J., & Chamberlain, L. (2007). What is 'neuromarketing'? A discussion and agenda for future research. *International Journal of Psychophysiology*, *63*(2), 199–204. doi:10.1016/j.ijpsycho.2006.03.007 PMID:16769143

Lusk, J. L. (2012). The political ideology of food. *Food Policy*, *37*(5), 530–542. doi:10.1016/j.foodpol.2012.05.002

Lusk, J. L., Crespi, J. M., Cherry, J. B. C., McFadden, B. R., Martin, L. E., & Bruce, A. S. (2015). An fMRI investigation of consumer choice regarding controversial food technologies. *Food Quality and Preference*, *40*, 209–220. doi:10.1016/j.foodqual.2014.10.005

Ma, Q., Wang, X., Shu, L., & Dai, S. (2008). P300 and categorization in brand extension. *Neuroscience Letters*, *431*(1), 57–61. doi:10.1016/j.neulet.2007.11.022 PMID:18155837

Mostafa, M. M. (2010). Brain processing of vocal sounds in advertising: A functional magnetic resonance imaging (fMRI) study. *Expert Systems with Applications*, *39*(15), 12114–12122. doi:10.1016/j.eswa.2012.04.003

Mostafa, M. M. (2012). Brain processing of vocal sounds in advertising: A functional magnetic resonance imaging (fMRI) study. *Expert Systems with Applications*, *39*(15), 12114–12122. doi:10.1016/j.eswa.2012.04.003

Ohme, R., Reykowska, D., Wiener, D., & Choromanska, A. (2010). Application of frontal EEG asymmetry to advertising research. *Journal of Economic Psychology*, *31*(5), 785–793. doi:10.1016/j.joep.2010.03.008

Padmala, S., & Pessoa, L. (2008). Affective learning enhances visual detection and responses in primary visual cortex. *The Journal of Neuroscience*, *28*(24), 6202–6210. doi:10.1523/JNEUROSCI.1233-08.2008 PMID:18550762

Petty, R. E., & Cacioppo, J. T. (1983). The role of bodily responses in attitude measurement and change. In J. T. Cacioppo & R. E. Petty (Eds.), *Social Psychophysiology: A Sourcebook*. New York: Guilford Press.

Plassmann, H., Ambler, T., Braeutigam, S., & Kenning, P. (2007). What can advertisers learn from neuroscience? *International Journal of Advertising*, *26*, 151–175.

Plassmann, H., Kenning, P., & Ahlert, D. (2007). Why companies should make their customers happy: The neural correlates of customer loyalty. *Advances in Consumer Research—North American Conference Proceedings, 34*.

Plassmann, H., O'Doherty, J., Shiv, B., & Rangel, A. (2008). Marketing actions can modulate neural representations of experienced pleasantness. *Proceedings of the National Academy of Sciences of the United States of America*, *105*(3), 1050–1054. doi:10.1073/pnas.0706929105 PMID:18195362

Reimann, M., Schilke, O., Weber, B., Neuhaus, C., & Zaichkowsky, J. (2011). Functional magnetic resonance imaging in consumer research: A review and application. *Psychology and Marketing*, *28*(6), 608–637. doi:10.1002/mar.20403

Rilling, J. K., & Sanfey, A. G. (2011). The neuroscience of social decision-making. *Annual Review of Psychology*, *62*(1), 23–48. doi:10.1146/annurev.psych.121208.131647 PMID:20822437

Rossiter, J. R., Silberstein, R. B., Harris, P. G., & Nield, G. A. (2001). Brain-imaging detection of visual scene encoding in long-term memory for TV commercials. *Journal of Advertising Research*, *41*(2), 13–21. doi:10.2501/JAR-41-2-13-21

Schaefer, M., Berens, H., Heinze, H. J., & Rotte, M. (2006). Neural correlates of culturally familiar brands of car manufacturers. *NeuroImage*, *31*(2), 861–865. doi:10.1016/j.neuroimage.2005.12.047 PMID:16487728

Schaefer, M., & Rotte, M. (2007). Favorite brands as cultural objects modulate reward circuit. *Neuroreport*, *18*(2), 141–145. doi:10.1097/WNR.0b013e328010ac84 PMID:17301679

Solnais, C., Andreu-Perez, J., Sánchez-Fernández, J., & Andréu-Abela, J. (2013). The contribution of neuroscience to consumer research: A conceptual framework and empirical review. *Journal of Economic Psychology*, *36*, 68–81. doi:10.1016/j.joep.2013.02.011

Stewart, D. W. (1984). Physiological measurement of advertising effect: An unfulfilled promise. *Psychology and Marketing*, *1*, 43–48. doi:10.1002/mar.4220010105

Stewart, D. W. (1985). Differences between basic research and the validation of specific measures: A reply to Weinstein et al. *Psychology and Marketing*, *2*(1), 41–49. doi:10.1002/mar.4220020105

Vecchiato, G., Astolfi, L., De Vico Fallani, F., Cincotti, F., Mattia, D., Salinari, S., & Babiloni, F. et al. (2010b). Changes in brain activity during the observation of TV commercials by using, GSR and HR measurements. *Brain Topography*, *23*(2), 165–179. doi:10.1007/s10548-009-0127-0 PMID:20033272

Vecchiato, G., De Vico Fallani, F., Astolfi, L., Toppi, J., Cincotti, F., Mattia, D., & Babiloni, F. et al. (2010a). The issue of multiple univariate comparisons in the context of neuroelectric brain mapping: An application in a neuromarketing experiment. *Journal of Neuroscience Methods*, *191*(2), 283–289. doi:10.1016/j.jneumeth.2010.07.009 PMID:20637802

Vecchiato, G., Toppi, J., Astolfi, L., De Vico Fallani, F., Cincotti, F., Mattia, D., & Babiloni, F. et al. (2011). Spectral EEG frontal asymmetries correlate with the experienced pleasantness of TV commercial advertisements. *Medical & Biological Engineering & Computing*, *49*(5), 579–583. doi:10.1007/s11517-011-0747-x PMID:21327841

Wallis, J. D. (2007). Orbitofrontal cortex and its contribution to decision-making. *Annual Review of Neuroscience*, *30*(1), 31–56. doi:10.1146/annurev.neuro.30.051606.094334 PMID:17417936

Weinstein, S., Drozdenko, R., & Weinstein, C. (1984). Brain wave analysis in advertising research. *Psychology and Marketing*, *1*(3-4), 83–96. doi:10.1002/mar.4220010309

Yılmaz, B., Korkmaz, S., Arslan, D. B., Güngör, E., & Asyalı, M. H. (2014). Like/dislike analysis using EEG: Determination of most discriminative channels and frequencies. *Computer Methods and Programs in Biomedicine*, *113*(2), 705–713. doi:10.1016/j.cmpb.2013.11.010 PMID:24326336

Young, C. (2002). Brain waves, picture sorts®, and branding moments. *Journal of Advertising Research*, *42*(4), 42–53. doi:10.2501/JAR-42-4-42-53

Zhao, R., & Zhong, S. (2015). Carbon labelling influences on consumers' behaviour: A system dynamics approach. *Ecological Indicators*, *51*, 98–106. doi:10.1016/j.ecolind.2014.08.030

This research was previously published in Applying Neuroscience to Business Practice edited by Manuel Alonso Dos Santos, pages 49-62, copyright year 2017 by Business Science Reference (an imprint of IGI Global).

Chapter 14
A Neuromarketing Perspective on Measuring Marketing Influence at the Unconsciousness Level

Ioana Iancu
Babeş-Bolyai University, Romania

ABSTRACT

In a context characterized by an inflation of marketing messages, it is imperious to understand how consumers succeed in making the buying decision. Starting by briefly describing the structure and the role of the brain and the differences between consciousness and unconsciousness, the paper aims to investigate the way neuromarketing can help in comprehending the feelings of the consumers, the way products or services match the consumers' needs, and the way companies can discover the insights of decision-making process. This paper can be perceived either as a guide for the companies that aim to find more on the way people manage information and make decisions or as a comprehensive description on human being marketing behavior that can serve both business, academic environments and consumers.

INTRODUCTION

In a context characterized by a dynamic market, by increasingly sophisticated consumers and by an immense amount of information, the way companies work in order to accomplish individuals' expectations become a complex issue. Therefore, the main questions that can be asked are what consumers really feel and think on a certain product or service, how that product or service should match the consumers' needs, and to what extend companies can discover the insights of decision-making process.

Considering the above background and the need to deeply enter the secrets of consumers, the present paper aims to theoretically investigate the role of unconsciousness on the individuals' decisions at the marketing level. In this respect, issues like how the brain works and neuromarketing are some of the main themes assessed.

DOI: 10.4018/978-1-5225-5478-3.ch014

This paper can be perceived either as a guide for the companies that aim to find more on the way people manage information and make decisions or as a comprehensive description on human being marketing behavior that can serve both business and academic environments and consumers. First, using the insights of this paper companies can develop instruments in order to find more valuable information on what exactly consumers' want and need. Second, academics can use the information as an inspiration source for their further experiments and in order to improve the already used instruments. Finally, consumers can find information about themselves and about the way brain works and, thus, uses it either in their shopping behavior or on other social decisions.

Starting by analyzing insights of the brain and the role of consciousness and unconsciousness, the paper emphasizes the way the modern marketing can complete traditional marketing by shifting the focus from the product to the complexity of the consumer. Thus, the core concept of this paper is neuromarketing. Based on this concept, the main discussed areas are related to the instruments of neuromarketing, the way it offers extra information comparing with the traditional research methods, its entrepreneurship applicability and its ethical perspectives.

BACKGROUND

An Insight of the Brain

Although representing only 2% of the total body mass, the brain is responsible for all the marketing behavior and it consumes approximately 20% of the body energy (Morin, 2011, p.134). Renvoisé and Morin (2007) consider that the simplified structure of the brain is the following one: the reptilian brain or R-complex (responsible with instinct), the middle brain or the limbic system (responsible with emotions), and the new brain or neo-cortex (responsible with rational thinking). Being entities for which the survival process is the most important one, the human behavior is mostly controlled by the reptilian brain (Morin, 2011, p.134). In comparison with the new brain that is slow in reaction, it effort itself in order to answer to a certain stimulus, it is intelligent, conscious and partially controllable, the reptilian brain is 500 million years old, is the first that reacts to incentives, it is always on, it is unconscious and incontrollable (Renvoisé and Morin, 2007). In the same respect, emphasizing that there are a lot of aspects related to impressions, intuitions and decisions that are far to be deeply understood, Kahneman (2011) splits the brain in two parts: System 1 (reptilian brain) and System 2 (new brain). The two concepts have been initially used by Keith Stanovich and Richard West. System 1 is described as being automatic, permanent, quick, effortless, and involuntary. In contrast, System 2 is the part of the brain that pays more attention to the mental effort, it can realize complex calculus, and it is associated with decision and with concentration (Kahneman, 2011).

The description of the parts of the brain and the way they work and react is important in the marketing context, mainly considering that almost 80% of the information received by a consumer is internalized at the unconsciousness level (Morin, 2011, p.134). Put it differently, although the cognitive functions of the brain are treasured, in a context in which individuals receive more that 10.000 messages per day, a large part of this information is lost unless it addresses the reptilian brain (Morin, 2011, p.135). Therefore, one can conclude that a message in general and a marketing message in particular are efficient mainly if it contains stimuli that can activate mainly the instinctual and unconscious part of the brain, namely the reptilian brain. Based on this issue, this chapter is going to underline the domain of neuromarketing and its several techniques already used by companies or academics, or techniques and experiments that can be used in order to gain business efficiency and knowledge.

Consciousness and Unconsciousness

The evolution of the techniques studying the marketing influence at the brain level (neuromarketing) is mainly due to the fact that researchers have realized that the process of decision making is not a rational one and that it is impossible to make an optimal rational decision in which all the alternatives are taken into consideration (Ramsøy, 2014). Although there is a part of the literature emphasizing that consumers' decisions are conscious and deliberated processes, another part of the literature debates the role of unconsciousness within the decision making process (Tom et al., 2007). For instance, in a study, thirsty consumers that have been exposed to smiley individuals have consumed higher quantities of juice and have been willing to pay more for the consumption in comparison with the same type of persons exposed to sad individuals (Winkielman, Berridge, & Wilbarger, 2001 in Tom et al., 2007). In the same respect, Tom et al. (2007) claims that consumers pay more on brands they like than on alternative brands. They also talk about two types of effects: *mere exposure effect* implying that a repeated or simple exposure to a certain stimulus results in developing an affective positive reaction on the stimulus; the *endowment effect* implies that a consumer perceives an object as being more valuable if he/she possesses that object. (Tom et al., 2007, pp.118-119). While the first effect increases the preference for a certain object (affection without cognitive mediation), the second effect increases the value of the object (cognition without affective mediation) (Tom et al., 2007, p.119).

Thus, the decisions are made based on two processes: a quick unconscious process and a rapid decision experience. Ramsøy (2014) considers that it is necessarily to understand both conscious and unconscious brain. Conscious events are those brain activities individuals can report with accuracy, in optimal conditions and with a minimum distraction. In contrast, unconscious events are those one knows that exist but there is a lack of ability to develop an accurate report on them (Baars, Ramsøy, & Laureys, 2003, p.672). While the unconscious processes imply limited and rigid processing of information, the conscious processes are associated with a much more vast and flexible processing of information, the effect on thinking and behaving being deeper (Shanahan & Baars, 2005 in Ramsøy & Skov, 2014, p.1).

Shiv and Fedorikhin (1999, pp.288-289) talk about the affective-cognitive model of behavior and consider that, in the context of exposure to certain alternatives during a task, there are two processes that can take place: an affective one and an cognitive one. While it is likely that fist one happens automatically, the second one is likely to be controllable focusing on the consequences of the chosen alternative. Moreover, while the consumer decision is likely to be based on affective aspects when the resources processing is constraint, the lack of constraints implies rather cognitive reactions (Shiv & Fedorikhin, 1999, pp.288-289).

Cognitively speaking, the objectives are mentally represented as schemas, similar with stereotypes or attitudes. Once having these constructs activated, the actions associated with them will be realized (e.g. the objective of saving money will lead to the behavior associated with it) (Bargh, 1990 in Chartrand et al., 2008, p.190). Another part of the literature sustains that in order to accomplish an objective, this must be associated with positive feelings (Custers & Aarts 2005 in Chartrand et al., 2008, p.190). Within this context, the activation of the objectives can be done either consciously, when the individual develops an objective and act in order to accomplish it, or unconsciously, when an objective has been frequently associated with a certain situation, that objective being automatically activated when the circumstances of that situation are recognized (Bargh, 1990 in Chartrand et al., 2008, p.190).

Dijksterhuis, Smith, van Baaren, and Wigboldus (2005 in Simonson, 2005) claim that there are a large number of decisions taken at the unconsciousness level and they are strongly affected by the environment factors. For instance, an adult in a supermarket that sees a child running is likely to unconsciously buy something that reminds him/her about childhood (Dijksterhuis, Smith, van Baaren, & Wigboldus, 2005 in Simonson, 2005, p.212). However, Simonson (2005) defend the theory that stresses that individuals are rather conscious when they make buying decisions, the above mentioned example being rather an accident than a behavioral model.

Bargh (2002, p.280) is interested in finding to what degree the individual is conscious about what influences him/her during buying behavior and to what degree he/she can control it. He stresses that there are two ways in which unconscious elements can be activated: either subliminal, when the stimuli are not accessible at the conscious level, or supraliminal, the individual being conscious about the stimuli but not about their effect (Bargh, 2002, p.282). For instance, there are a large amount of individuals that buy tickets to Super Bowl only in order to see the commercials. Thus, they know they are influenced, but they do not know which the effect of this influence is (Bargh, 2002, p.283).

Chartrand et al. (2008) aim to analyze whether random exposure to certain incentives (for instance, certain prints) can activate different buying objectives and whether it can influence the final decision. In the buying behavior context, there are a lot of studies showing that the individual is directed on accomplishing the objectives, these ones being the key motivational element within the decision making process (Chartrand et al., 2008, p.189). However, there is a smaller amount of literature talking about the factors that determine the choice of those particular objectives and their accomplishment. Thus, in contrast with the traditional research that stresses that individuals are perfectly conscious about their objectives regarding the choices, the newer literature talks about the possibility that the objectives can be unconsciously chosen (Chartrand et al., 2008, p. 189). Chartrand et al. (2008), based on several experiments, draw the conclusion that a subliminal exposure to certain brands (either associated with luxury products or cheaper products) activates objectives associated with these brands (e.g. the objective of buying a luxury product, or the objective of buying a cheaper product).

In the context of exposure to a certain stimuli, Berkowitz (1993 in Shiv & Fedorikhin, 1999, pp.279-280) talks about three types of processes. First, the information regarding the stimulus is perceived in an automatic associative process, affective sometime, that happens before the cognitive processes of evaluation, interpretation, schemas development, attributions and strategy. Second, the information on the incentive is the subject of a deliberative and cognitive process that can improve or aggravate action tendencies. Third, the affective significance of the results following a cognitive process can result in more complex affective reactions and in slower action tendencies in comparison with preliminary affective reactions, the deliberative process making the decision slower.

Considering this context, a study presents the following experiment. The individuals have been asked to memories either a number of 7 digits or a number of 2 digits, this task being described as very important. While the subjects' attention has been focused on learning the numbers, the subjects have been asked to choose between a very appealing but unhealthy chocolate cake and a very healthy bun non-appealing fruit salad. The results shows that the subjects are inclined to choose the unhealthy cake when they have a difficult task, as memorizing a 7 digits number (Shiv and Fedorikhin, 1999, p.279). Put it differently, in the case in which the rational part of the brain is occupied, the intuitive and impulsive brain has a stronger influence on behavior (Kahneman, 2011). Thus, in the context of a binary choice where one of the alternatives (chocolate cake) is much more associated with affection and less with cognition in comparison with the second alternative (fruit salad), the decision depends very much both on the level of processing the resources of the task and on the impulsive behavior (Shiv & Fedorikhin, 1999, p.288).

Another experiment was conducted in the same room in which the tea is usually served at the British University. Every individual had the freedom to put in a box a certain amount of money they considered to be enough for their consumption. Although there was a list with suggested prices, it was not mandatory to respect it. The experiment implied as well the existence of banner above the list with prices for 10 weeks time period, every week the image on the banner being changed. There were used images either with flowers, or with eyes part of the head of men and women. There was no explanation given for these pictures. The results showed that in the days in which there was shown eyes, the contribution for the consumed quantity increased almost three times than in the days in which the image represented flowers. Thus, it can be said that images with people watching the consumers, either they are happy, aggressive or neutral, can influence the level of obedience and the quantity of the given money (Kahneman, 2011).

The above-described literature emphasizes the role of the brain and of unconsciousness within the consumer behavior. At the same time, it underlines the need to deeply asses this behavior with other instruments than traditional ones. Therefore, neuromarketing might be a domain that helps in better understanding the influences of different incentives on the consumers' decision process.

A GENERAL PERSPECTIVE ON NEUROMARKETING

Victoria Phan (2010, p.14) claims that advertising becomes extremely advanced from a scientifically point of view. In this context and in a strong relation with the subliminal messages and the role of emotions within the decision making process, neuromarketing is a relatively new domain that describes clinical information on the way the brain functions and explains the consumer's behavior based on neurological activity (Fugate, 2007, p.385). Being a combination between neuroscience and marketing, neuromarketing has appeared in 2002 and its instruments seems to be more and more used in the business environment (Morin, 2011, p.131). In 2004, The Economist has stressed that Jerry Zaltman is the first one that proposed, in the mid '90, the use of imagistic technology in marketing (Wilson et al., 2008, p.390). However, another source stresses that the concept of neuromarketing has been invented in 2002 by the professor Ale Smidts from the Erasmus University and it has been recognized two years later at the first conference on neuromarketing (Phan, 2014, p.15). American companies as Brighthouse or SalesBrain are the first ones that have offered neuromarketing services (Morin, 2011, p.132).

Thomas Ramsøy (2014) claims that there is difference between *neuromarketing* and *consumer neuroscience*. While the first uses neuroscience having the commercial goal to improve the consumer's knowledge and the marketing effects, the second one is rather an academic exercise meant to understand consumer's psychology and behavior (Ramsøy, 2014).

Neuromarketing is a research domain that helps gather relevant information in the responses of the brain to different stimuli from marketing (Renvoisé´ and Morin, 2007 in Murphy et al., 2008, pp.293-294). The goal of neuromarketing is to offer objective information on the consumer's perceptions without registering his or her subjective answers (Murphy et al. 2008, pp.293-294). Lee, Broderick, and Chamberlain (2007 in Wilson et al., 2008, p.390) consider that neuromarketing implies the application of neuroscience methods and instruments in order to analyze and understand human behavior in its relation with the market and the changes within it. At the same time, neuromarketing uses neuro-imaging techniques in order to efficiently sell products (Lee et al., 2007, p. 200).

Thomson (2003 in Wilson et al. 2008, p.390) describes neuromarketing as a domain in which the methods specific for the neurological domain are used in advertising. In the same respect, Lee, Broderick, and Chamberlain (2007 in Wilson et al. 2008, p. 390) consider that neuromarketing implies the application of neuroscience methods in analyzing and understanding the human behavior in its relationship with the market and the changes within it. Thus, one can claim that neuro-imaging technology will contribute to the understanding of the way individuals take decisions and of the methods used by specialists to influence those behaviors (Wilson et al. 2008, p.390).

In addition, neuromarketing is perceived as being useful within the process of developing a new model of consumerism behavior. Thus, based on the neuromarketing instruments, the practitioners have access to valid data on consumers and the consumers can learn to know themselves better and to better understand their decision making process (Butler, 2008, pp.416-417). In the '50, two researchers from the McGill University conclude that within our brain there is an area that can be named the "center of pleasure". Based on this discovery, it can be said that humans are motivated by things that make them feel better, even when it comes to shopping (Nobel, 2013). However, although individuals can say to a certain degree what they want, what they like or how mush they would pay for a certain product, they cannot clearly say why. This is the context in which neuromarketing brings the necessary instruments that can help in better understanding the human behavior (Nobel, 2013).

While the instruments of neuronal imaging are widely used in economics (neuroeconomics), they are relatively slowly adopted in marketing. Lee *et al.* (2007) offers several explanations for that. First, neuroscience can be an intimidating domain in an academic context. Second, considering the costs implied for purchasing the necessary technology, universities cannot afford to empirically study this domain (Lee *et al.*, 2007, p. 199). Finally, neuromarketing raises a large amount of discussions regarding the search for the buy button in the brain of the consumers (Lee *et al.*, 2007, p.199).

The actors that can be involved in the neuromarketing research are advertising agencies and marketing departments (that aim to present their products, design, campaigns, to measure the individuals' reactions regarding the efficiency of promoted messages), media departments (that aim to determine the best media space for a campaign and the mix of efficient strategies), companies (that aim to understand the consumers' preferences), strategists from the political domain (that aim to evaluate the communication and debate styles), or film producers (that aim to take decisions regarding the best actors to use, the best end of a movie, the best scenes of a trailer in order to get the highest level of efficiency) (Bercea 2015). Thus, neuromarketing instruments can help in gathering relevant information mainly before a product is being launched on the market.

The first research in the neuromarketing field has been published by McClure, Li, Tomlin, Cypert, L. M. Montague and P. R. Montague in *Neuron* in 2004 (Morin, 2011, p.132). The research is conducted by using a set of blind and anonymous tests and passive experiments using Functional Magnetic Resonance Imaging (fMRI). The authors aim to investigate both the neuronal response that can be correlated with the preference for Coca-Cola or Pepsi, and the influence of brand image on the decision making process (McClure *et al.*, 2004, p.379). The research starts from three main research questions. The first one refers to the cognitive and neurological responses in the context in which the two soft drinks are tasted without knowing the brand. The second analyzed issue refers to the cognitive and behavioral influence in the context in which the subject knows the brand of the tasted soft drink. Based on these two aspects, the third one refers to a possible correlation between the behavioral preference and the neurological response measured with fMRI (McClure et al. 2004, p.380).

The researchers claim that the two soft drinks included in the experiment, Coca-Cola and Pepsi, are very similar from the point of view of chemical composition, the second one being only a little bit sweeter. However, the study concludes that the cultural messages, related to the brand, correlated with the content can influence the way individuals perceive a product (McClure et al. 2004, p.379). Thus, knowing the brand of the soft drink, especially when it comes to Coca-Cola, biases the consumers' preferences. When the subjects do not know which soft drink they are consuming, a large majority declares they prefer, at the taste level, Pepsi. In this particular context, limbic system of the brain, the one responsible with emotions and instinctual behavior is activated. In contrast, when the subjects know which of the two soft drinks they are consuming, a large majority prefers Coca-Cola. In this case, a different part of the brain, namely the frontal cortex, responsible with executive functions, with thinking and planning, is activated (McClure et al. 2004). Put it different, although they like better the Pepsi taste, the subjects claim they prefer Coca-Cola, mainly developing this response on the brand image rather than on taste (Phan, 2010, p.15). The insight of this conclusion is that a commercial or a set of commercials have the capacity to make an individual act in contradiction with what their bodies are actually asking for (Phan, 2010, p.16). Thus, there is the capacity of certain external factors, as brand, brand reputation, social desirability, to influence the consumers' preferences and their buying behavior.

The Instruments of Neuromarketing

The instruments used in neuromarketing can be grouped in at least three categories (Zurawicki, 2010 in Bercea, 2012). First, there are the instruments used to register the metabolic activity of the brain (functional Magnetic Resonance Imaging - fMRI, Positron emission tomography - PET). Second, there are the instruments used in order to measure the electrical activity of the brain (electro-encephalography – EEG, magneto-encephalography (MEG), Transcranial magnetic stimulation - TMS, Steady State Topography - SST). Finally, there are the instruments that do no require the registration of brain activity. They refer either to facial reactions (Eye Tracking, Facial coding, Facial electromyography), or to body reactions (Physiological response measurement, Galvanic skin). The most used techniques are fMRI, EEG, and magneto-encephalography (Fugate, 2007, p.386) at the brain level, and Eye Tracking, Facial coding and galvanic skin at the body level (Bercea, 2012). Usually, there is a combination of techniques used within an experiment. These instruments are non-invasive (Morin, 2011, pp.132-133). The following part of the chapter will comprehensively describe the most important of them.

Electroencephalography (EEG), although being an older method, it is still considered to be efficient regarding the measurement of brain activity (Morin 2011, p.132). In the 17th century scientists have discovered that individuals are controlled by electrical impulses, the first electroencephalograph being built at the beginning of the 20th century and the first demo being conducted by Hans Berger (Ohme & Matukin, 2012).

Neurons, the cells responsible for the cognitive responses, are more than 100 billions. At the same time, there are trillions of synaptic connections that represent the base of the neuronal circuit. When an individual is exposed to a stimulus (e.g. a commercial) the neurons are activated and they produce small electricity. This electricity can be associated with different emotional states. In marketing, EEG data are obtained by using a helmet with electrodes that is placed on the head of the subject and that is able to make up to 10.000 registrations per second (Morin, 2011, pp.132-133). However, by using EEG the precise location in the brain in which the neurons have been activated cannot be detected, mainly when it is about deeper areas in the brain. EEG devices can measure electrical signals only at the surface of the brain. Moreover, although data obtained by using EEG are valuable in the case of measuring the effect of an ad, they are not sufficient to understand the entire cognitive process that determines a certain activity (Morin, 2011, p.133). Studies that use EEG are based on inputs (e.g. images) presented on the screen of a computer, and an experiment can take approximately 30 minutes (Ohme & Matukin, 2012).

Although it is not sensitive at the deepest brain structures, EEG has a high temporal resolution (milliseconds), which means that it can detect very succinct brain events. At the same time, EEG has a small spatial resolution, the obtained information depending on the number of electrodes on the helmet. The more electrodes, the better the spatial resolution is and the more information is obtained (Ariely & Berns, 2010, p.288). One of the most common measurements regards the left-right asymmetry of the frontal EEG, meaning the measurement of alpha coefficient (alpha band 8-13 Hz) (Ariely & Berns, 2010, p.288). There are several types of electrical signals measured with EEG. For instance, while delta signals are register in lethargy state, alpha signals are registered in relaxation state, and beta signals are registered in conscious activity state (Ohme & Matukin, 2012).

The first studies in the psychology using EEG have been conducted around 1979. Davidson has been one of the first cognitive specialists that have studied the relationship between emotions and electrical signals in the brain. There are studies that have proved that the electrical models are located in the frontal part of the brain. For instance, while alpha signals in the left frontal lobe indicate positive emotions, electrical activity in the right part of the frontal lobe is usually correlated with negative emotions, with the desire to leave an action (Morin 2011, pp.132-13). By using EEG, there can be detected the scenes of a commercial that have the highest emotional intensity, or there can be evaluated the reaction of a subject to different elements of an ad (e.g. image, sound, words, special effects etc.). Moreover, there can be predicted to what degree a song will be liked by the public, a strategy will be efficient, a packaging will be preferred, or a story will be powerful (Ohme & Matukin, 2012).

Ohme and Matukin (2012), in one of their studies, about the analysis of *Balls* commercial for Sony Bravia LCD. The ad has been launched in November 2005 and won Gold Lion at Cannes Lions International Advertising Festival in June 2006 and Grand Prix at Midsummer Awards in London in June 2006 (the original commercial can be seen at https://www.youtube.com/watch?v=R8WkIC7bMnE). The special element of the ad is a number 250.000 colored real balls that are launched to freely jump on San Francisco streets (Ohme, & Matukin, 2012). Based on this commercial, the authors are interested in understanding which is the elements that have made this commercial so appreciated and which are the scenes with the highest emotional appeal. Thus, completing a traditional research the authors have conducted a neuromarketing study using EEG on a number of 45 subjects. One of the most intense emotional scenes is the moment in which a frog comes into play (17th second of the ad), this scene being filmed by chance. Moreover, the last part of the commercial in which informational aspects on the product are presented is considered to have a stronger emotional effect than the first part of the ad in which Sand Francisco city is visualized (Ohme & Matukin, 2012). Interestingly, in comparison with the neuronal results, at the declarative level (focus groups), the respondents claim that the most pleasant images are that of the balls falling. The research also tries to determine the importance of the music in this commercial (the song “Heartbeats”, singer - José González, composer - The Knife) and its impact on the entire ad. Thus, when the commercial runs without music, the impact is lower (less positive emotions) than in the case of the ad with music. The same thing happens in the case in which the song is presented without images.

Although in this case the image and the sound create a necessary unity, there are neuromarketing studies that claim that it is not a rule that this mix is efficient (Ohme & Matukin, 2012).

Magnetoencephalography (MEG) exists from the middle of the '60s and has the role of measuring the magnetic field of the brain and of developing the magnetic map of it (Morin 2011, p.134). MEG has a temporal and spatial resolution better than EEG. It is happening due to the fact that the magnetic field is less biased by the cranium and its artifacts in comparison with the electrical field (Ariely and Berns 2010, p.288). However, MEG is limited to measuring brain activity only on its surface. Thus, it is not a proper method for measuring brain activity in the sub-cortical areas (Morin 2011, p.134). The literature talks about studies that use MEG and that have succeeded in finding a correlation between certain frequency bands and certain controllable cognitive tasks (e.g. recognizing objects, accessing verbal memory, remembering specific events etc.). Thus, MEG is rather recommended in measuring activities that imply specific tasks than for exploratory experiments that imply investigating cognitive (cortical) or emotional (sub-cortical) functions (Morin 2011, p.134).

Functional Magnetic Resonance Imaging (fMRI) is used in measuring the blood flow in the brain. When the neurons are activated by a stimulus (e.g. commercial), they need energy that is transported in blood. In marketing studies using fMRI, the acronym BOLD (Blood Oxygen Level Dependant) is used. It refers to the oxygen quantity in blood. All the studies that use fMRI consider that a change in BOLD signal represent a valid measurement of neuronal activity (Morin 2011, p.134). Although the resolution in the care of fMRI is better than in the case of EEG and the deepness of activity measurement is higher, the first method is slower. There is a lateness of several seconds between the moment in which the neuron is activated and the moment in which a change in BOLD is remarked (Morin 2011, p.134). Moreover, although fMRI has the advantage of offering information from the profound areas of the brain, certain regions (e.g. frontal orbital cortex) are affected by artifact signals that can bias the data (Ariely and Berns 2010, p.288).

In the case of fMRI, the researchers can isolate systems of neurons associated with different functions of the brain. An example used in the literature is that of an individual that watches a print commercial. In this particular case, a part of the 125 millions of visual neuronal receptors in each eye are activated. The neuronal signals are transmitted to that part of the brain responsible with the pupils and that initiate the eye movement on the print. Other signals are issued, through the optical nerves, to the opposed part of the brain in such a manner that the left side of the print is perceived by the right hemisphere and the right part of the print is perceived by the left hemisphere (Dubuc 2007 in Wilson 2008, p.391). Thus, by using fMRI, it is possible to detect neuronal activity associated with the sight and with the cognitive and affective responses related to the print (Wilson 2008, p.391).

In the context in which the isolation of the neuronal systems formed of 100 billions of neurons in the human brain is a complex task, fMRI is the tool capable of locating the active systems by comparing the images in the moment the brain is active with the images in the moment in which the brain is not active (Wilson 2008, p. 391). Put it different, first the brain is scanned when it is not running the task and then the brain is scanned while developing the task. Thus, the activity in the brain is detected and it is interpreted by taking into account the lateness of the data, the movements of the head, the breathing or the heart beating (Wilson 2008, p.392). There is the need of involving many subjects in order to compare the activated locations in the brain (Wilson 2008, p.392).

Simply put, the brain activity follows the following pattern: the signals are transmitted from a neuron to the other through neurotransmitters, through synapse to the receptors (Carey 2005 in Wilson 2008, p.391); the synaptic activity of the neurons' network generates blood flow in that particular region (Raichle & Mintun 2006 in Wilson 2008, p.391); the blood brings a supplementary quantity of oxygen and hydrogen that increase the magnetic field during scanning, field that is detected by fMRI device (Matthews & Jezzard 2004 in Wilson 2008, p.391).

Although fMRI can be used in gathering data based on visual, aural or gustative stimuli, the scanning process can cost even 1000$ a day and there are needed at least 20-30 subjects (Karmarkar in Nobel, 2013). However, there is research that can predict the future popularity of a certain product. Such an example is a study conducted by Emory University in which the specialists succeed in correlating certain neuronal activities with the musical success. For instance, the song Apologize from OneRepublic was very well perceived both in scanning and in tops (Berns & Moore in Nobel, 2013).

Besides the neuro-imaging techniques, neuromarketing uses instruments that measure body reactions. Thus, *eye tracker* measures the way the subject looks at a certain stimulus, the level of pupil dilatation, where he/she is looking and for how long (Bercea 2013). The aim of the analysis is that of assessing the visual perception, the cognitive intention, the interest in and the importance given to an observed object (Zurawicki 2010 in Bercea 2013). Moreover, the measurement of the physiological response to certain incentives is a method through which there can be obtained data related to emotional effects of a certain stimulus at the heart level, blood pressure, skin conductivity (e.g. sweating), stress hormones in saliva, or muscles contractions of the face (Bercea 2013).

Although the potential of neuromarketing is vast, Lee *et al.* (2007) consider that the main aspect the neuronal imaging is focusing on is branding advertising, and consumer behavior. For instance, while EEG can be used in order to asses the impact of certain visual inputs, MEG can be used in order to understand the way cognitive and affective commercials generate different reactions in the brain. Another marketing specific issue that can be analyzed using neuronal imaging is the decision behavior of the consumer (for instance, the difference between predictable choices and unpredictable ones based on gender) (Lee *et al.*, 2007, p.201). Murphy et al. (2008, p.295) claims that by combining these methods, the value of the data increases.

NEUROMARKETING AS A COMPLETION FOR THE TRADITIONAL RESEARCH METHODS

Although marketing is perceived as a deleterious domain, it helps matching the products with the consumer through guiding the design and presentation of the product in such a way in order to correspond with individuals' preferences (Ariely & Berns, 2010, p.284). Traditional research methods used in marketing (opinion survey, focus group, interview etc.) tries for a long time now to predict the efficiency of marketing campaigns. However, considering that emotions are a strong and complex mediator of the way consumers process and understand messages, these traditional methods usually fail to develop valid results (Morin, 2011, p.132). Some researchers wrongly assume that individuals are capable to describe their cognitive processes, although they have a large amount of subconscious elements. Moreover, there are reasons for which consumers prefer to bias the offered information, as the incentives, time constraints, interviewees' pressure etc. (Morin, 2011, p.132).

As Plato claims, an individual is formed of two major elements: reason and emotion. However, most of the research has been based the human behavior's explanations solely on rational judgment (The Economist, 2005 in Fugate, 2007, p.386). In contradiction, nowadays, based on social, technological and economic evolution, the buying decision is perceived as the result of the combination between rationality and emotions. Using neuromarketing instruments, Fugate (2007, p.386) claims that rational and emotional thinking are co-dependent. In the same perspective, Damasio stresses that emotions are responsible with attracting consumer's attention and with their motivation in order to focus the rationality on certain elements (Damasio in Coy, 2005 in Fugate, 2007, p.386). In contrast with the traditional research methods that focus on self-evaluation and on the ability and desire of the individual to give exact and correct answers, instruments used by neuromarketing can detect data that are not necessarily visible at the consciousness level (Petty and Cacioppo, 1983 in Lee et al., 2007, p.200). Schafer emphasizes that neuromarketing, in contrast with the traditional methods, has the ability to focus on specific elements belonging to unconsciousness within the buying behavior (Schafer, 2005 in Fugate, 2007, p.386).

The tradition research methods used in analyzing the consumer behavior do not efficiently succeed in describing the emotions, these descriptions being dependent on the desire and on the competency of the individuals to emit real data (Morin, 2011, p.131). Thus, neuromarketing completes the traditional methodology through collecting data especially at the brain level (Morin, 2011, p.132). Thus, neuro-imaging techniques offer a fascinating alternative to traditional research. These techniques offer the possibility to find valuable information on the way a message in perceived and processed at the unconsciousness level and on the future success of that message (Morin 2011, pp.132-133).

Usually, specialists use research methods as focus group, opinion survey or market tests. On one hand, while some of these methods (e.g. focus group or opinion survey) are relatively simple to apply and cheap, they imply a large amount of possible errors. On the other hand, while using these techniques one can obtain precise data (e.g. market tests), the costs of application are very high and the product must exist already on the market in order to develop the tests (Ariely and Berns 2010).

While the tradition research methods can analyze the reactions of the consumers only after the product or the service is released on the market, neuromarketing techniques are used both before and after releasing it. Thus, concept-products can be tested and eliminated if they seems un-attractive, a large amount of resources being saved or reused for more efficient products (Ariely and Berns, 2010). For instance, the willingness to pay has started to be measured using fRMI. A specific example is an experiment that implies snacks' consume. Thus, the willingness to pay is associated with activity in the part of the brain named frontal orbital cortex and pre-frontal cortex. The authors of the study claim that similar parts of the brain are activated when the subjects anticipate a pleasant taste, hear a nice music, receive money or experienced a social prize (Ariely and Berns, 2010, p.285). In conclusion, one can say that neuromarketing completes the traditional research methods by adding information regarding the reaction at the brain or body level (Morin, 2011, p.132).

Ariely and Berns (2010) consider that, beside that neuromarketing brings information that is not detectable by using conversational research methods, it is likely that using neuromatketing instruments might be a quicker and low cost alternative comparing to other methods including in the phases before the moment in which the product is distributed on the market. Specialists claim that individuals do not know to comprehensively articulate their preferences and that the human brain hide information related to the real preferences of the consumer. Neuromarketing, as a domain that has the capacity to detect this hidden information could be used both for stressing what individuals want and what they would buy (Ariely & Berns 2010, p.284).

However, there are common features both traditional research methods and neuromarketing have. First, regarding the similarities with the quantitative research methods, the most important ones are the following:

the measurement of beliefs, knowledge, attitudes, attention, memory or memories; the development and testing of the hypothesis and causal effects; the use, analysis, interpretation and triangulation of statistical data; the objective observation role of the researcher; the possibility to generalize data, although the sample is small in the case of neuromarketing studies (the brain mechanisms are similar from person to person), predictions development regarding the consume behavior (Bercea 2013). Second, regarding the similarities with the qualitative methods, neuromarketing has the following characteristics: the objective is that of gathering deep information through studying certain phenomena; a small number of needed subjects (15-20 individuals are enough in order to have internal validity) (Bercea 2013). Third, there are some similarities of neuromarketing with both types of research methods: the resulted data can be images, numbers, statistics; the main objectives are the description, the explanation, the prediction (similar with quantitative methods), and the exploration, the discovery and the construction (similar with qualitative methods); the subjects' behavior is analyzed in controlled conditions and due to technical evolution, there is the possibility to develop experiments with a natural context (e.g. supermarkets) (Bercea 2013).

Although using neuroscience instruments as a completion of the traditional methods is useful regarding the decisional process of the consumer, the literature reflects on several concerns regarding this domain (Yoon et al., 2012, p.484). Being a domain that implies a different way of thinking and new theoretical perspectives, one of the concerns regards the high costs implied for educating especially the students. At the same time, the needed infrastructure is costly. Another concern states the necessity to calibrate the practitioners' expectations regarding the way they perceive the efficiency of the neuromarketing results, this domain not being able to make perfect predictions on consumer choice (Yoon et al., 2012, p.484).

NEUROMARKETING AND ITS ENTREPRENEURSHIP APPLICABILITY

Due to market globalization, to technological evolution and to changes in the purchase and consumption behavior, the contemporary marketing is constantly challenged in adapting itself (Pop, Iorga, 2012, p.632). Thus, the need for interdisciplinary and the need of clarifying the way people make decisions are increasing concerns for both marketing theorists and practitioners. In this context, the neuromarketing with its specific tools is considered a viable solution for the companies in order to better understand costumers in respect to different domains, as corporate economy, environmental economy, human recourses, or customer relations management (Pop, Iorga, 2012, pp.633-634).

While the rational hypothesis of classical economy that states that people are rational and follow their own interests is not longer valid, more recent theories emphasize that people are intuitive, impulsive, their decisions being rather filled with emotional attachment (Pop, Iorga, 2012, p.634). Therefore, practitioners are challenged to understand the irrational behavior in order to correlate "processes, such decision-making, judgment, encoding memory or emotions with marketing concepts, such as positioning, brand loyalty and consumer reaction to marketing messages" (Perrachione, 2008 in Pop, Iorga, 2012, p.634).

Neuromarketing is contradictory perceived by the researchers. While some are consider neuromarketing as a science, as a valid field of research that goes beyond the simple application of neuroimaging tools in order to sell products (Lee et al., 2007 in Pop, Iorga, 2012, p.636), others consider it as only a simple business tool and label it as "consumer neuroscience" (Hubert and Kenning, 2008 in Pop, Iorga, 2012, p.636). Consumer neuroscience is defined as an interdisciplinary field (psychology, neuroscience and economics) that analyzes the way the brain is affected by advertising and marketing strategies (Lee et al. 2007 in Agarwal and Dutta, 2015, p.458). Interestingly, although consumer neuroscience and neuromarketing are often understood as being the same thing, consumer neuroscience is associated with academic research and neuromarketing with the practitioners (Plassmann et al., 2011 in Agarwal and Dutta, 2015, p.458). Other specialists stress the importance of neuromarketing as a mechanism that helps entrepreneurs to deliver the right message to the right people (Georges and Badoc, 2010 in Pop, Iorga, 2012, p.636). While, Constensen (2011 in Pop, Dabija and Iorga, 2014, p.28) claims that neuromarketing is a component of marketing by means of which the customer behavior can be understood by interpreting psychological and neuropsychological information, neuromarketing is believed to be helpful for entrepreneurs when making decisions in both their communication with consumers and in their communication with partners (Pop, Dabija and Iorga, 2014, p.28).

Marketing research is considered to be a very important part of the marketing efforts, although it is perceived as being costly and time consuming. In the same respect, accurate results are needed for having the most efficient marketing decisions (Egrie and Bietsch, 2013). Consumer satisfaction is one of the most significant factors that determine the success of a business mainly in the context of relationship marketing. Satisfaction implies both buying the same product/service again and recommending the product to other potential consumers. At the same time, satisfaction relates to both customers' judgment of the quality of the product and costumer's evaluation of the interaction experience with the brand provider (Dapkevicius and Melnikas, 2009, p.17). However, there are large amounts of unknown variables that can determine a decision-making. In this context, neuromarketing is considered a solution (Egrie and Bietsch, 2013). Google is one of the companies that use neuromarketing research. Google specialists are using neuroscience tools in order to learn new information about the way the consumers perceive different web pages and to help companies to have their webpage as the top pick. The main variables that can be tested with neuromarketing tools are mainly visual appeal and emotional engagement (Egrie & Bietsch, 2013).

Another interesting example is that belonging to the neuroscientists form the University of California, Los Angeles, that have scanned the brain of individuals while watching Super Bowl commercials (Suomala et al., 2012, pp.12-13). This type of expertise is very important mainly taken into consideration that there are enormous amount of money invested in launching these commercials during this American football event. The main results of the research emphasize that while a very few commercials stimulate empathy, a large number of other ads provoke anxiety, fear, and insecurity (Suomala et al., 2012, pp.12-13). These results can be considered as being valuable for any entrepreneur in his or her decision to further use or not similar messages to the target group of the company.

When talking about using neuromarketing tools indoor of a company, a problem that is frequently mentioned in the literature is the need for training. Neuromarketing implies interdisciplinary. Thus, a combination of backgrounds in statistics, psychology, computer science might be an advantage (Smidts et al., 2014, p.260). In the same respect, Smidts et al. (2014, p.263) discuss three technical advances in consumer neuroscience: continuing innovation in scanner strength, speed and stability in order to provide improvements in spatial and temporal resolution; advances in modeling of behavior in order to elucidate processes that generate choice; and advances in analysis of algorithms that can identify multivariate patterns of brain activity correlated with future behavior.

There is an increasing number of companies that use neuromarketing research. Some of the main domains are media, advertising, media planning, automotive, technology, online auction or food. In the same respect, several exact examples are Yahoo, Hyundai, Microsoft, EBay, Daimler. The main analyzed issues are the following: how the brain react in the morning hours, the reaction to advertising, the reaction to viewing a product, understanding human (emotional) interaction with computers, adjusting products or commercials etc. (Flores, Baruca and Saldivar, 2014, p.78).

The marketing entrepreneurs are increasingly adopting neuroscience techniques in understanding consumer behavior. Levalloise et al. (2013 in Smidts et al, 2014, p.260) is talking about an increase from 13 neuromarketing companies in 2008 to more than 60 in 2012. Nowadays, on the website of the Neuromarketing Science & Business Association (NMSBAc) are listed more than one hundred companies that provide neuromarketing services. In addition, there are research companies that acquire leading neuromarketing companies proving the value of this neuroscience approach. Such an example is Nielsen that has bought in 2011 Neurofocus, a very appreciated neuromarketing company (Smidts et al, 2014, p.260).

A business model is composed of three main stages: the development of *value proposition* (product design and development), the development of *value delivery* (the channels used in order to address the message to consumers), and the development of *value capture* (the pricing formulas needed in order to generate income) (Teece, 2010 in Kamp, 2014, p.426). A market failure can intervene in each of these stages. While the literature defines a market failure as a mismatch between supply and demand (Kamp, 2014, 425), neuromarketing is considered to be a domain that can eliminate a part of the problems. For instance, by using neuroscience tools, practitioners can better understand the demand characteristics and influence the investments in new products and services, in innovation or in entrepreneurial initiatives.

One of the reasons the new products introduced in the market fail is the mismatch between the attitude and actual purchase behavior of the consumer (Agarwal and Dutta, 2015, p.457). Neuroscience approaches have the ability to predict choices and thus, have large implications for marketers (Plassmann, 2015 in Agarwal & Dutta, 2015, p.458). In this respect, neuroscience tools can be applied for testing the branding and marketing strategies before being implemented and can choose the most emotional strategies by using a celebrity endorsement or an association with a social or environmental cause. All these perspectives can help both in designing an efficient campaign and in avoiding the waste of money on implementing inefficient strategies (Agarwal & Dutta, 2015, p.458).

Desiring to know how consumers select their products, companies invest significant amount of recourses in order to better understand the purchasing behavior and what people really like (Pop, Iorga, 2012, p.640). Ariely and Berns (2010 in Pop, Iorga, 2012, p.640) identify several areas in which neuromarketing finds relevant application for the entrepreneurs. The first area is related to *designing food and beverage products*. In this respect, variables as flavor, texture, looks are important features. For instance, Plassmann et al. (2008 in Smidts et al., 2014, p.259) used fMRI to assess the role of information about product quality, as the price, on the perception of the product in terms of post-consumption rationalizing or in terms of changes in actual taste perception. The results show that changing the price of otherwise identical wines can affect the interpretation of taste pleasantness during sampling sessions (Plassmann et al., 2008 in Smidts et al., 2014, p.259). The subjects rate the expensive wine as being much more pleasant (the part of the brain that experiences pleasure being activated) (Dapkevicius and Melnikas, 2009, 19). Booth and Freeman (2014) name several situations in which neuromarketing can be useful. For instance, while it is considered that slogans and headlines in ads for food can increase sensory appeal (Elder and Krishna, 2010 in Booth and Freeman, 2014, p.178), the names of foods can affect the neural response to aroma (de Araujo *et al.*, 2005 in Booth and Freeman, 2014, p.178).

The literature consider that ad testing is the most popular area of neuromarketing practical application, the brain reactions implying a valuable added information to the results from the traditional research, especially at the prediction level (Ariely and Berns, 2010 in Smidts et al, 2014, p.260). By using neuromarketing instruments, advertisers can collect and assimilate information on consumers at the personal level in their endeavor to create intimate and lasting relationships with a brand (Rapp et al., 2009, p.51).

The second area refers to *architecture for designing buildings*. Neuromarketing can be used in order to assess the perceptions on the building design by using virtual reality or augmented reality (Pop, & Iorga, 2012, p.640).

The third area discusses the *movie industry* and the role neuromarketing have in evaluating the efficiency of certain scenes or of using certain characters. The same idea fits very well for music industry or for commercial industry (the way certain ads are designed, the music they use, the way websites are organized etc.). Thus, aspects as emotions, interest, trust, loyalty can be measured (Pop, & Iorga, 2012, p.640). In this respect, Stallen et al. (2010 in Smidts et al., 2014, p.259) talks about the need of companies to evaluate the brain reaction on celebrity endorsement.

Moreover, Shiv and Yoon (2012 in Agarwal & Dutta, 2015, p.459) emphasize ways in which neuroscience can be efficient: "providing opportunities and guidelines to facilitate theoretical development; facilitating new empirical tests of standard theoretical claims; explanations for observed heterogeneity within and across consumer groups; and novel mechanisms for considering the physiological context and the role of numerous biological factors, including hormones and genes, on consumer preferences and decisions".

In the same respect, Fugate (2007) underlines entrepreneurial application of neuromarketing by developing several perspectives. The first one talks about testing advertising effectiveness. Although there is no direct connection between arousal and behavior or purchase intent, the advertising specialists can use brain scan to predict the strength of an ad recall or the emotional attachment with the ad (Fugate, 2007, p.387). The second perspective refers to testing product appeal, meaning the efficiency of the design o the product. The third perspective emphasizes celebrity endorsement. In this respect, studies have shown that dopamine and phenylethylamine flood the brains in the moments in which consumers see familiar faces. It is believed to have the same reaction when world of mouth is used in the case of recommendations from close friends or family (Fugate, 2007, p.388). The fourth perspective refers to logo and brand selection, higher level of brain activation signifying a higher level of involvement, of relevance and importance to the person (Fugate, 2007, p.388). The fifth perspective deals with media selection. Besides analyzing which channel is the most appropriate for releasing a message, the research can be enlarged to the way consumers scan and read text, ads, images of even web pages (Reynolds, 2006 in Fugate, 2007, p.389). In this case, while an efficient message for the right part of the brain would be short and frequent television messages, an efficient message for the left side should be print and complete with factual information (Fugate, 2007, p.389). Not recommending a special media, Renvoise and Morin (2005 in Fugate, 2007, p.389) claim that the delivered message should be more visual and less verbal.

Although neuromarketing is generally perceived as a promising domain, there are authors claiming that that there are biases that can be encountered and that can make the data redundant. Several examples are the following: when observed, individuals act and react differently than in a usual context, there are no strict rules about the relationship between brain activity and behavior, the variables of the life stage are much more influential on behavior than brain arousal (James, 2004 in Fugate, 2007, p.391). Thus, neuromarketing is considered very unlikely to solve the problems marketing is confronting.

Moreover, although neuromarketing can provide valuable insights, it is criticized for being conducted in artificial environment and has limitations regarding the applicability in the real world (Barkin, 2013, p.50).

In the same respect, Medina (2004 in Fugate, 2007, p.391) considers that the technology is still imprecise and that neuroscientific understanding is not yet sufficient to understand a complex issue as consumer behavior. Another important concern states that there is the inability to generalized data from a small sample to a wide and diverse population (Nature Neuroscience, 2004 in Fugate, 2007, p.391), an example here being the diverse understandings of humor in different cultures (Harris, 2006 in Fugate, 2007, p.391).

Neuromarketing and Ethics

Being perceived as a domain that aims to find the buy button in the brain, neuromarketing is the subject of numerous criticisms and ethical discussions (Lee *et al.*, 2007, p.199). Although the majority of the instruments used by neuromarketing are not invasive, the scientific community has suggested and already works on a neuroethical code that aims to protect the involved entities (Murphy et al., 2008, p.295). There are at least two aspects that can create the connection between neuromarketing and non-ethical behavior (Ramsøy, 2014). First, generally speaking, the companies are aware of the ethical issues, social responsibility actions (Corporate Social Responsibility) getting advantage of the neuromarketing results as much as possible. Second, there is a false impression that neuromarketing can discover the buy button in the consumer brain (Ramsøy, 2014).

Thus, the literature talks about the need to protect the vulnerable groups, as individuals with neurological diseases, with psychological disorders, children or other protected groups (Murphy et al., 2008, p.295). In the same respect, the emphasis is on the following aspects: the need to provide complete information regarding the privacy protection of the subjects, the experimental protocol, the risks implied, the use of data, the possibility of retreat from the experiment anytime and regardless of the reason, the correct presentation of information in mass-media, and the control over the offered incentives for participants (Murphy et al., 2008, p.298).

Bercea (2015) underlines three types of responsibilities when it comes for the ethics in neuroscience. First, it is the responsibility on the subject. The subjects needs to be informed regarding the entire research process, to have the benefit of finding out examples of similar research already developed and without being used technical language. Second, there is the responsibility on the consumer. It implies using the results of the experiments in such a manner that these to be useful for the consumer and emphasizing honesty. The consumer must have the final word in the process of marketing decision. Finally, the literature stresses the responsibility on the researcher. Thus, being a growing and developing domain, it is necessary that neuromarketing to be accompanied by an open ethical guide (Bercea 2015).

One of the issues that preoccupy the neuromarketing specialists is the need to eliminate de problem of reverse inference. Usually, based on a study that involves neuroscience instruments, the statement that is designed is the following one: if the cognitive process X takes place (for instance, computing the desire to pay), then the Z zone from the brain is activated. The reverse inference implies changing the two components of the statement. Empirically, it is impossible to say that activation in certain zone of the brain implies a certain unique cognitive process. A single area of the brain can be activated by a large number of distinct cognitive processes (Plassmann, Ramsøy, Milosavljevic, 2012, p.29). Considering this context, one way of controlling the bias is that of finding the measure in which the interest region is selectively activated by a specific mental process. Thus, if a region is activated by a large number of cognitive processes, the activation of the region is not significant in relationship with the cognitive process chosen. However, if a region is selectively activated nu a mental process, then it can be validly infer the relationship between the cognitive process and the activation of brain area, a certain selectivity factor being required to be computed (Poldrack, 2006 in Plassmann, Ramsøy, Milosavljevic, 2012, p.30).

An exploratory research conducted among neuromarketing specialists around the world investigates the level of interest the neuromarketing companies have in the ethical issues (Pop, Dabija and Iorga, 2014). The survey includes the participation of 52 companies from Europe, North America, South America and Asia. The surveyed subjects are policy makers (chief executive officers, administrative officers, chief neuromarketing officers and chief research officers). Based on this survey, the authors develop a ranking with the most important areas in which neuromarketing tool are used: video ads creation, creation of communication campaign, printed ads/banners creation, product design, packaging, online media campaigns, sales promotion, entertainment (Pop, Dabija and Iorga, 2014, p. 34). Regarding the ethical issues mentioned by the specialists, the research mentions the following: the need to transparently and precisely inform the subjects, the need to guarantee data privacy and security, the fear that a third party might have access to disturbing information about subjects, and the security of data processing methods (Pop, Dabija & Iorga, 2014, p. 34).

The Neuromarketing Science & Business Association (NMSBA) has an important role regarding the issue of ethics in neuromarketing. NMSBA is a global association composed of researchers interested in neuromarketing domain and that aims to discuss the latest research in the domain, to develop an international network of knowledge in the domain, and to improve the quality of the domain (NMSBAa). Among the most important aspects mentioned in the ethical code this organization is proposing are the following: the importance of respecting a high degree of research professionalism, of respecting the subjects implied, of providing correct, complete and honest information, of offering the subjects the possibility to leave the experiment anytime, of respecting the privacy of the data, and of cautiously using the data only with the subjects' agreement (NMSBAb). This, based on these aspects and on what Racine et al. (in Bercea, 2015), is stating, confidentiality and protecting the privacy of the subjects are the most important aspects related to the way neuromarketing as a domain is entitled to use the neuroscience instruments.

Although there are organizations that militate against neuromarketing by underlying that the use of medical instruments in marketing purpose can deteriorate the collective mental, the optimists are emphasizing that the results in this domain can be useful for the society. For instance, it can be mentioned the contribution of neuromarketing for the public services industry, this domain being capable of leading to efficient anti-drug or anti-smoking campaigns (Phan, 2010, p.16).

In addition to all the above mentioned shortcomings of neruomarketing, in the majority of the studies using neuroscience instruments, one of the selection criteria of the subjects is related to the hand they are using for writing. Thus, the individuals that are left-handed are excluded form the experiments (Willems et al., 2014). In the context in which the left-handed people differently perceive the stimuli, rather using the right part of the brain, it is consider that their inclusion into experiments bring too much variety of the data. However, Willems et al. (2014) advocate for the introduction of left-handed people into experiments buy invoking that these individuals represent a significant part of the population (approximately 10%) and that losing them implies losing the representativeness of the results. Moreover, it is probable that by including them, the scientists could find new and relevant information on the way the brain works (Willems et al., 2014).

CONCLUSION

Concluding, although being a theoretical paper, the present study intents to treat the issue of unconsciousness reactions in business and marketing environments from a practical perspective. The brain and the way it works become important variables in the study of decision making process. Therefore, the measuring task of these processes implies a deep and careful design. In this respect, neuromarketing, as a new domain, is proposed as a solution in order to improve the knowledge specialists have.

Finding relevant information directly from the human brain and using neuromarketing instruments in order to gather and evaluate influences to market incentives are not totally well perceive processes. However, as this paper clearly states, there is an increasing need to better understand human behavior in general and in marketing in particular. Moreover, by using neuro-imaging instruments in marketing, there is a win-win situation. While entrepreneurs can find relevant information in order to develop their marketing strategies, consumers can both find relevant information about their own behavior and learn how companies are trying to persuade them. In addition, this paper emphasizes that there is a large number of experiments that can be replicated and that can complete the traditional methods in order to increase the amount of knowledge on human behavior.

REFERENCES

Agarwal, S., & Dutta, T. (2015, December). Tanusree Dutta. Neuromarketing and consumer neuroscience: Current understanding and the way forward. *Decision, 42*(4), 457–462. doi:10.1007/s40622-015-0113-1

Ariely, D., & Berns, G. S. (2010, April). Neuromarketing: The hope and hype of neuroimaging in business. *Nature Reviews. Neuroscience, 11*(4), 284–292. doi:10.1038/nrn2795 PMID:20197790

Baars, B. J., Ramsøy, T. Z., & Laureys, S. (2003, December). Brain, conscious experience and the observing self. *Trends in Neurosciences, 26*(12), 671–675. doi:10.1016/j.tins.2003.09.015 PMID:14624851

Bargh, J. A. (2002, September). Losing Consciousness: Automatic Influences on Consumer Judgment, Behavior, and Motivation. *The Journal of Consumer Research, 29*(2), 280–285. doi:10.1086/341577

Barkin, E. (2013, July). The prospects and Limitations of Neuromarketing. *Customer Relationship Management.*

Bercea, M. D. (2012). *Anatomy of methodologies for measuring consumer behavior in neuromarketing research.* Retrieved January 2014, from http://www.lcbronline.com/index_files/proceedingsemc12/12emc023.pdf

Bercea, M. D. (2013). *Quantitative versus qualitative in neuromarketing research.* Munich Personal RePEc Archive. Retrieved June 2015, from http://mpra.ub.uni-muenchen.de/44134/

Booth, D. A., & Freeman, R. P. J. (2014). Mind-reading versus neuromarketing: How does a product make an impact on the consumer? *Journal of Consumer Marketing, 31*(3), 177–189. doi:10.1108/JCM-08-2013-0674

Butler, M. J. R. (2008). Neuromarketing and the perception of knowledge. *Journal of Consumer Behaviour, 7*(4-5), 415–419. doi:10.1002/cb.260

Chartrand, T. L., Huber, Shiv, & Tanner. (2008). Nonconscious Goals and Consumer Choice. *Journal of Consumer Research, 35.*

Dapkevicius, A., & Melnikas, B. (2009). Influence of Price and Quality to Customer Satisfaction: Neuromarketing Approach. Science – Future of Lithuania, 1(3).

Egrie & Bietsch. (2013). *Marketing Research Ethics: How Consumers feel about Neuromarketing.* Academic Press.

Flores, Baruca, & Saldivar. (2014). Is Neuromarketing Ethical? Consumers Say Yes. Consumers Say No. *Journal of Legal, Ethical and Regulatory Issues, 17*(2).

Fugate, D. L. (2007). Neuromarketing: A layman's look at neuroscience and its potential application to marketing practice. *Journal of Consumer Marketing, 24*(7), 385–394. doi:10.1108/07363760710834807

Kahneman, D. (2011). *Think fast and slow.* New York: Farrar, Straus and Giroux.

Kamp, B. (2014). Overcoming Market and Network Failures by Means of Embedded Business Model Innovation. *Boletin de estudios economicos, 69*(213), 423-442.

Lee, N., Broderick, A. J., & Chamberlain, L. (2007). What is 'neuromarketing'? A discussion and agenda for future research. *International Journal of Psychophysiology, 63*(2), 199–204. doi:10.1016/j.ijpsycho.2006.03.007 PMID:16769143

McClure, S. M., Li, J., Tomlin, D., Cypert, K. S., Montague, L. M., & Read Montague, P. (2004). Neural correlates of behavioral preference for culturally familiar drinks. *Neuron, 44*(2), 379–387. doi:10.1016/j.neuron.2004.09.019 PMID:15473974

Morin, C. (2011). Neuromarketing: The New Science of Consumer Behavior. *Soc, 48*(2), 131–135. doi:10.1007/s12115-010-9408-1

Murphy, E. R., Illes, J., & Reiner, P. B. (2008). Neuroethics of neuromarketing. *Journal of Consumer Behaviour, 7*(4-5), 293–302. doi:10.1002/cb.252

NMSBA. (n.d.a). *The Neuromarketing Science & Business Association.* Retrieved June 2015, from http://www.nmsba.com/

NMSBA. (n.d.b). *The Neuromarketing Science & Business Association.* Retrieved June 2015, from http://www.nmsba.com/Ethics

NMSBA. (n.d.c). *The Neuromarketing Science & Business Association.* Retrieved April 2016, from http://www.nmsba.com/neuromarketing-companies

Nobel, C. (2014). HBS Working Knowledge, Neuromarketing: Tapping Into the 'Pleasure Center' of Consumers. In Consumer Neuroscience and Neuromarketing (2nd ed.). Academic Press.

Ohme & Matukin. (2012). *Small Frog That Makes a Big Difference.* DOI: ..10.1109/MPUL.2012.2189169

Olteanu, B. (2015, January). Neuroethics and Responsibility in Conducting Neuromarketing Research. *Neuroethics.* doi:10.1007/s12152-014-9227-y

Phan, V. (2010). *Neuromarketing. Who Decides to Buy?* The Triple Helix.

Plassmann, H., Ramsøy, T. Z., & Milosavljevic, M. (2012). Branding the brain: A critical review and outlook. *Journal of Consumer Psychology*, *22*(1), 18–36. doi:10.1016/j.jcps.2011.11.010

Pop, Dabija, & Iorga. (2014). Ethical Responsibility Of Neuromarketing Companies In Harnessing The Market Research – A Global Exploratory Approach. *Ethical Responsibility of Neuromarketing Companies in Harnessing the Market Research – a Global Exploratory Approach. Organizations' responsibility to Consumers, 16*(35).

Pop, N. A. (2012). Ana Maria Iorga. A New Challenge For Contemporary Marketing – Neuromarketing. *Management & Marketing Challenges for the Knowledge Society*, *7*(4), 631–644.

Ramsøy, T. (2014). Consumer Neuroscience and Neuromarketing (2nd ed.). Academic Press.

Ramsøy, T. Z., & Skov, M. (2014). Brand preference affects the threshold for perceptual awareness. *Journal of Consumer Behaviour*, *13*(1), 1–8. doi:10.1002/cb.1451

Rapp, J., Hill, R. P., Gaines, J., & Mark Wilson, R. (2009, Winter). Advertising and Consumer Privacy. Old Practices and New Challenges. *Journal of Advertising*, *38*(4), 51–61. doi:10.2753/JOA0091-3367380404

Renvoisé, P., & Morin, C. (2007). *Neuromarketing: Understanding the "Buy Button" in Your Customer's Brain*. Thomas Nelson.

Shiv, B., & Fedorikhin, A. (1999, December). Heart and Mind in Conflict: The Interplay of Affect and Cognition in Consumer Decision Making. *The Journal of Consumer Research*, *26*(3), 278–292. doi:10.1086/209563

Simonson, I. (2005). In Defense of Consciousness: The Role of Conscious and Unconscious Inputs in Consumer Choice. *Journal of Consumer Psychology*, *15*(3), 211–217. doi:10.1207/s15327663jcp1503_5

Smidts, A., Hsu, M., Sanfey, A. G., Boksem, M. A. S., Ebstein, R. B., Huettel, S. A., & Yoon, C. et al. (2014). Advancing consumer neuroscience. *Marketing Letters*, *25*(3), 257–267. doi:10.1007/s11002-014-9306-1

Suomala, J., Palokangas, L., Leminen, S., Westerlund, M., Heinonen, J., & Numminen, J. (2012, December). Technology Innovation. *Management Review*.

Tom, G., Nelson, C., Srzentic, T., & King, R. (2007). Mere Exposure and the Endowment Effect on Consumer Decision Making. *The Journal of Psychology*, *141*(2), 117–125. doi:10.3200/JRLP.141.2.117-126 PMID:17479582

Willems, R. M., Van der Haegen, L., Fisher, S. E., & Francks, C. (2014). On the other hand: Including left-handers in cognitive neuroscience and neurogenetics. *Nature Reviews. Neuroscience*, *15*(3), 193–201. doi:10.1038/nrn3679 PMID:24518415

Wilson, R. M., Gaines, J., & Hill, R. P. (2008). Neuromarketing and Consumer Free Will. *The Journal of Consumer Affairs*, *42*(3), 389–410. doi:10.1111/j.1745-6606.2008.00114.x

Yoon, C., Gonzalez, R., Bechara, A., Berns, G. S., Dagher, A. A., Dubé, L., & Spence, C. et al. (2012). Decision neuroscience and consumer decision making. *Marketing Letters*, *23*(2), 473–485. doi:10.1007/s11002-012-9188-z

YoutubeR. S. (2015). Retrieved from https://www.youtube.com/watch?v=R8WkIC7bMnE

Youtube. (n.d.). Retrieved September 2015, from https://www.youtube.com/watch?v=IGQmdoK_ZfY

This research was previously published in Digital Entrepreneurship and Global Innovation edited by Ioan Hosu and Ioana Iancu, pages 151-171, copyright year 2017 by Business Science Reference (an imprint of IGI Global).

Chapter 15
Neuromarketing and the Potential Application of Scientific Methods in Measuring Consumer Behaviour

Cynthia A. Bulley
Central University, Ghana

Veronica Adu-Brobbey
University of Education – Winneba, Ghana

Esther O. Duodu
ABL Intel Consulting Ltd., Ghana

ABSTRACT

Consumer behaviour studies have taken a new turn. Marketers, economists and other consumer related disciplines are looking to science to accurately determine consumer behaviour. The purpose of this chapter is to provide insight into a burgeoning field of study, neuromarketing, documenting various research studies and applications of mechanisms in determining brain activities and other uses of science to benefit marketing research. Data for the study is derived from impartial cross-referencing of conceptual and empirical articles published in major journals. The application of neuroimaging technique in research have provided marketers with concrete evidence of brain activation that signal increased activities during stimulation (Lewis & Bridger, 2005; Rossiter et al., 2001). Further, the implication and causes of concern in using neuroscience methods in marketing are highlighted. Developing country studies on neuromarketing are examined to determine its application and use as a marketing research tool.

INTRODUCTION

The medical environment and its numerous usage of imaging have found its way in marketing. Medical imaging basically provides a visual scan (still pictures) or video of any part of the body for analysis. It noninvasively reproduce image of the internal aspects of the human body. And this aspect of science

DOI: 10.4018/978-1-5225-5478-3.ch015

research in areas such as neuroscience, behavioural economics, human psychology have brought to the fore methods that marketers can explore to arrive at reliable conclusions on human behaviour and reactions to products and services. Marketers have to deal with consumer behaviour especially the decision making process and activities of the human brain.

Neuromarketing is a discipline that applies a purely scientific process to measure and analyse consumer brain activity to determine decisions and choices made in buying products and services. It is a synthesis of neuroimaging techniques in neuroscience and marketing. The transdisciplinary nature of neuroscience provides a platform that allows marketers to better understand consumer decision making process and personal choices. Marketers are making use of this scientific process in monitoring and analysing consumers. Clinical information and observations about the functions and movements in the brain facilitates explanation of the human mind and thought process to explain consumer behaviour. It is all about reactions of neurons in our brain and other parts of our human body. Neurons are cells that form part of the human nervous system whose chief function is to send messages to and from the brain. Marketing stimuli is transferred through or along the neurons by means of electrical impulse. The nervous system is made up of the brain, spinal cord and other peripheral nerves. These are nerve cells named neurons and other supporting cells called glial cells. The neurons are nerve cells that act in unison in transferring electrical and chemical stimuli in or to the brain. Neurons are complex cells that are the pathway for transmitting information in the nervous system. There are three main neurons – the sensory neurons, the motor neurons and interneurons. The sensory neurons are attached to receptors that detect and respond to various stimuli internally and externally. The receptors are very sensitive to stimuli and other internal and external changes. For instance, variations in sound, light, and other chemical and mechanical stimuli affect hearing, vision, touch, skin responses (pain), smell and taste. The motor neurons are the chief controllers of human muscles and therefore are responsible for movements and behaviour (even speech). The interneurons are cells that particularly populate the human brain (almost 80%). Interneurons are responsible for the highest functions of the brain and support optimal processing of information of any stimulus (Rawlins, 2005; Bear, Connors & Paradisco, 1996). This preamble provides a basic understanding of activities and functions of aspects of neuroscience that play a role in consumer research and methods applied in marketing studies.

Consumer behaviour is embodied in the "dynamic interaction of affect and cognition, behaviour and the environment" that influences human decisions and choice (Schiffman & Kanuk, 2007). These reactions and interactions are a function of the brain (neurons in the nervous system) responding to internal and external stimuli. Marketing researchers are now looking to clinical information and measurement of brain functions and its mechanisms to explain behaviour and decision making. Though the method involves imaging and scanning equipment that may hinder frequent usage, the rise of neuromarketing firms like Neurofocus, Brighthouse, EmSense, SalesBrain, Sands Research, Beckley Cali, Mindlab International and Neurosense (now Nielson) are contributing to its popularity as a more reliable and concise technique of data collection especially on consumer behaviour. There is a platform online - Neuromarketing Business Association - that encourages research, research firms and others interested in this discipline to network.

Research interest in neuromarketing has grown exponentially and the potential application of such scientific methods in measuring consumer behaviour is a positive and necessary endeavour. The purpose of this chapter is to provide insight and document various research studies and application of mechanisms in determining brain activities and other uses of science to benefit consumer research. Selected research studies are analysed accordingly to highlight the current state of research and areas of further possible research to bring to the fore this burgeoning research method. The usage of such methods in developing countries in Africa is a question that needs to be examined to throw more light on the discipline and increase its future deployment as a method of choice in consumer studies.

The chapter is organised in five sessions. This section introduces the chapter focusing on a general overview of neuromarketing. The background information follows examining related works that provide the main foundation for the review (under quest for in-depth literature). This includes other subtopics relevant to building a body of knowledge on neuromarketing. An analytical section that reviews tools and techniques used in neuromarketing research follows with relevant findings, discussions and recommendations. This takes a look at what is going on in developing countries in terms of brain imaging techniques and consumer research. Furthermore ethical issues and its implications emerging from neuromarketing research are evaluated. Finally, future research directions and conclusion end the chapter.

THE NEUROMARKETING DISCIPLINE

Until recently, the discipline and study of consumer behaviour using neurons under neuromarketing has been conceptually under studied. The word neuromarketing is derived from neurology or neuroscience and marketing. Martinez (2011) attributes neuromarketing to the integration and usage of aspects of three disciplines – neurology, psychology and marketing in the processes and procedures of the discipline. Neuromarketing taps into neuroscientific methods of examining changes in the brain to understand business reaction and decision process. This view is shared by many in the field of marketing, psychology, economics and science (Fugate, 2007; Rawlins, 2005; DeFelipe and Farinas, 1992). Defining neuromarketing basically delineates the effect of neuroscience imaging on marketing subjects that is, consumers. It seeks to map out brain processes in response to stimulus, and choices and decisions made thereof. Calvert and Brammer's (2012) definition of neuromarketing dwell on applying neuroscientific techniques to meaure *unconscious* reactions and responses of consumers to marketing stimuli. It measures consumers' actual feelings.

Neuromarketing mainly focuses on the complex interactions in the brain. Neuroscience and marketing in this sense share a common concern that is, finding out how patients/consumers react to stimulus and decisions made in this regard. It measures consumer 'sensorimotor', cognitive and affective responses to marketing stimuli. Hence marketing researchers have defined neuromarketing in simple terms as the research activity that uses methods, techniques and tools from the field of neuroscience (Genco et al., 2013).

The word neuromarketing is attributed to Smidts, a professor of marketing research at the Rotterdam School of Management, Erasmus University. Although the works of Plato, Adam Smith, Rene Descartes, Engel, Kollart and Blackwell focused on rational thinking and decision making, laying foundation for the development of science to address everyday issues (Glimcher, 2003; Fugate, 2007). These philosophers examined the human mind, human behaviour and rational and emotional thinking (the "economic man" – Adam Smith). Further, mental activities in relation to the nervous system (Rene Descartes) and thought process were used to articulate deeper subconscious rational thinking. The relevance of these studies promoted scientific inquiry into neuroscience. Significant progress was made in the area of structure and functioning of the brain in relation to marketing efforts. However neuroeconomics began the scientific inquest followed by neuropsychology, then neuromarketing and neuroethics. The implication of neuromarketing is that there will be a balance between science and marketing in seeking and determining consumer subconscious motives. The scientific tools and techniques ensure a bias free and accurate measurement of human subconscious thinking and reasoning. Marketers can now reliably and empirically measure, observe and predict behaviour. In other words, quantitative data can be derived from constructing inferential models from consumer behaviour outcomes and observations using neuroscience techniques. Neuromarketing research is capable of determining emotional and rational thoughts which in fact depend on each other in the human mind (brain).

In neuromarketing studies, rational activity in the human brain (*Neocortex*) and the unconscious mind can be measured, evaluated and used to reveal human decisions. For instance, emotional impulse decisions (especially impulse buying decisions) can be activated because humans have three distinctive brain functioning system (Zarã & Tuţã, 2013; Renvoisé & Morin, 2005). The *Neocortex*, the *Limbic system* and the *Reptilian* brain. The *Neocortex* also known as *Neopallium* is typically peculiar to humans and it enables the most complex mental activity. It is the main rational and thinking aspect of the brain. The *Neocortex* has four other subdivisions – the *frontal*, the *parietal*, *occipital* and *temporal lobes*. The *frontal lobe* is located at the front and upper portions of the brain hemisphere (cortex), where decision making and other high level human analysis occur. It enables reasoning, judgement and language among others. The *parietal lobe* is at the upper but back portion of the brain. It responds to sensory stimuli (taste, touch, feelings – temperature) and may be an important area for neuromarketing research (Swenson, 2006). The *occipital lobes* can be found at the bottom back portion of the cortex and it controls sight (the eyes and seeing). It is responsible for rapidly processing visual data from the eyes. The *temporal lobe* is for hearing and listening. It processes auditory information. The *temporal lobe* is for processing what the ears absorb. It processes and recognises auditory information. The lobe is an integral part of the cortex that is, key to comprehension and understanding. It is associated with memory and speech (Renvoisé & Morin, 2005).

The *Limbic system* is the middle inner surface portion of the brain. It is made up of interconnected structures (lobes, cortex, nuclei and others). It adjusts autonomic responses especially in relation to internal and emotional aspect of behaviour. It is the emotional brain (for arousal, motivation and reinforcement behaviour) and performs other important functions linked to learning and memory. The *reptilian brain* is the upper brain stem and it is made up mainly of the fore brain. It is the central system that controls the autonomic nervous system and regulates the body's total functioning. Also known as the *R-complex,* it controls breathing, the heart rate, body temperature and the total balance of the human being. The *reptilian brain* is in charge of reflexive and instinctive behaviours therefore decision – making processing and rational decisions are processed in that part of the brain. Hence consumer behaviour and purchase decision processing and other impulses are the fundamental function of the brain.

The use of neuroscience to find out the actual insights and brain functions of consumers' provides marketers with a better tool to predict subsequent buyer behaviour. And this has been discussed at length by neuroscience, neuroeconomics, and neuromarketing researchers like Fugate (2007), and McDowell and Dick (2013). The argument that has been propounded is the view that consumers as rational human beings will seek their interest and therefore decisions made would follow rational thinking that examines all other available options in any situation. However, this kind of classic economics premise is not usually followed by consumers in critical decisions. Rather emotional intuitions overtake or sometimes are put aside in consumer decisions. Pop and Lorga (2012) and Genco et al. (2013) have on these grounds concluded that research point to' strong emotions' overtaking rational decisions made by consumers. The researchers compare and contrast the *intuitive consumer model* against the *rational consumer model* to emphasize on emotions playing a major role in decisions and behaviour. The intuitive consumer model emphasizes the use of "emotions, habits and experience" in consumer decision – making and choice rather than using rational logical evaluations. The rational consumer model places emphasis on rational thinking which drives decision – making, and product and service decisions instead of intuition (Du Plessis, 2011; Damasio & Bechara, 2005). The two models play a critical role in decision making but it is the symbiosis relationship that drives choices and decisions that is important to consumer researchers. Damasio (2006) and Damasio and Bechara (2011) have argued that human rational decisions actually rely on prior processing of intuitions and emotions. This is a fact that drives marketing researchers to want to know more about the brain and its functions to understand consumer perceptions, behaviour and decision – making.

Neuromarketing encourages researchers to conduct studies on consumer behaviour that are transparent to increase the 'generalisablity' of findings and understanding of consumer decisions. The sizable research and literature on neuromarketing creates a fundamental foundation for the full application of scientific methods in marketing research.

THE QUEST FOR MORE IN-DEPTH LITERATURE ON NEUROMARKETING

Marketing research methods have come a long way and any addition that would deliver concise and more accurate understanding of consumer behaviour should receive positive acclaim. The extensive research in the field of neuromarketing indicate that the scientific research method is going to expand the scope of the traditional consumer behaviour studies (Mucha, 2005; Lee, Broderick & Chamberlain, 2007; Lindstrom, 2009). Neuromarketing research avoids rational interpretation and applies scientifically derived results of the subconscious areas of the brain to explain consumer behaviour, decision making and choice. This makes the discipline a more precise research method for understanding human behaviour. The acceptance of neuromarketing methods requires extensive conceptual and theoretical analysis of research in the field. Further empirical company specific research using neuromarketing tools and techniques should be replicated to affirm its acceptance and usage. Empirical studies where specific cases were researched by neuromarketing companies are geared toward predicting consumer reactions and behaviour with specific products. These studies focus on particular methods and tools. Nielsen's Neuro-Focus Company uses a specially developed EEG device and other eye and skin monitoring tools in their studies. Other research firms use different metrics to measure brain and skin changes to determine consumer response to stimuli and subsequent choices made. The practical application of these methods and techniques will *revolutionalize* the field and promote its acceptance. The use of imaging techniques provides marketers with an alternative research method that can lead to further mind targeting communications. The question that needs to be addressed is that – is there adequate literature to move neuromarketing to the next level of development? There is the need for more literature especially on interpreting and reading results of neuromarketing tests.

Some sceptics like Booth and Freeman (2014) have challenged the interpretations of the brain imaging scans and reports on the grounds that it produces *abstract* reports that can only tell one brand from the other. Their argument is that external cues (marketing related) and product variables influence decisions. The researchers conclude that individual behavioural mapping based on psychological responses can be more beneficial, and cognitive responses can produce better results. The study on consumer *choice processes* examines the *conceptual-sensory interaction in a marketing place*. Further it goes to prove that individual cognitive responses provide the necessary cues for preferences and choice making. For instance, information provided to the consumer is analysed and the resulting decisions made reflects the interaction between that consumer and any cues (advertisements) they come into contact with during the process.

There is a dearth of research papers on different aspects of neuromarketing and the use of neuroimaging technique. Although such literature provides detailed understanding of the underlying concepts and processes, there is the need for more literature on the subject. This calls for more empirical studies that creatively use the scientific process, concepts and framework to make its application more beneficial for better interpretation and advancement of the field and method. Theorising in neuromarketing is largely a matter of relying on neuroscience, social psychology, behavioural and biological science, economics and marketing. It is important to note that neuroscience research seeks to develop scientific knowledge to a higher level to simplify complex problems. This means complex issues can be defined, explained, and used in resolving future problems. There is the need for in-depth literature to add to the current array of literature to facilitate development of models and strategies for the discipline.

REVIEW OF STUDIES IN THE FIELD OF NEUROMARKETING

This review outlines the direction of prior research to stimulate new and innovative ideas that will promote neuromarketing studies. The section pulls together disparate research studies to present some of the articles on the current state of knowledge in the field. The research studies were selected from well known databases such as "EBSCO host", "emerald", "science direct" "WileyInterscience" and Google scholar. For the purpose of this evaluation, publications between 1995 and 2015 were reviewed. The impartial cross – referencing of conceptual and empirical articles published in major journals were targeted and used. The articles selected were essentially based on its academic background and scientific grounding. It is important to emphasize that the journal articles chosen were not limited solely to one particular field but rather an interdisciplinary approach that is directed at marketing research was the focus. The review regrouped the articles into two based on the methodology applied. Literature review studies and empirical papers indicating the actual tools applied are examined. The major conclusion and authors (as shown in Table 1) summarises important objectives, research methods and primary areas emphasized in these articles. The publications also addressed other objectives. This is a semi-systematic inquiry and the main objective is to provide information.

In all the literature review articles, information provision discourse is paramount. This is interspersed with definitions, history, brain structure and attributes of neuromarketing and its relevance to consumer behaviour studies. Furthermore imaging techniques and tools are emphasized in the research papers (Perrachione & Perachione, 2008; Lee et al. (2007; Fugate, 2007; Pechman et al., 2005). Paramount in all the articles is the description and purpose of the field. In all the articles reviewed, neuromarketing has been described as a scientific field under neuroscience (Perrachione & Perrachione, 2008), as a research method in marketing (Fugate, 2007; Fisher, Chin & Klitzman, 2010; Wilson et al., 2008), as part of psychology under consumer behaviour and perception (Butler, 2008; Pechmann et al., 2005), as part of economics (Hubert & Kenning, 2008) and as a new academic field (Garcia & Saad, 2008; Godwin, Kothai & Saranya, 2012).

Wilson et al. (2008), da Rocha et al (2013) and Burgos – Campero et al. (2013) provide a brief report on neuromarketing methodologies that comprehensively provide a gist of tools that can be used in the field. The electro encephalography (EEG), functional magnetic resonance imaging (fMRI) and other

Table 1. Synopsis of neuromarketing publications

Author/Year/Journal/Country	Title/Major Objectives/Type of Paper	Methodology Applied	Major Conclusion
Perrachione, T. K. and Perrachione, J. R. (2008). Journal of Consumer Behaviour Vol. 7 no. 4/5 pp. 303-318. DOI: 10.1002/cb.253 USA	**Topic:** "Brains and Brands: Developing mutually Informative Research in Neuroscience and Marketing." **Objective:** "to develop a synthesis of the theoretical interests of marketing and neurology to bring out an educative study" **Type:** Research paper.	Literature Review	1. The need for collaboration between neuroscientists and marketers to 'align' this aspect of researching emphasised. 2. All marketing mix variables were on the whole more likely to have correlations with neuroscience methods but calls for a strong scientific mindset.
Lee, N.; Broderick, A. J. and Chamberlain (2007). International Journal of Psychophysiology (Science Direct) Vol. 63 pp 199-204. DOI: 10.1016/j.ijpsycho.2006.03.007 UK	**Topic:** What is Neuromarketing: Discussion and Agenda for Research. **Objective:** to provide knowledge of neuromarketing to go beyond application. **Type:** Research paper.	Literature Review	Issues of trust, pricing and negotiation should be given more attention in neuromarketing studies.
Fugate, D. L. (2007). Journal of Consumer Marketing, Vol. 24 (7) pp. 385-394. http:dx.doi.org/10.1108/073637607/0834807 accessed on June 17, 2015.	**Topic:** Neuromarketing: A layman's Look at neuroscience and its potential Application to Marketing Practice. **Objective:** Examine the origins of neuromarketing, its processes in simple terms and documenting findings for future research **Type:** General Review	Content Analysis of Literature	There is the need to operationalise neuromarketing experiments, and construct models to measure and determine brain activities' overall contribution to decision making.

Source: Bulley, Adu Brobbey and Duodu (2015).

imaging applications are presented describing the processes involved and the sample size that can be utilized in such research. The studies agree that small samples are necessary due to the process and the procedure of the neuroscience research method. Reports by Morin (2011), Lang, Forson and Rothschild were used in majority of these studies to illustrate the use of EEG and fMRI tests especially. Some of these tests indicate that despite millions invested in brand emphasis and all advertising campaigns, research methods applied do not present actual reports of emotions and the deciding factor when consumers are exposed to promotions. Hence neuromarketing research has come to bridge that gap to provide in-depth analysis of consumer unconscious reactions.

Various case specific studies done by Gladwell (2005), McClure, Li, Tomlin, Cupert, Montague and Montague (2004), Lindstom (2010) and others about the subconscious image effects on consumers provide the base studies for neuromarketing. Consumer preferences for Coca Cola and Pepsi and the effects of advertisements on the subconscious mind research was a major step forward in the advancement of the field (McClure, Li, Tomlin, Cupert, Montague and Montague, 2004; Gladwell, 2005). Lindstom's (2010) research on the effect of cigarette packs advertisements on consumers, and the media time slots survey horn in the use of neuromarketing research. This study provided more specific responses based on images from particular areas of the brain to determine positive and negative reactions, emotional effect and, final choice and purchase decisions. Research companies and industry empirical studies outnumber academic studies (per the review conducted). This can be attributed to the elementary stages of the method and the cost involved in conducing such survey.

In terms of theorising, consumer behaviour theories are the main driving force for quantitative studies. The synopsis of journal articles below delve into areas that touch on various theories. Leslie (2005) and others use cognitive and affective constructs of consumer behaviour which is rooted in psychology. Garcia and Saad (2008) dwell on *evolution* and *Darwinising* the neuromarketing methods. Resource – based theories featured in Burgos – Campero (2013) studies. Da Rocha et al (2013) focus on consumer satisfaction while Wilson and Gaines (2008) point out the ethical issues and challenges that arises from using neuroimaging methods. Concepts have been explored by all the studies examined (Table 1), and the issue of trust, cost, sampling and techniques, and tools to apply were major huddles that must be addressed in conducting research using neuromarketing methods. Literature reviews outnumber empirical studies (reflected in the analysed articles), and researchers such as Fugate (2007) have advocated that theoretical and qualitative studies should be forthcoming since that would highlight the field and encourage model development by marketers.

The first attempt at model development was first muted by Butler (2008). He integrated views from academic studies and applied company specific empirical research and reporting to derive his *model of Co-Production in Organisational Cognitive Neuroscience*. The model indicates the interconnection of scientific research, knowledge and practice and the different perceptions generated in the marketing field. It brings together the various interests, parties and the use of neuroscience knowledge. Waldman (2013) and Butler (2008) point to a strategic fit between academic research (scientific rigorous research), practical commercial agencies application and the clients' satisfaction with the reporting. And emphasis is placed on the scientific rigorous process done under ethical considerations. The model has been criticised for various reasons. Edwards (2013) pinpoints how in reality the *mental processes* that come into contact with stimulus to result in consumer behaviour are dynamic processes hence would be difficult to determine. This may cause over simplification of the reporting of the findings. Lindebaum and Zunndel (2013) and Becker (2013) in their critique concluded that neuroscience and neuromarketing may not be able to stand on its own as a field *theoretically or empirically* but more research needs to be done. Butler (2013) further re-examined the model and added that a behavioural model alongside the *Organisational Cognitive Neuroscience* would be the answer to the criticisms.

Two researchers whose systematic reviews proposed frameworks for brand psychology and preferences are Schimtt (2012) and Plassman, Ramsoy and Milosavljevic (2012). Plassman et al. (2012) used consumer decision – making construct developed by psychologist (Wirtz, Kruger, Scallon & Diener, 2003; Rangel, Camerer & Montague, 2008; Kahneman & Snell, 1992) to arrive at their *value signals important for brand decisions* construct. The construct in four stages indicates the components needed to form brand preference. The stages are a) *representation and attention*, b) *prediction of value*, c) *experience the value* and d*) remember the value and learn* respectively. The first stage tackles the information consumers are exposed to and which brand catches their attention. The identification of a particular brand follows a *prediction of the value* and qualities of the item. The purchase and experience precedes memorisation of the particular brand. The area of the brain where this process takes place is indicated. Various studies support every stage of this assertion (Payne, Bettman & Johnson, 2003; Koch, 2004; McClure et al., 2004; Plassman et al., 2008). The review concludes that consumer psychology is supporting neuroscience in an interdisciplinary way to promote theorising and improve neuromarketing studies. This is in line with Perrachionne's (2008) article which examines neuroscience concepts and modelling the process of neuromarketing research.

Perrachionne (2008) emphasizes *that location, connection and representation* are the main components in neuroscience that neuromarketing must focus on. The *location* that is, focusing on a particular region of the brain that indicates areas that is more metabolically active. *Connectivity* is the means by which neurons in the located cerebral areas of the brain react to process data while *representation* is the *encoding process* that stores the information being processed. Based on research in neuroscience, the brain functions and maps can provide such information and research must pinpoint the exact area that are active in decision making that must be analysed. The use of imaging provides the functioning and necessary information in particular areas of the brain and its interpretation for further analysis. These studies validate and give credence to neuromarketing's relevance and studies. However it is important to note that researchers must decipher and interpret results to arrive at credible reporting.

The articles point to a collaboration between scientists and marketing researchers to arrive at reliable results

Using Neuroscience in Marketing Surveys/Research

Even though in the 1990s Dawkins' "meme" and Zaltman's Metaphor Elicitation Technique were the first methods used for eliciting information to understand the consumer's mind, currently other more sophisticated methods abound. The "meme" are analogous to genes and unit information that is stored in the human brain is measured. Memetics is a field of study that seeks to empirically provide understanding of human behaviour and cultural process by measuring genes in the brain. Zaltman Metaphor Elicitation Technique is a research method that builds statements that are used to probe and identify deeper feelings that motivate consumer behaviour. There are different tools and methods in neuroscience that can be applied in marketing to determine consumer reactions and decisions making. Some of the tools and techniques include: functional Magnetic Resonance Imaging (fMRI), electroencephalography (EEG), magneto encephalography (MEG), Positron Emission Tomography (PET), Steady State Topography (SST), Transcranial Magnetic Stimulation (TMS), eye tracking, facial coding and other biometrics that observe and measure autonomic arousal and movements. The fMRI, EEG and SST measures activities in specific areas of the brain to determine its response to changes in the respondent's physiological state and the portion of the brain that is affected. Other responses and technique include heart beat, respiratory rate, skin responses (galvanic skin response - GSR) to observe consumers' reaction, thoughts and the decisions that follow. Figure 1 provides a grouping of the tools and techniques applied in neuromarketing research.

The neuroscience tools used in neuromarketing studies can be grouped into three based on the purpose for which it is used. The fMRI and PET record metabolic reactions in the brain. The process is able to localise and image the neural activity in the brain, measuring the changes that occur in the blood flow.

The EEG, MEG, SST and TMS monitor and document electric activity in the brain. The process involve using electrodes (EEG), electrical helmet (magnetometer), to measure the frequency of electrical current and the electro-chemical changes and signals (Figures 2, 3, 4a – d). The biometric and other body response tools are non – neurological tools that measure specific portions of the body like the skin, face eye to read its responses to stimuli.

Marketing research can make use of these tools in numerous studies These tools can be applied in brand differentiation determination and product perception research, advertising and promotion research, and other studies that evaluates emotions, interests, trust, customer loyalty, decision – making and choice. Neuromarketing is a field of inquiry that is growing. The scientific methods if applied and the corresponding results call for expert interpretation and marketing researchers would have to gear up to take up the challenge.

Figure 1. Neuromarketing tools and instruments applied in data collection

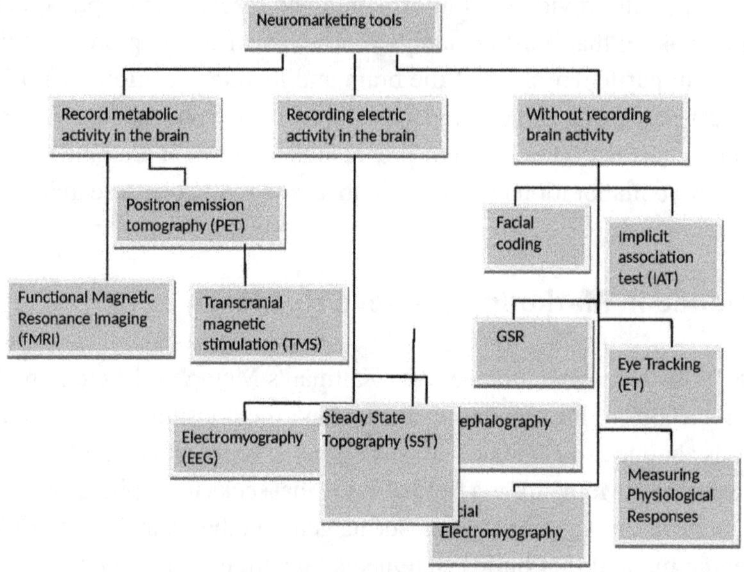

Marketing Research or Neuromarketing Research

Marketing research on consumer behaviour and decision making process have been reported to account for 80% of failures particularly in the effectiveness of advertising campaigns in terms of cost and new products and services (Morin, 2011; Pradeep, 2005). In conducting marketing research, respondents' willingness to provide truthful responses particularly in relation to feelings and reactions lie on the consumer. Evaluations of cognitive and affective processes are at a level that marketing researchers cannot determine but rather depend on consumers to provide answers. Marketing research methods like surveys, focus group discussions and interviews are not suitable for evaluating brain functions to determine feelings that can only be determined from the subconscious mind. Traditional marketing methods are not explicitly able to model what takes place in the brain or subconscious mind of the consumers when exposed to stimuli. Hence neuromarketing research which uses neuroscience imaging is able to map or scan consumer brain when activation takes place. Activation involves the introduction of any marketing related stimuli. The choice of methods in this sense sways toward meuromarketing. This demonstrates that neuroscience fills the gap in providing critical inputs for marketing research.

The question of samples and procedure needs to be addressed. Neuromarketing research uses smaller sample size and real time studies are recorded to provide a better understanding of the subconscious mind and its processes. The issue of sampling and the process is due to the equipment involved and the cost of imaging. Is marketing researchers ready to take on the challenge? Are consumers who mostly make up research respondents ready to go through such surveys? The respondents in most neural research cases may not be conscious of the actual process of the study though their consent is sought to avoid any contention. The non-invasiveness of neuromarketing methods allow the researcher to measure consumer physiological reactions. Direct benefits of neuromarketing research have been tabled by Zarã and Tuțã (2013) and these include consumer behaviour, decision process and choice particularly in field of marketing. The research method is able to process information using scientific tools to provide detailed knowledge about consumers.

Academic researches mostly dwell on qualitative and theorising with a few specific landmark studies while company specific surveys have made use of neuromarketing tools and methods. *Neurorelay* and the *Neuromarketing Science and Business Association* (NMSBA) have a list of neuromarketing research companies worldwide for 2014 and this is compiled to provide information on the tools available to these companies (the list is not as comprehensive since new companies are added from time to time). It provides the continent, country, the company and the technology available to the firms as listed below:

IMPLICATION OF NEUROMARKETING RESEARCH FOR CONSUMERS

The consumer in marketing is the prime focus of every marketing activity. The aim of neuromarketing research is to understand consumers, how and why they make buying decisions, the choices they make and the portion of the brain that support this process. The implication is that there will be a balance between science and marketing in seeking and determining what really is going on with consumers. However, the consumer may be inconvenienced in such surveys and ethical issues bothering on consumer rights and abuse may hinder the attractiveness of the method.

Table 2. A compilation of Neuromarketing companies and technology available

Continent	Country	Company	Technology Available
Europe	Milan, Italy	Itol Lab	EEG, Eye tracking
	Milan, Italy	Behaviour and Brain Lab	EEG, Eye tracking, GSR, Heart rate, respiration rate, Electromiography
	Milano – Modena, Italy	Fondazione Organismo di Ricera & Innovazione	EEG, Eye tracking
	Milan, Italy	SR Labs	Eye tracking
	Milan, Italy Barcelona, Spain	SalesBrain	Pain Probe (voice layering analysis) Brand Names: NeuroMap™, PAIN Probes™
	Milan, Italy	NeuroBiomarketing	EEG, EMG, ECG, GSR, Eye tracking
	Rome, Italy	Brain Signs	EEG, Heart rate, GSR, Eye tracking
	Rome, Italy	Brain Trends	EEG, Thermography, fMRI, Eye tracking, 3D Virtual Reality
	Trento, Italy	NeurExplore	EEG, GSR
	Marsa Malta	AAT Research Limited	EEG, GSR, Eye tracking, Virtual reality tracking
	Bruxelles, Belgium	Brain Impact	fMRI, EEG, Eye tracking, IAT priming test
	Heverlee, Belgium Amsterdamn, The Netherlands	Haystack	EEG, GSR, Eye tracking
	Amsterdam, The Netherlands Paris, France	Derval Research	Biometrics Brand Names: Hormonal Quotient®, Sensory Lab®, Sensory GeoMap®, Persona Quest™, Endocrinography™
	Amsterdamn, The Netherlands Stuttgart, Germany	Neurensics	fMRI, Brand Names: 3D Brain Rating™, 3D Mind Mapping™
	Berlin, Osnabrueck	Eye Quant	Eye Tracking

Nevertheless, the information from neuromarketing research using scientific tools provides a bias free and accurate measurement of human subconscious thinking and reasoning. And marketers can now reliably and empirically measure, observe and predict behaviour. The *Pepsi Challenge* by McClure et al. (2004) demonstrates the potential to use external information to trigger particular choices and brand decisions.

From 2005 to date, the advertising industry has moved to the next level in conducting consumer behaviour studies. Neuroscience techniques are supporting the industry in its efforts to critically examine consumers and brand engagement. In addition, promotional campaigns and other communications that call for more effective and detailed knowledge of consumer preferences and decision making can benefit from neuromarketing studies. The industry mainly uses sensory tools like the fMRI, EEG and Eye tracking tools in measuring consumer response to advertising message. Overall, marketing and public policy stand to benefit from the implication of neuromarketing research.

Ethical Concerns and Its Implications

Many concerns have been raised by consumer advocates about the intrusiveness of neuromarketing techniques. The first issue raised is the privacy and rights of consumers. The research method requires the consent of consumers and fundamentally consumer advocates could argue on the evasiveness of the procedures and interpretation of the results. The potential abuse of consumer rights and consent in acquiring information from brain activities to interpret behaviour and the end use of the data which influence decisions is the main issue. And this may lead to a breach of confidentiality of thought that can be challenged legally. Moreover neuromarketing methods and techniques may be criticised because it induces and interprets consumer behaviour at a subconscious level within the brain. Arguments developed by consumer and ethical advocates are on the basis of circumventing consumer rational decision and choice. The key issue (especially with advertising) is the potential to mislead consumers into believing that a purchase decision is a rationally motivated choice when in fact there may be a stimulant influencing their subconscious impulses and thinking. This means marketers and advertisers could go beyond inquiry to unlawfully coercing consumers into believing that they desire or want particular products and services for which they do not need. Based on this premise, the vulnerable populace could be disadvantaged and consumer laws need to be tightened or promulgated to protect consumers. The European Union's *Unfair Commercial Practices Directives* and other country-specific consumer protection laws need to be pronounced particularly if the neuromarketing technique should gain universal application. In the United States, the Federal Trade Commission Act, Section 5 protects consumers against *unfair and deceptive trade practices* and the Act has spelt out offenses that could emanate from trade practices against consumers especially failure to provide needed information. Neuromarketing studies could face various claims as the method override consumer rational decision – making process in favour of impulse propensity in brain activity in purchase choice. In Africa, the African Yamoussoukro decision on consumer protection, Article 9.6, ensures that consumer rights are protected. But by and large some African countries do not have any consumer protection laws or regulations (Consumers International, 2015). The fundamental rights of consumers are entrenched in other country-specific equality and unfair practices laws and acts. And investigating trade practices particularly marketing and advertising issues may not be given much attention or dealt with sceptically. In 2008, South Africa passed its consumer protection law, Act 68, and the number one declaration spells out equal rights to the consumer and protects against *discriminatory marketing practices* and privacy. Under this act, managing neuromarketing ethics issues adequately would not be a challenge.

Another ethical controversy is the assumption that firms could identify and spark brain mechanisms that would influence choices and purchase behaviour. This clearly would expose consumer thoughts and could be a dangerous direction ethically. Companies would be able to take advantage of the process and knowledge for commercial purposes (Javor et al., 2013; Morin, 2011; Senior & Lee, 2008; Fugate, 2007; Lee et al., 2007). There is the need for principles guiding the use of neuromarketing research. The global acceptance of a regulatory body like the Neuromarketing Science and Business Association (NMSBA), as the overall professional and trade group for protecting the interest of the field and consumers would be the way forward.

The third ethical challenge has to do with reliability of neuromarketing research results. This aspect of ethics questions the degree of predictability of the results of neuroimaging scans and reports, and the potential to misuse the information derived. Critics foresee a future of marketing research manipulating and coercing consumers to take decisions and make choices that promote particular products or services and thus are skeptical about the new field. There is the need to impose standards, boundaries and code of ethics for which researchers would have to abide by at all times.

The use of high and inflexible technological machines like the MRI scanners (as in Figure 2) and imaging instruments (Figure 3 and 4a - c) may cause serious setback. These imposing instruments could affect research respondent in different ways. The nature and size of the medical instruments used in research may deter respondents or intimidate and influence them. The machines and gadgets may scare research respondents. The laboratory setting also may pose serious problems that may influence responses. Is it ethically right to subject consumers to stressful procedures in the name of research and solving problems? Research respondents should be briefed on the instruments and techniques of data collection to overcome this challenge.

Are ethical concerns looking for regulations to guide and monitor neuromarketing application or restrictions that would prevent negative illegitimate practices? Ethics in neuromarketing should address the procedural and interpretational implications and the social ramification that could emerge from negative practices. Consumers must be informed about the purpose of the data collection, the process and potential end use of the information. This has been articulated by several researchers (Wilson, Gaines and Hill, 2008; Lee et al., 2010). Neuromarketing studies as it stands must focus on the advantages of using neuroscience research methods while solving business problems. As a discipline and research choice, neuromarketing studies have been criticised but in all, its reliability, benefits and measures make it more popular. In view of the fact that its indirect measurements addresses marketing and brand choice questions better than ordinary consumer surveys it must be encouraged under strict ethical considerations.

Developing Country Studies on Neuromarketing

In the last decade, information technology and medicine have made strides in imaging techniques that have moved diverse fields like psychology, marketing and engineering into using this scientific method in resolving and delving into the consumer mind to find reasons for decisions and choices made. Currently, the human brain activities' reaction to external stimuli can be captured to know what drives consumer behaviour. Championing research and the field of neuromarketing are studies from more advanced countries like the United States, Europe and Asia. These countries have more advanced markets and products. Industry and company promotions are geared toward consumers who have a myriad of product and service to choose from to satisfy their needs. Products and service communications are moving toward investigating consumer decision process and what drives certain brand choices. And neuroscience methods aptly provide the process for digging deeper to understand consumer thought process and decision making. Indeed researchers and communication agencies are seeking information to comprehend and find ways to coerce and encourage buying. Studies emerging from scholars from developing countries are limited to particular countries that are more technology advanced. Neuromarketing research from developing countries search yielded 523, 000 results (Google search as at 4/6/2015) but further probing revealed that most of these studies are not from African countries. Developing countries are making their mark in the field in terms of interest and academic research but empirical studies are lacking as per review for this chapter. Industry specific consumer advertising and brand research by international marketing firms in some developing countries are introducing this innovative research procedure gradually into their studies. Associations and groups are sensitising and providing knowledge on theory and practice. The Neuromarketing Science and Business Association (NMSBA) for example, is a professional trade group for disseminating information about the field. The association has established code of ethics and definition of neuromarketing terms for use by its members. Currently, there are 40 chairs globally (Appendix) with South Africa (SA) being the only African country. South Africa (SA) is now taking bold steps in neuromarketing research with the *beer and mind control* and other *consumer responses to advertising* studies (Moerdyk, 2013; Neurogadget, 2012). Two notable neuromarketing firms in South Africa are HeadSpace SA and Neural Sense, SA. The South Africa Marketing Research Association (SAMRA) in 2013 formalised neuromarketing research through its maiden conference that year to provide more insight into burgeoning field. du Plessis of Millward-Brown, SA spearheaded the introduction and sensitisation key note discussions at the conference. Other developing countries have their fair share of theoretical reviews. The challenge will be the cost of conducting neuromarketing surveys and availability of tools. Research companies so far are going global and taking on jobs that require the use of scientific tools and the advantages of this new field will endear itself to every country.

The question that must be addressed is that - Is Africa ready for the challenging? Future research on neuromarketing in Africa particularly will provide the answer to address this question.

FUTURE RESEARCH DIRECTIONS AND CONCLUSION

The knowledge of the world around us depends on the human thinking and perception. Neuroscience is breaking grounds in connecting research and practice with other important fields that study the human being. Consumer behaviour studies are moving toward the mapping of the thought process. And further studies could explore what consumers are thinking and what drives that process. The potential of neuromarketing lies in its ability to go beyond determining the effect of marketing campaigns and brands on consumers and the decisions they make thereafter to predicting future choices.

Neuromarketing is concern with generating knowledge which is verifiable by observation (direct or indirectly), or experiences. The tools and techniques are rooted in neuroscience and measurements derived indirectly provide responses that marketing researchers can use to gain a better understanding about consumer attitudes, behaviour, decision making and choice. These tools measure portions of the brain and its response to stimuli. Theories and methodologies of cognitive neuroscience is utilised by researchers to measure marketing activities. The meaningful interpretation of these test are the dilemma that marketers will need to sharpen to accurately present the factual information. This calls for special training to equip and prepare for in-depth interpretation of scientific imagining and other methods of neuromarketing. From all indications, neuromarketing advocates are increasing in number but the emphasis on the benefit of neuroscience is promoting the field.

The major controversy with the field has to do with intrusion on consumers and the ability to influence consumer decision making and choice. Also the invasive tools and techniques can be misused by marketing researchers to their advantage. Another challenge with neuromarketing research is adequately interpreting information from neuroimaging (fMRI, EEG and others). Ethical questions may be the biggest challenge that neuromarketing would have to address for the field to flourish. But all in all, consumer laws, advocates and associations would be able to protect consumer interest and guide best practices.

From the review, it can be posited that theoretical studies in neuromarketing is trending. The effort at developing models that can be used in this new territory is grow rapidly. In all, 120 articles were selected for review based on neuromarketing specifically. The convenient sampling method was applied in this case. The articles selected had 95% (114) being theoretical based review and 5% (6) were empirical. Empirical documentation studies abound but it is more on industry specific studies for organisations done by marketing research firms. This makes Garcia and Saad (2008) assertion that all empirical surveys are 'one-shot empirical studies' focusing on a single research worth reviewing. The NMSBA is gradually building a base that would monitor neuromarketing practice and activities.

Consumer behaviour and consumers are the focus of neuromarketing practice and research but this chapter delimits itself to the thematic issues relating to the topics discussed. In regard to the articles reviewed, it is important to note that it is not a representative of articles and empirical research of the whole arena of studies on neuromarketing but rather an optimization that encouraged an extensive analysis of the articles chosen.

In conclusion, it can be said that despite concerns raised by sceptics about the intrusiveness of the methods of enquiry on consumers, neuromarketing techniques will go a long way in providing more accurate information. Neuroimagining as a survey instrument cannot easily be over turned, but the question of acquiring some of the equipment may be a disadvantage to its usage. It is an important area of study and its application and tools for monitoring, measuring and possibly manipulating and knowing what is on the minds of consumers is the next colossus in marketing. Neuromarketiing research is here and the opportunity to determine consumer behaviour and measure actual responses is the way forward.

REFERENCES

Bear, M. F., Connors, B. W., & Paradisco, M. A. (1996). *Neuroscience: Exploring the Brain*. London: Williams & Wilkins.

Booth, D. A., & Freeman, R. P. J. (2014). Mind-reading versus Neuromarketing: How Does a Product make an impact on the Consumer? *Journal of Consumer Marketing*, *31*(3), 177–189. doi:10.1108/JCM-08-2013-0674

Butler, M. J. (2008). Neuromarketing and the Perception of Knowledge. *Journal of Consumer Behaviour*, *7*(4-5), 415 – 419.

Butler, M. J. (2013). Operationalising Interdisciplinary research – a model of Co-Production in Organisational Cognitive Neuroscience. *Frontiers of Human Neuroscience, 7*, 720. Retrieved on 4/06/2015 from: www.ncbbi.nlm.nih.gov/pmcparticles/PMC3893564/

Consumer Protection Act. (2009). *I know my rights. Do you know Yours*. The Department of Trade and Industry, Republic of South Africa. Retrieved on 20/06/2015 from http://www.consumer_protection_pocketguide.pdf

Consumers International. (2015). *Consumer Rights*. Retrieved on 13/06/2015 from http://www.consumersinternational.org/who-we-are/consumer-rights

Dowouna-Hammond, M. C., Atuguba, D. R., & Nabila, D. L. (2006). Legal and Policy Framework for Consumer Protection in Ghana. Accra: Friedrich Ebert Foundation, Ghana.

Du Plessis, E. (2011). *The Branded Mind* (1st ed.). London, UK: Kogan Page Limited.

Edwards, P. (2013). *Neuroscience and Reductionism: Some Realist Reflections*. Birmingham Business School. Discussion Paper Series, 2013 – 2008.

Eze, A. G. (2012). Consumer Rights as Constitutional Rights: A comparative analysis of some selected jurisdictions. *Nnamdi Azikiwe University Journal of International Law and Jurisprudnce*, 184 – 196.

Fisher, C. E., Chin, L., & Klitzman, R. (2010). Defining Neuromarketing: Practices and Professional Challenges. *Harvard Review of Psychiatry, 18*(4), 230 – 237. 10.3109/10673229.2010.496623

Glimcher, P. (2003). *Decisions, Uncertainty and the Brain: The Science of Neuroeconomics*. London: MIT Press.

Hubert, M. (2010). Does Neuroeconomics give new impetus to Economic and Consumer research? *Journal of Economic Psychology, 31*(5), 812 – 817. 10.1016/j.joep.2010.03.009

Hubert, M. & Kenning, P. (2008). A current Overview of Consumer Neuroscience. *Journal of Consumer Behaviour, 7*(4-5), 272 – 292.

Javor, A., Koller, M., Lee, N., Chamberlain, L. & Ransmayr, G. (2013). Neuromarketing and Consumer Neuroscience: Contributions to neurology. *BMC Neurology, 13*(1), 1-12. 10.1186/1471-2377-13-13

Kahneman, D., & Snell, J. (1992). Predicting a Changing Taste: Do people know what they will like? *Journal of Behavioral Decision Making, 5*(3), 187–200. doi:10.1002/bdm.3960050304

Karmarkar, U. R. (2011). Note on Neuromarketing. Harvard Business School.

Koch, C. (2004). *Quest for Consciousness: A Neurobiological Approach*. Englewood, CO: Roberts & Company Publishers.

Lee, N., Broderick, A. J., & Chamberlain, L. (2007). What is “Neuromarketing”? A discussion and agenda for future research. *International Journal of Psychophysiology, 63*(2), 199 – 204. 10.1016/j.ijpsycho.2006.03.007

Lee, N., Butler, M., & Senior, C. (2010). The Brain in Business: Neuromarketing and Organisational Cognitive neuroscience. *Journal of Marketing, 49*, 129–131. doi:10.1007/s12642-010-0033-8

Lewis, D. & Briader, D. (2005). Market Researchers make increasing use of Brain Imaging. *Advances in Clinical Neuroscience and Rehabilitation*, *5*(3), 35.

Lindebaum, D. & Zundel, M. (2013). Not quite a revoluntion: Scrutinizing Organisational Cognitive Neuroscience in Leadership Studies. *Human Relation, 66*, 857 – 877. 10.1177/0018726713482151

Martinez, P. (2012). *The Consumer Mind: Brand Perception and the Implication for Marketers* (1st ed.). London, UK: Kogan Page Limited.

McClure, S. M., Li, J., Tomlin, D., Cypert, K. S., Montague, L. M., & Montague, P. R. (2004). Neural correlates of Behavioral preferences for culturally familiar drinks. *Neuron*, *44*(3), 379–387. doi:10.1016/j.neuron.2004.09.019 PMID:15473974

Moerdyk, C. (2013). *Synopsis of Neuromarketing. South Africa Marketing Association* (SAMRA). Retrieved on 20/06/2015 from: www.erikdup.com

Morin, C. (2011). Neuromarketing: The New Science of Consumer Behaviour. *Society*, *48*(2), 131 – 135. 10.1007/s12115-010-9408-1

Neurogadget. (2012). *HeadSpace Agency brings neuromarketing Studies to South Africa*. Retrieved on 20/06/2015 from: http://www.neurogadget.com/2012/08/28/headspace-agency-brings-neuromarketing-studiesto-SA/4603

Payne, J. W., Bettman, J. R., & Johnson, E. J. (1993). *The Adaptive Decision Maker*. Cambridge, UK: Cambridge University Press. doi:10.1017/CBO9781139173933

Plassman, H., O'Doherty, J., Shiv, D., & Rangel, A. (2008). Marketing actions can modulate neural representations of experienced pleasantness. *Proceedings of the National Academy of Sciences of the United States of America*, *105*(3), 1050–1054. doi:10.1073/pnas.0706929105 PMID:18195362

Rangel, A., Camerer, C., & Montague, P. R. (2008). A Framework for studying the Neurobiology of Value-based Decision making. *Nature Reviews. Neuroscience*, *9*(7), 545–556. doi:10.1038/nrn2357 PMID:18545266

Rawlins, N. (2005). The Nervous System. In M. Hewstone, F. D. Fincham, & J. Forster (Eds.), *Psychology* (pp. 42–72). Blackwell Publishing.

Schaefer, M., Berens, H., Heinze, H. J., & Rotte, M. (2006). Neural correlates of culturally familiar brands of car manufacturers. *NeuroImage*, *31*(2), 861–865. doi:10.1016/j.neuroimage.2005.12.047 PMID:16487728

Senior, C., & Lee, L. (2008). A Manifesto for Neuromarketing Science. *Journal of Consumer Behaviour*, *7*(4-5), 263–271. doi:10.1002/cb.250

Waldman, D. A. (2013). Interdisciplinary Research is the key. *Frontiers in Human Neuroscience*, *7*, 5. doi:10.3389/fnhum.2013.00562 PMID:24062664

Wilson, M., Gaines, J., & Hill, R. (2008). Neuromarketing and Consumer Freewill. *The Journal of Consumer Affairs*, *42*(3), 389–410. doi:10.1111/j.1745-6606.2008.00114.x

Wirtz, D., Kruger, J., Scollon, C. N., & Diener, E. (2003). What to do on spring break. *Psychological Science, 14*(5), 520–524. doi:10.1111/1467-9280.03455 PMID:12930487

Zaltman, G. (1997). Rethinking Marketing Research: Putting people back in. *Journal of Marketing Research, 34*(4), 424 – 427. 10.2307/3151962

Zaltman, G. (2003). How Consumers think: Essential Insights into the Mind of the Markets. Boston: Harvard Business School Press.

KEY TERMS AND DEFINITIONS

Brain: A soft organ made up of blood vessels, interconnected nerves (neurons) and fluid - filled ventricles.

Galvanic Skin Response (GSR): Also known as skin conductance. It is a method applied to measure the skin responses to stimuli.

Glial: Cells that support neurons in their tasks and contribute to the development of the nervous system and its activities.

Receptors: Special spots or areas in the nervous system (like the brain) where neuron transmitters have their greatest effects.

This research was previously published in the Handbook of Research on Consumerism and Buying Behavior in Developing Nations edited by Ayantunji Gbadamosi, pages 263-282, copyright year 2016 by Business Science Reference (an imprint of IGI Global).

APPENDIX

Figure 2. The Tamale Hospital in Ghana's MRI Scanner
Courtesy www.ghanahealthnet.com

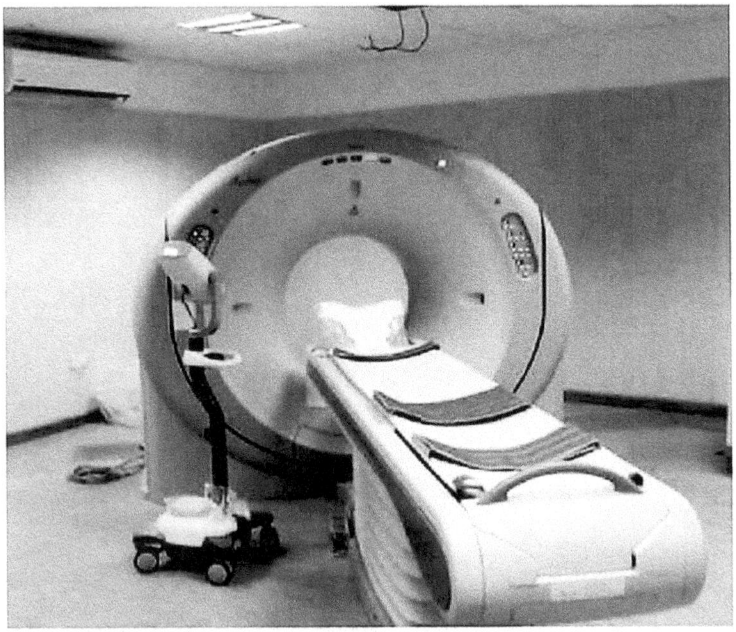

Figure 3. The Tamale Hospital in Ghana's computed tomography
Courtesy www.ghanahealth.com

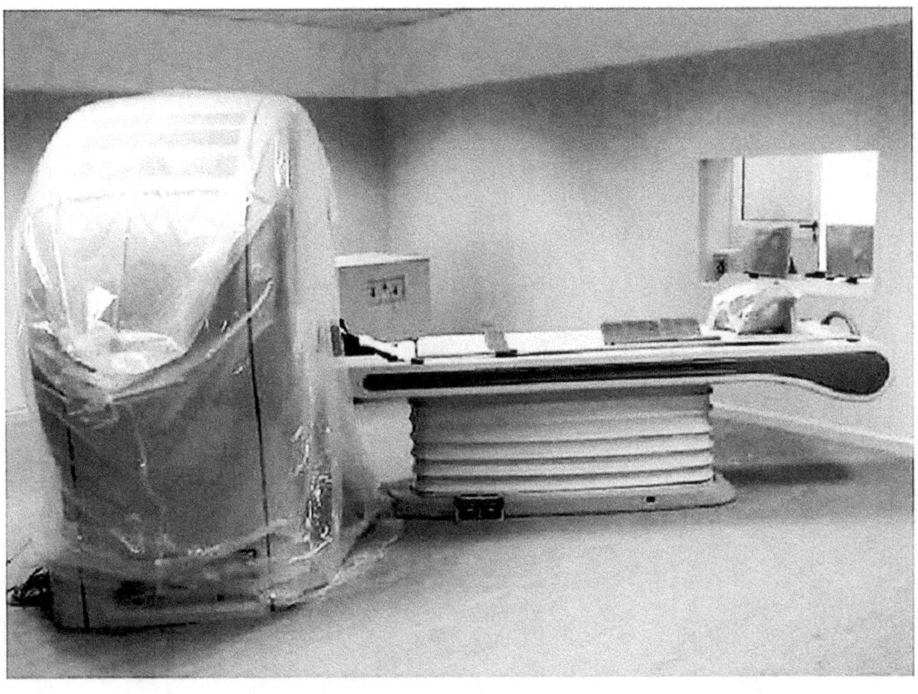

Figure 4. TNS Emor, Neuromarketing presentation: a) eye-tracking device, b) EEG, c) GSR
Courtesy of Kaidi Reedi

(a) (b) (c)

Chapter 16
Assessing Consumer Reactions with Neuroscientific Measurements

Christopher Rumpf
German Sport University Cologne, Germany

Christoph Breuer
German Sport University Cologne, Germany

ABSTRACT

Positive consumer reactions to corporate marketing activities are regarded a key driver of business success. Since consumer reactions occur to a large extent on non-conscious levels, traditional market research approaches provide limited insights into the consumer's perceptions and intentions. This chapter demonstrates how neuroscientific measurements can contribute to a deeper understanding about critical processes in the consumer's "black box". Two generic approaches will be outlined: Whereas brain imaging techniques create pictures reflecting brain activity in response to marketing stimuli, psychophysiological methods assess body signals as correlates of neural activity. The chapter provides a general understanding about the meaningful application of neuroscientific measurements in consumer research and presents a critical reflection on the opportunities and challenges of different neuroscientific measurements.

INTRODUCTION

Consumer research currently relies on a limited set of measurement techniques. To date, quantitative surveys and qualitative interviews are still the most widely used approaches to assess consumer reactions to any kind of marketing activities. However, it is quite obvious that people can only answer questions regarding their consciously reflected attitudes and intensions. Given that consumer reactions – to a large extent – occur on a non-conscious level, traditional questioning techniques might only capture the "tip of the iceberg". Moreover, survey data are vulnerable to serious biases (Gregg & Klymowsky, 2013) – such

DOI: 10.4018/978-1-5225-5478-3.ch016

as the social desirability bias (i.e. respondents may dissimulate their true attitude), self-enhancement bias (i.e. respondents wish to flatter themselves), and self-ignorance bias (i.e. respondents unwittingly express attitudes they do not have).

Against this backdrop, there is an urgent need to apply more advanced research methods which adapt to the complex nature of consumer behavior. In recent years, there was a rapid development of a new integrative discipline which has been labeled neuro-economics, neuro-marketing, or consumer neuroscience. The overarching aim of this emerging discipline is to transfer the insights and methodology from neurology to the study of consumer behavior in order to gain a deeper understanding about brand perceptions and consumer intentions. More generally speaking, neuroscientific measurements in consumer research are used to identify brain areas and cortical processes which are engaged in consumer reactions and economic decision making.

Current studies in the field of consumer neuroscience are concerned with a variety of marketing topics, such as shopping behavior (e.g. Knutson, Rick, Wimmer, Prelec, & Loewenstein, 2007), brand and product perception (e.g. McClure, Li, Tomlin, Cypert, Montague, & Montague, 2004; Deppe, Schwindt, Kugel, Plassmann, & Kenning, 2005) and advertising effects (e.g. Klucharev & Fernandez, 2005). One of the key finding from studies in the field of consumer neuroscience relates to the neuronal correlates of economic decision-making. For instance, Deppe et al. (2005) show how emotions influence consumer choices in a shopping situation. In a similar vein, McClure et al. (2004) demonstrate the effect of brand perception on brain activity and brand preferences. Further, the investigations of Plassman, Kenning, and Ahlert (2007) reveal that consumer loyalty is associated with activity changes in the neural reward system.

There is a set of instruments taken from neuroscience and medical research which has the potential to gauge data automatically, that is, without any verbal response from the participant (Green & Holbert, 2012). Such neuroscientific measurements can be classified in two categories: Brain imaging and psychophysiological techniques. Brain imaging techniques are used to depict brain activities in terms of spatial and temporal dimensions. In this regard, metabolic approaches measure substances which relate to neural activities (e.g. hemoglobin), whereas psychophysiological approaches directly track the outcome of neural activities (e.g. electric potential). In contrast to brain imaging techniques, psychophysiological methods capture body signals in the autonomic nervous system. Skin conductance, facial expressions, heart rates and eye movements serve as important indicators of neural consumer reactions (Kenning, 2014).

In the chapter's main body, several methods adapted from neuroscience, medical science and applied psychology will be critically discussed. After explaining the underlying concepts of each measurement approach the added value for consumer research will be outlined.

BRAIN IMAGING TECHNIQUES

Brain imaging techniques are used in consumer research to study psychological concepts like attention, affect, memory and desirability (Venkatraman et al., 2015). For example, based on a brain imaging study it was found that certain brain areas associated with pleasure, self-identification and rewards, strongly respond to well-known brands, whereas other parts in the brain associated with displeasure were evoked by unfamiliar brands (Hubert, & Kenning, 2008).

Different brain imaging techniques can be used in consumer research. In this chapter, the focus is on a direct and an indirect approach to visually depict brain activity. As an indirect approach, Functional

Magnetic Resonance Imaging (fMRI) measures metabolic processes that indirectly reflect neural activity, whereas Electroencephalography (EEG) – a direct approach – captures electrical potentials along the scalp as a direct result of neural activity.

fMRI: A WINDOW ON BRAIN ACTIVITY

The fMRI technique assesses changes in blood flow associated with neural activity in specific areas of the brain. More precisely, fMRI examines the changes in brain responses through recording BOLD (B - Blood, O – Oxygenation, L – Level, D – Dependent) activity response. To understand which areas of the brain are activated in particular consumer situations, the participant typically performs a cognitive task. The task could include a choice decision, for example, choosing between two different rewards such as $20 now or $35 in 10 days. While the participant mentally performs the task, an fMRI scanner determines which brain area requires more oxygenated blood, and thus, is more involved in the decision process (Yoon, Gonzalez, & Bettman, 2009).

In an empirical study on the role of product packaging in consumer decision making, four male and seven female participants took part in an fMRI experiment. To control for individual differences a homogenous sample was recruited aged 18 to 26 years. The participants were exposed to 30 different FMCG packages. Each product package was pre-tested whether it was perceived as highly attractive, less attractive or perceived as neutral. After the fMRI scanning, the participants were requested to fill out a questionnaire containing questions about the attitude toward the three main functions of packaging, that is, protection, information, and attention attraction at the point of sale. The study results show that attractive and unattractive package designs cause different cortical activity changes. Activity changes were detected in the occipital lope which is associated with visual processing and attention. Further, cortical areas involved with reward processing were increasingly activated when exposed to attractive packages. Even though the study results have an explorative character, they help to understand why attractive product packages gain more attention in a shopping situation, and thus, are more likely to be picked (Stoll, Baecke & Kenning, 2008).

Another fMRI study was conducted to investigate whether semantic judgements about products are processed similarly to semantic judgements about persons. The study engaged twenty-five healthy and right handed participants for the fMRI scanning. The participants were between 18 and 23 years old. During five runs of the fMRI session, the participants were exposed to trait adjectives which were to be evaluated whether they fit certain target brands. In each run, the participants were exposed to five target brands which were presented next to one trait adjective. The study results suggest that the judgements about products and persons are processed in different ways. The evaluation of neural correlates via fMRI session of product versus human descriptor judgments showed greater activation in the medial prefrontal cortex regions for person judgements. In response to product stimuli, activation was greater in the left inferior prefrontal cortex, an area of the brain recognized to be involved in object processing. These results are putting under question the general assumption in marketing science that brands can possess human-like traits (Yoon, Gutchess, Feinberg, & Polk, 2006).

Deppe et al. (2005) investigated how consumer decisions are influenced by implicit, that is automatic, memory processes. Twenty-two healthy participants were engaged in the study. The participants were instructed to imagine that they were purchasing coffee (female participants) or beer (male participants) in a shop. While their brain was scanned by fMRI, the participants had to choose between two different consumer brands. Visual pairs of brands were shown during this task. The participants were instructed to spontaneously choose between the brands (within 3 seconds) by mentally speaking the selected brand name. Changes of brain activity comparing decisions in the presence or absence of a specific target brand were detected by fMRI. If the target brand was favored by the participant (i.e. first choice brand), increased activity was found in brain areas associated with processing emotions and self-reflection during decision making. Other brain areas which are involved in reasoning and working memory were less activated. The authors conclude that the first-choice brand automatically evokes a positive state that becomes a "beacon of incentive" to select this brand over alternative options (Deppe et al, 2005).

Compared with alternative techniques in consumer neuroscience, fMRI creates very precise images of neural activity in the brain. As it is sensitive to the BOLD signals, fMRI is suitable to measure activity both in the cortical surface and deep brain structures. Thus, it enables researchers to draw well-founded conclusion on the engagement of certain brain areas in consumer reactions. Based on its high spatial resolution, fMRI is suited to study a variety of consumer reactions such as affect, memory and desirability. Since the processes underlying consumer affect are difficult to assess with traditional measurements, the benefits of fMRI are very obvious in this regard. Particularly, neural activity in the amygdala – which plays a key role in emotional processing – has been a focus in consumer neuroscience. Furthermore, fMRI research on ad encoding shows that the activation of the hippocampus – which is engaged in storing and connecting memories – is stronger for brands that were remembered by the consumer (Venkatraman et al., 2015). Moreover, it is know from fMRI research that activation in the ventral striatum – which is a part of the reward system – serves as a valid indicator of subsequent brand choice (Berns and Moore, 2012)

By contrast, fMRI studies have been criticized due to some technical and economic issues. Since the BOLD response to neural activity (i.e. flow of oxygen-rich blood into the activated brain area) occurs with a delay of approximately 2-5 seconds, the temporal resolution of fMRI needs to be taken into consideration when planning the study design. Therefore, fMRI studies are typically limited to simple designs with only a few stimuli presented in a given time frame. As a consequence thereof, the validity of fMRI data might be lower than desired and several studies require replication in the future (Plassman et al., 2007).

Another criticism relates to the costs of fMRI scanning compared with traditional methods. As a rule of thumb, fMRI scans cost US$500 to US$1'000 per lab hour. Given these high costs, fMRI studies are typically based on small sample sizes (n<20), and thus, lack statistical power and raise concerns about false positive results (Kenning, 2014).

ELECTRICAL RESPONSES FROM THE BRAIN

Electroencephalography (EEG) records electrical activity along the scalp produced by the firing of neurons within the brain. Thus, EEG is capable to detect cortical responses within experimental settings. Since neurons communicate with each other by passing electrical signals along the connecting fibers, low-voltage potentials can be picked up by highly sensitive electrodes placed on the scalp (Luck, 2014).

Typically, EEG systems consist of an electrode cap, an amplifier and a recording device. The electrodes embedded in an electrode cap need to touch the surface of the head to pick up electrical signals. The number of electrodes ranges between higher resolution systems (128 electrodes) to lower resolution systems (4 electrodes). Each electrode measures electrical potential in a different area of the brain so that activity can be distinguished between brain cortexes. The electrical signals are processed to the amplifier which converts the signals into digital codes to be visualized and stored as brainwaves by the recording device.

A recent EEG study in the field of consumer research was conducted by Vecchiato et al. (2014). They explored the cognitive and emotional changes during the exposure of TV advertisements. The experiment engaged 28 healthy participants (12 females and 16 males) who were given the task to follow a twenty minute documentary in which two commercial breaks were inserted by the experimenter. The commercials were shown after five and fifteen minutes. Each interruption was formed by six advertisement spots with a duration of thirty seconds. The advertisement spots were unknown to the participants prior to the study and represented six product categories: Perfume, general consumption, banking, sport, telephone, and clothing. With regard to perfumes, advertisements for the luxury brands "Cartier" and "Prada" were presented. Compared to male participants, females were found to be significantly more aroused during certain scenes of the "Cartier"-spot, but less for the "Prada" spot. In particular, a dance scene in the "Prada" commercial aroused higher emotion with men while the dance scene in the "Cartier" spot created higher emotional response among women. The study results show how EEG results can be used to create more customized spot elements with regard to a specific target group (Vecchiato et al., 2014).

In general terms, findings from EEG studies provide marketers with the opportunity to better understand hidden processes in the consumer's mind. Most frequently, EEG technology has been employed to study the implicit (i.e. non-conscious) consumer reactions to marketing communication such as advertising spots. However, the interpretation of EEG recordings can be problematic or even misleading. This is because EEG is a mixed up conglomeration of dozens of different neural sources of activity, making it difficult to isolate individual neuro-cognitive processes. Embedded within the EEG are the neural responses associated with specific sensory, cognitive, and motor events which are difficult to extract from the overall EEG. The study of event-related potentials (ERP) can overcome this issue. ERPs are defined as changes in EEG recordings due to a randomly presented event, that is, a sound such as a brand jingles or images such as band logos (Green & Holbert, 2012).

In ERP study designs, a trigger signal is send to the recording device to be stored along the EEG data, whenever a brand stimulus is presented to the participant. Since the stimulus-induced brain signal is too small within the much larger EEG, the stimulus is frequently presented to the participant (15-25 times). The EEG waveforms are then averaged across many trials. This procedure leads to an averaged ERP waveform for the stimulus at each electrode whereas the brain activities that were unrelated to the stimulus will cancel out each other (Luck, 2014).

ERP data are visualized as a series of positive and negative peaks that vary in polarity, amplitude, and duration as the waveform unfolds overtime. The observed ERP waveform can be defined as a representation of the changes in scalp recorded voltage over time that reflect the sensory, cognitive, and affective processes elicited by a stimulus (Hajcak, Weinberg, MacNamara, & Foti, 2012).

One of the major strength of EEG is the high temporal resolution (i.e. the degree to which a single object can be separated in time from another). In terms of spatial resolution (i.e. the degree to which a single object can be separated in space from another), however, EEG is limited to detect brain activity in superficial cortex layers which makes it less sensitive to neural response in deeper brain areas such as the hippocampus and the amygdala (Xue, Chen, Lu, & Dong, 2010).

PSYCHOPHYSIOLOGICAL TECHNIQUES

Psychophysiological techniques have aroused increasing interest by marketing researchers and practitioners. In the context of consumer research, psychophysiological methods are mostly used to examine the emotional reaction of consumers to marketing stimuli. Such methods can be employed to test, for instance, the effectiveness of an advertisement spot in terms of valence and arousal (Kivikangas et al., 2011).

SKIN CONDUCTANCE AS A MEASURE OF AROUSAL

Electrodermal Activity (EDA), also known as the measurement of skin conductance, assesses the change in electrical properties of the skin in response to the emission of sweat (Lajante, Droulers, Dondaine, & Amarantini, 2012). Sweat glands are involved in emotion-evoked sweating and cover the whole body. The highest density of sweat glands can be found on the palms and on the soles of the feet. Since increasing sweat secretion indicates a higher degree of activation in the autonomic nervous system, EDA can be used in consumer behavior research as a measure of arousal (Poels & Dewitte, 2006). For this purpose, electrodes measuring conductance are placed on the participant's palms.

In a recent literature review, Lajante et al. (2012) report that EDA has been applied in more than 20 studies in the field of consumer research, most frequently used to analyze the effectiveness of marketing communication activities. In one of the first EDA studies in the context of advertising, Belch, Holgerson, Belch, and Koppman (1982) investigated the consumer response to sexual content in ads. Participants were exposed to three different sets of advertisements including partially nude, fully nude, and fully dressed up models. The study results reveal that the use of suggestiveness in advertisements elicits strong physiological and cognitive reactions, with different effects for male and female consumers. Whereas females reacted much stronger towards suggestive ads than men, both opposite sex and same sex nude advertisements elicited strong physiological reactions (Belch et al., 1982).

In an exploratory study, Lajante et al. (2012) tested thirty participants (22 females and 8 males) on their degree of emotional activation in response to seven distinct commercials. Each commercial was presented in a randomized order for 30 seconds each. Data analysis reveals significant differences in skin conductance, that is, emotional activation between the different commercials. The authors conclude that EDA means a valid approach to evaluate emotional reactions triggered by marketing stimuli. However, Lajante et al. (2012) also report complications in the analysis if the frequency of responses is high.

As it indicates the emotional reaction of consumers triggered by marketing stimuli, EDA bears the potential to inform researchers and marketing decision makers regarding emotionally appealing content. Moreover, rational marketing decisions on design elements such as color, size and message can be based on EDA data. However, it needs to be considered that electrical properties of the skin are not capable of reflecting brain activity. Furthermore, EDA results are not specific with regard to the valence of emotions (i.e. positive vs. negative). Finally, Poels and Dewitte (2006) mention that the measurement and interpretation of skin conductance data requires a high level of analytical expertise. Otherwise, there is a high risk of producing invalid data and misleading implications.

STUDYING FACIAL EXPRESSIONS

Another psychophysiological measurement technique captures facial expression to analyze the behavioral response of consumers to marketing stimuli. Facial expressions have been frequently used to assess the state of basic emotions such as joy, disgust and anger (Poels & Dewitte, 2006). Based on the Facial Action Coding System (FACS), psychologists have studied facial expressions indicating the emotional stimulus response.

Debaix (1995) conducted one of the largest FACS studies with 228 participants to compare verbal emotional responses and coding of facial expressions to a set of commercials. Thirteen commercials were divided into three sets out of which one was the known advertisement for a familiar brand, six were unknown advertisement for a familiar brand, and six were unknown advertisement for unfamiliar brands. The results showed that emotional responses stemming from verbal measures correlate with ad liking and brand attitude whereas facial expressions measured by FACS did not. As a conclusion, the authors argued that the changes in muscular activity evoked by marketing stimuli are often too delicate to be measured by FACS.

Since more sensitive measures are needed for detecting facial reactions to marketing stimuli, facial Electromyography (EMG) can be seen as a beneficial advancement of FACS and more accurate measure of facial expressions. It has proven to register facial muscle activity when no changes in facial expressions were found through FACS. In facial EMG, two muscles in the human face are of importance: the corrugator and zygomatic muscle. The corrugator muscle is located above the eyebrow and contractions in this muscle are involved in frowning. The zygomatic muscle is located around the cheeks and controls smiling. Electrodes that register muscle contractions are placed on these two muscles (Bolls, Lang, & Potter, 2001).

In an EMG study 64 students were recruited by Lang et al. (1993). The participants were either assigned to a positive or negative picture group. When exposed to positive pictures the zygomatic muscles was increasingly activated whereas the corrugator muscle showed higher activation when participants saw negative pictures. The authors concluded that facial EMG can be applied in consumer research as a valid measure of valance (Lang et al., 1993).

To date, a number of follow-up studies employed facial EMG to measure the emotional consumer reactions through facial expressions. For example, Bolls, Lang, and Potter (2001) found that the zygomatic muscles activity was larger in response to radio commercials with positive emotional tone and the corrugator muscle activity in commercials with a negative emotional tone.

Despite its given strength, EMG measurements are faced with the problem of ecological validity, since facial EMG is usually measured in lab settings and the placement of electrodes on the face can bias the participant's behavior. The lab environment can make the participants more aware of her or his facial expression, and thus, lead to inaccurate results.

SIGNALS FROM THE CARDIOVASCULAR SYSTEM

Another approach to understand emotional consumer reactions is based on heart rates (HR). The speed of HR is able to indicate various phenomena such as attention, arousal, or valence. In physiology research, heart rate is typically operationalized as the number of milliseconds since the previous heart rate. Poels and Dewitte (2006) suggested that the variations in the speed of HR occur due to either arousal or attention when a consumer is exposed to marketing stimuli such as commercials.

Since the interpretation of HR data in a consumer research context is based on phasic (i.e. short term) and tonic (i.e. long term) changes, it is important to identify which changes refer to attention and which changes refer to arousal. Lang (1990) identified phasic changes in HR relating to attention whereas tonic changes indicate a shift in arousal.

Furthermore, HR can also provide insights into the valence of emotional responses. Valence is defined as a negative or positive psychological value associated with a person, event, object, or situation. In relation to neutral stimuli, positive and negative stimuli evoke a phasic decrease in HR. While positive stimuli cause an increase in tonic HR, it decreases in response to negative stimuli. These general insights from psychophysiology can be used in consumer behavior to better understand the emotions of people evoked by marketing stimuli (Poels & Dewitte, 2006).

HR is not necessarily measured at the breast, but also at more convenient places like the wrist to minimize potential interference effects. All in all, the measurement of HR means one of the most simple and cost-efficient ways to assess psychophysiological reaction evoked by marketing activities or brand stimuli.

Despite its simplicity and its valid indication of attention, arousal and valence, the measurement of HR comes with some problems and limitations. For example, a consumer who is exposed to a highly exciting sport broadcast would devote a significant share of attention to sports action, and thus, produce a phasic declaration in HR. At the same time, the excitement of the sport broadcast can evoke a high arousal level leading to an increase in tonic HR. Given that the sport broadcast creates negative emotions, the HR can decrease indicating negative valence. Due to the ambiguity of HR data leading authors suggest to use HR measurement in combination with other neuroscientific techniques such as EEG or EDA (Poels & Dewitte, 2006).

EYE-TRACKING: LOOKING THROUGH CONSUMERS' EYES

One of the most popular psychophysiological techniques in consumer research is called eye-tracking. The fundamental principle of eye-tracking is to measure the movement of the consumer's eyes by recording the position of the pupils with high frequency as horizontal and vertical coordinate values. Since the focus of the gaze can change not only due to the movement of the eyeballs, but also through head movements, modern systems measure both the pupil position and the position of a reflection point on the cornea of the eye (Duchowski, 2007).

On the basis of the spatial relation between the position of the pupils and the position of the corneal reflection on the eye, eye-tracking systems can calculate the direction of the gaze independently of head movement. For this purpose, infrared light is beamed from a fixed lighting source onto the test person's face, while a special infrared camera records the reflected light. On the recorded video image, eye-tracking systems can identify the participant's pupils as dark circles, since the beamed light is absorbed by the pupils. As opposed to the pupils, the cornea reflects the infrared light very strongly, which is why the corneal reflection appears as a bright point on the video image and is identified by the system (Holmqvist et al., 2011).

The application of the corneal reflection method allows for the precise recording of the direction of consumer's eyes. However, this method does not make it possible per se to determine whether a person's gaze falls on brand messages in the environment. To record gaze contact with brand messages, so-called areas of interest (abbreviated as AOI) must be marked within the scene video. Due to the short length of fixation (100-400 milliseconds), this marking should be done frame-by-frame, whereby the manual work process can be supported by the automatic algorithms of modern computer programs. After the complete marking of AOIs, it is possible to automatically evaluate how long and frequently a consumer's gaze enters into a certain brand message in order to draw conclusions about the capture of attention (Rumpf & Breuer, 2014).

Eye-tracking provides valuable insights into the effectiveness of visual marketing communication, particularly if the method is used within a controlled experimental research design. Through the systematic manipulation of visual properties of brand stimuli (e.g. color, form, motion), cause-effect-relationships can be assessed. In a recent study on consumer attention for sponsor brands in a sport context, the impact of color variables and different types of animation were tested in a lab. Sponsor signage was systematically manipulated with regard to brand color and animation and then virtually embedded into sport broadcasts. It was found that light and high-contrast colors as well as certain types of animation (e.g. blinking) facilitate the visual perception of brands during TV sport consumption (Breuer & Rumpf, 2015).

Such evidence-based findings enable researchers and managers to better understand the drivers of successful brand communication in emotionally-loaded environments. In another eye-tracking study it was found, for example, that brand placements during highly arousing scenes are less likely to be visually processed and therefore branding objectives might be missed. This is most probably caused by the fact that brand stimuli in more exciting environments have to compete with a higher number of more intense stimuli to be perceived. Therefore, it is argued that a moderate level of surrounding intensity should be preferred for brand placements (Rumpf, Noel, Breuer, & Memmert, 2015).

SHORTCOMINGS AND FUTURE DIRECTIONS

In the main body of this chapter, several neuroscientific measurement techniques have been analyzed regarding their contribution to understanding consumers' brand perceptions and reactions to marketing stimuli. In the remainder, the shortcomings and potential future directions of neuroscientific measurements will be discussed before drawing a conclusion. In this regard, the cost-benefit ratio, technical limitations, and ethical concerns are to be briefly discussed.

Neuroscientific measurements are still more expensive and less accessible compared to traditional techniques. As a consequence thereof, sample sizes are typically small and research designs are held simple. In a similar vein, it needs to be pointed out that most neuroscientific studies are conducted inside the lab, whereas applications in a real-world environment are rare. These issues put under question the validity, and thus, the benefit of neuroscientific measurements for marketing decision-makers. Two major advancements are expected in the future: First, due to technological development and increasing demand the costs for neuroscientific instruments will decrease which will make such instruments more affordable and accessible for consumer research. This trend can be observed, for example, with regard to EEG and eye-tracking systems which are now available at reasonable costs. Second, there is an ongoing trend of wearable neuroscientific technology, for example, eye-tracking glasses, EMG or EEG systems which work wireless and allow test persons to move naturally in a real-world environment. Such wearable gear offers researchers new opportunities with regard to more advanced study designs and natural test settings, for example, in-store tests.

Furthermore, the measurement of brain activity is still limited in terms of temporal resolution (i.e. the degree to which a single object can be separated in time from another) or spatial resolution (i.e. the degree to which a single object can be separated in space from another). Despite major improvements over the last decade, there is still no technique which excels the others in terms of validity. Consequently, leading researchers suggest pursuing convergent validity by linking various approaches. Future research is therefore expected to make more use of meaningful combination of neuroscientific measurement techniques. For example, leading researchers in neuroscience already make use of an integration approach where fMRI data (due to its high spatial resolution) and EEG data (due to its high temporal resolution) are recorded simultaneously to achieve both a high temporal and spatial resolution (e.g. Huster, Debener, Eichele, & Herrmann, 2012).

Finally, neuroscientific studies have been faced with ethical concerns. Critics claim that medical instrumentations such as fMRI which have been developed for pathologic diagnosis are now misused for commercial purposes. Moreover, some media reports have created fear in the general public that neuroscientific measurements enable companies to scan private thoughts and to manipulate buying behavior in the future (Javor et al., 2013). In response to this issue, it will be increasingly important to better inform consumers about the purpose and limitations of neuroscientific measurements and to clear up several myths circulating about consumer neuroscience.

CONCLUSION

Understanding the consumer's reaction to marketing activities is essential to gain business success in competitive markets. In times of cluttered media environments and in light of the increasing popularity of indirect marketing forms such as sponsorship, social media, or ambient marketing, most consumer reactions are assumed to occur on non-conscious levels. Since traditional market research techniques simply capture the consumer's consciously reflected attitudes and intensions, they provide limited understanding about the success of up-to-date marketing actions.

Over the last two decades, neuroscientific measurements have been successfully applied in testing consumer behavior theories and providing new insights for business practice. Since consumer behavior is influenced by neurobiological processes, neuroscientific approaches mean a meaningful methodological enhancement to the ordinary research inventory in market research and marketing science (Kenning, 2014).

In this chapter, a variety of brain imaging and psychophysiological measurement approaches have been explained and discussed. The focus was on the application and benefits for consumer research. It can be concluded that each of the analyzed techniques delivers a contribution to a deeper understanding of consumer behavior in general and consumer reactions to marketing stimuli in particular. Due to some critical limitations, however, neuroscientific measurements should be regarded as a valuable supplement rather than a replacement of the traditional measurement approaches. If applied appropriately, neuroscientific measures can substantially help marketing practitioners to better explain and foresee the success of their marketing actions. From a scientific perspective, neuroscientific measures are beneficial to test and extent marketing theories, and to enhance the understanding about the consumer "black box".

REFERENCES

Belch, M., Holgerson, B., Belch, G., & Koppman, J. (1982). Psychophysiological and cognitive responses to sex in advertising. *Advances in Consumer Research. Association for Consumer Research (U. S.), 9*, 424–427.

Berns, G. S., & Moore, S. E. (2012). A neural predictor of cultural popularity. *Journal of Consumer Psychology, 22*(1), 154–160. doi:10.1016/j.jcps.2011.05.001

Bolls, P. D., Lang, A., & Potter, R. F. (2001). The effects of message valence and listener arousal on attention, memory, and facial muscular responses to radio advertisements. *Communication Research, 28*(5), 627–651. doi:10.1177/009365001028005003

Breuer, C., & Rumpf, C. (2015). The impact of color and animation on the sports viewers' attention for televised sponsorship signage. *Journal of Sport Management, 29*(2), 170–183. doi:10.1123/JSM.2013-0280

Deppe, M., Schwindt, W., Kugel, H., Plassmann, H., & Kenning, P. (2005). Nonlinear responses within the medial prefrontal cortex reveal when specific implicit information influences economic decision making. *Journal of Neuroimaging, 15*(2), 171–182. doi:10.1111/j.1552-6569.2005.tb00303.x PMID:15746230

Derbaix, C. M. (1995). The impact of affective reactions on attitudes toward the advertisement and the brand: A step toward ecological validity. *JMR, Journal of Marketing Research, 17*(4), 470–479. doi:10.2307/3152182

Duchowski, A. (2007). *Eye-tracking methodology, theory and practice* (2nd ed.). London: Springer.

Green, S., & Holbert, N. (2012). Neuroimaging and marketing research. In R. J. Kaden, G. Linda, & M. Prince (Eds.), *Leading edge, marketing research–21st-century tools and practices* (pp. 252–272). Thousand Oaks, CA: Sage Publications. doi:10.4135/9781452240558.n13

Gregg, A. P., & Klymowsky, J. (2013). The implicit association test in market research: Potentials and pitfalls. *Psychology and Marketing, 30*(7), 588–601. doi:10.1002/mar.20630

Hajcak, G., Weinberg, A., MacNamara, A., & Foti, D. (2012). ERPs and the study of emotion. In E. S. Kappenman & S. J. Luck (Eds.), *The Oxford handbook of Event-Related Potential component* (pp. 441–472). New York, NY: Oxford University Press.

Holmqvist, K., Nyström, M., Andersson, R., Dewhurst, R., Jarodzka, H., & van de Weijer, J. (2011). *Eye-tracking: A comprehensive guide to methods and measures*. New York, NY: Oxford University Press.

Hubert, M., & Kenning, P. (2008). A current overview of consumer neuroscience. *Journal of Consumer Behaviour, 7*(4-5), 272–292. doi:10.1002/cb.251

Huster, R. J., Debener, S., Eichele, T., & Herrmann, C. S. (2012). Methods for simultaneous EEG-fMRI: An introductory review. *The Journal of Neuroscience, 32*(18), 6053–6060. doi:10.1523/JNEUROSCI.0447-12.2012 PMID:22553012

Javor, A., Koller, M., Lee, N., Chamberlain, L., & Ransmayr, G. (2013). Neuromarketing and consumer neuroscience: Contributions to neurology. *BMC Neurology, 13*(1), 13. doi:10.1186/1471-2377-13-13 PMID:23383650

Kenning, P. (2014). *Consumer Neuroscience*. Stuttgart, Germany: Kohlhammer.

Kivikangas, J. M., Chanel, G., Cowley, B., Ekman, I., Salminen, M., Jarvela, S., & Ravaja, N. (2001). Review on psychophysiological methods in game research. *Journal of Gaming and Virtual Worlds*, *3*(3), 181–199. doi:10.1386/jgvw.3.3.181_1

Klucharev, V., Fernandez, G., & Smidts, A. (2005, September). *Why celebrities are effective: Brain mechanisms of effective advertising*. Presented at the *3rd annual meeting of the Society for Neuroeconomics*. Kiawah Island, SC.

Knutson, B., Rick, S., Wimmer, G. E., Prelec, D., & Loewenstein, G. (2007). Neural predictors of purchases. *Neuron*, *53*(1), 147–156. doi:10.1016/j.neuron.2006.11.010 PMID:17196537

Lajante, M., Droulers, O., Dondaine, T., & Amarantini, D. (2012). Opening the "black box" of Electrodermal Activity in consumer neuroscience research. *Journal of Neuroscience, Psychology, and Economics*, *5*(4), 238–249. doi:10.1037/a0030680

Lang, A. (1990). Involuntary attention and psychophysiological arousal evoked by structural features and mild emotion in TV commercials. *Communication Research*, *17*, 275–299. doi:10.1177/009365090017003001

Lang, P. J., Greenwald, M. K., Bradley, M. M., & Hamm, A. O. (1993). Looking at pictures: Affective, facial, visceral, and behavioral reactions. *Psychophysiology*, *30*(3), 261–273. doi:10.1111/j.1469-8986.1993.tb03352.x PMID:8497555

Luck, S. J. (2014). *An Introduction to the Event Related Potential Technique* (2nd ed.). Cambridge, MA: MIT Press.

McClure, S. M., Li, J., Tomlin, D., Cypert, K. S., Montague, L. M., & Montague, P. R. (2004). Neural correlates of behavioral preference for culturally familiar drinks. *Neuron*, *44*(2), 379–387. doi:10.1016/j.neuron.2004.09.019 PMID:15473974

Plassman, H., Kenning, P., & Ahlert, D. (2007). Why companies should make their customers happy: The neural correlates of customer loyalty. *Advances in Consumer Research. Association for Consumer Research (U. S.)*, *34*, 735–739.

Poels, K., & Dewitte, S. (2006). *How to capture the heart? Reviewing 20 years of emotion Measurement in Advertising* (Working Paper No. MO 0605). Retrieved from10.2139/ssrn.944401

Rumpf, C., & Breuer, C. (2014). Predicting the attentional effect of sport sponsorship information as an innovative evaluation approach. In M. Alonso (Ed.), *Strategies in sports marketing: Technologies and emerging trends* (pp. 238–254). Hershey, PA: IGI Global.

Rumpf, C., Noël, B., Breuer, C., & Memmert, D. (2015). The role of stimulus intensity and working memory capacity in the consumer's processing of brand information in entertainment media. *Psychology and Marketing*, *28*(8), 764–770. doi:10.1002/mar.20816

Stoll, M., Baecke, S., & Kenning, P. (2008). What they see is what they get? An fMRI-study on neural correlates of attractive packaging. *Journal of Consumer Behaviour*, *7*(4-5), 342–359. doi:10.1002/cb.256

Vecchiato, G., Maglione, A. G., Cherubino, P., Wasikowska, B., Wawrzyniak, A., & Latuszynska, A. et al.. (2014). Neurophysiological tools to investigate consumer's gender differences during the observation of TV commercials. *Computational and Mathematical Methods in Medicine*, 1–12. PMID:25147579

Venkatraman, V., Dimoka, A., Pavlou, P. A., Vo, K., Hampton, W., Bollinger, B., & Winer, R. S. et al. (2015). Predicting advertising success beyond traditional measures: New insights from neurophysiological methods and market response modeling. *JMR, Journal of Marketing Research*, *52*(4), 436–452. doi:10.1509/jmr.13.0593

Xue, G., Chen, C., Lu, Z., & Dong, Q. (2010). Brain imaging techniques and their applications in decision making research. *Xin Li Xue Bao. Acta Psychologica Sinica*, *42*(1), 120–137. doi:10.3724/SP.J.1041.2010.00120 PMID:20376329

Yoon, C., Gonzalez, R., & Bettman, J. R. (2009). Using fMRI to inform marketing research: Challenges and opportunities. *Marketing Research*, *46*(1), 17–19.

Yoon, C., Gutchess, A. H., Feinberg, F., & Polk, T. A. (2006). Comparing the structures of brand and human personality: A functional magnetic resonance imaging study. *The Journal of Consumer Research*, *33*, 31–40. doi:10.1086/504132

KEY TERMS AND DEFINITIONS

Brain Imaging: A set of neuroscientific techniques which are either sensitive to metabolic substances related to neural activities (e.g. hemoglobin), or to the outcome of neural activities (e.g. electrical potential).

Electrodermal Activity (EDA): A psychophysiological method which is used to assess changes in electrical properties of the skin in response to the emission of sweat.

Electroencephalography (EEG): A neuroscientific method which records electrical activity along the scalp produced by the firing of neurons within the brain.

Electromyography (EMG): A method which measures frowning and smiling via electrodes placed on two muscles in the human face: The corrugator muscle (involved in frowning) and the zygomatic muscle (controls smiling).

Eye-Tracking: A psychophysiological approach to study the allocation of attention by measuring the direction of the test person's gaze.

Functional Magnetic Resonance Imaging (fMRI): Brain imaging technique which assesses changes in blood flow associated with neural activity in specific brain areas.

Heart Rates (HR): The number of milliseconds between two heart rates indicating various phenomena such as attention, arousal, or valence.

Psychophysiology: The study of body signals in the autonomic nervous system through indicators such as skin conductance, facial expressions, heart rates, and eye movements.

This research was previously published in Applying Neuroscience to Business Practice edited by Manuel Alonso Dos Santos, pages 1-17, copyright year 2017 by Business Science Reference (an imprint of IGI Global).

Chapter 17
Upgrading Marketing Research:
Neuromarketing Tools for Understanding Consumers

Anka Gorgiev
The University of Sheffield International Faculty, City College, Greece

Nikolaos Dimitriadis
The University of Sheffield International Faculty, City College, Greece

ABSTRACT

The following chapter focuses on the recent developments in the marketing industry resulting from the rise of neuromarketing. It offers analysis of the problems initiated by the traditional marketing research approaches and how these problems are being resolved by implementation of neuromarketing techniques. Moreover, both physiological and neurological measures are explained, providing advantages and disadvantages of all available neuromarketing tools that may be used in overcoming previously mentioned problems. Finally, the authors provide initial insights from a small scale qualitative research conducted among representatives from all segments of marketing industry.

INTRODUCTION

Since the first recognitions of marketing as an independent field, it has always had the reputation of a dynamic area. The reason for such perception lays in its core function - to satisfy the needs of customers. The entire marketing field is designed in such manner to adapt to the changing needs and habits of the customers, regardless if the changes come from within the field or from neighboring disciplines. It is within the duty of marketing to satisfy the most recent needs, wants and desires of consumers by uncovering their emerging preferences, attitudes, interests and opinions.

The most recent trends in this respect are coming from the developments in behavioral economics, social psychology and neuroscience (Genco, Pohlmann & Steidl, 2013). These fields enable marketers to comprehend consumer behavior not as an action, but as a reaction to certain underlining processes.

DOI: 10.4018/978-1-5225-5478-3.ch017

They also take into consideration the individual as a member of a complex social system, rather than just considering one of the two constructs. And finally, it is becoming clear to scientists that consumers are not as conscious of their decisions as it was previously considered.

BACKGROUND

Generally speaking, behavioral science promises to provide the knowledge necessary to predict future behaviors. In the realm of economics, the entire system of economic theories has been based on the *Homo Economicus* assumption, which implies that people make rational and conscious decisions (Rubinson, 2010; De Vroey, 2009; Morgan, 2006). However, for some time now, it has been clear that people do not follow the rational economic postulates; rather, important ingredients in human decision making are emotions (DellaVigna, 2009; Lee, Amir & Ariely, 2009; Damasio, 2005). Empirical findings in this area are challenging many rules-of-thumb and establishing new confidence in the knowledge of human behavior.

Social psychology complements this knowledge by adding the social influencers into the equation. It has been a common understanding that people's answers, thoughts, choices, preferences and even attitudes and beliefs change depending on the fact if they are alone or in presence of others. Apart from the fact that acknowledging this phenomenon can have significant value to the marketing professional, understanding of such observation contributes vastly to the holistic construct of human behavior.

If behavioral economics and social psychology represent the front office of the behavioral studies, then neuroscience dominates the back office, by supplying the knowledge about the neural activities involved in particular behaviors. Brain science has received significant popularity in the recent years, mainly due to the two high profile projects that have budget of $2.3 billion since they joined forces (Reardon, 2014). The BRAIN (Brain Research through Advancing Innovative Neurotechnologies) Initiative in USA and the Human Brain Project in Europe are expected to provide detailed map of the brain and its neural network in a computerized simulation (Blau, 2013; Wolf & Morrissey, 2013).

While all these findings impose new perception of the world, they also call for new methodologies and tools to be used in order to investigate this new world. Put together, at the intersection of these fields is the newly formed neuromerketing area that can help marketing academics, researchers and practitioners to overcome the problems and get closer to reaching the ultimate goal in marketing - delighting the consumers - and have more effective approach towards issues of societal significance.

MAIN FOCUS OF THE CHAPTER

Problems With Traditional Marketing Research Tools

It is widely known that the holy trinity of research tools in traditional marketing consists of surveys, interviews and focus groups. Majority of the marketing knowledge that exists at this point is result of one of the three tools, or their combination. And it seems that they have shown to be quite useful for the purpose intended. In the previous marketing paradigms, with focus on products or sales, they were appropriately designed to achieve the necessary goals. However, as consumers and their needs become more sophisticated and more informed, new tools need to be employed in order to achieve more complex goals.

The three traditional marketing research methods are solely based on the self-reporting measures. This involves consumer to provide the necessary information based on which marketing theory or marketing strategy will be based upon. As a consequence, all the trust and confidence in the correctness of the information is leveraged towards the consumer. Nevertheless, there have been instances when inconsistencies appeared between consumer indications and implementation results. Numerous such examples exist within the 80% failure rate in new products. In order to demystify this, recent findings from behavioral economics, social psychology and neuroscience have brought to light activities that take place within a person and filter the inputs considered on both conscious and subconscious level, until this person provides information when asked a question (Ohme, 2009). These findings can explain why previously mentioned inconsistencies happen in the first place, but they also bring into questioning any result of traditional marketing research (Fugate, 2007).

The biggest concern regarding the three research instruments usually employed in marketing research is the response bias, when participants misrepresent the true answers (Martin, Engelland & Collier, 2011). Sometimes the failure to tell the truth can be deliberate, while sometimes people are simply not aware of the truth behind the questions asked. Genco, Pohlmann & Steidl (2013) have provided a list of all response biases previously studied by psychologists that affect marketing research, including both intentional and unintentional ones.

Agreeableness Bias - It is in human nature to try to find ways to be likable to those a person is interacting with. And in order to be likable to the person asking the question, people tend to provide answers they believe will satisfy the others (Detric, 2010). In trying to be agreeable with the person asking the questions, people provide incorrect answers, which distort the results of the study. However, traditional marketing research tools are designed in a way that makes it difficult to distance the researcher from the respondents, especially when interviews and focus groups are employed.

Social Desirability Bias - The phenomenon of socially desirable answers has been known to marketers for a long time (Nancarrow, Brace & Wright, 2001). It reflects the tendency of people to provide answers that are considered socially acceptable. This is especially the case with brands that are associated with particular target group characteristics that some might consider negative, but still buy the brand; for example, people are sometimes ashamed to admit that they consume brand that is believed to be associated with low income target group, as this might imply they have financial problems. Due to such beliefs, respondents sometimes give false answers which result in incorrect assumptions for the topic in question.

Knowledge Exposure Bias - Nobody wants to be caught not knowing the answer to the question they have been given. People would rather make something up to conceal the fact that they are not informed on particular topic or simply never even thought of it. In this way, they are not exposed in a negative way in front of the researcher. For example, this bias can be found in research investigating characteristics of a product or service, where participants would rather name some attributes than admit they are ignorant about it.

Misinformation Bias - Opposite from knowledge exposure bias is the misinformation bias. It can occur when participants in a research do not feel comfortable taking part and are deliberately giving incorrect answers. This can happen with people that are overwhelmed with inquiries to participate in studies and as a sign of protest provide answers that are opposite from their beliefs or attitudes. Such bias has been noticed among political voters in the United States.

Memory Bias - While previous biases involve providing incorrect answers consciously, memory bias represents involuntary misrepresentation of the respondent's beliefs and attitudes. Memory bias occurs when a participant is asked to recall some previous experiences (Singh & Churchill, 1986). Scientific findings reveal that creation of memories is a complex process that can be easily distorted in the process of retrieval, either by the researcher or other immediate factors. For example, memories that are emotionally relevant to the respondent are found to be carefully stored and easily retrieved, compared to memories lacking such component.

Emotion (and Preference) Access Bias - Since it has become clear that emotions play important part in building relationships between brand and consumers, researchers have included evaluation of emotional states in their studies. However, people do not always have clear understanding of their emotional state. Rather, in order to provide adequate answer to such questions, they start analyzing relevant situation. And by analyzing and rationalizing emotional state, the 'emotional' part gets distorted. For example, such bias can occur in projects studying how much consumers like or love a product or a brand. As it is the case with memory bias, regardless of the answer given at the time, the respondent will believe he/she answered truthfully.

Prediction Bias - Many research studies are conducted in order to develop estimations for the future as precise as possible. In such cases, consumers are asked to predict their future behavior in particular conditions. And even though they provide honest answer regarding how they expect to behave in future, this cannot be accurate. In order to predict their own future behavior, research shows that people analyze their past behavior and emotional experience in such behavior, and based on this data they predict how they will feel in the future and, thus, how they will behave (Genco, Pohlmann & Steidl, 2013). Like any prediction based on many unknown factors, answers provided by consumers represent an educated guess. And as such, can have significant consequences on research results and actions that are based on those results. For example, in market research studies consumers might be asked whether they would buy a particular product at a particular price; they analyze the feasibility of the price based on their previous purchases, disposable income, comparison with other products, and similar factors. However, they cannot predict the circumstances of their future purchase when exposed to this product, such as unforeseen costs or lifestyle changes.

Design-wise, self-reporting measures provide a lot of loose space where response biases can appear and distort the results of the research in question. Even though this is the case, there are ways to either eliminate or reduce the opportunity for the bias occurrence. This primarily involves the acknowledgement and understanding of these biases, and extensive training and experience with techniques designed to avoid them. And even then a researcher cannot be completely confident that all the loose ends have been tied.

Neuromarketing Tools and Measures as a Potential Solution

Based on previous discussion, traditional marketing research tools seem to be unreliable source of information due to the response biases, especially when used in market research as a base for formulation of consumer strategies and tactics. Therefore, in order to preserve the core function of the practice, marketers are in need of the tools that would allow access to true underlining needs, preferences, attitudes, interests and opinions of their customers, which new technologies are promising to deliver.

Advances in science have revealed tools that can be used in indentifying, monitoring and interpreting physiological and neurological signals from human body (Narvainen et al, 2013; Ohme et al, 2007). Human body is designed in such a way that brain controls all body movements, regardless if these movements are conscious or unconscious. Nevertheless, unconscious signals from the body represent a reliable source of information as they are unbiased and cognitively unfiltered response from the body, and thus not manipulated. They allow marketers to interpret messages from consumers beyond their cognitive comprehension. And as a result, marketers can use these signals in an effort to understand the effect stimuli has on a person.

Physiological body signals represent unconscious body movements that result from exposure to a particular stimuli. Physiological measures that are being used in marketing research involve study of facial expressions, tracking of eye movement, interpretation of heart rate and respiration, electrodermal activity, as well as the behavioral response times. Apart from interpreting signals from the body, marketers can use signals from the brain to understand reactions to a stimuli. These signals can either be of electrical, which is measured with EEG, or represent changes in blood flow in the brain, which is measured with fMRI.

Regardless of their nature, both physiological and neurological signals provide valuable resource in understanding human behavior and underlying processes influencing resulting actions (Bercea, 2012). In that respect, marketing is just one of the fields where these novel technologies have found its purpose.

Facial Expressions

Emotions experienced by a person are usually represented on one's face and they are difficult to conceal. They represent response to one's emotional state, which is why the interpretation of facial expressions in considered a very useful tool in understanding human emotions. The most prominent person in the area of facial expression research is certainly Dr. Paul Ekman. In late 1960s, Dr. Ekman studied the nonverbal behavior of the Fore tribe in Papua New Guinea (www.paulekman.com). As the result of this research, Dr. Ekman, together with his colleagues, identified six universal emotions, each represented in the distinctive form of universal facial expressions: happiness, sadness, surprise, fear, anger and disgust. Remarkable discovery is that these emotions appear to be manifested in the same way in people from different cultural backgrounds (Ekman, 1992). The research has identified patterns in certain muscle movements, such as frown or smile that points towards specific emotions.

Dr. Ekman and his colleagues observed movements of facial muscles for each of the emotion and were able to code each expression based on muscles involved in its making. Each muscle movement represents one action unit (AU). Under these assumptions, the researchers have developed Facial Action Coding System (FACS) (Ekman & Friesen, 1978). FACS keeps track of facial expressions by interpreting the micro-movements of facial muscles (Bove, 2008; Craig et al, 2008). According to Cohn, Ambadar & Ekman (2006), FACS distinguishes between 9 AUs in upper and 18 AUs in lower face and 5 miscellaneous AUs, that can be combined with 14 head and 9 eye movements, as well as "*9 action descriptors, 9 gross behaviors, and 5 visibility codes*" (p. 205). When analyzing facial expressions, experts view the video containing person(s) in question and carefully record the micro-movements of facial muscles. Combining the activation of different action units, experts gain insight into which emotion is being experienced by the person.

Facial action coding system represents important tool for interpreting human emotions as it is completely based on the knowledge of human face anatomy and understanding of the nervous system. Due to this, there is strong support towards the premise that facial expressions provide reliable representation of emotions experienced. Put in marketing context, facial expressions have be very insightful indication of consumer emotions, especially in real-life circumstances. Nevertheless, since facial expressions are partly consciously controlled, sometimes consumers might experience emotions that do not surface, and thus fail to be captured in the form of facial expressions. In addition, interpretation of facial expressions requires expertly trained professional who performs these analysis manually. Apart from low availability of such experts, it implies the difficulty for conducting analysis in real-time, significant time consumption, and costly set-up. There are some softwares available to marketers that can speed up the analysis of facial expressions and can cost less, but they seem to be less reliable.

Facial coding is very useful in analyzing advertisements prior to their broadcasting, as it can help companies identify emotional responses to their messages. It can also interpret consumer experience in whichever situation that is being recorded. In addition, there are softwares used today that can connect to customers' webcams and record their online experience. Of course, such analysis are conducted after the fact and can be used for improving marketing activities in future.

In an effort to overcome some of the limitations imposed by FACS, marketers use Facial Electromyography (EMG). EMG also interprets facial expressions, but it does so by measuring the electrical activity of facial muscles that are being active (Rasch & Teichert, 2013). This tools enables the researchers to identify even the smallest muscle movement that otherwise can be invisible to the naked eye in FACS technique. This applies especially to the unconscious movements that occur when specific emotions are experienced, even prior to their manifestation of the face. Electromyography can record muscle movements in real time and with high precision. However, it requires the participant to wear specially designed sensors on his face during the research, which makes it very difficult for usage in consumers' natural environment.

Due to its technical requirements, EMG is suitable only for laboratory settings. Similarly to FACS, it can provide significant insights when used in the analysis of marketing messages.

Eye Tracking

Human brain heavily relies on sight as the primary source of information and stimuli from the external environment. Research shows that visual inputs represent the most reliable source of information about our external environment. The evolutionary response to this phenomenon manifests in the existence of an entire brain area solely dedicated to receiving, interpreting and storing visual stimuli. Having in mind its importance to any individual, it is natural to consider it in trying to gather insight behind the human behavior.

First and foremost, tracking eye movements has provided useful way in identifying focus of attention (Becker, 2013; Casarotti, 2012). It is in human nature, or rather it is an involuntary reflex, to direct sight towards the source that grabs human attention. When a person is looking at something, it can be concluded that this stimulus in of his interest. The longer the fixation on a stimulus, the greater the interest for this stimulus; the reason behind it can be either the difficulty of processing the input or simply the high interest. Apart from fixations, the movement in-between fixations can have a significance to a researcher. If viewer is showing rapid shifts in fixations, it may indicate that the viewer has lost interest in the material viewed or that he is having difficulty following and comprehending the material. Moreover, the shifts that are inconsistent with the intended flow of the material may indicate the necessity to adapt it accordingly.

The pattern of eye movement has shown to be very useful in the analysis of websites. The fixations indicate the content that is considered more interesting to the viewer. For example, some research has shown that in an online behavior the majority of attention is given to the photos and titles compared to the textual content (Djamasbi, Siegel, Skorinko & Tullis, 2011). These findings can help website editors in managing their online content in order to make it more attractive for the target group.

Secondly, eye tracking can be used in identifying the occurrence of blinking, which has shown to be correlated with the attention and engagement (Mauri, 2011). More precisely, suppressed levels of blinking indicate higher level of attention and engagement, as in situations requiring higher concentration people tend to gather as much input as possible; in contrast, blinking at higher rate is present in tasks that require less attention. Similarly, blinking rate can indicate emotional state of the viewer. As blinking represent's our defense mechanism, its frequent occurrence can indicate negative emotions (Genco, Pohlmann & Steidl, 2013).

The blinking rate can be used as complementary analysis of eye movements to indicate sources of attention in exposure to marketing content. They can also be employed over longer periods of time in order to observe consumers' adaptation to a message or the effectiveness of pre-exposure (priming).

Finally, the most interesting and insightful information in eye movement analysis can be gained from the pupils. The changes in size of pupils represent unconscious actions, usually as a response to emotional reaction (Mauri, 2011). More precisely, increase in pupil size happens when a person experiences emotional arousal. However, analysis of pupil dilation cannot indicate whether the change is size is due to negative or positive emotions; it can only indicate an emotional response. Apart from emotional arousal, pupil dilation has been recorded in instances when participants in research have been involved in performing difficult tasks or tasks that required higher attention (Steinhauer, Siegle Condray & Pless, 2004). Having in mind its sensitivity to emotional and cognitive stimuli, pupil dilation represents very reliable measure for stimuli of such nature. And the best part is - the analysis can be performed in real time. Based on the dilation changes, marketers can instantly adapt content in question.

As one of the most reliable indicators of emotional state, analysis of eye movements, including the fixations, blinking rates and pupil dilations, have become very popular among marketers nowadays. With increasing spread, its cost becomes lower, which makes this tool more and more accessible for market research purposes. Nevertheless, due to the eye sensitivity to changes in brightness (especially pupils), this tool should be used primarily in evaluating print or web content rather than videos.

Respiration, Heart Rate and Blood Pressure

Everyone is familiar with what lie detector is, at least from the commercial movies involving government agencies. Lie detector or polygraph is a machine designed to monitor body functions throughout the sessions of questioning. The entire technique is based on establishing the referent point for physiological impulses, such as blood pressure, heart rate and respiration, and observe their change when a particular question is attended. For many years this technique has been used by governments throughout the world in pursue of truth. Neuromarketers nowadays use the same principle when conducting research of their own.

Going deeper into the brain, there are certain primary body functions, such as breathing and heart activity, which are controlled by the brain without any conscious effort (Afifi & Bergman, 2005). This implies that signals received through these functions represent true body responses, as they cannot be controlled or managed consciously. Given the situation, brain manages these activities based on the assessment of external environment, thus suppressing or increasing the rate of these actions.

According to De Melo, Kenny & Gratch (2010), there are specific patterns of respirations that are associated to specific emotional states. For example, extensive research has shown that deeper breathing at faster pace reflects excitement, but also anger and fear. Such pattern is present when brain is experiencing fight-or-flight effect. On the other hand, fast and shallow breathing reflects either anxiety or a situation when higher attention is needed. When a person is breathing slowly with deep intakes of air, it can mean that he is very relaxed or sad. Interestingly, the same slow and deep respiration pattern has been identified when a person is experiencing positive emotions, such as joy. There have been numerous studies conducted in an effort to identify emotional states through the respiratory patterns and the findings can be very useful to neuromarketers trying to understand the audience's feelings (Philippot, Chapelle & Blairy, 2002). Before new research findings are available, respiration *per se* might not be the best tool to use to identify particular emotions. Rather, at this point it should be used primarily as a measure of arousal, as it can be especially insightful when employed in pre-screening of marketing messages.

Hearth function is controlled by the brain in the same way as respiration, which is through the autonomic nervous system. Based on the inputs received from the external environment and brain's interpretation of these inputs, both hearth rate and blood pressure can show oscillations.

There are a lot of research studies investigating heart function and its relationship with emotional experiences, but there are many inconsistent results. In a study exploring the relationship between emotions and music, participants had lower heart rate when exposed to unpleasant music, while their heart rate increased when listening to music they liked (Sammler, 2007). On the other hand, there are studies that report increased heart rate when a person is experiencing positive emotions and decreased heart rate during negative emotions. Having this in mind, heart rate might not be the most reliable tool in interpreting emotional valance, but with high assurance it can identify the presence of emotional response. There are even indications that heart rate can be used in indicating attention or cognition levels, with heart rate decreasing at times of higher attention efforts.

In addition to heart rate, numerous clinical studies had blood pressure in focus, reporting increased activity when participants are experiencing negative emotions, while positive emotions are usually accompanied by decrease in blood pressure (Sundararajan, 2010). Apart from emotions, blood pressure seems to be highly sensitive to memory, as well. However, the results from this tool are considered unreliable, since they can differentiate based on the placement of the measurement object on participant's arm. As such, they are only used in combination with other techniques identifying the presence of emotion, but not its valance.

There is no question whether measures of heart rate, blood pressure and respiration can be useful in marketing research. Nevertheless, the interpretations of body responses using previously mentioned tools are still unreliable. As future research demystifies further these autonomic inputs, they will be more and more applicable when trying to understand emotional states of target group, especially in pre-launch market testing research.

Electrodermal Activity

Similarly to previously discussed body functions, perspiration is also controlled by the autonomic nervous system, and as such cannot be consciously manipulated (Lajante et al, 2012). As brain controls the glands producing the sweat, some of the emotional responses of human body result in the production of sweat on the skin surface, causing sweaty palms. In essence, electrodermal activity represents measure of electrical characteristics of the skin as the response to the presence or absence of sweat (Lajante, Droulers, Dondaine & Amarantini, 2012).

On the contrary to heart rate and respiration, though, significance of perspiration in marketing research has been acknowledged for thirty years now, while this tool has been used in marketing research since 1960s. In one of the first studies of such kind, the researchers were investigating correlation between the skin response and the feeling of warmth among participants exposed to advertising stimuli (Aaker,

Stayman & Hagerty, 1986). Accompanied with numerous other similar studies, measuring electrodermal activity became reliable tool for marketers due to its sensitivity to stimuli originating from advertising messages (Kroeber-Riel, 1979).

Reactions like this enable marketers to identify emotions experienced by consumers, even if they fail to report it. Electrodermal activity is measured using galvanic skin response (GSR) or skin conductance response (SCR). Sensors that are connected with the galvanometer are placed on participant's palms and fingers; these sensors report changes in electrical conductance of the skin, which can be monitored on the galvanometer. Studies conducted so far have found the connection between perspiration and arousal and stimulation (Kreibig, 2007). This means that when GSR is higher, it is safe to conclude that the participant is experiencing emotional arousal due to the exposure to presented stimuli (Gakhal & Senior, 2008). In addition, Caffyn (1964) found that results captured by GSR can even differentiate between print and TV ads.

Electrodermal activity represents a highly sensitive measure to changes in affective states. Some even consider it to be among the most reliable in capturing unconscious bodily signals (LaBarbera & Tucciarone, 1995). As such, they are considered a good measure of arousal. However, the studies conducted using GSR only managed to conclude whether there is an emotional reaction to stimuli, but not its directionality. EDA measures cannot be used to determine whether the consumer likes or dislikes the stimuli; it can be only used to see whether there is a reaction of some kind and how strong this reaction is. Moreover, marketers wanting to employ this measure in their research need to find experts and highly trained personnel who know how to handle galvanometer and read the galvanic skin responses. Even though more than twenty studies have been conducted in the past thirty years, still there is no standardized methodology or analysis of the signals used in interpreting result (Lajante, Droulers, Dondaine & Amarantini, 2012).

This tool still provides better indication of consumer emotional stimulation than any self-reporting method. But, the complexity involved with interpreting results prevents it from being primary source of market inputs. Nevertheless, when coupled with other tools, electrodermal activity can provide very insightful information to marketers. In a study conducted by Hensel (1970), it was found that high GSR and pupil dilation indicate higher receptiveness of the message, which ultimately leads to higher retention. Following this example, marketers can use galvanic skin response or skin conductance response measures to validate results gathered otherwise and support findings of other neuromarketing tools.

Behavioral Response Time

Human memory is one of the most complex cognitive functions by itself. However, it also represents a foundation for higher level cognitive functions, such as learning, decision making, language, etc (Del Missier et al, 2013; Flight, 2011; Ratnayake, Broderick & Mitchell, 2010; Jefferies, Frankish & Noble, 2009). People receive inputs from external environment through their senses. These inputs are then processed and stored in human memory. Brain creates mental representations as a way of storing all this information. This means that there are specific mental representations for each attitude, emotion, person, etc. When trying to retrieve some information, human brain searches a database of mental representations trying to find the correct one. As a way to facilitate this process, human brain creates associations among specific objects, words, pictures or sounds and attitudes, situations or emotions. Stronger the association, faster the retrieval or recall. Due to such diligent mechanism, these associations are of crucial importance to marketing professionals.

Attitudes represent one of the most intriguing concepts to contemporary marketers, since attitudes play an important role in the formation of opinions and preferences, which are crucial for brand and product consumption, and significantly affect human behavior. Explicit attitudes represent those attitudes that are created consciously and on purpose. For example, after watching a documentary on fast food, a person develops a strong opinion that fast food is not healthy. As such, explicit attitudes are controlled by a person and can easily self-reported during a research. Implicit attitudes, on the other hand, represent opinions, preferences and feeling that are developed based on previous experiences. Due to their nature, they cannot be consciously controlled, and thus cannot be correctly self-reported, but still significantly influence behavior (Dimofte, 2010). In the case of fast food, a person might believe that such food is not healthy, but he might go and eat it, regardless. For example, after many years of consuming Coca-Cola, a person might read about negative effects of this beverage and decided to stop drinking it; however, this would not stop him to feel excited and smile while watching Coca-Cola's Christmas commercial with polar bears or Santa Claus.

People are exposed to a large number of stimuli on daily basis. Some of these stimuli do not affect us at all, some are accommodated consciously and a person purposely redirects his behavior, and there are stimuli that affect behavior of a person without him even realizing it. Such phenomenon happens throughout the process of priming. Priming can be described as a process of influencing behavior by introducing a stimuli that will subconsciously trigger an idea associated with this stimuli; this can further activate a physical response and even behavior. In recent years, there have been numerous studies investigating priming and the effects of influence without awareness, providing very interesting results. For example, handing a warm cup of coffee to a business partner may make him friendlier and less aggressive during negotiations or exposing a person to Apple logo might trigger him to perform better on creative tasks. Research has revealed two types of priming: semantic and affective. Semantic priming reflects triggering a certain mental representation by using words, while affective priming involves triggering a feeling by using a prime of the same valance as the target feeling (Davis & Herr, 2014; Yi, 1990).

In order to be able to implement priming in marketing campaigns, marketers need to evaluate the strengths of associations between the stimuli with desired response or behavior. This can be achieved with measuring time needed to respond to particular concept. Depending on the research requirement, there are three techniques available: semantic priming, affective priming and implicit associations test. All three have the same research mechanism. The participant is exposed to paired concepts, either words or pictures, one after another only fraction of seconds apart. Based on the previously predetermined criteria, participant needs to make a decision of choosing one of the two concepts, usually by clicking on the corresponding button. Response latency, or time needed to respond can reveal whether and to what extend the stimuli has activated mechanisms necessary for desired behavior (Mast & Zaltman, 2005). Shorter time needed to respond indicates stronger association between the stimuli and the target concept.

The most common behavioral response method used among marketers is implicit associations test (IAT), which evaluates implicit attitudes towards brands. More importantly, IAT can provide comparative information when used to study two brands. However, IAT seems to be more complex that semantic and affective priming describe previously, as it can employ both techniques as part of the methodology (Genco, Pohlmann & Steidl, 2013; Gregg & Klymowsky, 2013; Gregg et al, 2013; Berger & Fitzsimons, 2008).

Behavioral time response studies represent very useful tool, since they allow marketers to observe behavioral responses that are not controlled by the participant. As discussed on several occasions previously, this is very important in evaluating marketing outputs. Priming and IAT are one of the most available neuromarketing tools, as they can be performed by the researcher alone. And most importantly, these studies can be conducted with minimal costs, which is usually not the case with neuromarketing tools. However, not all research questions can be answered through time response studies. The researcher needs to consider carefully all the aspect of this approach before conducting the experiment. Otherwise, the researcher might have serious problems with the validity of the results.

Blood Flow in Brain

All the tools discussed previously are designed to study the autonomic signals coming from the brain and influencing behavior subconsciously. The reason why there are tools interpreting every such signal is because scientists have always been interested in uncovering what is happening in the so-called 'black box'. For the past 22 years, there is the opportunity to look directly into the brain and observe how it functions using magnetic resonance imaging (Bandettini, 2012).

Magnetic resonance imaging (MRI) is a tool designed to develop images of the brain activity by measuring the blood flow. This approach is very often used in medicine to analyze the irregularities in the brain functioning and structure. Neuroimaging tools work on the principle of identifying blood oxygenation level dependent (BOLD) signal changes in brain as an indication of a neural activity (Parens & Johnston, 2014). The oxygenated parts are then identified by the magnetic field developed by the fMRI machine. Activation of a particular part of the brain requires higher supply of oxygen compared to other brain parts, and thanks to the increased level of oxygen, blood develops stronger magnetic field that is depicted on fMRI images.

Scientists from other disciplines that require brain activity analysis use functional magnetic resonance imaging (fMRI), as it allow them to see the brain activity while engaging participants in different tasks, such as watching videos, pictures or listening to music. The function of fMRI is based on the same principles as the regular MRI, but it provides the opportunity to the researcher to expose the participant to the wide range of stimuli. As a consequence, analysis of the fMRI images can provide more reliable results that are specific to the particular circumstances studied. For example, fMRI can show brain activity of a participant watching TV commercial or listening to a song.

Advances in studying the anatomy of human brain have indicated patterns in the activation of specific brain areas when performing specific tasks. As a result, there are indicative results that certain brain parts are activated when a person is experiencing pain, or love, or fear, or when a person is retrieving memories or talking. This properties of fMRI are very attractive to marketing researchers trying to uncover the significance of stimuli by identifying which part of the brain is engaged in such situation (Craig et al, 2012; Kenning, Plassmann & Ahlert, 2007).

fMRI has become very popular among neuromarketing researchers due to its ability to provide clearer understanding of the brain mechanisms behind certain behaviors, opinions and preferences. However, its usages cannot be widespread yet, because such studies involve very high costs. There is a small number of fMRI machines, since they cost several million dollars per unit, and only expertly trained personnel can operate them. This has resulted in limited availability of this research tool to neuromarketers.

Apart from the logistics problems, fMRI has its own technical shortcomings. fMRI creates images of the brain and it is very successful in depicting spatial resolution of brain activity. However, the images are not available instantly as the stimuli is introduced; rather, there is a time lapse between the capturing and producing the image of approximately six seconds (Genco, Pohlmann & Steidl, 2013). And even capturing a change in neural activity, researchers cannot still be sure why it happened. With future research, it is expected that this limitation will be eliminated. Also, in order to be able to receive a clear image, the participants are not supposed to move while in fMRI scan. This forbids neuromarketers the access to brain activity of participants in more natural consumption environment.

In addition to fMRI scans, positron emission tomography (PET) scans also measure brain activity based on the blood flow changes. Before the measurement, participants are being injected with radioactive isotopes. The levels of these isotopes increase in the parts of the brain that are being activated and they can be captured on an image. PET scan has technical advantages and disadvantages over an fMRI scan. Nevertheless, due to its invasive nature, it is not expected to become a tool neuromarketers would consider for their consumer research purposes.

Electrical Activity in Brain

While fMRI measures changes in the blood flow in the brain, Electroencephalography (EEG) measures the electrical activity of the brain (Morin, 2011). Brain is designed as a network of neurons that communicate among each other constantly. Individual brain cell, neuron, transmits information to its neighboring cell by "firing", or rather releasing electrical potential. EEG can capture whenever a brain cell fires. Participants being monitored under the EEG have electrodes placed on the top of their scalp in the form of a cap or a headset. The headset is connected to a computer operating a specific program designed to interpret input coming from the participant. The electrodes in headset can measure any change in the electrical potential within the brain. Scientists believe that whenever there is increased frequency of neuron firing, the brain has been engaged.

EEG has a wide spectra of inputs it can measure. Apart from just measuring signals from the brain, it can identify which signals are caused by which stimuli. Due to its very high temporal resolution, EEG can provide analysis of the event-related potentials, brain signals that happen as a response to particular event (Ohme, Matukin & Pacula-Lesniak, 2011). In addition, as different brain signals emit different frequency corresponding to different brain states, scientists have found a way to interpret this frequency inputs. EEG studies can reveal the strength of signals involved in performing certain mental activity and the parts of brain that are used in performing this activity.

Researchers have also found the way to apply the knowledge of frontal asymmetry paradigm in EEG frequency analysis. As the theory suggests, when people are motivated by desirable outcomes (approach motivation), there seems to be higher activation of the right brain hemisphere; on the contrary to that, behavior evoked by aversive outcomes (avoidance motivation), shows correlation with the activation of the left brain hemisphere (Pizzagalli, Sherwood, Henriques & Davidson, 2005). Now, neuromarketers can use EEG to determine consumer motivation for their brand or product and use this information in developing more effective marketing strategies (Ravaja, Somervuori & Salminen, 2013; Ohme, Reykowska, Wiener & Choromanska, 2010).

EEG represent highly valuable tool to marketers and neuromarketers. As a research tool, EEG can provide information whether and which part of the stimuli causes increased brain activity, thus indicating increased attention, arousal or even surprise (Silberstein & Nield, 2008). More importantly, all this information can be available almost instantly, which makes EEG highly reliable. Nevertheless, EEG has very limited spatial resolution. Since electrodes are placed on the surface of the scalp, it can only measure signals that originate closely to the electrodes and not the signals that occur deep within the brain. This permits researchers from identifying the involvement of the precise parts of the brain as a response to stimuli. Moreover, the electrodes used to measure electrical signals are very sensitive to body movements, which means that if a participant is not still during research, the results might be corrupted.

There are efforts nowadays to develop wireless set that can monitor brain activity in natural consumption setting, but their reliability is still to be scientifically and empirically proven.

Impact and Challenges

It has been shown that at certain times consumers do not understand their needs or are not comfortable at expressing them (Nancarrow, Brace & Wright, 2001). In such cases, neuromarketing tools can provide valuable resource in interpreting body and brain signals, which are so strong that they cannot be consciously controlled, but they still influence human behavior. These tools enable marketing professionals the access to the underlining truth behind people's thoughts, feelings, attitudes and preferences.

Neuromarketing is sometimes disputed due to potential ethical issues of taking advantage of the "black box" and manipulating people into doing things beyond their free will. Nevertheless, such accusations have not been proven to this point. Even though people are not purely rational human beings, they are not slaves of consumerism either. The fact that sometimes consumers are not consciously aware of their needs does not mean they can be forced into choices. The subconscious mind, impulses and instincts are there to protect them from such instances.

Actually, neuromarketing research potential and the principles behind it, provide a great hope for solving serious societal issues. Understanding the brain mechanisms behind receiving, processing, memorizing, recalling and learning messages can be very useful in constructing the messages that need to reach great masses. Public campaigns can benefit greatly from neuromarketing, from constructing to testing and broadcasting campaigns. In addition, neuromarketing can contribute to better understanding of the consumerism as a behavioral aspect. There is a large set of data collected in commercial purposes mainly that points towards behavioral patterns observed over many years. This input can complement many studies in the areas of neuroscience and neurology, as well as psychology and other behavioral sciences (Javor, Koller, Lee, Chamberlain & Ransmayr, 2013).

Nevertheless, neuromarketing tools still represent novel technologies, which implies that there is not sufficient empirical evidence in order to make strong conclusion about the findings (Ariely & Berns, 2010). More so, it is believed that neuromarketing tools are very costly to obtain and require personnel with specialized training to handle them correctly. And this is true for some of the specialized tools, like fMRIs, EEGs and FACS. However, there are those tools that can deliver information about the subconscious signals, but do not require a fortune; these include time-response studies or even behavioral experiments.

Whatever the research budget available, the choice on the tools to use should depends on the research question and what is being investigated. Sometimes traditional tools are enough and neuromarketing tools simply do not justify the costs; but there are also times that require more advanced techniques and technologies to be used in solving the problem, which is when neuromarketing tools are irreplaceable. As a result, many scientists and marketing professionals still believe that neuromarketing tools are most effectively used in combination with the traditional marketing research techniques (Bercea, 2012; Fugate, 2007).

FUTURE RESEARCH DIRECTIONS

Small Scale Primary Research

Neuromarketing as a sub-discipline has been thriving for more than 10 years now (McClure, Li, Tomlin, Cypert, Montague & Montague, 2004). However, it seems that that the proliferation it experienced so far is nothing compared to what is expected of neuromarketing in the future.

In order to gain initial insights regarding neuromarketing in South East European region, the authors have conducted a small scale qualitative research. The data was collected using in-depth unstructured interviews, as such approach allows the identification of major topics and important themes (Arskey & Knight, 1999). In total, the authors conducted interviews with representatives from three companies operating in marketing industry. One company in the sample represents a neuromarketing consulting agency from Greece. Two participants from this company were interviewed, the director of neuromarketing and strategic communication and the market research and neuromarketing consultant. The second company provides market research services at the global level. The participant from this company holds the position of the regional managing director responsible for Adriatic region. The third company represents a big advertiser operating in the sector of financial services in Serbia, as a member of a regional group. The interviewee is the marketing and communications director in the company. The maximum variation sampling method was used to choose these individuals. This approach allowed the researchers to investigate the variations in the major topics depending on the industry segment (Patton, 1990).

The aim of this research was to start a conversation about neuromarketing that will include all players in marketing industry and acquire opinions and personal observations to complement the existing knowledge. Based on the qualitative data collected, several conclusions have been made.

Neuromarketing has low awareness, but high persuasive effect - As representatives from neuromarketing agency have argued, neuromarketing as a discipline is still experiencing low awareness among the marketing professionals. However, they claim that after marketers are being presented with the benefits and characteristics of neuromarketing, they realize the opportunities it can offer. As all the participants were aware of neuromarketing prior to this discussion, their opinions did not differ; they all believe neuromarketing is providing valuable insights into human behavior, and as such, can bring the consumer and the company closer.

Neuromarketing as a way of thinking and new research technique – The most interesting discussion was held regarding the ultimate impact of neuromarketing. On one hand, there are those who believe that neuromarketing will enrich the pool of market research tools and complement the traditional techniques. On the other hand, there are those who believe that neuromarketing represents a way of thinking that can be implemented even when neuromarketing research tools are not present. Regardless of how marketing professionals look at neuromarketing, they all seem to agree that neuromarketing input, primary or secondary, does make significant difference on the end result.

The most effective way to promote neuromarketing is through case studies – As the participants indicate, there are examples of neuromarketing studies being conducted in the South East European region. The most common ones involve using EEG and eye tracking systems, as well as the analysis of emotions. Even though there are not many, these studies represent the strongest persuasion tactic available at this stage. Nevertheless, all the studies need to be carried out according to strict protocol and thoroughly documented, so that science behind them cannot be disputed.

This small scale qualitative research represents only the first step towards realizing the true significance of neuromarketing among industry professionals in South East European region. Future investigation in this area is necessary and highly anticipated; there are still many questions to be answered. But there is a hope that neuromarketing will upgrade the entire marketing knowledge and will close the gap between the consumers and the companies.

CONCLUSION

This is the age of the brain. Many resources are being dedicated to uncovering the power behind the brain mechanisms and its impact on the world. It can be argued that the twenty first century is the period of brain paradigm. The knowledge from neuroscience is entering all aspects of business and science, from medicine and psychology to urbanism, economics and law. Luckily enough, marketing is not an exception.

Marketing research is changing and marketing professionals can no longer ignore it. Neuroscience tools have already found its application in marketing research and neuroscience knowledge is changing the perception and mechanisms of marketing. However, it is only the transition period right now. This means that the innovative techniques have not reached full confidence and mature usage levels. As such, they should be used in combination with traditional marketing techniques; in doing so, marketing professionals will be able to enrich their data and incrementally accept neuromarketing inputs.

REFERENCES

Aaker, D. A., Stayman, D. M., & Hagerty, M. R. (1986). Warmth in Advertising: Measurement, Impact, and Sequence Effects. *The Journal of Consumer Research, 12*(4), 365–381. doi:10.1086/208524

Afifi, A., & Bergman, R. (2005). *Functional Neuroanatomy: Text and Atlas*. New York, NY: McGraw-Hill Medical.

Ariely, D., & Berns, G. S. (2010). Neuromarketing: The Hope and Hype of Neuroimaging in Business. *Nature Reviews. Neuroscience, 11*(4), 284–292. doi:10.1038/nrn2795 PMID:20197790

Arksey, H., & Knight, P. (1999). *Interviewing for Social Scientists*. London, UK: Sage.

Bandettini, P. A. (2012). Twenty Years of Functional MRI: The Science and the Stories. *NeuroImage, 62*(2), 575–588. doi:10.1016/j.neuroimage.2012.04.026 PMID:22542637

Becker, S. (2013). Why You Cannot Map Attention: A Relational Theory of Attention and Eye Movements. *Australian Psychologist, 48*(6), 389–398. doi:10.1111/ap.12028

Bercea, M. D. (2012). Taking the Pulse of the Market: A Consumer's Decision-Making Algorithm Using Insights from Neuromarketing Research. *Proceedings of the Neuropsychoeconomics Conference* (pp. 42).

Berger, J., & Fitzsimons, G. (2008). Dogs on the Street, Pumas on Your Feet: How Cues in the Environment Influence Product Evaluation and Choice. *Journal of Marketing Research, 45*(1), 1–14. doi:10.1509/jmkr.45.1.1

Blau, J. (2013). Europe's Billion-Euro Brain Research. *Research Technology Management, 56*(3), 8–10.

Bove, J. (2008). What Makes Shoppers Tick? *Display & Design Ideas*, *20*(3), 76–78.

Caffyn, J. (1964). Psychological Laboratory Techniques in Copy Research. *Journal of Advertising Research*, *4*(4), 45–50.

Casarotti, M., Lisi, M., Umiltà, C., & Zorzi, M. (2012). Paying Attention through Eye Movements: A Computational Investigation of the Premotor Theory of Spatial Attention. *Journal of Cognitive Neuroscience*, *24*(7), 1519–1531. doi:10.1162/jocn_a_00231 PMID:22452561

Cohn, J., Ambadar, Z., & Ekman, P. (2006). In J.A. Coan, & J.B. Allen (Eds.), *The Handbook of Emotion Elicitation and Assessment*. New York, NY: Oxford University Press.

Craig, A. W., Loureiro, Y. K., Wood, S., & Vendemia, J. M. (2012). Suspicious Minds: Exploring Neural Processes during Exposure to Deceptive Advertising. *Journal of Marketing Research*, *49*(3), 361–372. doi:10.1509/jmr.09.0007

Craig, S. D., D'Mello, S., Witherspoon, A., & Graesser, A. (2008). Emote Aloud during Learning with Autotutor: Applying The Facial Action Coding System to Cognitive-Affective States During Learning. *Cognition and Emotion*, *22*(5), 777–788. doi:10.1080/02699930701516759

Damasio, A. R. (2005). *Descartes' Error: Emotion, Reason, and the Human Brain*. New York, NY: Penguin Books.

Davis, D. F., & Herr, P. M. (2014). From Bye to Buy: Homophones as a Phonological Route to Priming. *The Journal of Consumer Research*, *40*(6), 1063–1077. doi:10.1086/673960

De Melo, C. M., Kenny, P., & Gratch, J. (2010). Influence of Autonomic Signals on Perception of Emotions in Embodied Agents. *Applied Artificial Intelligence*, *24*(6), 494–509. doi:10.1080/08839514.2010.492159

De Vroey, M. (2009). A Marshall-Walras Divide? A Critical Review of the Prevailing Viewpoints. *History of Political Economy*, *41*(4), 709–736. doi:10.1215/00182702-2009-038

Del Missier, F., Mäntylä, T., Hansson, P., Bruine de Bruin, W., Parker, A. M., & Nilsson, L.-G. (2013). The Multifold Relationship between Memory and Decision Making: An Individual-Differences Study. *Journal of Experimental Psychology. Learning, Memory, and Cognition*, *39*(5), 1344–1364. doi:10.1037/a0032379 PMID:23565790

DellaVigna, S. (2009). Psychology and Economics: Evidence from the Field. *Journal of Economic Literature*, *47*(2), 315–372. doi:10.1257/jel.47.2.315

Detric, P. (2010). Demand Effects on Positive Response Distortion by Police Officer Applicants on the Revised NEO Personality Inventory. *Journal of Personality Assessment*, *92*(5), 410–415. doi:10.1080/00223891.2010.497401 PMID:20706927

Dimofte, C. V. (2010). Implicit Measures of Consumer Cognition: A Review. *Psychology and Marketing*, *27*(10), 921–937. doi:10.1002/mar.20366

Djamasbi, S., Siegel, M., Skorinko, J., & Tullis, T. (2011). Online Viewing and Aesthetic Preferences of Generation Y and the Baby Boom Generation: Testing User Web Site Experience through Eye Tracking. *International Journal of Electronic Commerce*, *15*(4), 121–158. doi:10.2753/JEC1086-4415150404

Ekman, P. (1992). An Argument for Basic Emotions. *Cognition and Emotion*, *6*(3), 169–200. doi:10.1080/02699939208411068

Ekman, P., & Friesen, W. V. (1978). *Facial Action Coding System*. Palo Alto, CA: Consulting Psychologists Press.

Flight, M.H. (2011). Memory: The Foundations of Accurate Recall. *Nature Reviews. Neuroscience*, *12*(7), 372–373. doi:10.1038/nrn3055 PMID:21610731

Fugate, D. L. (2007). Neuromarketing: A Layman's Look at Neuroscience and Its Potential Application to Marketing Practice. *Journal of Consumer Marketing*, *24*(7), 385–394. doi:10.1108/07363760710834807

Gakhal, B., & Senior, C. (2008). Examining the Influence of Fame in the Presence of Beauty: An Electrodermal 'Neuromarketing' Study. *Journal of Consumer Behaviour*, *7*(4/5), 331–341. doi:10.1002/cb.255

Genco, S., Pohlmann, A., & Steidl, P. (2013). *Neuromarketing for Dummies*. Mississauga, ON: John Wiley & Sons Canada.

Gregg, A. P., & Klymowsky, J. (2013). The Implicit Association Test in Market Research: Potentials and Pitfalls. *Psychology and Marketing*, *30*(7), 588–601. doi:10.1002/mar.20630

Gregg, A. P., Klymowsky, J., Owens, D., & Perryman, A. (2013). Let Their Fingers Do The Talking? Using The Implicit Association Test in Market Research. *International Journal of Market Research*, *55*(4), 2–14. doi:10.2501/IJMR-2013-013

Hensel, J. (1970). *Physiological Measures of Advertising Effectiveness: A Theoretical and Empirical Investigation* [Doctoral Dissertation]. Ohio State University, School of Business. Retrieved from https://etd.ohiolink.edu/

Javor, A., Koller, M., Lee, N., Chamberlain, L., & Ransmayr, G. (2013). Neuromarketing and Consumer Neuroscience: Contributions To Neurology. *BMC Neurology*, *13*(13), 1–12. PMID:23383650

Jefferies, E., Frankish, C., & Noble, K. (2009). Lexical Coherence in Short-Term Memory: Strategic Reconstruction or "Semantic Glue"? *Quarterly Journal of Experimental Psychology*, *62*(10), 1967–1982. doi:10.1080/17470210802697672 PMID:19255945

Kenning, P., Plassmann, H., & Ahlert, D. (2007). Applications of Functional Magnetic Resonance Imaging for Market Research. *Qualitative Market Research*, *2*(2), 135–152. doi:10.1108/13522750710740817

Kroeber-Riel, W. (1979). Activation Research: Psychobiological Approaches in Consumer Research. *The Journal of Consumer Research*, *5*(4), 240–250. doi:10.1086/208736

LaBarbera, P. A., & Tucciarone, J. D. (1995). GSR Reconsidered: A Behavior-Based Approach to Evaluating and Improving the Sales Potency of Advertising. *Journal of Advertising Research*, *35*(5), 33–53.

Lajante, M., Droulers, O., Dondaine, T., & Amarantini, D. (2012). Opening the "Black Box" of Electrodermal Activity in Consumer Neuroscience Research. *Journal of Neuroscience, Psychology, and Economics*, *5*(4), 238–249. doi:10.1037/a0030680

Lee, L., Amir, O., & Ariely, D. (2009). In Search of Homo Economicus: Cognitive Noise and the Role Of Emotion In Preference Consistency. *The Journal of Consumer Research*, *36*(2), 173–187. doi:10.1086/597160

Martin, W. C., Engelland, B. T., & Collier, J. E. (2011). Assessing the Impact of Acquiescence Response Bias on Marketing Data. *Marketing Management Journal*, *21*(1), 31–46.

Mast, F. W., & Zaltman, G. (2005). A Behavioral Window on the Mind of the Market: An Application of the Response Time Paradigm. *Brain Research Bulletin*, *67*(5), 422–427. doi:10.1016/j.brainresbull.2005.06.004 PMID:16216689

Mauri, M., Cipresso, P., Balgera, A., Villamira, M., & Riva, G. (2011). Why is Facebook so successful? Psychophysiological Measures Describe A Core Flow State While Using Facebook. *Cyberpsychology, Behavior, and Social Networking*, *14*(12), 723–731. doi:10.1089/cyber.2010.0377 PMID:21879884

McClure, S. M., Li, J., Tomlin, D., Cypert, K. S., Montague, L. M., & Montague, R. P. (2004). Neural Correlates Of Behavioral Preference For Culturally Familiar Drinks. *Neuron*, *44*(2), 379–387. doi:10.1016/j.neuron.2004.09.019 PMID:15473974

Morgan, M. S. (2006). Economic Man as Model Man: Ideal Types, Idealization And Caricatures. *Journal of the History of Economic Thought*, *28*(1), 1–27. doi:10.1080/10427710500509763

Morin, C. (2011). Neuromarketing: The New Science of Consumer Behavior. *Symposium: Consumer Culture in Global Perspective.*

Nancarrow, C., Brace, I., & Wright, L. (2001). Tell Me Lies, Tell Me Sweet Little Lies: Dealing With Socially Desirable Responses In Market Research. *Marketing Review*, *2*(1), 55–55. doi:10.1362/1469347012569427

Narvainen, J., Pulkkinen, K., Kuoppa, P., Tarvainen, M., Lankinen, M., & Lapveteläinen, A., & ... Karhune, L. (2013). Probing Psychophysiology by Pictorial Stimuli of Emotional Content: General Patterns and Personality Effects. *Neuropsychoeconomics Conference Proceedings* (p. 25).

Ohme, R. (2009). Special Topics Session: How Neurophysiological Research Contributes to Advertising Research. *American Academy Of Advertising Conference Proceedings* (pp. 55-57).

Ohme, R., Kowalski, J., & Reykowska, D. (2007). Exploring Consumers' Mind Via EEG And EMG Activity. *Neuropsychoeconomics Conference Proceedings* (p. 26).

Ohme, R., Matukin, M., & Pacula-Lesniak, B. (2011). Biometric Measures For Interactive Advertising Research. *Journal of Interactive Advertising*, *11*(2), 60–72. doi:10.1080/15252019.2011.10722185

Ohme, R., Reykowska, D., Wiener, D., & Choromanska, A. (2010). Application of Frontal EEG Asymmetry To Advertising Research. *Journal of Economic Psychology*, *31*(5), 785–793. doi:10.1016/j.joep.2010.03.008

Parens, E., & Johnston, J. (2014). Neuroimaging: Beginning to Appreciate its Complexities. *The Hastings Center Report*, *45*(2), s2–s7. doi:10.1002/hast.293 PMID:24634082

Patton, M. (1990). *Qualitative Evaluation and Research Methods*. London, UK: Sage.

Philippot, P., Chapelle, G., & Blairy, S. (2002). Respiratory Feedback in the Generation of Emotion. *Cognition and Emotion*, *16*(5), 605–627. doi:10.1080/02699930143000392

Pizzagalli, D. A., Sherwood, R. J., Henriques, J. B. & Davidson, R. J. (2005). Frontal Brain Asymmetry and Reward Responsiveness. *Psychological Science (Wiley-Blackwell), 16*(10), 805-813.

Rasch, C., & Teichert, T. (2013). Assessing Consumers' Preferences with Facial Electromyography - Insights from a Discrete Choice Experiment. *Neuropsychoeconomics Conference Proceedings*, 54.

Ratnayake, N., Broderick, A. J., & Mitchell, R. C. (2010). A Neurocognitive Approach to Brand Memory. *Journal of Marketing Management*, *26*(13-14), 1295–1318. doi:10.1080/0267257X.2010.522531

Ravaja, N., Somervuori, O., & Salminen, M. (2013). Predicting Purchase Decision: The Role of Hemispheric Asymmetry over the Frontal Cortex. *Journal of Neuroscience, Psychology, and Economics*, *6*(1), 1–13. doi:10.1037/a0029949

Reardon, S. (2014). Brain-Mapping Projects to Join Forces. *Scientific American*. Retrieved from http://www.scientificamerican.com/article/brain-mapping-projects-to-join-forces/

Rubinson, J. (2010). What Behavioral Economics Can Teach Marketing Research. *Journal of Advertising Research*, *50*(2), 114–117. doi:10.2501/S0021849910091269

Sammler, D., Grigutsch, M., Fritz, T., & Koelsch, S. (2007). Music and Emotion: Electrophysiological Correlates of the Processing of Pleasant and Unpleasant Music. *Psychophysiology*, *44*(2), 293–304. doi:10.1111/j.1469-8986.2007.00497.x PMID:17343712

Silberstein, R. B., & Nield, G. E. (2008). Brain Activity Correlates of Consumer Brand Choice Shift Associated with Television Advertising. *International Journal of Advertising*, *27*(3), 359–380. doi:10.2501/S0265048708080025

Singh, S. N., & Churchill, G. A. Jr. (1986). Using the Theory of Signal Detection to Improve Ad Recognition Testing. *JMR, Journal of Marketing Research*, *23*(4), 327–336. doi:10.2307/3151808

Steinhauer, S. R., Siegle, G. J., Condray, R., & Pless, M. (2004). Sympathetic and Parasympathetic Innervation of Pupillary Dilation during Sustained Processing. *International Journal of Psychophysiology*, *52*(1), 77–86. doi:10.1016/j.ijpsycho.2003.12.005 PMID:15003374

Sundararajan, M. (2010). Physiological Emotions and Entrepreneurial Decisions. *Global Business and Management Research*, *2*(4), 310–322.

Wolf, L., & Morrissey, S. (2013). Brain Initiative Unveiled. *Chemical and Engineering News*, *91*(14), 9–9.

Yi, Y. (1990). Cognitive and Affective Priming Effects of the Context for Print Advertisements. *Journal of Advertising*, *19*(2), 40–48. doi:10.1080/00913367.1990.10673186

ADDITIONAL READING

Achrol, R., & Kotler, P. (2012). Frontiers of the Marketing Paradigm in the Third Millennium. *Journal of the Academy of Marketing Science*, *40*(1), 35–52. doi:10.1007/s11747-011-0255-4

Ariely, D. (2008). *Predictably Irrational: The Hidden Forces That Shape Our Decisions*. New York, NY: HarperCollins.

Ariely, D. (2009). The End of Rational Economics. *Harvard Business Review*, *87*(7/8), 78–84. PMID:19968059

Bargh, J. A. (2005). Bypassing the Will: Toward Demystifying the Nonconscious Control of Social Behavior. In R. R. Hassin, J. S. Uleman, & J. A. Bargh (Eds.), *The New Unconscious* (pp. 37–61). New York, NY: Oxford University Press.

Berkman, E. T., & Lieberman, M. D. (2011). What's Outside the Black Box?: The Status of Behavioral Outcomes in Neuroscience Research. *Psychological Inquiry*, *22*(2), 100–107. doi:10.1080/1047840X.2011.550182 PMID:21984865

Causse, M., Sénard, J.-M., Démonet, J. F., & Pastor, J. (2010). Monitoring Cognitive and Emotional Processes through Pupil and Cardiac Response during Dynamic versus Logical Task. *Applied Psychophysiology and Biofeedback*, *35*(2), 115–123. doi:10.1007/s10484-009-9115-0 PMID:19816770

Du Plessis, E. (2011). *The Branded Mind: What Neuroscience Really Tells Us about the Puzzle of the Brain and the Brand*. London, UK: Kogan Page Limited.

Eser, Z., Isin, F., & Tolon, M. (2011). Perceptions of Marketing Academics, Neurologists, and Marketing Professionals about Neuromarketing. *Journal of Marketing Management*, *27*(7/8), 854–868. doi:10.1080/02672571003719070

Fitzsimons, G. M., Chartrand, T. L., & Fitzsimons, G. J. (2008). Automatic Effects of Brand Exposure on Motivated Behavior: How Apple Makes You "Think Different". *The Journal of Consumer Research*, *35*(1), 21–35. doi:10.1086/527269

Fortunato, V. C. R., Giraldi, J. D. M. E., & De Oliveira, J. H. C. (2014). A Review of Studies on Neuromarketing: Practical Results, Techniques and Limitations. *Journal of Management Research*, *6*(2), 201–220. doi:10.5296/jmr.v6i2.5446

Genco, S., Pohlmann, A., & Steidl, P. (2013). *Neuromarketing for Dummies*. Mississauga, ON: John Wiley & Sons Canada.

Gladwell, M. (2007). *Blink: The Power of Thinking without Thinking*. New York, NY: Little, Brown and Company.

Kahneman, D. (2011). *Thinking, Fast and Slow*. New York, NY: Farrar, Straus, and Giroux.

Kim, K., Hayes, J. L., Avant, J., & Reid, L. N. (2014). Trends in Advertising Research: A Longitudinal Analysis of Leading Advertising, Marketing, and Communication Journals, 1980 to 2010. *Journal of Advertising*, *43*(3), 296–316. doi:10.1080/00913367.2013.857620

Lindstrom, M. (2008). *Buy-ology*. New York, NY: Doubleday.

Maison, D., Greenwald, A. G., & Bruin, R. H. (2004). Predictive Validity of the Implicit Association Test in Studies of Brands, Consumer Attitudes, and Behavior. *Journal of Consumer Psychology*, *14*(4), 405–415. doi:10.1207/s15327663jcp1404_9

Micu, A., & Plummer, J. T. (2010). Measurable Emotions: How Television Ads Really Work. *Journal of Advertising Research*, *50*(2), 137–153. doi:10.2501/S0021849910091300

Plassmann, H., Ramsoy, T. Z., & Milosavljevic, M. (2011). Branding the Brain: A Critical Review and Outlook. *Journal of Consumer Psychology*, 1–19.

Pradeer, A. K. (2010). *The Buying Brain: Secrets for Selling to the Subconscious Mind*. Hoboken, NJ: John Wiley & Sons.

Sinha, P., Sharma, A., & Singh, H. (2012). Prediction for the 2012 United States Presidential Election Using Multiple Regression Model. *Journal of Prediction Markets*, *6*(2), 77–97.

Stoll, M., Baecke, S., & Kenning, P. (2008). What They See is What They Get? An fMRI Study on Neural Correlates of Attractive Packaging. *Journal of Consumer Behaviour*, *7*(4-5), 342–359. doi:10.1002/cb.256

KEY TERMS AND DEFINITIONS

Attitude: Evaluation of objects, people, situation, etc., that has significant influence on one's behavior. Attitudes can be either positive or negative, present for a long time or changeable by present circumstances and experiences.

Behavioral Economics: Interdisciplinary field that studies how emotions, social environment, and psychological factors influence economic decisions.

Bias: Strong opinion about certain people or objects, even when contrary evidence is present. They are formed as a result of judgment heuristics, the shortcuts created by brain in making decisions.

Neuromarketing: Emerging independent discipline that studies actual behavior of consumers.

Social Psychology: Scientific field that studies how social environment influences individual's behavior and vice versa.

This research was previously published in Trends and Innovations in Marketing Information Systems edited by Theodosios Tsiakis, pages 337-357, copyright year 2015 by Business Science Reference (an imprint of IGI Global).

Chapter 18
Neuroscience Applications on the Assessments of TV Ads

Tuna Çakar
Acıbadem University, Turkey

Kaan Gez
Anatolian University, Turkey

ABSTRACT

The progress in neurotechnologies has enabled a potentially better and cheaper analysis for the neural signals not limited to medical applications but influencing several fields from marketing to economics and law to ethics. Since the main targets have been to understand the brain mechanisms better as well as providing useful applications specifically regarding the sector-specific interest, one related application has been about the assessments of TV ads as a complementary and more objective tool than traditional methods that rely on the verbal self-reports and interviews that could be speculative and misleading depending on the given context. For assessing several TV ads within a shorter duration, the use of neuroscientific methods has attracted much interest. This chapter will focus on the current practices with the given constructs for the TV ad research specifically in relation to the practices such as attention, emotional engagement, individual preference, and market success.

INTRODUCTION

Advertising is the structured and composed impersonal communication of information, usually paid for and usually persuasive in nature, about products (goods, services and ideas) by identified sponsors through various media (Arens, 2006). Advertisements are generally presented through media channels such as magazines, newspapers, the Internet, radio, outdoor, and television. Although advertisements on television have lost their influence to some extent, it is undeniable that these types of advertisement still continue to have an impact on the general marketing strategies of companies. Marketing research is the planning, collection, and analysis of data relevant to marketing decision-making and the communication of the results of this analysis to management (McDaniel & Gates, 2013). Advertising as a discipline in social science also uses research methods familiar to social sciences. As such methods, either qualitative

DOI: 10.4018/978-1-5225-5478-3.ch018

or quantitative, depend on statements, a subject may give misinformation intentionally or unintentionally related to various factors. In this context, marketing research professionals have demanded supplementary methods that preferably rely on objective tools such as neuroscientific measurement tools.

Therefore, applied neuroscience especially focuses on neurophysiological and biometric responses of the body as a complementary tool for advertising research. Neuromarketing research mainly aims to associate the activities within the neural system (brain and the whole body) with consumer behavior (Hubert, 2008). Even though data collection via neuroimaging seems closer to quantitative approaches, neuromarketing research, including biometric research, shows that it also possesses certain features in common with qualitative methods. In terms of advertising research and assessment, besides bearing a resemblance to quantitative and qualitative approaches in traditional methods, this application-based form of neuroscience is a relatively new approach that has gradually become prominent. Marketing research methods are traditionally based on several quantitative and qualitative methods by which the introspective and verbal outputs are confronted. Despite the fact that these methods are frequently used for sector-specific applications, there has been a trend towards the use of more objective methods, especially in the last decade (Ariely & Berns, 2010). The main motivation has arisen in relation to the need for accessing hidden information that could be described as unconscious information that is even not penetrable by the individual herself (McClure et al., 2004; Dijksterhuis 2004). The second motivation has been related to the possibility of providing cheaper and quicker settings for the assessments so that they could be automatized and serviced at a speed able to would satisfy the sector-specific demand (Ariely & Berns, 2010). These initial motivations resulted in the establishment of more than 300 companies by which different applications have been developed for sector-specific use within the application domain named as neuromarketing. This interdisciplinary field is at the intersection of various fields including marketing, economics, psychology, and neuroscience. There are many applications developed regarding the scientific findings in the academic literature.

One such application based on the use of these neuroscientific methods has been developed for the assessment of TV advertisements. The assessment of TV ads poses a significant challenge, since they are composed of dynamical and complicated bundles of information packages that are influenced by the affect of brand image, the content of the ad and emotional as well as cognitive impact during the ad. These neural assessments are currently used as complementary methods in the sense that the TV ads are examined via the use of neuro-application before they are released to the market. Or they are reprocessed with the deliberate evaluation of the TV ad in the process of optimization. The most widely used applications such as the detection of attention, emotional engagement, and the stress level of the participants during the watching a TV ad (Ambler et al., 2000). The pros and cons of these measurements will be addressed in the second section of this chapter. On the other hand, the most frequently used methods are Electroencephalography (EEG/ERP), functional Magnetic Resonance Imaging (fMRI), Positron Emission Tomography (PET), functional Near Infrared Spectroscopy (fNIRS), Eye-tracking (E-T), Facial Coding (F-C), Skin Conductance Resistance (SCR) and Heart Rate (HR). These are considered as the main measurement tools derived or directly adopted from medical devices that have been shown to provide enough scientific reliability for such applications. There are several reasons for the increasing demand and the use of these methods. These will be briefly explained in the first part of this chapter. Interestingly enough, the academic studies on various subdomains shed light on the evaluation processes of TV ads, revealing that cultural differences appear to be highly significant regarding the ad content (Vecchiato et al., 2011). Different segments of a TV ad could be assessed via the different objective methods (Ohme et al., 2008). The market success of the movies as well as the individual preferences related to these could be predicted via the assessment of their fragments (Boksem & Smidts, 2015; Venkatraman et al., 2015). This chapter will cover the main academic studies and findings that suggest a potential sector-specific use. Meanwhile, the need for an increase in academic studies supporting the current approaches is also indicated in order to achieve a higher level of scientific reliability in the middle and long-terms. Thus, this chapter includes the review of the current empirical literature on the assessment of TV ads as well as providing the necessary comparative analysis in relation to the traditional research methods. In other words, this chapter aims to provide a critical bridge between the academic studies on one side and the sector-specific applications on the other, as well as underlining the most frequent uses of these methods for the assessments of TV ads. More interestingly, the debatable issues and future prospects regarding the research agenda will be considered in the last part of this chapter.

BACKGROUND

The first advertisements appeared in the commercial activities since of the early civilizations including the Ancient Egyptian, Ancient Greek, and Ancient Romean. The noisy advertisers in the *souks* performed every kind of marketing activity. The first written records of advertising practice date from Ancient Egypt (Yilmaz, 2014). Until the industrial revolution, the main criterion related to marketing activities was limited to the amount of sales generated. After the industrial revolution, the sales-oriented marketing approach was replaced by the customer-oriented approach. In addition, new domains for advertisements such as television were introduced. The first TV ad consisted of a spot announcement accompanied by the image of watch on July 1, 1941. During 1960s, the creative stage of ad production has appeared and by 1980s, advertising involved an integrative process across different disciplines. During recent decades, social media have provided a huge and alternative platform for new advertisement strategies.

The concept of advertisement has undergone considerable transformations regarding its form and structure, especially in the last decade. Advertisements can be defined as strategic ways of convincing the potential clients towards a definite product or service through the use of mass communication tools by an identified agent (Arens, 2008). Printed advertisements such as newspaper ads target the visual system, whereas radio ads target the auditory system-and TV ads stimulate interaction through both of these modalities. Beyond these features, TV ads are tested regarding the basic constructs such as attention, desirability, memory and recognition, since these require high amount of investment to be produced and released on television (Belch & Belch, 2003). TV ads basically serve to convince potential clients of the value of the product or service concerned (Richards & Curran, 2002).

Research on TV commercials has become one of the most controversial issues, since they include a process of artistic creation, marketing strategy and technical components (Belch & Belch, 2003). Moreover, the use of qualitative and quantitative marketing methods is considered to provide a more scientific dimension in the sense that the analyses could be performed via more objective methods. On the other hand, advertising research is inevitably interdisciplinary, being related to various different fields including marketing, psychology, sociology, anthropology and statistics. These components tended to increase the complexity and uncertainty of advertising research. Moreover, advertising campaigns are expensive for the companies and there is a steady increase during recent decades that has been projected to continue in the next three years (See Figures 1 and 2). This increase indicates the importance of more efficient advertising strategies.

Figure 1. Distribution of advertising spending worldwide in 2013 and 2018, by medium
Source: Statista, 2015

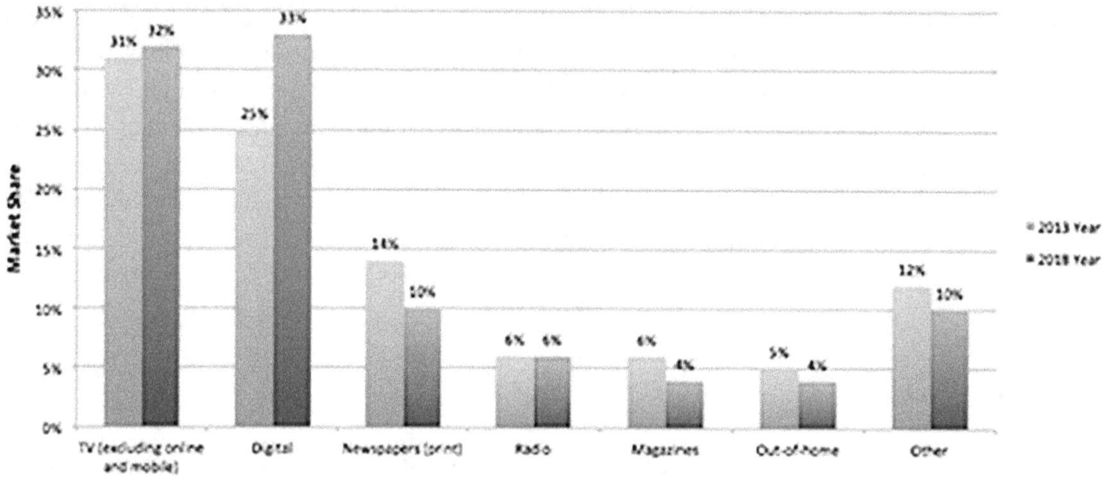

Figure 2. Total media ad spending worldwide, 2014-2019 (billion $)
Source: eMarketer, Sep 2015

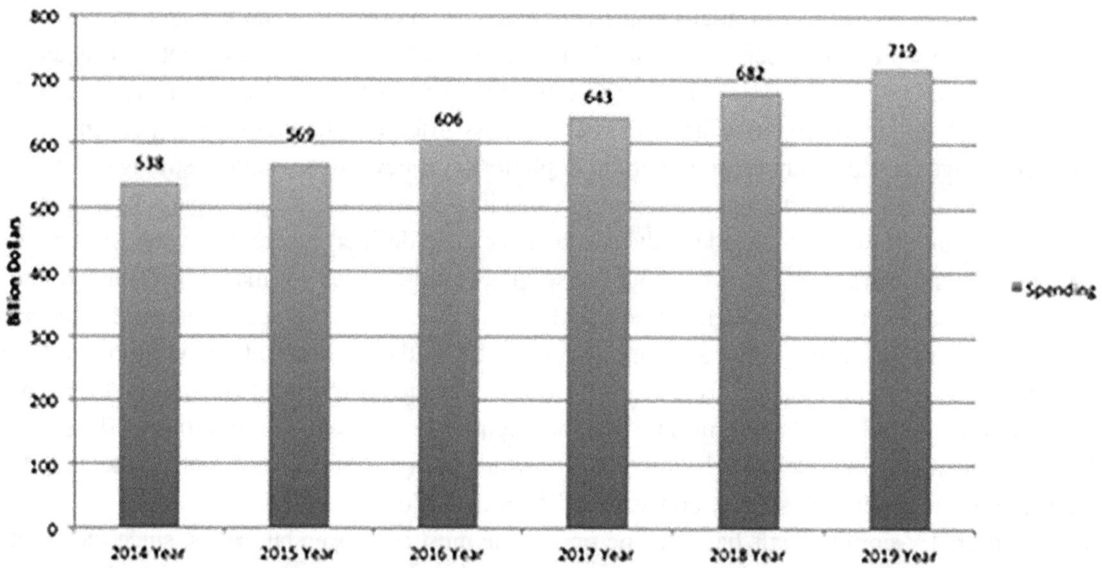

TV ads could positively be assessed based on features such as selectivity, impact, creativity, prestige and social dominance. However, there are also negative aspects such as high production cost (up to one million dollar budgets), high airtime cost, limited selectivity, brevity, and clutter. Moreover, the long-term influence of TV ads is another critical issue mainly due to the short display duration., These ads are presented several times in order to compensate for their short duration but they must not bore the target audience. Thus, there are several pre-test and post-test phases before and after the release of a TV ad in order to assess its effectiveness. Until the early 2000s, traditional qualitative and quantitative methods were practiced intensively in spite of their limitations. However, the limitations of these traditional methods paved the way for the development of more objective and reliable methods especially for, but not limited to, TV ad assessments (Ariely & Berns, 2010). In this respect, the potential use of neuroscientific methods, namely neuromarketing, has caused an increasing interest on the part of academics and marketing professionals. Thus, the major issue has become to provide new methods that are reliable and potentially more objective, especially regarding the sector-specific demand from the companies – including multinationals– for assessing TV ads and other domains of marketing (Morin, 2011). Neuromarketing has been accepted as an applied field of neuroeconomics in which the neuroscientific methods as well as the paradigms are applied to marketing stimuli to test their success in a specific dimension or its overall effectiveness (Lee et al., 2007; Fugate, 2007).

To sum up, TV ads are still crucial communication means for presenting products and services to potential clients. Despite this fact, there are also potentially negative aspects, mainly referring to their high production and airtime costs as well as the increasing dominance of social media. Thus, the TV ads need to be produced very effectively and efficiently. Research methods developed for the assessments of TV ads have thus acquired a higher level of importance. The next part of this chapter covers the traditional methods before introducing, the widely used neuroscientific methods for the assessment of TV ads. Following that is a description of common measures and practices among the sector-specific and academic studies.

Traditional Methods in Advertising Research

The traditional methods used in advertising research mainly rely on the qualitative and quantitative methods that have widely been used in social sciences. The qualitative methods generally aim to reflect the emotions, thoughts, and attitudes of the target population towards the presented TV ad but they carry the risk of being speculative. On the other hand, the quantitative methods focus on acquiring data through questionnaires that have very strict guidelines for data acquisition and data analysis procedures. There are various measures developed for the evaluation of advertising effectiveness that are generally referred to as ad-related measures such as liking, familiarity, relevance and informativeness. This part introduces the most common methods used during the research process (Patton, 2014).

Focus Groups

These are qualitative research methods utilizing groups composed of 6 to 8 members from a specific socioeconomic segment and guided by an educated specialist as the moderator. These conversations might take from one to three hours. The members are encouraged to declare their opinions about the presented TV ads freely. However, there is the issue of a potentially dominant view among the group members and the members with the counter views might tend to avoid declaring opinions contradicting the dominant view. Thus, there appears the potential problem for generalizability of these "common" opinions (Rubin & Babbie, 2013). On the other hand, these conversations might highly fruitful and creative sessions that provide enriching useful (*or rich*) information, especially about developing novel concepts and ideas for future marketing strategies in relation to the in-group dynamics and interactions. Despite the fact that the focus group method seems to differ from the other methods, it benefits from the deeper level of information obtained, which could be used as guidance for the questionnaires that will be developed (Morgan, 1997). The focus group method requires careful planning of the research aim, the development of questions, the participant profile and their invitation before the actual work of data collection. The workload is even more difficult after the focus group, including the allocation, encryption and analysis of the obtained data (Kruger & Casey, 2014). Since there are no strict rules or guidelines provided for focus groups, this relative flexibility enables fruitful discussions. The participants might be encouraged to think freely during the discussions (Patton, 2014).

In-Depth Interviews

Interviews are one of the most common methods for collecting data that could be directed to a specific subpopulation. There are three types of interviews in the academic literature:

1. Informal conservational interviews,
2. General interview guide approach, and
3. Standardized open-ended interviews.

These interviews are performed one-by-one with the members of the target group, and thus the potential social pressure observed in the focus groups can be eliminated. These interviews are generally performed within a fully- or semi-structured fashion depending on the desired framework. The fully-structured interviews do not allow the interviewers to ask questions different from the prepared questions whereas the semi-structured interviews give space for additional questions. Despite the fact that this method provides a valuable source for TV assessments, it is expensive and time-consuming, which is a crucial issue in cases when the advertisement must be released or optimized urgently. The strong sides of the in-depth interviews include the relative flexibility of the respondents, their nonverbal behavior, the quick responses and the control over the environment. Moreover, the behavior of the participants, including the eye movements and facial expressions can be recorded and analyzed as supportive tools for the encryption of the verbal statements. Since the interviewer actively participates in this process, she can ensure that the full information is gathered. Moreover, the interviewer can use different means of communication during the data collection, including graphs or other illustrations in order to provide a richer framework for the participants. However, it is a quite vulnerable issue that the participants could easily be manipulated even by the order of the questions, social status or the facial expressions of the interviewer. Moreover, in-depth interviews also have the problem of generalizability as in the focus group research method. This method has also the negative aspects such as the length of the procedure, the lack of privacy, the absence of standardization in terms of questions and the potential influence of the interviewers. There are also crucial drawbacks that might occur during the process of data collection and analyses. The interviewers might derive inappropriate and biased conclusions that are not consistent with the statements of the participant.

Questionnaires

Questionnaires have been the most widely used quantitative method specifically for the TV ad assessments. The questionnaires are structured and prepared in very high detail in order to infer the potential attitude and behavior of the target group towards a given case such as a TV ad that will be released soon (Stack, 2010). It is highly crucial to set the sample from the target population correctly as well as to determine how the questions will be prepared (Gut & Marsh, 2011). First of all, the questions should be easily understandable by the target group. The questionnaires have been practiced through different means like face-to-face, telephone, and mail depending on the physical and time limitations. Nowadays, Internet communication such as email has been preferred due to its practicality and flexibility (Rubin & Babbie, 2013). The questionnaires are also considered to be highly practical regarding the integration of computer technologies that very quickly enable statistical analysis using software packages. Thus, it is easier to manage large numbers of responses and promptly produce the final output of the research. Despite the fact that the questionnaires are highly preferred and practiced during advertising research, there are several drawbacks, such that the participants could be bored while responding to the test items that might affect their responses. Moreover, the participants need to rely on their introspection during the evaluation of the presented items and weight these items with respect to the given response scale. Additionally, there is the critical issue of manipulation by the coordinator or even by the order of the presented questions. Thus, the design of the test items is one of the highly significant issues to which careful attention must be given. Otherwise, the research findings inevitably face the potential problems related to reliability and validity. Moreover, the process of preparation of the questionnaires has been criticized to be artificial and superficial with not being able to refer to social processes but limited to introspective deliberations of the participants. Among all of these positive and negative aspects, it is highly possible that this method will still be used in the long-term due to its practicality and cost-effectiveness.

Neurophysiological and Biometric Methods

The increasing use of neuroscientific methods has initiated a long-lasting effect on various fields including economics, ethics and marketing. The application of these neuroscientific methods have directly been adapted from the neuroscience literature thus they provide scientific validity and reliability as research framework (Hammou et al., 2013). The methods are generally categorized into two main groups with respect to their data-source: whether they can obtain data directly from the brain such as analyzing the brainwaves (EEG/ERP) or they can have indirect measures of the brain activations such as skin conductance resistance (SCR) or eye-movements. The first one will be referred as neurophysiological methods and the second one as biometric methods throughout this chapter. In this part, the most common methods used for TV ad assessments will be explained very briefly.

The prices of the necessary devices for using these methods highly vary depending on their technical specifications and limitations. For instance, an EEG/ERP device can be found with a price of 100 USD up to more than 100.000 USD with respect to its sampling frequency, channel number and the software package for data acquisition and data analysis. The prices generally tended to decrease in the last decade, since the use of open source (free software) is preferred as the possible integration with cheaper devices has become possible. For instance, there are eye-tracking devices with the prices down to 100 USD with no software package developed for high level of data analyses. However, the users might integrate their own software and reduce the overall cost. Thus, the new trend enables the researchers and the companies to acquire these devices much more easily than before. However, there appears the issue of developing software for the data analyses including the artifact removal steps. It is a highly crucial fact to remove the artifact components from the raw data in order to provide a set of reliable and valid data analyses. Otherwise, it might result with misleading outputs that could include excessive amount of internal and external noise. Occasionally, it has been a useful idea to set up short control experiments just before each of the experimental setups in order to validate how clean and usable the acquired data is. One could find more detailed information about the technical specifications of these methods in Chapter XX in this book.

Main Empirical Findings and Neural Constructs

There are various neurophysiological methods used during the advertisement assessments through the use of neurophysiological and biometric methods. In this part, the best-known and accepted practices will be introduced with a special emphasis on the relevant methods and very brief technical explanations. The well-known practices include the measurements of attention, cognitive engagement (interest), emotional engagement (affect), memory, and market success. The most common methods used are EEG/ERP, eye-tracker, fMRI, SCR/GSR and HR. There are also supplementary methods such as MEG that is highly expensive and relatively new technologies such as fNIRS. The specific use of one of these methods generally depends on the purpose and the budget of these projects. Accordingly, there are two main domains of research with respect to the aim of the projects. First, there are academic studies that are performed mainly for the reason to understand and to demonstrate the functioning one of the mechanisms related to advertising research. These research-based approaches generally involve controlled experiments. There are several institutions including Advertising Research Foundation (ARF) that provide funding for this kind of fundamental research. This type of academic research is generally referred to as consumer neuroscience that covers a relatively huge domain of applications from logo-tests to taste buds. Secondly, there are the sector-specific studies (or semi-academic studies) that rely on the academic findings but mainly interested in performing a research project in order to fulfill the commercial request of one of its clients. The client may ask for the neuroassessment of one of their TV ads while providing comparison with the TV ads of one or two of the competitors. This second one is the direct exemplification of the applied neuroscience on business practice and is specifically called neuromarketing in this text.

Attention Level

The attention level of the viewers has been considered to be a significant indicator for the success of a TV ad, since attention has been defined as the selective cognitive process by which the agent focuses on a particular aspect of the environment while ignoring the rest. The attention level of the viewers has been investigated through several experimental studies via the use of eye-tracker, EEG/ERP, Heart Rate (Deceleration), and Respiration Index (Milosavljevic et al., 2012; Wedel & Pieters, 2008; Foxe & Snyder, 2011; Lang et al., 1999). The studies via eye-tracker has increased especially in the last decade mainly because of the reason that the results are generally robust, easy to interpret and the commercial software packages enable the researcher to perform the analyses relatively easily. Since the outputs of the eye-tracker demonstrate the areas that the participants looked at, it is relatively common and easier to inspect through the looking patterns of the participants during watching a specific TV commercial for instance. However, the results of an eye-tracking study are evaluated to be limited and should be supported with the other methods in order to have inspection for the direction and intensity of affect of the viewers. Rather these E-T outputs including fixation counts and dwell time are beneficial to provide quantitative information as independent indicators of the attention levels of the participants (Milosavljevic et al., 2012). There is a considerable consensus on the view that the higher fixation counts and the longer dwell time refers to the higher level of attention by the participant (Wedel & Pieters, 2008). For instance, a TV ad containing a specific scene with a target product could be found to be ineffective, since none of the participants looked at this product of interest at that specific scene. Or a celebrity presenting a novel product might be suspected to have a negative impact on the product because of the possible fact that she could take more attention than the product itself that could be argued relying on the eye-tracking results. Thus, potentially valuable and useful information for the sector-specific projects that might not possibly be scientifically interesting could be obtained via the use of eye-tracker outputs. On the other hand, there are empirical studies with EEG/ERP showing that low alpha oscillatory activity mainly among the occipital lobes is correlated with the attention level of the participants regarding their roles in suppressing the distractors that acquired a critical role in selective attention (Klimesch, 1999; Foxe & Snyder, 2011; Vecchiato et al., 2011). To measure the attention level of a participant while watching a TV ad through the use of EEG/ERP method, it is necessary to cut the whole ad into segments of 100 milliseconds and calculate the power spectral density of the low alpha band for each temporal segment. The predefined segments that have excessive artifacts might be replaced by the average values of the whole TV ad. If there are too many segments labeled as artifact (more than a threshold level, for instance 30% of the whole commercial) then it might be a better idea to remove this data set from the further steps of analyses. It is also significant to consider the individual differences regarding the oscillatory activity.

Interest Level (or Cognitive Engagement)

One of the potentially useful information is related to understanding the interest level of the ad viewers, since the interest level could be correlated with the purchasing tendency on one hand and the suitableness of the brand-product in general on the other. The empirical findings in the relevant academic literature have indicated that the increase in theta activity specifically from the frontal lobes (mainly F3, Fz, and F4) could be used as "a signature of successful working memory manipulation" (Itthipuripat et al., 2012). This increase among the frontal lobes has potentially been associated with the consumers' interest levels depending on their profile, being a green (GR) consumer (or not) in a recent EEG study by Lee et al. (2014). The participants composed of GR and Non-GR consumers were asked to view TV advertisements of environment-friendly and non-environment-friend products. According to their results, the ad messages related to green products cause higher frontal theta activations for GR consumers than Non-GR consumers. This finding by Lee et al. (2014) has been reported as a potential neural indicator of cognitive engagement as fulfilling one's environment-friendly life view. This potential effect might not be limited to environment-friendly aspect of the products but rather from a broader perspective might be related to the products or services fulfilling one's personal values (as in this case) as opposed to the unsuitable ones. For a set of brand-loyal customers regarding the Brand A, a similar activation in neural patterns might be expected to occur as opposed to the ad of its main competitor, Brand B. Thus, the theta oscillations during the view of a TV ad might reflect the cognitive aspects of the viewer for the presented products, specifically in relation to the personal values. This has been a quite interesting finding that needs to be further investigated through many more experimental studies. However, the possible impacts of personal values on products should not be limited within the scope of cognitive processing. For instance, a person actively protesting the products of fur will not only have decline in the cognitive engagement with respect to her interest but could be argued to be affected highly on the emotional side rather than cognitive side. Therefore, it is also important to understand the intertwined interaction between emotion and cognition.

Emotion, Affect, and Positiveness

Measuring the affect of the viewers in terms of a specific TV ad has attracted attention for two reasons. First, it has been of interest for marketing professionals and researchers to know how a TV ad will be perceived that could provide insight for the positivity of the target group. Secondly, it has been significant to provide scene-by-scene (moment-by-moment) analyses for the affect of the viewers while watching a TV ad. This is especially important to grasp the specific scenes causing significant decline as well as increase in emotional engagement. Affect has mainly been evaluated within two dimensions:

1. Valence, and
2. Arousal.

Thus, it has been crucial to understand the emotional tendency of their potential customers as well as having an objective tool for a possible optimization process of the whole ad (Silberstein, & Nield, 2012).

There are several different techniques that have been actively used for detecting affect level as a part of research on TV ads. Implicit Association Task (IAT) has been accepted as a reliable indicator of affect in the sense that people tend to respond more quickly to the brand-adjective couples that they agree with as opposed to disagreed ones. Despite the fact that there is considerable degree of individual differences, the normalization protocols adjust the obtained values and provide a reasonably convincing framework to rely on. Pupil size has also been accepted as a reliable measure for the degree of affect in the sense that the pupil size tends to decrease as the emotional engagement increases (Graur & Siegle, 2013). However, it is also important to mention that pupil size is very sensitive to various other external and internal factors such as color contrast. Thus, it is crucial to provide baseline for correct measure of affect especially during dynamic settings as in TV ad watching. Skin conductance reaction (SCR or Galvanic Skin Response) has been demonstrated to be in relation with the level of arousal generally irrespective of the valence of the stimulus (Lang et al., 1999). In other words, the increase in level of skin conductance resistance has been considered to be independent indicator of affect levels specifically in the arousal dimension (Ohme et al., 2009; Wang et al., 2011; Potter & Bolls, 2012). Facial Coding Systems, on the other hand, have intensively been used for measuring emotion through the facial expressions, since these expressions are accepted as the direct measures of emotional categories. Despite the universal versus culture specific facial expressions debate, there is general consensus for the main emotional categories. This is a widely preferred method due to the fact that it has no direct contact with the participants as well as there are automatic measurement systems even accessible through internet (McDuff et al., 2014). The study by McDuff et al. (2014) has demonstrated that smile detection algorithms have classified with 81% accuracy if the participants have a tendency to watch an ad again that is accepted as a strong predictor of marketing success.

Among the brain imaging methods, EEG/ERP has been the most widely used technique to assess the affect level of the viewers during watching TV ads. The common analysis method is measuring the frontal alpha asymmetry over the channels among the frontal lobe (mainly F4-F3 and F8-F7 channel couples). The empirical evidence shows that the frontal alpha asymmetry (specifically dominance of left over right hemisphere) is correlated with the affection level of the participant that is also referred to as "positiveness" (Davidson et al., 1990; Ohme et al., 2008, 2010; Vecchiato et al., 2011). To be more specific, the higher the frontal alpha asymmetry (as calculated in power spectral density for alpha band and specifically left over right) the more positive is the participant's attitude towards the displayed advertisement. Thus, the studies mainly focused on plotting the affect of the participants in relation to the scene with an analysis on millisecond level, since TV ads are highly dynamic. In addition, Ohme et al. (2010)'s study has indicated that it might be possible to provide an extensive analysis of a TV ad in order to suggest the affect level during different phases of an advertisement. They have compared three consecutive creative executions of three TV ads of Sony Bravia. They have selected TV ads of the same brand in order to isolate the possible impacts of brand-perception. These particularly selected TV ads are Balls (2005), Paints (2006) and Play Doh (2007). The analyses have been performed separately for different parts of these ads such that the emotional and informational parts. Contrary to the current understanding, the informational parts such as the scenes related to product-benefit and product itself have been observed to cause approach-behavior (regarding the higher level of positiveness) more than the emotional parts specifically for the scenes of "Balls". Despite this piece of information seems to be a highly interesting and beneficial one, it is also important to underline that this might be related to the specific target group.

Lastly, the empirical findings obtained via fMRI have indicated that the neural activation of amygdala is directly related to the affection of the viewer. The earlier neuroimaging studies with fMRI have focused on the processing of negative emotions. And the academic literature has initially been dominated by the studies indicating that the amygdala is responsible for the processing of negative emotions. However, the findings from the relatively recent studies have indicated that the amygdala is also responsible for the processing of positive emotions (Pessoa & Adolphs, 2010). The results of the fMRI study have shown that for the cases having higher degree of likability, there is significantly higher level of neural activation among the right amygdala as well as right dorsolateral prefrontal cortex (dlPFC) and ventromedial prefrontal cortex (vmPFC) (Venkatraman et al., 2015). Moreover, heart rate deceleration has been shown to be an independent measure for affect level (Lang et al., 1999; Potter & Bolls, 2012; Wang et al., 2011; Venkatraman et al., 2015). The empirical findings from the academic literature demonstrate that the instantaneous decreases in heart rate (that could be calculated with instantaneous derivation of the data sample) might be used as an indicator for the affect level during the TV ad watching. It is then important to provide Z-score analysis of the whole output. The final output can be used for a further moment-by-moment analysis for the whole TV ad preferably accompanied by the results from the eye-tracker.

Memory

One of the leading interests has been related to the assessment of the TV ads in relation to how well it is encoded and thus retrieved in the future, since it has been a constructed measure for evaluating the middle and long-term effects of TV ads. The studies utilizing traditional methods have shown that there are familiarity and recognition effects for the assessments of the TV advertisements. In these studies, the participants are asked to recall or recognize the target among several distractor items (Du Plesis, 1994). Better memory is generally accepted as an indicator for better ad processing, however, it is also difficult to keep track of and dissociate these processes via the use of traditional methods related to the complexity of the mental processes.

The methods including fMRI and fNIRS have strong limitations for the temporal resolution especially for the facts that the main interest is related to watching a video or a TV ad. For this kind of stimuli, there is the risk of accumulation of the neural signals related to low temporal resolution and thus, there appears a debatable issue of providing a robust algorithm for a scene-by-scene analysis. On the other hand, the gap in temporal resolution has been compensated via the use of methods such as EEG/ERP.

A couple of such studies performed with EEG/ERP by Vecchiato mainly indicating the role of affect and positivity through the calculation of Frontal Alpha Asymmetry (FAA). Additionally, they have investigated the role of cognitive and affective processes related to memory. The results of the EEG/ERP study by Vecchiato et al. (2011) have indicated that theta, alpha and gamma oscillations dominantly in the left hemisphere could be related to the memory processes, since these neural patterns are observed among the participants who remembered the ads as opposed to the brain activations of the participants who forgot these TV ads. On the other hand, a series of experimental studies have indicated that the background music is processed earlier than the other elements within a TV ad (Fraser & Bradford, 2013). The background music has also been shown to cause a decrease in recall performance possibly indicating that the distractors might act as potentially crucial factors towards the memory-related processes (Fraser & Bradford, 2013). Moreover, the presented product orientation in the advertisements, directly related to our mental simulations of the product at target, has also been found to be important for our decisions and actions (Elder & Krishna, 2012).

Predicting Individual Preference

One of the most practical applications of the neuroscience methods will most likely be in relation to the prediction of individual preferences. The traditional methods obtain the individual preference generally through the application of questionnaires on a point or order scale. There are several occasions in which the traditional methods fail to discriminate two or more different products or ads because of similar scores on average that cannot be differ at a statistically significant level. The very basic claim is that the neuroscores might be helping to discriminate among very similar options that could not be discriminated by the results of traditional research methods. Accordingly, the findings from the fMRI study by Venkatraman et al. (2015) has indicated that the individual preferences have been related to the activation of the ventromedial prefrontal cortex (vmPFC) potentially regarding its role in brain reward system. Thus, the main idea is that the participants might be willing to prefer the items that have been more rewarding to them. Similarly, a very recent EEG/ERP study by Boksem and Smidts (2015) has indicated that the individual preferences could be predicted via the estimation of activity on beta band specifically over the frontopolar channels. In that study, the participants watched movie fragments (selected from the International Movie Database, IMDB) while their brain activities were recorded via EEG/ERP equipment and the participants were asked to order these movies regarding their preference just after watching these fragments. The relevant link between the individual preferences and beta oscillations among the frontopolar channels has been explained with the role of prefrontal regions, particularly the beta activity, in human brain reward system (Boksem & Smidts, 2015). To put it simply, the accumulated scientific evidence has suggested that the increase in neural activity in the reward system possibly causes higher level of positive affect that result with the higher likelihood for individual preference. However, it is also highly crucial to underline that the link between preference and purchasing is not still that clear. In other words, having grasped the individual preferences by neuroscientific methods does not guarantee that these preferences will turn into actions in real-world situations. Thus, it is also significant to attempt to provide framework that would relate these findings with the actual behavior. The evidence from other domains such as the neural correlates of purchasing also provides supportive evidence for the preference and even highlights the possible roots for predicting the actual behavior in the next decades (Knutson et al., 2007; Mitsuda et al., 2012; Cakir et al., 2015).

Market (Population-Wide) Success

One of the other prioritized targets of the whole marketing research is to predict the market success of the products or ads especially before their release to the market. This is the most desired scenario for the companies to able to decide to release the product or service depending on these quantitative predictions. However, no application has yet been developed regarding both the traditional and neuroscientific methods, to predict the market success reliably before the release of the TV ad. One of the neuroimaging studies performed with fMRI has claimed to predict the market-level success of 30-second long TV ads through the application of a unique experimental protocol. The experimental results demonstrated that the neural activity increase in ventral striatum demonstrates the real-world success beyond the elasticities in advertisements. In other words, the market-success of the advertisements could successfully be predicted via the observation of neural activations among the ventral striatum (VS) that is one of the integral parts of the brain reward system. In addition, the results of the EEG/ERP study by Boksem and Smidts (2015) have indicated that the population-wide success of the movies could be predicted by the neural correlates of the participants while they are watching the trailers of these movies. More specifically, the oscillations in gamma bands are positively correlated with the market success that is directly inferred from the results of US box office (it is referred as population-wide success in this article). Although these findings are argued to have direct implications for the marketing research, it is much more complicated to evaluate the consequences of a TV ad), since there are various potentially significant factors influential on the market success of the TV ads and the related product or service. However, it is highly remarkable that the brain oscillations within a specific band could be providing more successful predictive information for the market success than the self-reported measures. Thus, in near future, it might be the case that the potential market success of a TV ad could be provided before its release in television. Thus, the companies will be able to increase the effectiveness of their marketing strategies simply by eliminating the TV ads that are classified as unsuccessful relying on the obtained neuroscores. Although this sounds a very useful tool in practice, it should also be mentioned that there is need for more scientific studies to support and validate these findings in the middle and long-terms.

DISCUSSION

The progress in neurotechnologies (especially in brain imaging technologies as well as the recent advances in analysis methods) and the potential applications in marketing have resulted in many different initiatives but especially have given rise to the potentially objective tools for providing measures for the TV ads since this has been a huge industry involving sums of several hundred billions USD per year. Despite the fact that neurotechnologies have provided a variety of reliable tools for the marketing professionals regarding their decisions related to marketing strategies, such as selecting the more suitable option among two (or more) alternatives or deciding on the scenes to be excluded through a TV ad optimization process, there are also several debatable issues that have emerged especially regarding the commercial activities in the field. For instance, is it possible to determine which algorithm works better? Or how is it possible to distinguish between the main factor and the potentially irrelevant ones. This part is devoted to the discussion of the most relevant issues especially regarding the business practices.

Scientific and Statistical Validity of the Findings

There are several neuromarketing companies established especially in the last decade that have actively been providing service in several domains of marketing research. A considerable proportion of these companies aim to develop their own algorithms as a part of their R&D programs. However, there is also the issue of validity and reliability for the measures provided by these developed algorithms. There are important questions at this point: Which one can be evaluated to be the most successful? Since these algorithms are not publicly shared, except the fully accessible scientific projects, due to confidentiality issues, it is not possible to observe and test the validity of these algorithms. With respect to this, it might be safer to have these developed algorithms based on, or at least in line with, the main empirical findings in the academic literature. However, the current academic literature is still limited and requires further exploratory research to establish a more solid basis. Moreover, there are ethical issues to be discussed related to these outputs such that the clients are presented the report of the findings from the research. The results section of these reports may (or may not) include the details of the statistical analyses (including standard deviations and errors) performed. Thus, there is the critical issue of statistical validity of a reported finding indicating a difference if it implies statistically significant difference or not with standard error and deviation values. Hereby, the extended transparency of the research findings appears as a necessary issue for providing a reliable medium for communication between parties.

Differentiating Between Potential Causes

Beyond the use of any well-developed algorithm to assess the neurophysiological data, it is also challenging to find valid and convincing explanations for the obtained scores. For instance, a decrease in attention level could be caused by several different factors while watching a TV ad (Ohme et al., 2009). First of all, it is important to find out the precise timing of the initiation of the decrease. The lady in the ad scene might be evaluated as incompatible with the product or the background color might not be appropriate or the onset of vocal sounds could be distracting. Determination of the specific cause of a positive or negative trend regarding the neuroscores is one of the most significant aspects of this neuro-assessment process (Smidts et al., 2010, 2014). In this respect, the use of various measurement tools simultaneously -especially the integration of the eye-tracker system- is crucial to observe the looking patterns of the participants. Thus, one might be able to argue that the change is more likely to happen because of a specific feature in the scene that is (or not) attended by the viewers. The more professional and structured projects involve the testing phase of the suggested changes. For instance, the neurophysiological tests might have resulted with the claim that background music or vocal sound within a scene is not appropriate. Then the client might be convinced to provide these suggested changes and thereafter may request a test of the newer version of the ad. Thus, the clients can have the opportunity to observe the final version with the guidance of neuroassessments performed. Similar projects are performed for the optimization processes of the TV ads mainly for reducing the length to 30 seconds. Moreover, it is also crucial to combine the outputs of these neuro-assessments with the outcomes of traditional methods. This combination of mutual information properly is a crucial aspect of the marketing research on TV ads in order to have more powerful as well as fruitful outputs. Regarding a project, the obtained neuroscores might be reflecting an increase in attention related to a specific scene such as a balloon going up. However, it is not possible to infer why these participants have a general tendency in attention increase and follow-up in-depth interviews might reflect that specific cause. There might appear challenging cases that the results of neuroassessments alone could be conflicting and contradictory. For this reason, the integration of traditional and neuroscientific methods has generally been the preferred option during the last decade.

Which Algorithm Presents the Best Output?

Not limited to scene-by-scene analysis, one of the other neuromarketing applications on TV ads is to provide overall neuroscores for the tested ads. Thus, it appears the case that several TV ads could be given neuroscores that would inevitably enable them to order from the best to the worst. Hereby, it should be clearly declared that there is yet no universal (perfect) algorithm to give the most suitable neuroscores for the TV ads. Thus, different algorithms might end up with considerably different scores for the same test items, in this case TV ads. The first algorithm may end up with the result that Ad A has a higher attention score than Ad B. Another algorithm may end up with contradictory results for the attention score, promoting Ad B over Ad A. Even without the use of neuroscores, it is already a controversial and difficult issue to compare two different TV ads on different levels and attempt to conclude that one is better than the other especially for the cases that these ads are from different brands. The process of generating neuroscores is confidential for the moment, as mentioned above, and the clients have to rely on the algorithms developed by the companies. As mentioned above, they are not shared with the public except the scientifically published ones (as in conferences, journals and books) that occur only for the minority of the whole bundle of projects in business practices. However, there are recent trends for sharing the databases related to brain imaging research. In near future, it might be the case that this kind of knowledge will be freely accessed and so shared as public knowledge (as open source programs that could be accessed by anyone). In other words, several companies might become able to use these analyses tools easily may be with the contribution of their consultants.

Towards a More Objective Assessment of TV Ads

Despite the existence of obstacles to the provision of more objective ways for TV ad assessments, the current empirical evidence in the relevant literature indicates that there is considerable potential to provide reliable tools in the middle and long-terms (Venkatraman et al., 2015). However, it is also crucial to establish a multi-faceted process of dialogue among all parties. First of all, the underlying mechanisms of these neuroassessment applications should be explained to potential clients mainly including marketing professionals. Secondly, related to the first point, it is more likely that these methods will not replace the traditional methods, but the application of these methods in collaboration is expected to be more beneficial and helpful for explaining the facts. It should also be remarked that the use of different methods might cost more for the clients but it might be necessary to obtain more reliable analyses for the final output. On the other hand, there is also another highly controversial issue regarding the occasional cases when the results from traditional and neuroscientific methods contradict each other. These situations occasionally happen during the sector-specific studies and an urgent demand from the client forces the researcher to decide on one of the options in order to move onto the next step. Despite the fact that the neuroassessments can be considered superior in terms of objectivity, it is generally the case that the results, either from traditional methods or neuroscientific methods are evaluated with respect to their reasonableness. To be clearer, the accounts that are more reasonable and accountable are generally preferred by marketing professionals and researchers. This exemplary situation demonstrates that there will possibly be necessity for the deliberative and intuitive judgments of the professionals during the projects especially with the sharp deadlines to have guidance through the fuzzy or contradictory parts of the actual findings.

CONCLUSION

In conclusion, the frequent use of neuroscientific methods in the assessments of TV ads has provided a considerable hope of more objective evaluations and decisions for better marketing strategies. Beside the issue of providing objectivity through these evaluations, it is also argued that the assessments and reporting in general will be quicker, more robust and may be cheaper than the traditional methods, since the experiments require a considerably smaller number of participants when compared to the requirements of the traditional methods (Ariely & Berns, 2010). The application of various neurophysiological methods during watching TV ads is used for the assessments of the constructs including attention, emotional engagement, and individual preferences of the participants. These parameters have direct applications for the decisions on the preferable scenes during an optimization process of a TV ad. However, it is also worth mentioning that these are, at least for the moment, possibly limited measures that rely on the algorithm of the developers (despite the fact that they could be based on scientific findings). Although the relevant academic literature is currently limited, the number of scientific studies and business practices are increasing exponentially. Thus, the scientific validity and reliability might not be controversial at least in the middle term.

REFERENCES

Ambler, T., Ioannides, A., & Rose, S. (2000). Brands on the brain: Neuro-images of advertising. *Business Strategy Review, 11*(3), 17–30. doi:10.1111/1467-8616.00144

Arens, W. F. (2006). *Contemporary Advertising*. New York: McGraw-Hill.

Ariely, D., & Berns, G. S. (2010). Neuromarketing: The hope and hype of neuroimaging in business. *Nature Reviews. Neuroscience, 11*(4), 284–292. doi:10.1038/nrn2795 PMID:20197790

Asp, E., Manzel, K., Koestner, B., Cole, C. A., Denburg, N. L., & Tranel, D. (2012). A neuropsychological test of belief and doubt: Damage to ventromedial prefrontal cortex increases credulity for misleading advertising. *Frontiers in Neuroscience, 6*(100), 1–9. PMID:22787439

Bailey, K. (2007). *Methods of Social Research*. New York: The Free Press.

Belch, G. E., & Belch, M. A. (2003). *Introduction to Advertising and Promotion Management*. Academic Press.

Boksem, M. A. S., & Smidt, A. (2015). Brain responses to movie trailers predict individual preferences for movies and their population-wide commercial success. *JMR, Journal of Marketing Research, 52*(4), 482–492. doi:10.1509/jmr.13.0572

Bolls, P. D., Lang, A., & Potter, R. F. (2001). The effects of message valence and listener arousal on attention, memory, and facial muscular responses to radio advertisements. *Communication Research, 28*(5), 627–651. doi:10.1177/009365001028005003

Cacioppo, J. T., Petty, R. E., & Quintanar, L. R. (1982). Individual Differences in Relative Hemispheric Alpha Abundance and Cognitive Responses to Persuasive Communications. *Journal of Personality and Social Psychology, 43*(3), 623–626. doi:10.1037/0022-3514.43.3.623 PMID:7131245

Davidson, R. J., Ekman, P., Saron, C. D., Senulis, J. A., & Friesen, W. V. (1990). Approach-Withdrawal and cerebral asymmetry: Emotional expression and brain physiology. *Journal of Personality and Social Psychology, 58*(2), 330–341. doi:10.1037/0022-3514.58.2.330 PMID:2319445

Du Plessis, E. (1994, May-June). Recognition Versus Recall. *Journal of Advertising Research*, 75-91.

Elder, R. S., & Krishna, A. (2012). The "Visual Depiction Effect" in Advertising: Facilitating Embodied Mental Simulation through Product Orientation. *The Journal of Consumer Research, 38*(6), 1–17. doi:10.1086/661531

Fraser, C., & Bradford, J. A. (2013). Music to Your Brain: Background Music Changes Are Processed First, Reducing Ad Message Recall. *Psychology and Marketing, 30*(1), 62–75. doi:10.1002/mar.20580

Fugate, D. L. (2007). Neuromarketing: A layman's look at neuroscience and its potential application to marketing practice. *Journal of Consumer Marketing, 24*(7), 385–394. doi:10.1108/07363760710834807

Fugate, D. L. (2008). Marketing services more effectively with neuromarketing research: A look into the future. *Journal of Services Marketing, 22*(2), 170–173. doi:10.1108/08876040810862903

Godwin, B. J., & Saranya, V. P. (2012). Braining the brands: Ethnographic assessment on neuromarketing with reference to ultimate ad campaign. *Asia Pacific Journal of Marketing and Management Review, 1*(2), 117–125.

Guth, D. W., & Marsh, C. (2011). *Public Relations: A Value Driven Approach*. New York: Pearson.

Hammou, K. A., Galib, H., & Melloul, J. (2013). The contributions of neuromarketing in the marketing research. *Journal of Management Research, 5*(4), 20–33. doi:10.5296/jmr.v5i4.4023

Hansen, F., & Christensen, S. R. (2007). *Emotions, advertising and consumer choice*. Copenhagen Business School Press.

Hazlett, R. L., & Hazlett, S. Y. (1999). Emotional response to television commercials facial EMG vs. self-report. *Journal of Advertising Research*, (April), 7–23.

Heath, R., & Nairn, A. (2005). Measuring affective advertising: Implications of low attention processing on recall. *Journal of Advertising Research, 45*(2), 269–281. doi:10.1017/S0021849905050282

Hubert, M. (2010). Does neuroeconomics give new impetus to economic and consumer research? *Journal of Economic Psychology, 31*(5), 812–817. doi:10.1016/j.joep.2010.03.009

Hubert, M., & Kenning, P. (2008). A current overview of consumer neuroscience. *Journal of Consumer Behaviour, 7*(4-5), 272–292. doi:10.1002/cb.251

Kemp, E., Bui, M., & Chapa, S. (2012). The role of advertising in consumer emotion management. *International Journal of Advertising*, *31*(2), 339–353. doi:10.2501/IJA-31-2-339-353

Kenning, P. (2008). What advertisers can do and cannot do with Neuroscience. *IJA. Advertising & The Brain Special Issue*, *27*(3), 472–473.

Kenning, P., Deppe, M., Schwindt, M., Kugel, H., & Plassman, H. (2009). The good the bad and the forgotten – an fMRI study on ad liking and ad memory. *Advances in Consumer Research. Association for Consumer Research (U. S.)*, *36*, 4.

Klimesch, W. (1999). EEG alpha and theta oscillations reflect cognitive and memory performance: A review and analysis. *Brain Research. Brain Research Reviews*, *29*(2-3), 169–195. doi:10.1016/S0165-0173(98)00056-3 PMID:10209231

Knutson, B., Rick, S., Wimmer, G. E., Prelec, D., & Loewenstein, G. (2007). Neural Predictors of Purchases. *Neuron*, *53*(1), 147–156. doi:10.1016/j.neuron.2006.11.010 PMID:17196537

Kruger, R. A., & Casey, M. A. (2014). *Focus Groups: A Practical Guide for Applied Research*. Academic Press.

Langleben, D. D., Loughead, J. W., Ruparel, K., Hakun, J. K., Busch-Winokur, S., Holloway, M. B., & Lerman, C. et al. (2009). Reduced prefrontal and temporal processing and recall of high "sensation value" ads. *NeuroImage*, *46*(1), 219–225. doi:10.1016/j.neuroimage.2008.12.062 PMID:19457412

Lee, N., Broderick, A. J., & Chamberlain, L. (2007). What is neuromarketing? A discussion and agenda for future research. *International Journal of Psychophysiology*, *63*(2), 199–204. doi:10.1016/j.ijpsycho.2006.03.007 PMID:16769143

McClure, S. M., Li, J., Tomlin, D., Cypert, K. S., Montague, L. M., & Montague, P. R. (2004). Neural correlates of behavioral preference for culturally familiar drinks. *Neuron*, *44*(2), 379–387. doi:10.1016/j.neuron.2004.09.019 PMID:15473974

McDuff, D., Kaliouby, R. E., Senechal, T., Demirdjian, D., & Picard, R. (2014). Automatic measurement of ad preferences from facial responses gathered over the internet. *Journal of Image and Vision Computing*, *32*(10), 630–640.

Morgan, D. L. (1997). *Focus Groups As Qualitative Research*. New York: SAGE Publications, Inc. doi:10.4135/9781412984287

Morin, C. (2011). Neuromarketing: The new science of consumer behavior. *Society*, *48*(2), 131–135. doi:10.1007/s12115-010-9408-1

Morris, J. D., Klahr, N. J., Shen, F., Villegas, J., Wright, P., He, G., & Liu, Y. (2009). Mapping a Multidimensional Emotion in Response to Television Commercials. *Human Brain Mapping*, *30*(3), 789–796. doi:10.1002/hbm.20544 PMID:18286463

Mostafa, M. M. (2012a). Brain processing of vocal sounds in advertising: A functional magnetic resonance imaging (fMRI) study. *Expert Systems with Applications*, *39*(15), 12114–12122. doi:10.1016/j.eswa.2012.04.003

Mostafa, M. M. (2012b). The persistence of memory: An fMRI investigation of the brain processing of surrealistic imagery in advertising. *Journal of Marketing Communications*, 1–19.

Ohme, R., & Matukin, M. (2012). A small frog that makes a big difference: Brain wave testing of TV advertisements. *IEEE Pulse*, *3*(3), 28–33. doi:10.1109/MPUL.2012.2189169 PMID:22678837

Ohme, R., Matukin, M., & Pacula-Lesniak, B. (2011). Biometric measures for interactive advertising research. *Journal of Interactive Advertising*, *11*(2), 60–72. doi:10.1080/15252019.2011.10722185

Ohme, R., Reykowska, D., Wiener, D., & Chormanska, A. (2009). Analysis of Neurophysiological Reactions to Advertising Stimuli by Means of EEG and Galvanic Skin Response Measures. *Journal of Neuroscience, Psychology, and Economics*, *2*(1), 21–31. doi:10.1037/a0015462

Ohme, R., Reykowska, D., Wiener, D., & Chormanska, A. (2010). Application of frontal EEG asymmetry to advertising research. *Journal of Economic Psychology*, *31*(5), 785–793. doi:10.1016/j.joep.2010.03.008

Patton, M. Q. (2014). *Qualitative Research & Evaluation Methods: Integrating Theory and Practice*. New York: SAGE Publications, Inc.

Plassmann, H., O'Doherty, J. P., Shiv, B., & Rangel, A. (2008). Marketing actions can modulate neural representations of experienced pleasantness. *Proceedings of the National Academy of Sciences of the United States of America*, *105*(3), 1050–1054. doi:10.1073/pnas.0706929105 PMID:18195362

Plassmann, H., Venkatraman, V., Huettel, S., & Yoon, C. (2015). Consumer neuroscience: Applications, challenges, and possible solutions. *JMR, Journal of Marketing Research*, *52*(4), 427–435. doi:10.1509/jmr.14.0048

Potter, R. F., & Bolls, P. D. (2012). *Psychophysiological measurement and meaning: cognitive and emotional processing of media*. New York: Routledge.

Richards, J. I., & Curran, C. M. (2002). Oracles on "Advertising": Searching for a Definition. *Journal of Advertising*, *31*(2), 63–77. doi:10.1080/00913367.2002.10673667

Rossitier, J. R., Silberstein, R. B., Harris, P. G., & Nield, G. (2001). Brain-imaging detection of visual scene encoding in long-term memory for TV commercials. *Journal of Advertising Research*, *41*(2), 13–21. doi:10.2501/JAR-41-2-13-21

Rothschild, M. L., & Hyun, Y. J. (1990). Predicting Memory for Components of TV Commercials from EEG. *The Journal of Consumer Research*, *16*(4), 472–478. doi:10.1086/209232

Rubin, A., & Babbie, E. R. (2013). *Research Methods for Social Work*. Cengage Learning.

Salomons, T. V., Johnstone, T., Backonja, M. M., & Davidson, R. J. (2004). Perceived controllability modulates the neural response to pain. *The Journal of Neuroscience*, *24*(32), 7199–7203. doi:10.1523/JNEUROSCI.1315-04.2004 PMID:15306654

Silberstein, R. B., & Nield, G. E. (2012, May-June). Measuring emotion in advertising research. *IEEE Pulse*.

Smidts, A., Hsu, M., Sanfey, A. G., Boksem, M. A. S., Ebstein, R. B., Huettel, S. A., & Yoon, C. et al. (2014). Advancing consumer neuroscience. *Marketing Letters, 25*(3), 257–267. doi:10.1007/s11002-014-9306-1

Stacks, D. W. (2010). *Primer of Public Relations Research*. The Guilford Press.

Thornton, J., & Rossiter, J. R. (2004). Predicting the effectiveness of anti-speeding TV advertisements by skin conductance response. *Proceedings of the Australian and New Zealand Marketing Academy Conference* (pp. 1-7).

Vance, K., & Virtue, S. (2011). Metaphoric advertisement comprehension: The role of the cerebral hemispheres. *Journal of Consumer Behaviour, 10*(1), 41–50. doi:10.1002/cb.345

Vecchiato, G. et al. (2011). On the use of EEG or MEG Brain Imaging Tools in Neuromarketing Research. *Computational Intelligence and Neuroscience*. doi:10.1155/2011/643489

Vecchiato, G., Kong, W., Maglione, A., & Wei, D. (2012). Understanding the impact of TV commercials: Electrical neuroimaging. *IEEE Pulse, 3*(3), 42–47. doi:10.1109/MPUL.2012.2189171 PMID:22678840

Vecchiato, G., Toppi, J., Astolfi, L., De Vico Fallani, F., Cincotti, F., Mattia, D., & Babiloni, F. et al. (2011). Spectral EEG frontal asymmetries correlate with the experienced pleasantness of TV commercial advertisments. *Medical & Biological Engineering & Computing, 49*(5), 579–583. doi:10.1007/s11517-011-0747-x PMID:21327841

Venkatraman, V., Dimoka, A., Pavlou, P. A., Vo, K., Hampton, W., Bollinger, B., & Winer, R. S. et al. (2015). Predicting Advertising Success Beyond Traditional Measures: New Insights from Neurophysiological Methods and Market Response Modeling. *JMR, Journal of Marketing Research, 52*(4), 436–452. doi:10.1509/jmr.13.0593

Wang, Z., Lang, A. N., & Busemeyer, J. R. (2011). Motivational Processing and Choice Behavior During Television Viewing: An Integrative Dynamic Approach. *Journal of Communication, 61*(1), 71–93. doi:10.1111/j.1460-2466.2010.01527.x

Yılmaz, A. (2014). *Yüzyılı Dönüştüren Yaratıcılar*. Literatürk Academia.

Yoon, C., Gonzalez, R., Bechara, A., Berns, G. S., Dagher, A. A., Dubé, L., & Spence, C. et al. (2012). Decision neuroscience and consumer decision making. *Marketing Letters, 23*(2), 473–485. doi:10.1007/s11002-012-9188-z

Zaltman, G. (1997). Rethinking Market Research: Putting People Back in. *JMR, Journal of Marketing Research, 34*(4), 424–437. doi:10.2307/3151962

This research was previously published in Applying Neuroscience to Business Practice edited by Manuel Alonso Dos Santos, pages 231-256, copyright year 2017 by Business Science Reference (an imprint of IGI Global).

Chapter 19
Marketing Meets Neuroscience:
Useful Insights for Gender Subgroups During the Observation of TV Ads

Patrizia Cherubino
IULM University, Italy & BrainSigns srl, Italy

Giulia Cartocci
Sapienza University, Italy & BrainSigns srl, Italy

Arianna Trettel
BrainSigns srl, Italy

Dario Rossi
Sapienza University, Italy & BrainSigns srl, Italy

Enrica Modica
Sapienza University, Italy & BrainSigns srl, Italy

Anton Giulio Maglione
Sapienza University, Italy & BrainSigns srl, Italy

Marco Mancini
BrainSigns srl, Italy

Gianluca Di Flumeri
Sapienza University, Italy & BrainSigns srl, Italy

Fabio Babiloni
Sapienza University, Italy & BrainSigns srl, Italy

ABSTRACT

In this chapter, findings of an experiment aimed to investigate cognitive changes of cerebral activity during the observation of TV commercials will be presented. In particular, it has been recorded Electroencephalographic data (EEG) from a group of 24 healthy subjects during the observation of a series of TV advertisements. The group was divided by gender (male, female). Comparisons of cerebral index previously defined have been performed to highlight gender differences between scenes of interest of a specific TV commercials and the influence of the speaker's gender on the subgroups perception. Findings show how EEG methodologies could be used to obtain information not obtainable otherwise with verbal interviews. These cerebral index could help to analyze the perception of TV advertisements according to the target consumer's gender.

DOI: 10.4018/978-1-5225-5478-3.ch019

INTRODUCTION

Goal of any advertising campaign is to convey a specific message and reach a specific audience. Optimize advertising investments getting a good effectiveness of their communication is one of the main objectives of a company. It is known how the big companies invest a significant portion of their budget to promote their products or to improve their image and one of the most used channels is the television, the most effective communication medium to reach the greatest number possible of potential customers.

Recently, a rapidly growing approach within consumer research has developed under the label of "consumer neuroscience." Its goal is to use insights and methods from neuroscience to enhance the understanding of consumer behavior. The application of neuroscience to consumer psychology has gained popularity over the past decade in academic research and business practice. In fact, the goal of consumer neuroscience is to adapt methods and theories from neuroscience combined with behavioral theories to develop a neuropsychologically sound theory to understand the consumer behavior.

The birth of the field of consumer neuroscience has generated wide-ranging, ongoing debates of whether this hybrid field benefits its parent disciplines (consumer psychology and neuroscience) and, within them, what forms these benefits might take (Ariely & Berns, 2010; Kenning & Plassmann, 2005; Lee et al., 2007; Plassmann et al., 2007).

In these last years, findings from the consumer neuroscience experiments deconstruct the picture of perfectly rational humans, which are deliberating their choices by weighting costs and benefits until a deliberative equilibrium is reached. Although humans are definitely capable of conscious deliberation, many, if not most economically relevant decision processes are characterized by certain other features: first, they rely on automatic, fast and effective cognitive processes, which are not under direct volitional control (Bargh & Chartrand, 1999). Second, they are under the influence of unrecognized and finely tuned affective mechanisms, which often play a decisive role in action (Damasio et al., 1996; Davidson et al., 1999; Panksepp et al., 2004). Third, many of these processes have been shaped by evolution in order to serve social purposes (Adolphs, 2003; Cacioppo et al., 2002).

The term "neuromarketing" refers to practitioner and commercial interest in neurophysiological tools, such as eye tracking, skin conductance, electroencephalography (EEG), and functional magnetic resonance imaging (fMRI), to conduct company-specific market research. Neuromarketing has received considerable attention in the corporate world, and the growth of neuromarketing companies over the last decade has been impressive (Plassmann, 2012). More properly, neuromarketing can be defined as the field of study that applies the methodologies of the neuroscience to analyze and understand the human behavior related to market and economic exchanges (Lee et al., 2007). Hence, the contribution of neuroscientific methods becomes significant for the knowledge of the human behavior in the marketing scope. Moreover, another interesting issue is overcoming the dependence from the verbal answering nowadays used on testing subjects in traditional marketing researches where insights and indicators depend on the good-faith and accuracy of the experimental subject reporting his own sensations and opinion to the experimenter. Instead, the use of the brain imaging technique can distinguish the subject's cognitive and emotional experiences (verbally expressed during the interviews) from the activations of cerebral areas related to different, and unconscious, mental states. Interesting experimental evidences suggest that the use of the brain imaging, in a near future, could be placed side by side to classical tests today largely used in the marketing sciences (Vecchiato et al., 2013).

The idea behind the present chapter is to illustrate the potential of EEG techniques when applied to the analysis of brain activity related to the observation of TV commercials. In particular, purpose of the present research is to describe how by using appropriate statistical analysis it could be possible to recover significant information about scalp areas engaged by particular scenes inserted within the TV commercial analyzed. In the last five years, the authors conducted and published several studies showing how it is possible to detect hidden signs of the memorization process and emotional engagement, such as pleasantness perceived while watching an advertisement, by employing neurophysiological recordings.

In particular, the researchers were interested in analyzing the brain activity occurring during the "naturalistic" observation of commercial ads intermingled in a random order in a documentary. The aim was to recreate, as much as possible, an ecologic approach to the task, in which the observer is viewing the TV screen without particular goals in mind. In fact, the subjects were not instructed at all on the aim of the experiment, and they were not aware that an interview about the proposed TV commercials would be generated at the end of the task.

In particular, the experimental questions are the following:

1. Is it possible to track the cerebral activity and the emotional state, frame by frame, of subjects watching TV commercials?
2. Are visual and auditory features of commercial advertisements classifiable in "pleasant" and "unpleasant" on the basis of central and autonomic nervous system activity variations?

Background

In the last years there has been an increased interest in the use of brain imaging techniques for the analysis of brain responses to the commercial advertisements or for the investigation of the purchasing attitudes of the subjects (Ioannides et al., 2000; Knutson et al., 2007; Astolfi et al., 2008; Morris et al., 2009; Vecchiato et al., 2014; Cherubino et al., 2016). The interest is justified by the possibility to correlate the particular observed brain activations with the characteristics of the proposed commercial stimuli, in order to derive conclusions about the adequacy of such ad stimuli to be interesting or emotionally engaging. Standard marketing techniques, so far employed, involved the use of an interview and the compilation of a questionnaire after the exposition to novel commercial ads before the massive launch of the ad itself (ad pre-test). However, it is now recognized that often the verbal advertising pre-testing is flawed by the respondents' cognitive processes activated during the interview, being the implicit memory and subject's feelings often inaccessible to the interviewer that uses traditional techniques (Zaltman, 2003). In addition, it was also suggested that the interviewer on this typical pre-testing interviews has a great influence on what respondent recalls and on the subjective experiencing of it (Mc Donald, 2003; Franzen & Bouwman, 2001). Taking all these considerations in mind, researchers have attempted to investigate the signs of the brain activity correlated with an increase of attention, memory or emotional engagement during the observation of such commercial ads. Researchers within the consumer neuroscience community promote the view that findings and methods from neuroscience complement and illuminate existing knowledge in consumer research in order to better understand consumer behaviour (e.g., Ambler et al., 2000; Klucharev et al., 2008).

It is very well known that the hemodynamic measurements of the brain activity allow a level of localization of the activated brain structures on the order of cubic mm, being capable to detect activations also in deep brain structures such as amigdala, nucleus accumbens etc etc. However, the lack of time resolution, due to the delay of the cerebral blood flow's increment after the exposition to the stimuli, make the fMRI unsuitable to follow the brain dynamics on the base of its subseconds activity. The use of EEG allows following the brain activity on a millisecond base, bearing in mind that the recorded EEG signals are mainly due to the activity generated on the cortical structures of the brain. In fact, the electromagnetic activity elicited by deep structures advocated for the generation of emotional processing in humans is almost impossible to gather from usual superficial EEG electrodes (Nunez, 1995; Urbano et al., 1998). However, nowadays high-resolution EEG (hrEEG) technology has been developed to enhance the poor spatial information content of the EEG activity in order to detect the brain activity with a spatial resolution of a square centimeters and the unsurpassed time resolution of milliseconds (Nunez, 1995; Bai et al., 2007; He et al., 1999; Dale et al., 2000; Babiloni et al., 2005).

It has underlined that the positive or negative emotional processing during the observation of the commercial ads is an important factor for the formation of stable memory traces (Kato et al., 2009). A central brain region for emotional responses is the amygdala. In particular, it is involved in the processing of negative emotions and unknown stimuli, as well as in aversive responses to inequity (Rilling & Sanfey, 2011). Another key emotion-related region for the negative emotions is the insula (or insular cortex) (Preuschoff et al., 2008; Knutson & Bossaerts, 2007).

Hence, it became relevant to infer the emotional engage of the subject by using indirect signs for it. In fact, indirect signs of emotional processing could be gathered by picking variations of the activity of the anatomical structures linked to the emotional processing activity in humans, such as the activity of sweat glands on the hands and/or the variation of the heart rate (Baumgartner et al., 2006). Such gathering could be performed also along with the tracking of the variations of the pre-frontal and frontal cortex (PFC and FC respectively) (Davidson & Irwin, 1999).

In particular, by monitoring autonomic activity using devices able to record the variation of the skin conductivity (Galvanic Skin Responses, GSR) and the heart rate (HR) it is possible to assess the "internal" emotional state of subjects, measuring variables of the autonomic nervous system. In fact, GSR activity is actually viewed as a sensitive and convenient measure of indexing changes in sympathetic arousal associated with emotion, cognition and attention (Critchley, 2002). Studies using functional imaging techniques (Critchley, 2002; Nagai et al., 2004) have related the generation and level of electrodermal activity to specific brain areas. These specific regions are the ventromedial prefrontal cortex, orbitofrontal cortex, left primary motor cortex, and the anterior and posterior cingulate, which have been shown to be associated with emotional and motivational behaviours. Such as findings indicate the close association between peripheral and central measures of arousal, re-emphasise the close connections between electrodermal activity, arousal, attention, cognition and emotion. In addition, the link between the HR or the Heart Rate Variability (HRV) and the sympatho/vagal balance has been also suggested in literature (Malik et al., 1996; Malliani, 2005; Montano et al., 2009).

Matching the information conferred by the GSR and the HR signals the Emotional Index (EI) has been defined. The EI can vary between -1 and +1. Negative and positive values of the EI are related to negative and positive emotions, respectively, spanning the whole affect plane (for a review, Mauss & Robinson, 2009). As far as the construction of such variable is concerned, the authors refer to the affect plane where the coordinates of a point in this space are defined by the HR (horizontal axis) and the GSR (vertical axis). Several studies have highlighted that these two autonomic parameters correlate with valence and arousal, respectively (Astolfi et al., 2009; Vecchiato et al., 2013).

The PFC region is structurally and functionally heterogeneous but its role in emotion is well recognized (Davidson, 2002). EEG spectral power analyses indicate that the anterior cerebral hemispheres are differentially lateralized for approach and withdrawal motivational tendencies and emotions. Specifically, findings suggest that the left PFC is an important brain area in a widespread circuit that mediates appetitive approach, while the right PFC appears to form a major component of a neural circuit that instantiates defensive withdrawal (Davidson, 2000; 2004).

Information about the brain activity can be obtained by measuring different physical variables arising from brain processes, such as the increase in consumption of oxygen by the neural tissues or a variation of the electric potential over the scalp surface. All these variables are connected in direct or indirect way to the neural ongoing processes and each variable has its own spatial and temporal resolution. Hence, the different neuroimaging techniques are then constrained by their spatio-temporal resolution, depending on the monitored variables. Human neocortical processes involve temporal and spatial scales spanning several orders of magnitude: from the rapidly shifting somatosensory processes characterized by a temporal scale of milliseconds and a spatial scale of few square millimeters, to the memory processes, involving time periods of seconds and spatial scale of square centimeters. Today, no neuroimaging method provide a spatial resolution on a millimeter scale and a temporal resolution on a millisecond scale at the same time.

Electroencephalography is an interesting technique characterized by a high temporal resolution, on the millisecond scale, adequate to follow a particular kind of brain activity (to deepen this topic, far from the goal of the present book, Michel & Murray, 2012; Hallez et al., 2007; Michel et al., 2004; for a review). Unlikely, this technique has a relatively modest spatial resolution, beyond the centimeter, because of the inter-sensor distances and the fundamental laws regulating the electromagnetism (Nunez, 1995). The simultaneous activation of an entire population of neurons can generate an electric signal detectable on the head surface with electrodes placed on the scalp. In order to estimate the cortical activity, the EEG signal has to be measured in different scalp sites; the most common measurement system is the international montage 10-20. Generally, the standard EEG analysis, using 20-30 electrodes, brings to a spatial resolution of about 5÷7 cm. It might be argued that such spatial resolution could be not enough to solve the scientific issues related to standard clinical investigations. In fact, researchers have developed a body of techniques called "high resolution EEG techniques" that allow to move up to 2 cm of spatial resolution with the use of 64-128 electrodes (Nunez, 1995). However, a low spatial resolution of 5-7 cm is sufficient in the case experiments are aimed to differentiate the activity of the right and left PFC as in a typical "neuromarketing" experiments (Davidson et al., 1994, Vecchiato et al., 2014). In the following, it well be provide an example of the use of the unbalance for the activity of left and right prefrontal cortex for the analysis of a TV advertising.

METHODOLOGY OF A TYPICAL NEUROMARKETING EXPERIMENT

In the investigated ecologic situation, as Figure 1 shows, the experimental subjects are asked to comfortably have a seat in front of a computer screen by means of which a documentary is presented. A train of two commercial breaks, inserted within it, constitute the stimuli of interest. The cerebral signals gathered during the observation of the documentary will be used to generate a suitable baseline for each subject's

Figure 1. Picture shows the experimental setup of our neuromarketing studies. The subject is seated in front of a computer screen on which a neutral documentary is running. We inserted a commercial break within, consisting of 12 TV advertisements, among which was randomly inserted our target advertisement.

cerebral activity. Performing the experiment, subjects are not aware that an interview would be held at the end of the movie. They are simply told to pay attention to what they watch. In the interview, subjects are asked to recall the commercial clips they remember, tell if they are consumers or not of the products advertised and express the degree of pleasantness perceived. In such a way, the signals of the recorded population is divided into different datasets that will be compared for analysis.

Thanks to the high resolution EEG techniques it is possible to track subject's brain activity during visualization of the commercials: in such manner it is possible to obtain a global measure of the reconstructed cortical signals by means of a simple graphic tool which allows to distinguish the activity of different cortical areas.

During the all length of the stimuli presentation biometric signals can be recorded by different devices, allowing for instance the study of the EEG signal variation, the HR variability and the GSR alteration in response to the stimuli exposure.

The biometric EEG, HR and GSR recordings returned neurometric indexes linked to the variation of the memorization, attention, pleasantness and emotional involvement of each analyzed person. Variation of such indexes along the adopted time segments returned information about the ad perception, second by second, and for conceptual sequence, on the total sample of recorded subjects and on sub-samples of such group.

For more technical details concerning the EEG recordings and analysis please read below in the experimental section reported.

GENDERS' DIFFERENT REACTION TO TV ADS

The personal reaction to external stimuli is highly influenced by the personal condition and the emotional sensitivity of the subject (Allen & Kline, 2004). Furthermore, despite the human capacity of conscious deliberation, many economically relevant decisions rely on automatic, fast and effective processes, which are not under the direct volitional control (Bargh & Chartrand, 1999). Many of these processes have been shaped by evolution in order to serve social purposes (Adolphs, 2003; Astolfi et al., 2011; Cacioppo, 2002) while decision making and evaluation in economic contexts are influenced by mechanisms dedicated to social interaction. For instance, in men, the observation of different cultural objects associated with wealth and social dominance (e.g. sports cars) elicits activation in reward-related brain areas (Erk et al., 2002; Braeutigam, 2005).

The peculiar targeting of a spot has been also exploited in the field of health promotion and consciousness. An example of this is the case of the human papilloma virus (HPV), which infection is associated with cervical cancer in women, and which knowledge was increased among students watching HPV informative TV commercials (Sandfort and Pleasant, 2009); or the increased handwashing behavior reported by women in Ghana reached by TV and radio commercials concerning this hygiene topic (Scott et al., 2008); or also the targeted online, radio and print advertisement for tobacco control in young and middle-aged males subjects (Clayforth et al., 2014).

Additionally to the specific targeting of a campaign, the peculiar creativity ("the story", the concept) displayed in the advertisement, represents a key feature in determining the effectiveness of the commercial. This statement appears extremely clear in the evidence that male adolescents showed a consistent preference for beer ads with sports content in comparison to ads without the sport subject (Slater et al., 1996).

In this regard, the prefrontal cortex (PFC) appears to play a pivotal role in a larger overall circuit involved in emotional and motivational processes (Davidson, 2002). In fact, the frontal electroencephalographic (EEG) alpha rhythm activity is frequently used to detect inter - and between - subjects asymmetries in cortical activation. Hemispheric asymmetries in the alpha frequency range over prefrontal electrodes have become widely accepted as a correlate of approach and withdrawal related motivation in basic research (Harmon-Jones et al., 2010; Price et al., 2012; Briesemeister et al., 2013). According to the ''Approach/Withdrawal''(AW) model, analysis of the EEG power spectrum suggests a different lateralization for the anterior cerebral hemispheres in approach and withdrawal motivational tendencies and emotions (Coan & Allen, 2004).

According to the aforementioned theoretical framework, the expectation is that: the alpha EEG frontal asymmetry would reflect the appreciation of TV commercials, showing major (minor) activation on the left frontal lobe (right frontal lobe) during the observation of pleasant (unpleasant) scenes. In other words, the activity imbalance expressed by the AW index would report higher values for pleasant/engaging scenes and lower values for the unpleasant/disengaging ones. Decreased alpha activity (or higher alpha desyncronization) indicates an increase in allocation of cortical activation (Allen & Kline, 2004; Davidson, 1988). So, calculating the frontal alpha asymmetry by subtracting the right-hemispheric electrodes data minus their left-hemispheric electrodes counterparts - ln(R)-ln(L) - (Allen & Kline 2004; Davidson et al., 1979), positive frontal asymmetry values will indicate larger relative right-hemispheric power, which corresponds to larger cortical activation in the left hemisphere, and viceversa. Specifically, a relative power suppression of the alpha rhythm across the left PFC is associated with a propensity to approach toward a stimulus, while the relative power suppression of the alpha rhythm across the right PFC is associated with a propensity to withdraw from a stimulus.

The study of the engagement/emotional impact generated by TV commercials has been investigated by EEG alpha frontal asymmetry and showed that the observation of two versions of the same advertisement, produced significantly different approach reactions (Ohme et al., 2009). The same authors, Ohme and colleagues (2010), studied the approach reaction, as indexed by the frontal alpha asymmetry, to 4 scenes belonging to three alternative commercials of the same product (a flat screen TV), each characterized by a different kind of creative solution. The scenes were considered the "emotional part" of the commercials under evaluation, which had in common the same "informational part" concerning the product and the brand. Results showed that for the emotional part there was a statistical significant difference in the approach reaction towards the three commercials, but not for the informational part. Furthermore, difference in cerebral activity coding the pleasantness can be also observed by event-related potentials (ERPs) analysis. In particular, Handy and colleagues (Handy et al., 2010) proved that visuocortical processing evoked an increase of the early positive component (named P1), at central and parietal sites, along with an increase of the later negative component (named N2), at parietal and occipital sites, related to the observation of disliked logos. Taken together, these mentioned examples bring evidences of the feasibility of linking some properties of the collected EEG rhythms during the observation of TV advertisements with the overt preferences of the observers in terms of emotions.

The importance of more "objective" instruments to measure the appreciation of a product is witnessed by the evidence that at least the 70% of the new products launched worldwide (including cars, shoes, clothes, etc.), tested by traditional techniques with questionnaires or psychological interviews, fail within the first six months (Vecchiato et al., 2014). This happens because people do not (or are not able to) report precisely their internal perceptions when interviewed concerning product experiences or TV commercials.

Gender is one of the most commonly used variables in marketing and consumer research. Behavioral consequences of the exposure to commercial, for instance the food intake after watching food-related advertisements, is already object of research, showing higher vulnerability in women for eating snack food when exposed to food commercials (Anschutz et al., 2011). It is worth noticing that such result is also not merely a difference between women and men, but the modulation operated by the transportability within each sex (Wonderlich-Tierney et al., 2013). However, the effect of the consumer's gender in commercials perception still needs investigation. Some evidences suggest a different perception between women and men, in general, with an higher approach motivation in case of an advertisement of a typical feminine item (e.g. a perfume), but mainly circumscribed to particular scenes (e.g. dancing) of the video, and not providing univocal data if different videos are considered (Vecchiato et al., 2014). When the advertised product is an item often considered as masculine (e.g. a car), becomes evident the influence of the peculiar creativity of the commercial in order to produce efficient advertising also towards women. In particular, evaluating the emotional response, as indexed by heart rate (HR) and Galvanic Skin response (GSR) variations, women showed an emotional profile presenting a major activity in the first section of the commercial, characterized by the presence of a famous actress and three pretty little girls, probably due to the identification by the women sample with the actress and a sort of maternal reaction toward the children; on the other hand men's emotional profile was predominantly activated in the final section, corresponding to the technical presentation of the product, probably due to the higher level of reward elicited in men by the product, in comparison to women (Vecchiato et al., 2013 pp.101-106). The sum of these evidences suggest the importance to further the investigation of gender-related commercials perception by cerebral and time varying techniques, such as EEG, able to measure the approach/withdrawal motivation along the development of the different scenes composing a TV commercial.

MATERIAL AND METHODS

Participants: 24 healthy subjects were enrolled for the study and paid for their performances. Due to technical problems 18 subjects' recordings were included in the analysis, 8 men (27.25 ± 3.58) and 10 women (27.5 ± 2.88), age ranging from 22 and 33 years old have been enrolled in the study. Informed consent was obtained from each subject after the explanation of the study, which was approved by the local institutional ethic committee. The experiment was conducted following the principles outlined in the Declaration of Helsinki of 1975, as revised in 2000.

The protocol has been reported in the Figure 1.

After watching the video, for each commercial three images were shown to participants, and they were asked to rate the pleasantness they would have assigned to each advertisement, ranging from 1 (total dislike) to 10 (total like).

The advertised target commercial was a phone company promotion concerning the advantages of the optical fibre technology. The advertisement was chosen because of the great number of successive scenes displayed, approximately twelve. Some of those could be assumed to be "masculine appealing", other "feminine appealing" and other "neutral appealing". The selected scenes were:

- Fireworks (neutral)
- Baby girl (feminine)
- Soccer players (masculine)
- Brand (product).

Additionally, the present advertisement presented the alternation of two speakers: a man and a woman. This feature rendered the spot suitable also for the investigation of the speaker's gender perception in the participants gender subgroups.

The target advertisement was developed by the famous Italian phone company Tim and its synopsis was the development of new technologies enabling people to communicate ever better and the acme is represented by the possibility of connections offered by the use of the optical fibre.

EEG RECORDINGS AND SIGNAL PROCESSING

Specifically, subjects were comfortably seated on a reclining chair, in an electrically shielded, dimly lit room. A 10-channels EEG system (BrainAmp, Brainproducts GmbH, Germany) was used to record the electrical potentials, according to an extension of the 10-20 international system. Recordings were initially extra-cerebrally referred and then converted to an average reference off-line. The EEG activity has been collected at a sampling rate = 256 Hz while the impedances kept below 5kΩ. Each EEG trace was then converted into the Brain Vision format (BrainAmp, Brainproducts GmbH, Germany) in order to perform signal preprocessing such as artefacts detection, filtering and segmentation. Raw EEG traces were first band-pass-filtered (high pass = 2 Hz; low pass = 47 Hz), and the independent component analysis (ICA) was then applied to detect and remove components due to eye movements and blinks that could afflict the estimation of the EEG power spectra on the prefrontal areas. In addition, the Individual Alpha Frequency (IAF) has been calculated for each subject in order to define the band of interest according to the method suggested in the scientific literature (Klimesch, 1999).

Figure 2. Frame and scenes segments of the TV commercial

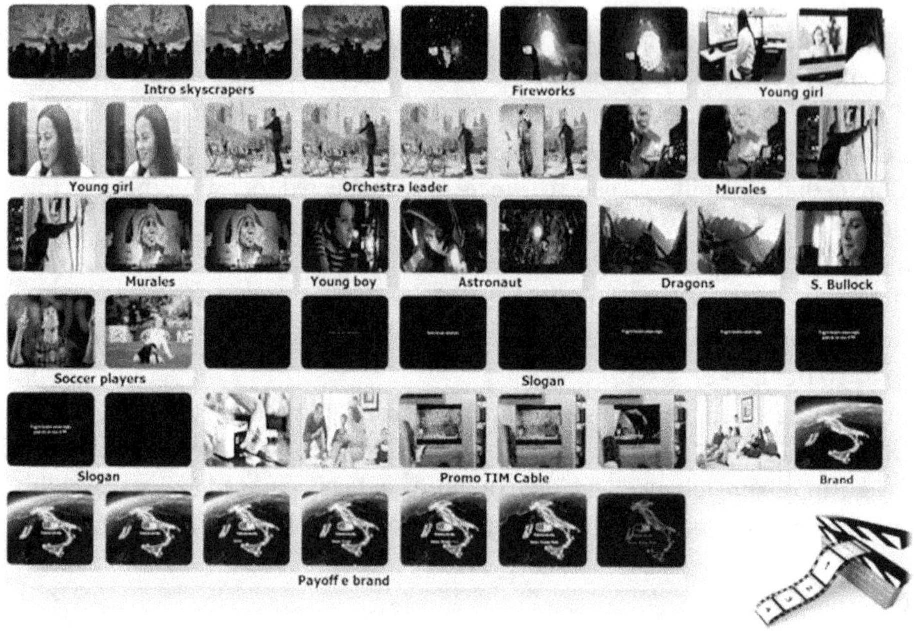

These EEG traces were then segmented to extract the cerebral activity during the observation of the TV commercials and the one associated to the documentary (baseline period). Then, this dataset has been further segmented into one-second length trials. Only artefacts-free trials have been considered for the following analysis.

Different indexes related to the perception of the TV commercial advertisings have been estimated. In particular, the index related to the Approach/Withdrawal theory previously proposed by Davidson and colleagues is calculated (Davidson et al, 1994, Vecchiato et al, 2012).

Each EEG trace is band pass filtered in order to isolate the spectral components in the alpha band from the whole EEG spectrum. The filtered traces have been used to calculate the Global Field Power (GFP) (Lehmann et al., 1980). Frontal electrodes, since a clear role of the frontal areas has been depicted for the investigated phenomena (Erk et al., 2002; Braeutigam, 2005; Vecchiato et al., 2011; Knutson et al., 2007) have been used to compute the GFP indexes in this study, by selecting the electrodes F7, F3, Fp1, Fz, Fp2, F4, F8 of the 10-20 International System. Such selection was performed to evaluate the Approach/Withdrawal Index (AW index).

The formula that defines the AW index is the following:

$$AW = GFP_{\alpha\_right} - GFP_{\alpha\_left} \quad (1)$$

where the GFP_{α_right} and GFP_{α_left} stand for the GFP calculated among right (Fp2, F4, F8) and left (F7, F3, Fp1) electrodes, in the alpha band, respectively. The waveform of AW cerebral index has been estimated for each seconds and then averaged for all the duration of the stimuli. The AW index was then standardized according to the baseline EEG activity acquired at the beginning and at the end of the experiment. Positive AW values mean an approach motivation toward the stimulus expressed by the subject, while negative AW values a withdrawal tendency.The AW index was in fact normalized returning a z-score values across all the experiment for each subject. In fact, such index has been defined by taking into account the frontal EEG asymmetry's theory by Davidson and coworkers.

AUTONOMIC RECORDINGS AND SIGNAL PROCESSING

The autonomic data, in particular the Galvanic Skin Response (GSR) and the Heart Rate (HR) were recorded by the NeXus-10 (Mindmedia, The Netherlands) instrument with a sampling rate of 128 Hz. Disposable electrodes provided by the Mindmedia company were applied to the subject's wrist to collect cardiac activity. The Pan-Tompkins algorithm (Pan & Tompkins 1985) has been used to obtain the HR signal. Skin conductance was acquired by the constant voltage method (0.5 V). The electrodes were attached, on non-dominant hand, to the palmar side of the middle phalanges of the second and third fingers of the participant, following published procedures (Boucsein et al. 2012). The tonic component of the skin conductance (Skin Conductance Level, SCL) was obtained using LEDAlab software (Benedek & Kaernbach 2010). In order to match GSR and HR signals producing a monodimensional variable which returns the emotional state of subjects the Emotional Index (EI) definition has been employed (Vecchiato et al. 2014). For the plotting of GSR and HR the authors referred to the circumplex model of affect plane (Russell & Barrett 1999; Posner et al. 2005), where the coordinates of a point in the space are defined by the HR (horizontal axis) to describe the valence and by the GSR (vertical axis) to describe the arousal phenomena (Mauss & Robinson 2009 for a review). The interpretation of the EI implies that the higher the value the more positive the emotion experienced by the subject and vice versa.

STATISTICAL ANALYSIS

Scenes for the statistical analysis of both the indices investigated in the experiment (AW and EI), repeated measures ANOVA has been performed on the average of the seconds of each of the four selected scenes. The between factor "gender" having 2 levels: men and women; and the within factor "scene" having 4 levels: fireworks, baby girl, soccer players, brand.

Speaker T test was performed to investigate the differences in the perception of the speaker'gender in the two experimental gender subgroups (men and women). This analysis was conducted for both the indices (AW and EI), with between factor "gender" (2 levels: men and women) and within factor "speaker" (2 levels: man speaker, woman speaker).

RESULTS AND DISCUSSION

Frontal Alpha Band GFP Asymmetry (AW Index)

Scenes The profile of the curves representing the average AW values reported by men and women for each scene composing the TV advertisement, showed a similar trend in the approach and withdrawal motivation between the two experimental subgroups. The women group expressed positive values, suggesting a slight higher approach motivation than men kept along all the spot development, even if not reaching the statistical significance (gender effect F=1.65 p=0.21).

The effect of the factor scenes didn't produce a significant effect between the subgroups considered (F=1.66 p=0.19), with the men group expressing less variable AW values among the scenes than women, even though without reaching a statistical significant difference in the comparison between scenes within each group (men: F=0.84 p=0.48; women: F=2.52 p=0.09. The effect of the interaction gender*scene did not reach the statistical significance (F=0.20 p=0.89).

Speaker Analyzing the comparison of the AW values in response to the Men Speaker there was a statistically significant increase in the men group in comparison to the women group (t=-2.43 p=0.03), but there was no difference in response to the Woman Speaker (t=0.16 p=0.88) (Fig.3). There wasn't any statistically significant effect of the gender (F=1.28 p=0.27), of the speaker (F=0.73 p=0.40) nor the interaction between the gender of the subjects and the speaker's one (gender*speaker F=1.92 p=0.18). The analysis of the difference in the speaker's gender perception within each group didn't reveal any significance (women: t=0.01 p=0.99; men: t=1.27 p=0.23).

Autonomic Data (Emotional Index)

Scenes The factor scenes produced a significant effect (F=9.55 p<0.01) and there was also a significant interaction of the factors gender*scenes (F=6.43 p<0.01).

Performing a group by group (men versus women) comparison for each of the investigated scenes, a statistical significant difference has been highlighted for the soccer players scene (t=-2.85 p=0.01) and for the brand scene (t=2.22 p=0.04) (Figure 4).

Considering the different perception of the scenes within each group there wasn't any significance for the women group (F=0.87 p=0.47), but in the men group a statistical significance has been shown (F=8.5 p<0.01), being the soccer players scenes the most emotionally involving in comparison to all the other scenes.

Figure 3. AW index expressing the GFP frontal alpha band asymmetry. The AW index showed a statistically significant difference between the groups for the Man Speaker (unpaired t-test p=0.03).

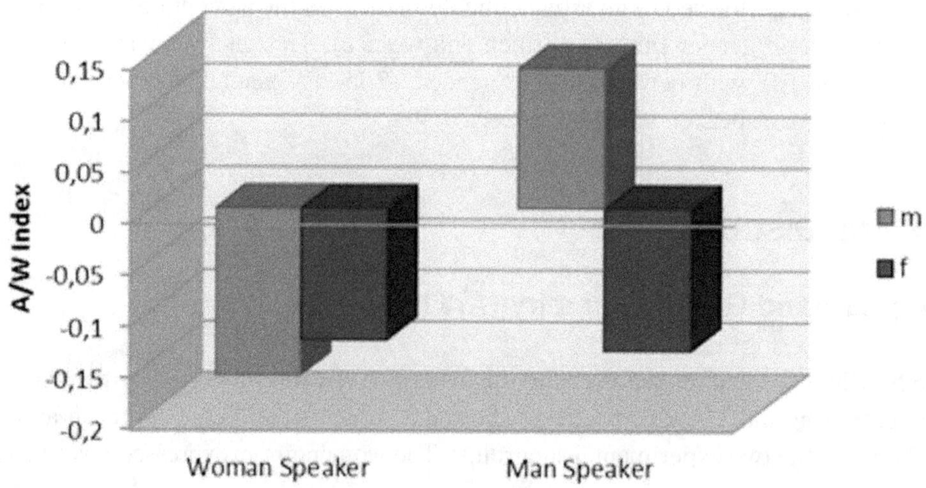

Figure 4. Emotional index (EI) expressing emotional involvement of the subjects as returned by the HR and GSR signals processing. The EI showed a statistically significant difference between the groups for the soccer players scene (unpaired t-test p=0.01) and the brand scene (unpaired t-test p=0.04).

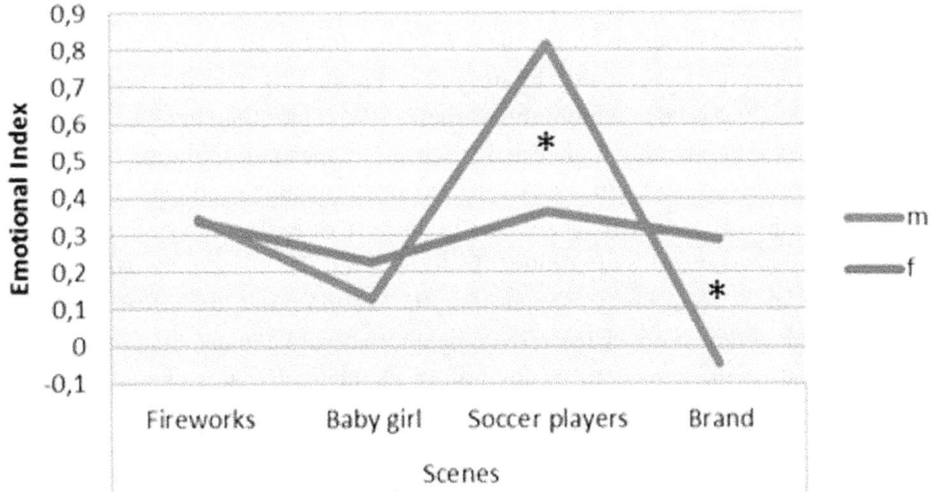

Speaker There was a statistically significant effect of the gender (F=6.93 p=0.02), furthermore women subjects expressed a higher emotional reaction toward the woman speaker than men sample (gender*speaker F=6.21 p=0.02). No statistical significant effect has been found for the factor speaker per se (F=0.15 p=0.07). In addition, analyzing the effect of the speaker's gender within each group no differences have been reported (women: F=3.61 p=0.09; men: F=3.19 p=0.12)

No significant difference was shown by the comparison between the means of the pleasantness rating provided by the men (mean 6.12 ± s.d. 1.05) and by the women group (mean 6.37 ± s.d. 0.92), t=0.55 p=0.59.

DISCUSSION

The different perception reported by men and women in the experiment conducted suggested: 1) the targeting of scenes on the basis of their content (e.g. sport subjects) ; 2) a different reaction to auditory (feminine or masculine voice of the speaker) features of the commercial.

Concerning the first point, in support of this hypothesis of a sex-based dimorphic interest, there are behavioral evidences based on the rating assigned to men's, women's, girls', and boys' magazines, the most preferred advertisement for women was pleasant and active, while for men unpleasant and active (Whissell & McCall, 1997). Furthermore, considering a very specific example as the influence of gender on the impact of food advertising, results are not univocal and need further investigation. In particular, studies found an increase in women food consumption, and in men food intake as a consequence of the exposure to commercials about food, but also other researchers didn't find any difference between men and women in the food intake while watching food-advertising (Mills et al. 2013). In addition to behavioral results, there are magnetoencephalographic evidences of an enhanced negativity on the occipito-temporal areas activity (early posterior negativity, EPN), considered a neural signature of motivated attention, expressed by female participants in response to feminine stimuli (shoes) and on the contrary in response to masculine stimuli (motorcycles) by male subjects (Junghöfer et al., 2010). Another evidence is represented by the higher AW highlighted by women during the all length of a female perfume commercial, a product typically addressed to women (Vecchiato et al., 2014). Additionally, the same researchers found that after an analysis of different scene, the AW values for the intro were higher for men than for women, this could explain the significantly higher value reported by men for the first scene, the baby cry scene, that theoretically should rise more the women's AW values than the men's ones. The cerebral response to infants in fact, produces a higher activation in the mesocorticolimbic system of women in comparison to men (Proverbio et al., 2011). A very recent study suggested also the importance of the creativity for the targeting of an advertisement also in the case of typical masculine products, e.g. cars, and the possible different perception of auditory features, even the silence, between male and female observers (Cartocci et al. 2016 in press). Concerning the second point, the advertisement presents an interesting auditory characterization, characterized by alternatively a male and a female voice speaker. A previous study based on the behavioral evaluation of the characteristics of male and female voices during and advertisement, did not show any significant effect of the sex of the speaker or the receiver on the perceived credibility of the voice (Chebat et al., 2007). Finally, biosignals analysis showed to be able to confer more information, concerning the different perception between groups, than behavioral data, which did not highlight any difference in the pleasantness perception of the advertisement.

FUTURE RESEARCH DIRECTIONS

Although gender differences have been historically taken into account by traditional marketing research, the issue of the present chapter witnesses, also from an electroencephalographic point of view, the importance of the knowledge of the desired (or undesired) target of TV ads. It has been shaping a "sex-dimorphic classification of the processing of consumer goods" (Junghöfer at al., 2010). Finally aiming to address the public of interest for the message, delivered by appropriate creative solutions. Gender issues represent a field of research in neuromarketing requiring deeper evaluation of the different perception of TV ads, and it would be worthy to extend the study of the different reaction also to commercials with no "product" subjects, such as public service announcements.

CONCLUSION

The sum of the present evidences state the importance of the creativity in advertising, whose peculiar features could address the product mainly towards a particular population. Furthermore, present results are coherent with studies that highlighted the criticality of the creative/story telling part of commercials (Ohme et al., 2009; 2010). In particular, this matter could be mostly of interest for the section preceding the presentation of the product, as a feasible resource to elicit the engagement in users. In order to attract the selected target population, the link between behavioral response and EEG activity could be used to generate a metric helping in selecting scene that could fit with the possible appreciation response of the group. This information could also be used "a posteriori" to redraw partially the advertising in order to increase the appearance of the ''like'' parts while depressing the ''dislike'' parts.

ACKNOWLEDGMENT

The present research was partially supported by the Minister of Foreign Affair under the research program between Italy and China "NeuroPredictor" as well as by the European Project "Smoke Free Brain" under the Horizon2020 research framework. The support of BrainSigns srl is also gratefully recognized.

REFERENCES

Adolphs, R. (2003). Cognitive neuroscience of human social behaviour. *Nature Reviews. Neuroscience*, *4*(3), 165–178. doi:10.1038/nrn1056 PMID:12612630

Allen, J. J. B., & Kline, J. P. (2004). Frontal EEG asymmetry, emotion, and psychopathology: The first, and the next 25 years. *Biological Psychology*, *67*(1–2), 1–5. doi:10.1016/j.biopsycho.2004.03.001 PMID:15130523

Ambler, T., Ioannides, A., & Rose, S. (2000). Brands on the Brain: Neuro-Images of Advertising. *Business Strategy Review*, *11*(3), 17–30. doi:10.1111/1467-8616.00144

Anschutz, D. J., Engels, R. C. M. E., van der Zwaluw, C. S., & Van Strien, T. (2011). Sex differences in young adults' snack food intake after food commercial exposure. *Appetite*, *56*(2), 255–260. doi:10.1016/j.appet.2010.12.010 PMID:21172394

Ariely, D., & Berns, G. S. (2010). Neuromarketing: The hope and hype of neuroimaging in business. *Nature Reviews. Neuroscience*, *11*(4), 284–292. doi:10.1038/nrn2795 PMID:20197790

Astolfi, L., Fallani, F. D. V., Cincotti, F., Mattia, D., Bianchi, L., Marciani, M. G., & Babiloni, F. et al. (2009). Brain activity during the memorization of visual scenes from TV commercials: An application of high resolution EEG and steady state somatosensory evoked potentials technologies. *Journal of Physiology, Paris*, *103*(6), 333–341. doi:10.1016/j.jphysparis.2009.07.002 PMID:19619647

Astolfi, L., Soranzo, R., Cincotti, F., Mattia, D., Scarano, G., & Gaudiano, I., … Babiloni, F. (2008). Assessing the memorization of TV commercials with the use of high resolution EEG: a pilot study. *Conference Proceedings: ... Annual International Conference of the IEEE Engineering in Medicine and Biology Society. IEEE Engineering in Medicine and Biology Society. Annual Conference, 2008*. http://doi.org/ doi:10.1109/IEMBS.2008.4650025

Astolfi, L., Toppi, J., De Vico Fallani, F., Vecchiato, G., Cincotti, F., Wilke, C. T., & Babiloni, F. et al. (2011). Imaging the Social Brain by Simultaneous Hyperscanning During Subject Interaction. *IEEE Intelligent Systems*, *26*(5), 38–45. doi:10.1109/MIS.2011.61 PMID:22287939

Babiloni, F., Cincotti, F., Babiloni, C., Carducci, F., Mattia, D., Astolfi, L., & He, B. et al. (2005). Estimation of the cortical functional connectivity with the multimodal integration of high-resolution EEG and fMRI data by directed transfer function. *NeuroImage*, *24*(1), 118–131. doi:10.1016/j.neuroimage.2004.09.036 PMID:15588603

Bai, X., Towle, V. L., He, E. J., & He, B. (2007). Evaluation of cortical current density imaging methods using intracranial electrocorticograms and functional MRI. *NeuroImage*, *35*(2), 598–608. doi:10.1016/j.neuroimage.2006.12.026 PMID:17303438

Bargh, J. A., & Chartrand, T. L. (1999). The unbearable automaticity of being. *The American Psychologist*, *54*(7), 462–479. doi:10.1037/0003-066X.54.7.462

Baumgartner, T., Esslen, M., & Jäncke, L. (2006). From emotion perception to emotion experience: Emotions evoked by pictures and classical music. *International Journal of Psychophysiology: Official Journal of the International Organization of Psychophysiology*, *60*(1), 34–43. doi:10.1016/j.ijpsycho.2005.04.007 PMID:15993964

Benedek, M., & Kaernbach, C. (2010). A continuous measure of phasic electrodermal activity. *Journal of Neuroscience Methods*, *190*(1–5), 80–91. doi:10.1016/j.jneumeth.2010.04.028 PMID:20451556

Boucsein, W., Fowles, D. C., Grimnes, S., Ben-Shakhar, G., Roth, W. T., Dawson, M. E., & Filion, D. L.Society for Psychophysiological Research Ad Hoc Committee on Electrodermal Measures. (2012). Publication recommendations for electrodermal measurements. *Psychophysiology*, *49*(8), 1017–1034. doi:10.1111/j.1469-8986.2012.01384.x PMID:22680988

Braeutigam, S. (2005). Neuroeconomics—From neural systems to economic behaviour. *Brain Research Bulletin*, *67*(5), 355–360. doi:10.1016/j.brainresbull.2005.06.009 PMID:16216681

Briesemeister, B. B., Tamm, S., Heine, A., & Jacobs, A. M. (2013). Approach the Good, Withdraw from the Bad—A Review on Frontal Alpha Asymmetry Measures in Applied Psychological Research. *Psychology (Savannah, Ga.)*, *04*(03), 261–267. doi:10.4236/psych.2013.43A039

Cacioppo, J. T. (2002). *Foundations in Social Neuroscience*. MIT Press.

Cartocci, G., Cherubino, P., Rossi, D., Modica, E., Maglione, A.G., Di Flumeri, G., & Babiloni, F. (2016. (in press). Gender and age related effects while watching TV advertisements: An EEG study. *Computational Intelligence and Neuroscience*. PMID:27313602

Chebat, J. C., El Hedhli, K., Gelinas-Chebat, C., & Boivin, R. (2007). Voice and persuasion in a banking telemarketing context 1, 2. *Perceptual and Motor Skills*, *104*(2), 419–437. doi:10.2466/pms.104.2.419-437 PMID:17566432

Cherubino, P., Trettel, A., Cartocci, G., Rossi, D., Modica, E., Maglione, A. G., & Babiloni, F. (2016). Neuroelectrical Indexes for the Study of the Efficacy of TV Advertising Stimuli. In *Selected Issues in Experimental Economics* (pp. 355–371). Springer International Publishing. doi:10.1007/978-3-319-28419-4_22

Clayforth, C., Pettigrew, S., Mooney, K., Lansdorp-Vogelaar, I., Rosenberg, M., & Slevin, T. (2014). A cost-effectiveness analysis of online, radio and print tobacco control advertisements targeting 25-39 year-old males. *Australian and New Zealand Journal of Public Health*, *38*(3), 270–274. doi:10.1111/1753-6405.12175 PMID:24750454

Coan, J. A., & Allen, J. J. B. (2004). Frontal EEG asymmetry as a moderator and mediator of emotion. *Biological Psychology*, *67*(1-2), 7–49. doi:10.1016/j.biopsycho.2004.03.002 PMID:15130524

Critchley, H. D. (2002). Electrodermal responses: What happens in the brain. *The Neuroscientist*, *8*(2), 132–142. doi:10.1177/107385840200800209 PMID:11954558

Dale, A. M., Liu, A. K., Fischl, B. R., Buckner, R. L., Belliveau, J. W., Lewine, J. D., & Halgren, E. (2000). Dynamic statistical parametric mapping: Combining fMRI and MEG for high-resolution imaging of cortical activity. *Neuron*, *26*(1), 55–67. doi:10.1016/S0896-6273(00)81138-1 PMID:10798392

Damasio, A. R., Everitt, B. J., & Bishop, D. (1996). The Somatic Marker Hypothesis and the Possible Functions of the Prefrontal Cortex. *Philosophical Transactions of the Royal Society of London. Series B, Biological Sciences*, *351*(1346), 1413–1420. doi:10.1098/rstb.1996.0125 PMID:8941953

Davidson, R. J. (1988). EEG measures of cerebral asymmetry: Conceptual and methodological issues. *The International Journal of Neuroscience*, *39*(1-2), 71–89. doi:10.3109/00207458808985694 PMID:3290140

Davidson, R. J. (1994). Asymmetric brain function, affective style, and psychopathology: The role of early experience and plasticity. *Development and Psychopathology*, *6*(04), 741–758. doi:10.1017/S0954579400004764

Davidson, R. J. (2000). Affective style, psychopathology, and resilience: Brain mechanisms and plasticity. *The American Psychologist*, *55*(11), 1196–1214. doi:10.1037/0003-066X.55.11.1196 PMID:11280935

Davidson, R. J. (2002). Anxiety and affective style: Role of prefrontal cortex and amygdala. *Biological Psychiatry*, *51*(1), 68–80. doi:10.1016/S0006-3223(01)01328-2 PMID:11801232

Davidson, R. J. (2004). What does the prefrontal cortex "do" in affect: Perspectives on frontal EEG asymmetry research. *Biological Psychology*, *67*(1-2), 219–233. doi:10.1016/j.biopsycho.2004.03.008 PMID:15130532

Davidson, R. J., Abercrombie, H., Nitschke, J. B., & Putnam, K. (1999). Regional brain function, emotion and disorders of emotion. *Current Opinion in Neurobiology*, *9*(2), 228–234. doi:10.1016/S0959-4388(99)80032-4 PMID:10322186

Davidson, R. J., Schwartz, G. E., Saron, C., Bennett, J., & Goleman, D. J. (1979). Frontal versus parietal EEG asymmetry during positive and negative affect. In Psychophysiology (Vol. 16, pp. 202–203). Cambridge Univ Press.

Davidson & Irwin. (1999). The functional neuroanatomy of emotion and affective style. *Trends in Cognitive Sciences, 3*(1), 11–21.

Erk, S., Spitzer, M., Wunderlich, A. P., Galley, L., & Walter, H. (2002). Cultural objects modulate reward circuitry. *Neuroreport*, *13*(18), 2499–2503. doi:10.1097/00001756-200212200-00024 PMID:12499856

Franzen, G., & Bouwman, M. (2001). *The mental world of brands: Mind, memory and brand success.* Oxford, UK: World Advertising Research Centre.

Hallez, H., Vanrumste, B., Grech, R., Muscat, J., De Clercq, W., Vergult, A., & Lemahieu, I. et al. (2007). Review on solving the forward problem in EEG source analysis. *Journal of Neuroengineering and Rehabilitation*, *4*(1), 46. doi:10.1186/1743-0003-4-46 PMID:18053144

Handy, T. C., Smilek, D., Geiger, L., Liu, C., & Schooler, J. W. (2010). ERP evidence for rapid hedonic evaluation of logos. *Journal of Cognitive Neuroscience*, *22*(1), 124–138. doi:10.1162/jocn.2008.21180 PMID:19199410

Harmon-Jones, E., Gable, P. A., & Peterson, C. K. (2010). The role of asymmetric frontal cortical activity in emotion-related phenomena: A review and update. *Biological Psychology*, *84*(3), 451–462. doi:10.1016/j.biopsycho.2009.08.010 PMID:19733618

He, B., Wang, Y., & Wu, D. (1999). Estimating cortical potentials from scalp EEG's in a realistically shaped inhomogeneous head model by means of the boundary element method. *IEEE Transactions on Bio-Medical Engineering*, *46*(10), 1264–1268. doi:10.1109/10.790505 PMID:10513133

Ioannides, A. A., Liu, L., Theofilou, D., Dammers, J., Burne, T., Ambler, T., & Rose, S. (2000). Real time processing of affective and cognitive stimuli in the human brain extracted from MEG signals. *Brain Topography*, *13*(1), 11–19. doi:10.1023/A:1007878001388 PMID:11073090

Junghöfer, M., Kissler, J., Schupp, H. T., Putsche, C., Elling, L., & Dobel, C. (2010). A fast neural signature of motivated attention to consumer goods separates the sexes. *Frontiers in Human Neuroscience*, *4*, 179. doi:10.3389/fnhum.2010.00179 PMID:21079751

Kato, J., Ide, H., Kabashima, I., Kadota, H., Takano, K., & Kansaku, K. (2009). Neural Correlates of Attitude Change Following Positive and Negative Advertisements. *Frontiers in Behavioral Neuroscience*, *3*. doi:10.3389/neuro.08.006.2009 PMID:19503749

Kenning, P., & Plassmann, H. (2005). Neuroeconomics: An overview from an economic perspective. *Brain Research Bulletin*, *67*(5), 343–354. doi:10.1016/j.brainresbull.2005.07.006 PMID:16216680

Klimesch, W. (1999). EEG alpha and theta oscillations reflect cognitive and memory performance: A review and analysis. *Brain Research. Brain Research Reviews*, *29*(2), 169–195. doi:10.1016/S0165-0173(98)00056-3 PMID:10209231

Klucharev, V., Smidts, A., & Fernández, G. (2008). Brain mechanisms of persuasion: How "expert power" modulates memory and attitudes. *Social Cognitive and Affective Neuroscience*, *3*(4), 353–366. doi:10.1093/scan/nsn022 PMID:19015077

Knutson, B., & Bossaerts, P. (2007). Neural antecedents of financial decisions. *The Journal of Neuroscience*, *27*(31), 8174–8177. doi:10.1523/JNEUROSCI.1564-07.2007 PMID:17670962

Knutson, B., Rick, S., Wimmer, G. E., Prelec, D., & Loewenstein, G. (2007). Neural predictors of purchases. *Neuron*, *53*(1), 147–156. doi:10.1016/j.neuron.2006.11.010 PMID:17196537

Lee, N., Broderick, A. J., & Chamberlain, L. (2007). What is "neuromarketing"? A discussion and agenda for future research. *International Journal of Psychophysiology: Official Journal of the International Organization of Psychophysiology*, *63*(2), 199–204. doi:10.1016/j.ijpsycho.2006.03.007 PMID:16769143

Lehmann, D., & Skrandies, W. (1980). Reference-free identification of components of checkerboard-evoked multichannel potential fields. *Electroencephalography and Clinical Neurophysiology*, *48*(6), 609–621. doi:10.1016/0013-4694(80)90419-8 PMID:6155251

Malik, M., Bigger, J. T., Camm, A. J., Kleiger, R. E., Malliani, A., Moss, A. J., & Schwartz, P. J. (1996). Task Force of the European Society of Cardiology. Heart rate variability standards of measurement, physiological interpretation, and clinical use. *European Heart Journal, 17*(3), 354–381. doi:10.1093/oxfordjournals.eurheartj.a014868 PMID:8737210

Malliani, A. (2005). Heart rate variability: From bench to bedside. *European Journal of Internal Medicine, 16*(1), 12–20. doi:10.1016/j.ejim.2004.06.016 PMID:15733815

Mauss, I. B., & Robinson, M. D. (2009). Measures of emotion: A review. *Cognition and Emotion, 23*(2), 209–237. doi:10.1080/02699930802204677 PMID:19809584

Mc Donald, C. (2003), Hearth rate variability: from bench to bedside. *Eur J Intern Med, 16*(1), 12-20.

Michel, C. M., & Murray, M. M. (2012). Towards the utilization of EEG as a brain imaging tool. *NeuroImage, 61*(2), 371–385. doi:10.1016/j.neuroimage.2011.12.039 PMID:22227136

Michel, C. M., Murray, M. M., Lantz, G., Gonzalez, S., Spinelli, L., & Grave de Peralta, R. (2004). EEG source imaging. *Clinical Neurophysiology: Official Journal of the International Federation of Clinical Neurophysiology, 115*(10), 2195–2222. doi:10.1016/j.clinph.2004.06.001 PMID:15351361

Mills, S. D. H., Tanner, L. M., & Adams, J. (2013). Systematic literature review of the effects of food and drink advertising on food and drink-related behaviour, attitudes and beliefs in adult populations. *Obesity Reviews, 14*(4), 303–314. doi:10.1111/obr.12012 PMID:23297736

Montano, N., Porta, A., Cogliati, C., Costantino, G., Tobaldini, E., Casali, K. R., & Iellamo, F. (2009). Heart rate variability explored in the frequency domain: A tool to investigate the link between heart and behavior. *Neuroscience and Biobehavioral Reviews, 33*(2), 71–80. doi:10.1016/j.neubiorev.2008.07.006 PMID:18706440

Morris, J. D., Klahr, N. J., Shen, F., Villegas, J., Wright, P., He, G., & Liu, Y. (2009). Mapping a multidimensional emotion in response to television commercials. *Human Brain Mapping, 30*(3), 789–796. doi:10.1002/hbm.20544 PMID:18286463

Nagai, Y., Critchley, H. D., Featherstone, E., Fenwick, P. B. C., Trimble, M. R., & Dolan, R. J. (2004). Brain activity relating to the contingent negative variation: An fMRI investigation. *NeuroImage, 21*(4), 1232–1241. doi:10.1016/j.neuroimage.2003.10.036 PMID:15050551

Nunez, P. L., & Cutillo, B. A. (Eds.). (1995). *Neocortical dynamics and human EEG rhythms*. Oxford University Press.

Ohme, R., Reykowska, D., Wiener, D., & Choromanska, A. (2009). Analysis of neurophysiological reactions to advertising stimuli by means of EEG and galvanic skin response measures. *Journal of Neuroscience, Psychology, and Economics, 2*(1), 21.

Ohme, R., Reykowska, D., Wiener, D., & Choromanska, A. (2010). Application of frontal EEG asymmetry to advertising research. *Journal of Economic Psychology, 31*(5), 785–793. doi:10.1016/j.joep.2010.03.008

Pan, J., & Tompkins, W..J. (1985). A real-time QRS detection algorithm. *Biomedical Engineering, IEEE Transactions on*, (3), 230-236.

Panksepp, J., Nocjar, C., Burgdorf, J., Panksepp, J. B., & Huber, R. (2004). The role of emotional systems in addiction: A neuroethological perspective. *Nebraska Symposium on Motivation*, *50*, 85–126. PMID:15160639

Plassmann, H., Ambler, T., Braeutigam, S., & Kenning, P. (2007). What can advertisers learn from neuroscience? *International Journal of Advertising*, *26*(2), 151–175. doi:10.1080/10803548.2007.11073005

Plassmann, H., Ramsøy, T. Z., & Milosavljevic, M. (2012). Branding the brain: A critical review and outlook. *Journal of Consumer Psychology*, *22*(1), 18–36. doi:10.1016/j.jcps.2011.11.010

Posner, J., Russell, J. A., & Peterson, B. S. (2005). The circumplex model of affect: An integrative approach to affective neuroscience, cognitive development, and psychopathology. *Development and Psychopathology*, *17*(3), 715–734. doi:10.1017/S0954579405050340 PMID:16262989

Preuschoff, K., Quartz, S. R., & Bossaerts, P. (2008). Human Insula Activation Reflects Risk Prediction Errors As Well As Risk. *The Journal of Neuroscience*, *28*(11), 2745–2752. doi:10.1523/JNEUROSCI.4286-07.2008 PMID:18337404

Price, T. F., Peterson, C. K., & Harmon-Jones, E. (2012). The emotive neuroscience of embodiment. *Motivation and Emotion*, *36*(1), 27–37. doi:10.1007/s11031-011-9258-1

Proverbio, A. M., Riva, F., Zani, A., & Martin, E. (2011). Is It a Baby? Perceived Age Affects Brain Processing of Faces Differently in Women and Men. *Journal of Cognitive Neuroscience*, *23*(11), 3197–3208. doi:10.1162/jocn_a_00041 PMID:21557651

Rilling, J. K., & Sanfey, A. G. (2011). The neuroscience of social decision-making. *Annual Review of Psychology*, *62*(1), 23–48. doi:10.1146/annurev.psych.121208.131647 PMID:20822437

Russell, J. A., & Barrett, L. F. (1999). Core affect, prototypical emotional episodes, and other things called emotion: Dissecting the elephant. *Journal of Personality and Social Psychology*, *76*(5), 805–819. doi:10.1037/0022-3514.76.5.805 PMID:10353204

Sandfort, J. R., & Pleasant, A. (2009). Knowledge, attitudes, and informational behaviors of college students in regard to the human papillomavirus. *Journal of American College Health: J of ACH*, *58*(2), 141–149. doi:10.1080/07448480903221368 PMID:19892651

Scott, B. E., Schmidt, W. P., Aunger, R., Garbrah-Aidoo, N., & Animashaun, R. (2008). Marketing hygiene behaviours: The impact of different communication channels on reported handwashing behaviour of women in Ghana. *Health Education Research*, *23*(3), 392–401. doi:10.1093/her/cym056 PMID:18000025

Slater, M. D., Rouner, D., Murphy, K., Beauvais, F., Van Leuven, J., & Rodríguez, M. D. (1996). Male adolescents' reactions to TV beer advertisements: The effects of sports content and programming context. *Journal of Studies on Alcohol*, *57*(4), 425–433. doi:10.15288/jsa.1996.57.425 PMID:8776685

Urbano, A., Babiloni, C., Onorati, P., & Babiloni, F. (1998). Dynamic functional coupling of high resolution EEG potentials related to unilateral internally triggered one-digit movements. *Electroencephalography and Clinical Neurophysiology*, *106*(6), 477–487. doi:10.1016/S0013-4694(97)00150-8 PMID:9741747

Vecchiato, G., Cherubino, P., Trettel, A., & Babiloni, F. (2013). *Neuroelectrical Brain Imaging Tools for the Study of the Efficacy of TV Advertising Stimuli and Their Application to Neuromarketing*. Springer. Retrieved from http://link.springer.com/content/pdf/10.1007/978-3-642-38064-8.pdf

Vecchiato, G., Kong, W., Maglione, A. G., & Wei, D. (2012). Understanding the impact of TV commercials: Electrical neuroimaging. *IEEE Pulse*, *3*(3), 42–47. doi:10.1109/MPUL.2012.2189171 PMID:22678840

Vecchiato, G., Maglione, A. G., Cherubino, P., Wasikowska, B., Wawrzyniak, A., Latuszynska, A., & Babiloni, F. et al. (2014). Neurophysiological Tools to Investigate Consumer's Gender Differences during the Observation of TV Commercials. *Computational and Mathematical Methods in Medicine*, *2014*, 1–12. doi:10.1155/2014/912981

Vecchiato, G., Toppi, J., Astolfi, L., Fallani, F. D. V., Cincotti, F., Mattia, D., & Babiloni, F. et al. (2011). Spectral EEG frontal asymmetries correlate with the experienced pleasantness of TV commercial advertisements. *Medical & Biological Engineering & Computing*, *49*(5), 579–583. doi:10.1007/s11517-011-0747-x PMID:21327841

Whissell, C., & McCall, L. (1997). Pleasantness, activation, and sex differences in advertising. *Psychological Reports*, *81*(2), 355–367. doi:10.2466/pr0.1997.81.2.355 PMID:9354085

Wonderlich-Tierney, A. L., Wenzel, K. R., Vander Wal, J. S., & Wang-Hall, J. (2013). Food-related advertisements and food intake among adult men and women. *Appetite*, *71*, 57–62. doi:10.1016/j.appet.2013.07.009 PMID:23917064

Zaltman, G. (2003). *How Customers Think: Essential Insights Into the Mind of the Market*. Harvard Business Press.

KEY TERMS AND DEFINITIONS

Approach and Withdrawal Motivation: EEG asimmetry during emotional states. they are two motivational systems that organize behavior. one system (left pFC) involves behavior prompted by a possible desirable outcome, whereas the other (right pfc) involves behavior prompted by a possible aversive outcome.

Approach Withdrawal (AW) Index: The present index is calculated estimating the difference between the right and the left frontal EEG activity, as first described by Davidson and colleagues in 1994. Positive values of the index suggest an approach behavior toward a stimulus, vice versa negative values a withdrawal behavior.

Consumer Neuroscience: Approach within consumer research that adapt methods and theories from neuroscience combined with behavioral theories, models, and tested experimental designs from consumer psychology and related disciplined such as behavioral decision sciences to develop a neuropsychologically sound theory to understand consumer behavior.

Electroencephalography (EEG): The recording of electrical activity along the scalp. eeg measures voltage fluctuations resulting from ionic current flows within the neurons of the brain.

Emotional Index (EI): The index derived from the mathematical matching of the Galvanic Skin Response (GSR) and Heart Rate (HR) signals. The EI varies between -1 and +1. Negative and positive values of the EI are related to negative and positive emotions, respectively, spanning the whole affect plane (Mauss&Robinson, 2009).

Global Field Power (GFP): According to Lehmann, EEG potentials are recorded simultaneously from many electrodes on the human head surface, and this can be viewed as series of maps of the momentary spatial distributions of electric potential. The distribution varies in electric strength and the GFP constitutes a parametric assessment of map strength, computed as standard deviation of the momentary potential values.

Neuromarketing: A new field of marketing research that studies consumers sensorimotor, cognitive, and affective response to marketing stimuli.

Neuroscience: The scientific study of the nervous system. traditionally, neuroscience has been seen as a branch of biology. however, it is currently an interdisciplinary science that collaborates with other fields such as chemistry, computer science, engineering, linguistics, mathematics, medicine and allied disciplines, philosophy, physics, and psychology.

Subgroup: A portion of the experimental sample identifiable through measurable (e.g. age), observable (e.g. eye color) or reported (e.g. migraine) conditions and characteristics of the subjects assigned to the subgroup; in the present study subgroups assignment was determined by gender.

TV Commercials: 10- to 60-second commercials on network TV stations. It is about relaying a specific social, political or sales pitch in a limited amount of time, broadly ranging in between few seconds to several minutes.

This research was previously published in Applying Neuroscience to Business Practice edited by Manuel Alonso Dos Santos, pages 163-190, copyright year 2017 by Business Science Reference (an imprint of IGI Global).

Chapter 20
The Neuroscience of Social Television

Shaun A Seixas
Neuro-Insight, Australia

Geoffrey E Nield
Neuro-Insight, Australia

Peter Pynta
Neuro-Insight, Australia

Richard B Silberstein
Neuro-Insight, Australia

ABSTRACT

In a short few years, social media has become the dominant way in which we communicate with the outside world. It has become prevalent in almost every aspect of our daily lives, but one of the most significant changes social media has had, has been on the way we watch television. This phenomenon, known as dual screening, has caused some concern amongst marketers and advertisers, who believed that this behaviour was having an overall negative impact on consumer engagement with television. This chapter attempts to address some of these concerns by providing evidence obtained from the neurosciences and from a case study. The evidence we present in this chapter demonstrates the opposite effect, whereby social media can actually be used to enhance viewer engagement.

INTRODUCTION

The availability and widespread use of Internet connected devices has seen a fundamental shift in the way we consume media today. In particular, many of our interactions on the web are now driven around social media, which itself has become a revolution in the way we create and share content. As a result of both of these changes, there has also been a shift in way media is consumed as traditional media streams have fragmented. For marketers and advertisers this has not only meant an increase in the overall number of available consumer touch points but importantly a new avenue for direct interaction with the consumer.

DOI: 10.4018/978-1-5225-5478-3.ch020

One of the most interesting phenomena arising from the social media revolution is dual-screening, where television viewing which was traditionally a sole viewing experience, is now shared with a second Internet-enabled screen. This observation has been backed by a number of quantitative investigations that report widespread dual screening in the household (Deloitte, 2012; Google, 2012; Thinkbox, 2012). Of particular interest is live tweeting behaviour, which not only provides a means for the viewer to express their thoughts and emotions in real-time, but also allows marketers and advertisers access to a new source of consumer driven insight. Twitter now provides a list of best practices so TV show producers know how best to interact and 'engage' viewers and even more recently has joined Nielsen in providing a new TV ratings metric (Nielsen, 2013a; Twitter, 2014).

When you consider that the vast majority of televised content is now shared with a second screen, it can be somewhat difficult to determine what impact if any, dual screening has on the audience engagement. In particular, if we consider that 'engagement' is a finite resource then it is logical to conceive that dividing this resource amongst two different activities may ultimately result in a reduction in the amount of engagement that is directed to TV viewing. While this may be true simplistically, current research from the neurosciences alludes to a more complicated relationship.

A BRIEF INTRODUCTION TO THE HUMAN BRAIN

The human brain is a vastly complicated organ that serves many functions that range from the automatic control of our hearts, to the voluntary control of our feet as we walk. The brain itself is divided into two halves, called hemispheres, which are connected by a thick bundle of nerves fibres known as the corpus callosum. Each hemisphere is divided into separate anatomical and functions regions known as the lobes. The most common of these are the frontal, temporal, parietal and occipital lobes (see Figure 1).

The frontal lobe is the largest of all the lobes and is located at the front of each hemisphere. The frontal lobes subserves many functions ranging from memory encoding (Brewer, Zhao, Desmond, Glover, &

Figure 1. Lateral view of the cerebral lobes of the human brain. Adapted from Gray's Anatomy (Gray, 1918)

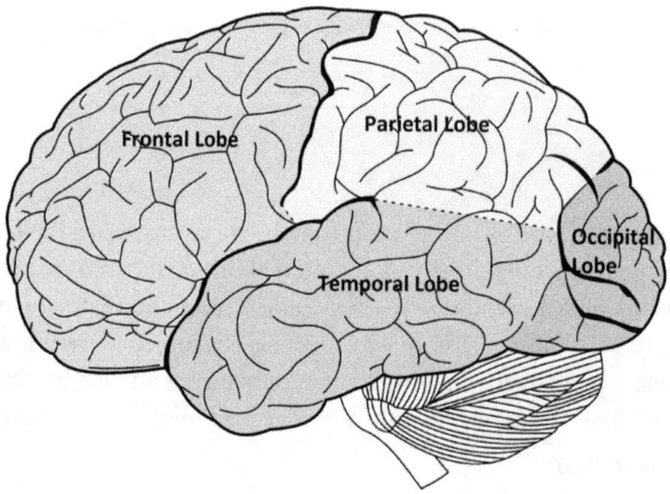

Gabrieli, 1998; Buckner, Kelley, & Petersen, 1999), voluntary motor control (Penfield & Boldrey, 1937; Penfield & Rasmussen, 1950), working memory (Miller & Cohen, 2001) and anticipation (Rolls, 2000). Traditionally seen as a classical sensory association area, the parietal lobe is also implicated in spatial ability, motor planning and attention processes (Bisley & Goldberg, 2010; Snyder, Batista, & Andersen, 2000). A prominent example of this occurs in visual sciences where the parietal lobe is referred to as the dorsal stream or "where" and "how" pathways (Goodale & Milner, 1992; Ungerleider & Mishkin, 1982). The parietal lobe is located adjacent to the frontal, temporal and occipital lobes. The temporal lobe is located towards the lateral side of the head, just beneath the parietal lobe. The lobe is involved in object recognition (Goodale & Milner, 1992; Ungerleider & Mishkin, 1982), facial processing (Sergent, Ohta, & MacDonald, 1992), memory (Maguire, Frith, Burgess, Donnett, & O'Keefe, 1998; Wood, Dudchenko, Robitsek, & Eichenbaum, 2000) and the understanding of language (Wernicke, 1970). Lastly, the occipital lobe is located at the very posterior of the head with its functioning primarily involved in vision.

ENGAGEMENT IS RELATED TO PERSONAL RELEVANCE

Engagement is a term that is widely and frequently used in media, most notably in the area of advertising research. The peak research body for advertising research, the Advertising Research Foundation (ARF) defines engagement as "turning on a prospect to a brand idea enhanced by the surrounding context". Using the findings obtained from the field of neuroscience, we suggest a much simpler definition. Specifically, we suggest that engagement occurs when a situation elicits a 'sense of personal relevance' and we will use this definition throughout the remainder of this chapter. This notion of personal relevance is backed by a number of observations of functional brain activity that have isolated an area within prefrontal region of the brain near the forehead known as Brodmann area 10 (BA10) that is believed to play a role in the processing of engagement.

A functional magnetic resonance imaging (fMRI) study by Phan et al. (2004) was one of the earliest investigations that reported activity related to the processing of personal relevance. In their study, the authors showed participants a series of pictures from the International Affective Picture System (IAPS) and asked them to rate the extent to which they associated with that picture. The activity in BA10 was not only highly correlated with the personal association score, the region also was more active when participants were engaged in appraising the self-relatedness of the images. In line with this approach, Heatherton and colleagues (2006) conducted an investigation that required their participants to evaluate a series of adjectives that either described themselves or a close friend, or to simply state if the adjective was in uppercase. The authors reported increased brain activity in BA10 when participants where evaluating statements pertaining to themselves, and a reduction in activity in the other two conditions which had less personal relevance to the individual. .

Additionally, a subsequent investigation by Abraham & von Cramon (2009) also reported relevance-related activations in BA10. In this investigation participants were asked to think of either a friend or a famous person while undergoing an fMRI scan. The results demonstrated greater activation in BA10 when the participants thought of their friend, a condition that can be argued is more personally relevant to the participant.

These are a few examples from a growing body of literature that have implicated BA10 in the processing of personal relevance or what we term as 'engagement'. From the above observations, it appears that the process is largely automatic and draws upon a number of different processes. These include the recall of explicit and implicit memories, reflection on past experiences and projecting oneself into the current context (Maddock, 1999).

LONG TERM MEMORY ENCODING

Engagement and long term memory (LTM) encoding are two automatic, interdependent neural processes that have a significant influence on our behaviour. Our memory is divided into two different systems comprising short term storage and long term storage. Short term memory, typically referred to as working memory, is a small and limited temporary storage of information which is represented through specific patterns of ongoing neural activity. In contrast, the long term storage (known as LTM) is the storage of information of more than one hour to decades. In regards to differences in brain activity, LTM information is stored in the form of changes in the strength of synaptic contacts (Hebb, 1949) through a process known as long term potentiation (Cooke & Bliss, 2006).

LTM is commonly categorised into explicit and implicit memories with different memory subtypes within each group. Explicit memories comprise episodic (memories of personal experiences) and semantic (memories of facts) memories, whereas implicit memories comprise procedural memories and priming. In this section we restrict ourselves to explicit memories. The form in which information is stored in LTM is influenced by many factors, an important one being the period elapsing since the memory was first encoded or stored. A considerable body of evidence points to long-term memories being initially stored in the hippocampus and then over a period of days being reorganised and redistributed within various regions of the cerebral cortex (Teng & Squire, 1999). The processes of engagement and LTM are closely linked to each other. When you consider the sheer amount of information that we are subjected to daily, it becomes apparent that only a few select memories are stored in LTM. What determines what is stored and what is forgotten? A potential explanation to this question may be engagement, which could mediate the strength of encoding processes. For example, taking an evolutionary approach, it is fair to say that the experiences that have survival or reproductive significance are the most valuable to the individual (LaBar & Cabeza, 2006). In this sense, LTM encoding is an indication that the "encoded experience" is of personal relevance to the individual, irrespective of whether the source of this personal relevance is consciously recognised.

In terms of LTM encoding, a number of functional neuroimaging investigations have reported an association between the LTM encoding process and activity in an area known as the lateral prefrontal cortex (PFC)(Brewer, et al., 1998; Buckner, et al., 1999). The link between lateral PFC activity and encoding has been demonstrated across a variety of experimental paradigms, which have revealed a number of noteworthy features of the PFC response. Firstly, PFC activation is related to encoding of the meaning rather than the surface features of stimuli (Fletcher, Frith, & Rugg, 1997; Fletcher, Shallice, & Dolan, 1998, 2000). Secondly, PFC activation occurs independently of the intention to memorise material (Buckner, et al., 1999; Demb et al., 1995). Lastly, PFC activation has been linked with encoding success (Wagner et al., 1998).

Based on these findings, the lateral PFC has been proposed to be intimately involved in the process by which LTM encoding occurs. In support of this view, converging evidence from event-related potential and functional brain imaging studies suggests that cortical activity at posterior regions of the frontal cortex and prefrontal cortex predicts the strength of LTM traces (Brewer, et al., 1998; Buckner, et al.,

1999). Furthermore, the nature of the material also influences the laterality of activation. In particular, the encoding of verbal material associated with left frontal activation while encoding non-verbal material such as unfamiliar faces has been associated with right frontal activation (Kelley et al., 1998; Sergerie, Lepage, & Armony, 2005).

In the field of advertising research LTM has been identified as a key measure in predicting future behaviour (Silberstein & Nield, 2008). High levels of LTM encoding are an indication that the message of the advertisement, whether it is a sale at Saks®, or the launch of a new Mustang®, is considered to be personally relevant by the brain, which in turn may drive or reinforce behaviour. If LTM encoding is also high during branding (a key indicator for advertising effectiveness), then this information is incorporated by the brain into future behaviours. Now it should be noted that we are not saying that high levels of branding will make you go out and buy that Mustang®, but rather LTM encoding reveals what factor may drive behaviour.

MEASURING AUDIENCE ENGAGEMENT IN A LIFE-LIKE SETTING

The majority of academic research in the area has used fMRI to investigate the regions that underpin the neural process of engagement and memory processing. While there are certainly strengths in using this imaging method- namely the high spatial resolution that allows researchers to investigate deeper cortical activity- there are a number of considerations that make this imaging method less than ideal for real-world audience measurement. The largest and most practical concern is that the setting in which an fMRI study takes place is very unrealistic and far removed from would be considered a "typical" home setting. Most MRI scanners are located within hospitals or Universities and require participants to view content whilst lying on their backs in a small confined space that can be very claustrophobic. Also, the specific characteristics of MRI scanners limit their utility. MRI scanners have poor temporal resolution, restricting their ability to detect rapid changes in brain activity associated with a rapidly changing stimulus that is typical of advertising material - such as a television commercial. Furthermore, the low signal to noise ratio of MRI scanners means responses to multiple presentations (repetitions) of a stimulus must be averaged to produce a reliable result. This is satisfactory in an academic research environment but becomes problematic when assessing a real life stimulus (such as a TV commercial), where it is important to understand the impact of a single viewing and, importantly, how responses to the stimulus (TV commercial) change from one viewing to the next.

Brain Imaging Using Steady State Topography

Steady State Topography (SST) is a non-invasive method of measuring the electrical activity of the brain (Silberstein et al., 1990). It uses a dim ongoing sinusoidal visual stimulus that is presented in the visual periphery while participants are performing a cognitive task, which in this case might be to sending out a tweet while watching a favourite TV program. The visual stimulus is present the whole time and elicits a small rhythmic sinusoidal response at the stimulus frequency that is termed the Steady State Visually Evoked Potential (SSVEP). The sinusoidal nature of the stimulus signal means the SSVEP response can be characterised entirely by its amplitude and the phase. By time-locking (or synchronising) the recorded data to the stimulus signal, the difference between the stimulus signal and the SSVEP can be derived and represented as a complex number. Variations in phase between the stimulus and SSVEP are related to the processing speed of the brain. For example, if there is a reduction in latency between the stimulus signal and SSVEP, then it is considered to reflect an increase in activity, while the opposite is true if there in an increase in latency (Silberstein, Ciorciari, & Pipingas, 1995; Silberstein, Line, Pipingas, Copolov, & Harris, 2000). SST has some unique features that make it ideal in conducting this type of market research. Firstly, the methodology has high temporal resolution and is able to track rapid changes in brain activity over an extended period of time. Secondly, the methodology is resistant to noise signals originating from muscle activity, blinks, eye movements as well as other forms interference which make it ideal when measuring the neural responses in dual screening. Lastly, the methodology has a high signal-to-noise ratio meaning it is possible to measure the response to a single viewing of an advertisement. It should be noted that other brain imaging methodologies require averaging responses from multiple viewings per participant in order to extract any meaningful data, which itself is problematic as you do not get a 'pure' response, nor one that is contextually relevant in a real world media environment. Presenting the brain with repeated or multiple exposures results in very specific changes in neural responses, namely a reduction over repeated trials. A practical example of this is to compare your viewing experience of the first time you saw an episode of your favourite show and contrast it with what your experience would be if you saw that very same episode again. For a more detail review of this phenomena known as repetition-suppression, please see Grill-Spector, Henson & Martin (2006).

INTERACTION ON SOCIAL MEDIA INCREASES VIEWER ENGAGEMENT

Recently, we conducted an investigation where we were interested in understanding the relationship that dual screening (with a particular emphasis on social media use) has on audience engagement during a live TV broadcast (Pynta et al., 2014). Given the high prevalence of Twitter feeds during a live broadcast TV programs, we chose X-Factor (Seven Network, Australia, under license from FremantleMedia Enterprises), a reality music competition that features a live broadcast a few nights a week. In addition, the show had a consistent format each night that it was aired which provided a great deal of homogeneity across the four day testing period. Furthermore, the TV show encouraged tweeting behaviour, by frequently displaying viewers' tweets during the episode. We chose to measure audience engagement using SST as it allowed for a continuous measurement of the entire viewing experience.

Study Design

The study comprised thirty-six participants aged from 18-31 (26 female & 10 males, M = 21.5) who were regular viewers of the program and were frequent users of Twitter or the interactive-television application Fango (Seven Network, Australia). We collected brain activity from these participants using a 24 channel electrode cap with electrodes position in accordance with the universal 10-20 system. Participants also wore a specially designed visor which generated a peripheral sinusoidal visual flicker that was present throughout the entire recording session. The participants watched 35 minutes of the X-Factor on their own 22-inch LCD monitor and were free to interact with Twitter or Fango using their own mobile devices or tablets as they pleased. During each session, we recorded the social media interactions of each viewer using a camera system that was synchronised to their SSVEP data, allowing each viewer's behaviour to be classified with a high degree of temporal resolution.

Participant viewing behaviour was classified into three distinct conditions which were directed around social media usage. Specifically, these were a "pre", "during" and a "post" condition. We defined the "pre" condition as any period of time when the sole focus was directed towards the television (Screen 1). The "during" condition was any period time when the show was still screening, but attention was now directed to the second screen (mobile phone or tablet). Lastly, the "post" condition was defined as any period of time after the interaction with the second screen had ended and attention was redirected towards the television show (Screen 1) once again.

Results

On average, each participant had 4.25 second screen interactions, which were distributed throughout the broadcast (a total of 153 Screen 2 interactions were made). The average length of these was 45.3 seconds (SD = 88 seconds) and importantly the majority of screen 2 interactions (76%) occurred during the program, with the remaining 24% extending into an ad break. In order to determine the levels of engagement during these interactions, we looked at SST engagement in the period prior to, including, and the following screen 2 events. In total, this window was 285 second long and comprised of a 120 second "Pre-period", a 45 second "Screen 2" ("During") period and a 120 second "Post-period".

The Effect of Screen 2 Engagement on Subsequent Screen 1 Engagement

To examine any changes in viewer engagement, we examined the mean levels of engagement across the Pre, During and Post conditions. Specifically, we divided the Pre and Post conditions into one minute segments and calculated the mean engagement for each of these periods (See Figure 2). When we exam-

Figure 2. Averaged brain activity corresponding to the different time periods within the broadcast. Adapted from Pynta et al., (2014).

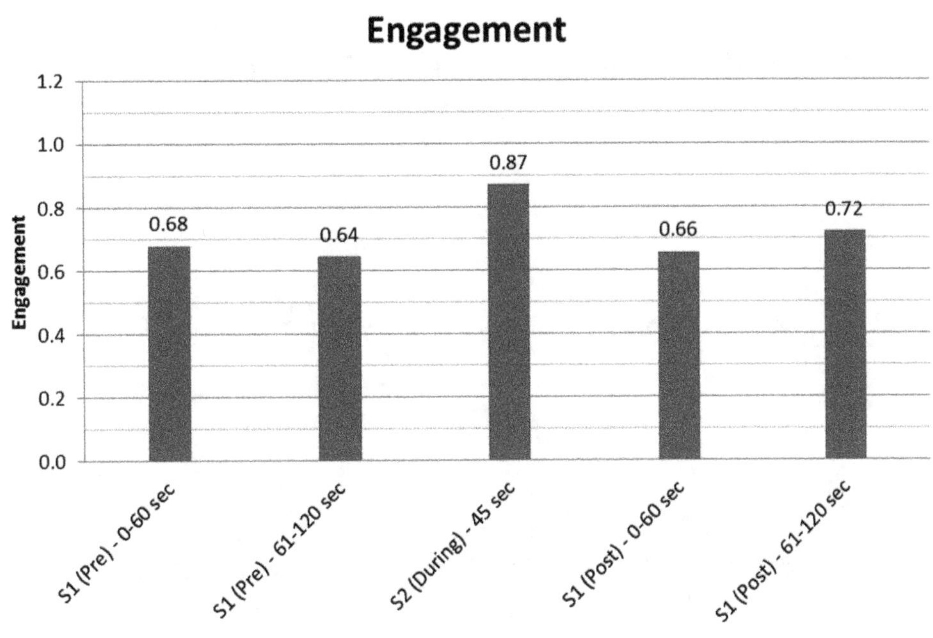

ined the results, what was immediately apparent was that the mean level of engagement was significantly higher while the viewers were engaged with Screen 2, than they were in the Pre-condition (Figure 2). Following the ending of the Screen 2 period, there was a small reduction in mean activity, following by a 9% increase in the last 60 seconds. There were no significant differences in the one minute periods on either side of the Screen 2 interaction, but interestingly the engagement time series (see Figure 3) showed that activity initially dropped below the mean of the first sixty seconds of the Post-period.

Long Term Positive Impact of Screen 2 Interaction

The results of the study were ground-breaking in that they showed interaction with social media while watching TV had a positive impact on the overall levels of program engagement. In particular, while tweeting there was an immediate increase in engagement. This is not surprising when you consider composing a tweet to share with a much wider audience is something we would assume to have a great deal of personal relevance to the composer. While this increase was not entirely unexpected, what was surprising was the neural response upon returning to the TV program – a 9% increase in engagement when compared to levels during the "pre" condition (See Figure 2).

While the results demonstrated the positive short term effects of tweeting, it would be interesting to know if this transient effect gave rise to a more enduring effect over the duration of the program. To answer this question, the authors drew upon the Neuro-Insight database of 150 prime-time television programs where typical viewing durations were 30 minutes or more. When examining the level of engagement over the total duration we found that, irrespective of genre, engagement would generally reduce after the first three minutes of the program. The nature of this reduction typically followed a linear

Figure 3. Time Series illustrating responses for Screen 1 – Screen 2 – Screen 1 during the "Pre" (0-120 seconds), "During" (120-165) & return back to Screen 1 "Post Period" (165-285 seconds). Adapted from Pynta et al., (2014).

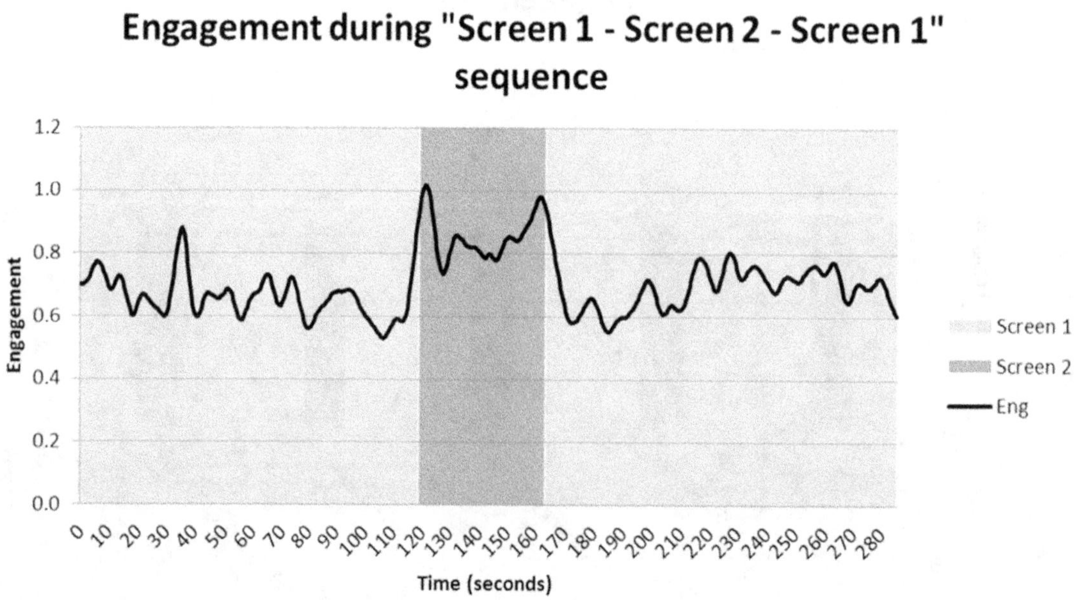

trend and resembled the physiological process of "habituation." A characteristic feature of habituation consists of elevated responses at the start of a period of sustained sensory stimuli followed by a reduction to a lower sustained value. We termed this effect "program habituation" and the neural behaviour demonstrated in the "Pre Period" (see Figure 2), is clearly indicative of this response. We subtracted the expected change in engagement over the program duration (based on our program database norms) in order to compensate the effects of habituation, which subsequently allowed us to examine the true impact of Screen 2 events. Screen 2 events reduced the effect of program habituation and maintained engagement at a higher level throughout the program, particularly towards the end. This meant that engagement was 26% higher than what could be expected for a primetime program and 23% higher than what could be expected for the reality TV genre. So while the 9% Pre-Post increase associated with a single tweet event is relatively small, these individual effects appear to accumulate creating a long term effect that is statistically significant when compared to our program database.

USING SOCIAL MEDIA TO INCREASE AUDIENCE ENGAGEMENT

By creating content that encourages viewers to engage with social media, content producers are ensuring that the level of program engagement within each viewer remains high. Because engagement and LTM process are closely linked, high levels of engagement are likely to result in high levels of LTM, which is extremely important in terms of advertising effectiveness.

What we have learnt while observing brain activity across an entire program is that engagement will generally reduce to some degree over the course of the program. This effect is irrespective of genre and resembles the physiological process of habituation. Interacting with social media appears to reduce the effects of habituation, by keeping viewers more engaged. In the specific case of the X-Factor, the rise in engagement may signify the participant's anticipation and excitement of having their tweet shown on live TV, which in turn has a flow on effect that lasts for the duration. Importantly, the effects of tweeting appear to be cumulative, as engagement by the end of the viewing period is 26% higher than would be expected from a prime-time program and 23% higher than would be expected from reality genre programs (source: Neuro-Insight). These results are also support similar results presented by Nieslen and Twitter which have reported a positive correlation between Twitter volumes and TV ratings (Nielsen, 2013a).

IMPLEMENTING SOCIAL MEDIA INTO A TV MARKETING STRATEGY

The results from the current study indicate that the growing practice of integrated sponsorships and branded elements in TV shows, especially shows optimised for social media, is likely to prove effective. While a traditional sponsorship will usually carry television commercials, many program formats will also include in-program branding, customised branded segments of 'behind-the-scenes' content &/or product usage. These additional sponsorship assets take on an even greater role with social TV users because they are seamlessly layered through the program – effectively forming part of the entertainment content. When we consider the fact that second screen usage tends to occur on average, throughout all stages of a the program (not just during commercial breaks), the heightened levels of engagement due to social activity would flow through to the elements within a program that carry branded entertainment beyond conventional commercials. This is mostly likely to be the case with competition based reality shows like the X-Factor where viewers have the opportunity to see their tweets on air, are encouraged to vote for their favourite contestants and interact with the shows' social media sites e.g. The X-Factor's Facebook page. The single largest analysis of Facebook users by SecondSync across Australia, UK and USA found a uniquely distinct pattern across competition based programs, where posts directly mapped to individual performances throughout the program (SecondSync, 2014).

Furthermore, as an advertiser, sponsoring a program that promotes social media interactivity into the viewing experience could prove to be a worthwhile exercise from an auditory branding standpoint. In other words, just because the social TV viewer doesn't 'watch' or see the sponsors' commercial or specific moments of brand integration within a program, doesn't necessarily mean that the brand linkage hasn't been successful. Brands that would traditionally use a mnemonic or a 'brand sound signature' during the branding sequence in a television commercial would definitely benefit from the execution of an auditory brand trigger e.g. The Intel Chime sound signature, verbal brand mentions by the programs' host or advertisements that feature jiggles that are strongly linked to the product or brand. We espouse this benefit purely from the fact that this paper has discussed the strong link between memory encoding & viewer engagement. In line with this fact that as engagement levels are significantly heightened during second screen usage, so too is the level of memory encoding. Brand-related stimulus – particularly those of an auditory nature – would likewise, be strongly encoded into long-term memory during second screen usage as well as during the longer-term period of increasing cumulative engagement during the program.

Finally from a broadcasters' perspective, the strategic implications of Social TV viewing are equally as significant as they are for advertisers. It's interesting to note the mounting evidence for the complimentary nature of TV viewing & social media since this study was conducted (November 2012). According to Twitter, the single strongest spike of Twitter activity in Australia for 2013, was the winner announcement of the X-Factor for 2013 (AdNews, 2013). A significant proportion of social activity is occurring around key television programs, Twitter confirming that 45% of their users engage in the platform while watching television (Broderick, 2014). In October 2013, Nielsen – the TV ratings provider – announced it was launching Nielsen Twitter TV Ratings (NTTR) in an effort to tap this synergistic relationship between the two platforms (Nielsen, 2013b) . Program creators & executive producers can be more confident in using social media as an enhancement to the TV viewing experience. For example, placing social media cues earlier in the program (as opposed to a prompt during the final sequences) can actually drive viewer engagement. Higher viewer engagement is likely to translate into increase the entertainment value of the program, open up potentially new avenues of program promotion as well as consolidating existing viewer loyalty.

CONCLUSION

The primary insight gained from this investigation is that interacting with social media while watching TV has a positive effect on viewer engagement. For marketers, any increase in viewer engagement will also be beneficial for advertising, as social media not only provides an additional consumer touch-point, but also has the ability to improve effectiveness of program sponsorships. Although we did not specifically test the hypothesis, content producers will need to find the right balance of integrating social media into their productions, as too much social TV interactions may in fact be counterproductive.

REFERENCES

Abraham, A., & von Cramon, D. Y. (2009). Reality=Relevance? Insights from Spontaneous Modulations of the Brain's Default Network when Telling Apart Reality from Fiction. *PLoS ONE*, *4*(3), e4741. doi:10.1371/journal.pone.0004741 PMID:19277108

AdNews. (2013). *Twitter and TV a match made in 140 characters*. Retrieved from http://www.adnews.com.au/adnews/twitter-and-tv-a-match-made-in-140-characters

Bisley, J. W., & Goldberg, M. E. (2010). Attention, Intention, and Priority in the Parietal Lobe. *Annual Review of Neuroscience*, *33*(1), 1–21. doi:10.1146/annurev-neuro-060909-152823 PMID:20192813

Brewer, J. B., Zhao, Z., Desmond, J. E., Glover, G. H., & Gabrieli, J. D. E. (1998). Making Memories: Brain Activity that Predicts How Well Visual Experience Will Be Remembered. *Science*, *281*(5380), 1185–1187. doi:10.1126/science.281.5380.1185 PMID:9712581

Broderick, T. (2014). *The Power of TV x Twitter*. Retrieved from https://blog.twitter.com/2014/the-power-of-tv-x-twitter

Buckner, R. L., Kelley, W. M., & Petersen, S. E. (1999). Frontal cortex contributes to human memory formation. *Nature Neuroscience*, *2*(4), 311–314. doi:10.1038/7221 PMID:10204536

Cooke, S. F., & Bliss, T. V. P. (2006). Plasticity in the human central nervous system. *Brain*, *129*(7), 1659–1673. doi:10.1093/brain/awl082 PMID:16672292

Deloitte. (2012). *The rise and rise of 'second screening'*. Retrieved from http://www.deloitte.com/view/en_GB/uk/industries/tmt/f0f3f07a77349310VgnVCM3000001c56f00aRCRD.htm

Demb, J. B., Desmond, J. E., Wagner, A. D., Vaidya, C. J., Glover, G. H., & Gabrieli, J. D. (1995). Semantic encoding and retrieval in the left inferior prefrontal cortex: A functional MRI study of task difficulty and process specificity. *The Journal of Neuroscience*, *15*(9), 5870–5878. PMID:7666172

Fletcher, P. C., Frith, C. D., & Rugg, M. D. (1997). The functional neuroanatomy of episodic memory. *Trends in Neurosciences*, *20*(5), 213–218. doi:10.1016/S0166-2236(96)01013-2 PMID:9141197

Fletcher, P. C., Shallice, T., & Dolan, R. J. (1998). The functional roles of prefrontal cortex in episodic memory. I. Encoding. *Brain*, *121*(Pt 7), 1239–1248. doi:10.1093/brain/121.7.1239 PMID:9679776

Fletcher, P. C., Shallice, T., & Dolan, R. J. (2000). "Sculpting the Response Space"—An Account of Left Prefrontal Activation at Encoding. *NeuroImage*, *12*(4), 404–417. doi:10.1006/nimg.2000.0633 PMID:10988034

Goodale, M. A., & Milner, A. D. (1992). Separate visual pathways for perception and action. *Trends in Neurosciences*, *15*(1), 20–25. doi:10.1016/0166-2236(92)90344-8 PMID:1374953

Google, I. S. B. (2012). *The New Multi-screen World: Understanding Cross-platform Consumer Behavior*. Retrieved from http://services.google.com/fh/files/misc/multiscreenworld_final.pdf

Gray, H. (1918). Anatomy of the human body (20th ed.). Philadelphia Lea & Febiger.

Grill-Spector, K., Henson, R., & Martin, A. (2006). Repetition and the brain: Neural models of stimulus-specific effects. *Trends in Cognitive Sciences*, *10*(1), 14–23. doi:10.1016/j.tics.2005.11.006 PMID:16321563

Heatherton, T. F., Wyland, C. L., Macrae, C. N., Demos, K. E., Denny, B. T., & Kelley, W. M. (2006). Medial prefrontal activity differentiates self from close others. *Social Cognitive and Affective Neuroscience*, *1*(1), 18–25. doi:10.1093/scan/nsl001 PMID:18985097

Hebb, D. O. (1949). *The Organization of Behavior*. New York: Wiley.

Kelley, W. M., Miezin, F. M., McDermott, K. B., Buckner, R. L., Raichle, M. E., Cohen, N. J., & Petersen, S. E. et al. (1998). Hemispheric specialization in human dorsal frontal cortex and medial temporal lobe for verbal and nonverbal memory encoding. *Neuron*, *20*(5), 927–936. doi:10.1016/S0896-6273(00)80474-2 PMID:9620697

LaBar, K. S., & Cabeza, R. (2006). Cognitive neuroscience of emotional memory. *Nature Reviews. Neuroscience*, *7*(1), 54–64. doi:10.1038/nrn1825 PMID:16371950

Maddock, R. J. (1999). The retrosplenial cortex and emotion: New insights from functional neuroimaging of the human brain. *Trends in Neurosciences*, *22*(7), 310–316. doi:10.1016/S0166-2236(98)01374-5 PMID:10370255

Maguire, E. A., Frith, C. D., Burgess, N., Donnett, J. G., & O'Keefe, J. (1998). Knowing where things are: parahippocampal involvement in encoding object locations in virtual large-scale space. *Journal of Cognitive Neuroscience, 10*(1).

Miller, E. K., & Cohen, J. D. (2001). An integrative theory of prefrontal cortex function. *Annual Review of Neuroscience*, *24*(1), 167–202. doi:10.1146/annurev.neuro.24.1.167 PMID:11283309

Nielsen. (2013a). *New Nielsen Research Indicates Two-Way Causal Influence between Twitter Activity and TV VIiewership*. Retrieved from http://www.nielsen.com/us/en/press-room/2013/new-nielsen-research-indicates-two-way-causal-influence-between-.html

Nielsen. (2013b). *Nielsen Launches 'Nielsen Twitter TV Ratings'*. Retrieved from http://www.nielsen.com/us/en/press-room/2013/nielsen-launches-nielsen-twitter-tv-ratings.html

Penfield, W., & Boldrey, E. (1937). Somatic motor and sensory representation in the cerebral cortex of man as studied by electrical stimulation. *Brain*, *60*(4), 389–443. doi:10.1093/brain/60.4.389

Penfield, W., & Rasmussen, T. (1950). *The Cerebral Cortex of Man: A Clinical Study of the localization of Function*. New York: Macmillan.

Phan, K. L., Taylor, S. F., Welsh, R. C., Ho, S. H., Britton, J. C., & Liberzon, I. (2004). Neural correlates of individual ratings of emotional salience: A trial-related fMRI study. *NeuroImage*, *21*(2), 768–780. doi:10.1016/j.neuroimage.2003.09.072 PMID:14980580

Pynta, P., Seixas, S. A. S., Nield, G., Hier, J., Millward, E., & Silberstein, R. B. (2014). The Power of Social Television: Can Social Media Build Viewer Engagement? A New Approach to Brain Imaging of Viewer Immersion. *Journal of Advertising Research*, *54*(1), 71–80. doi:10.2501/JAR-54-1-071-080

Rolls, E. T. (2000). The orbitofrontal cortex and reward. *Cereb Cortex*, *10*(3), 284–294. doi:10.1093/cercor/10.3.284 PMID:10731223

SecondSync. (2014). *How TV Drives Conversation on Facebook*. Retrieved from http://blog.shift.com/blog/2014/4/2/secondsync-how-tv-drives-conversation-on-facebook

Sergent, J., Ohta, S., & MacDonald, B. (1992). Functional neuroanatomy of face and object processing. A positron emission tomography study. *Brain*, *115*(Pt 1), 15–36. doi:10.1093/brain/115.1.15 PMID:1559150

Sergerie, K., Lepage, M., & Armony, J. L. (2005). A face to remember: Emotional expression modulates prefrontal activity during memory formation. *NeuroImage*, *24*(2), 580–585. doi:10.1016/j.neuroimage.2004.08.051 PMID:15627601

Silberstein, R. B., Ciorciari, J., & Pipingas, A. (1995). Steady-state visually evoked potential topography during the Wisconsin card sorting test. *Electroencephalography and Clinical Neurophysiology/Evoked Potentials Section, 96*(1), 24-35.

Silberstein, R. B., Line, P., Pipingas, A., Copolov, D., & Harris, P. (2000). Steady-state visually evoked potential topography during the continuous performance task in normal controls and schizophrenia. *Clinical Neurophysiology*, *111*(5), 850–857. doi:10.1016/S1388-2457(99)00324-7 PMID:10802456

Silberstein, R. B., & Nield, G. (2008). Brain activity correlates of consumer brand choice shift associated with television advertising. *International Journal of Advertising*, *27*(3), 359–380. doi:10.2501/S0265048708080025

Silberstein, R. B., Schier, M. A., Pipingas, A., Ciorciari, J., Wood, S. R., & Simpson, D. G. (1990). Steady-State Visually Evoked Potential topography associated with a visual vigilance task. *Brain Topography*, *3*(2), 337–347. doi:10.1007/BF01135443 PMID:2091717

Snyder, L. H., Batista, A. P., & Andersen, R. A. (2000). Intention-related activity in the posterior parietal cortex: A review. *Vision Research*, *40*(10-12), 1433–1441. doi:10.1016/S0042-6989(00)00052-3 PMID:10788650

Teng, E., & Squire, L. R. (1999). Memory for places learned long ago is intact after hippocampal damage. *Nature*, *400*(6745), 675–677. doi:10.1038/23276 PMID:10458163

Thinkbox. (2012). *Multi-screening encourages more TV and ad viewing*. Retrieved from http://www.thinkbox.tv/multi-screening-encourages-more-tv-and-ad-viewing

Twitter. (2014). Best Practices: Media.

Ungerleider, L. G., & Mishkin, M. (1982). Two cortical visual systems. In D. J. Ingle & M. A. Goodale (Eds.), *Analysis of Visual Behavior* (pp. 549–586). MIT.

Wagner, A. D., Schacter, D. L., Rotte, M., Koutstaal, W., Maril, A., & Dale, A. M. et al.. (1998). Building memories: Remembering and forgetting of verbal experiences as predicted by brain activity. *Science*, *281*(5380), 1188–1191. doi:10.1126/science.281.5380.1188 PMID:9712582

Wernicke, C. (1970). The aphasic symptom-complex: A psychological study on an anatomical basis. *Archives of Neurology*, *22*(3), 280. doi:10.1001/archneur.1970.00480210090013

Wood, E. R., Dudchenko, P. A., Robitsek, R. J., & Eichenbaum, H. (2000). Hippocampal neurons encode information about different types of memory episodes occurring in the same location. *Neuron*, *27*(3), 623–633. doi:10.1016/S0896-6273(00)00071-4 PMID:11055443

KEY TERMS AND DEFINITIONS

Dual-Screening: Watching television whilst also using a second Internet-enabled device, such as a mobile phone, tablet or laptop computer.

Engagement (Neuroscience): Engagement refers to a sense of personal relevance.

Long Term Memory (LTM): Information that is stored for a period of time of more than 30 minutes to decades. In the brain, LTM is stored in the form of changes in the strength of synaptic contacts.

Social Television: A television program that encourages the use of social media, such as Twitter or Facebook during the broadcast.

Spatial Resolution: Spatial resolution refers to the spatial detail of an image acquired by an imaging device such as a MRI.

Steady State Topography: Steady State Topography is a method for observing and measuring human brain activity that uses the presence of a dim sinusoidal visual flicker presented in the visual periphery. The sinusoidal flicker elicits an oscillatory brain electrical response known as the Steady State Visually Evoked Potential (SSVEP). Neural speed can be calculated by measuring the delay between the visual flicker and the SSVEP response.

Temporal Resolution: Temporal resolution refers to the ability to take an accurate measurement with respect to time.

This research was previously published in the Handbook of Research on Integrating Social Media into Strategic Marketing edited by Nick Hajli, pages 153-166, copyright year 2015 by Business Science Reference (an imprint of IGI Global).

Chapter 21
Measuring Cognitive and Emotional Processes in Retail:
A Neuroscience Perspective

Patrizia Cherubino
IULM University, Italy & BrainSigns srl, Italy

Anton Giulio Maglione
Sapienza University, Italy

Ilenia Graziani
BrainSigns srl, Italy & Sapienza University, Italy

Arianna Trettel
BrainSigns srl, Italy

Giovanni Vecchiato
BrainSigns srl, Italy & Sapienza University, Italy

Fabio Babiloni
BrainSigns srl, Italy & Sapienza University, Italy

ABSTRACT

The purpose of this chapter is to share scientific methods for the quantitative measurement of emotion through the recording of physiologic and cerebral variables of consumers in relation to advertising stimuli and during the purchase in the store. For this reason, the authors describe the way to estimate the emotion along the visit of a shop by using the approach-withdrawal index. It demonstrates how it is possible to describe the variation of the appreciation of a shop visit by two groups of persons. The specific contribution to the scientific literature is the use of such approach-withdrawal index and the estimation of the emotion linked with the visit of a large point of sale (e.g. a supermarket). The proper use of these methodologies can provide information related to cognitive and emotional aspects of persons involved in the appreciation of products in retail points of sale.

DOI: 10.4018/978-1-5225-5478-3.ch021

INTRODUCTION

It is well known from neuroscience that the cerebral systems related to emotion in humans play a vital role and often unconscious (that is subtracted from the conscious cognitive control) during the decisions that we generate every day. Very often the emotions perceived during sensory stimulation to which we are subjected in everyday life do not manifest themselves in our conscious control, while driving our behavior instead perceived as "spontaneous" or natural.

Often we are not able to rationally justify our purchasing behavior, which often differ greatly from our original intent in favor of a purchase "decided on the spot" on the basis of purely instinctive motivations. The reason for this behavior of "decision-making" seemingly out of control during the conscious rational choice of goods or services is based on the fact that the different emotional systems in the brain of humans can easily access to the centers of decision behavior without be subjected to the filter of the cognitive part of the brain. The emotion is therefore a very important factor in the decision to purchase a good or a service, as well as plays a vital role in the perception of a product. It therefore seems clear that the possibility to understand and to measure the emotion induced by an advertising, as well as to measure the emotion induced by the vision of a logo can be extremely important for a company that invests on advertising.

To measure the emotion induced by a product or an image in a consumer, however, can be a difficult exercise. In fact, there may be a problem of "self-perception" of the emotion itself: as we have already said a perceived emotion cannot simply become apparent to our conscious attention while instead is inducing a precise behavior. When this happens it is typical to collect verbal reports from the consumers such as "I cannot explain why I made this choice," or similar. Another problem related to the perception of the emotion in the consumers is related to the ability by the consumer to describe properly the emotion perceived during the observation of a product or a service. Another possible issue is related to the will of the interviewed to share with the others the own emotion related to the product or service consumption. It appears clear from these previous possible issues, that an objective measure of the emotion could be of some help in the evaluation of the correct proposition of goods and services in the retail to the consumers.

The technological and methodological innovations of this paper is that the emotions (and the related processes in the brain that occurs during such emotions) can be measured during the exposition of the subjects to the advertisements or during the purchase decision in the store. In addition, this can be performed in ecologic conditions close to what happens in the normal life. This is possible thanks to advanced technological devices able to measure brain and emotional activities outside of the scientific laboratories of the universities or the research centers.

The real advantage of the quantitative measure of emotional during the use of advertising in selected groups of users is the possibility to better modulate the advertising message on the point of sale recognizing its strengths and weaknesses. The relevance of this contribution for companies lies in the information that the quantitative measure of emotion:

1. It is possible out of scientific laboratories and universities;
2. It is able to operate effectively for the improvement of the advertising messages and the corporate communications;
3. Is able to provide information not directly obtainable by the processes of explicit verbalization of the respondents.

The measurement of emotions and the relative brain processes associated with the instinctive perception of advertising can thus provide additional evidence to the measurement of the efficacy of the advertising obtained via other typical methods of marketing research.

The purpose of this chapter is then to share scientific methods for the quantitative measurement of emotion through the recording of physiologic and cerebral variables of consumers in relation to advertising stimuli and during the purchase in the store. For this reason we describe the way to estimate the emotion along the visit of a shop by using the approach-withdrawal index, described in the following pages. It will be demonstrated how using such index it can be possible to describe the variation of the appreciation of a shop visit by two groups of persons. The specific contribution to the scientific literature is the use of such approach-withdrawal index and the estimation of the emotion linked with the visit of a large point of sale (e.g. a supermarket).

Background

In the last two decades, different techniques and devices have been developed to measure thoughts and feelings, by measuring the brain activity through the collection of the hemodynamic or electromagnetic signature of activated neural networks. Since last 10 years, unsolved problems and questions related to the evaluation of economic transactions, reached the neuroscience labs. Therefore, neuroscience researchers began to cooperate with economists in order to evaluate the brain activity during the generation of economic value judgments. What can cognitive neuroscience teach economics on the base of these last decades of common research? Cognitive neuroscience deconstructs the picture of perfectly rational humans, which are deliberating their choices by weighting costs and benefits until a deliberative equilibrium is reached. Although humans are definitely capable of conscious deliberation, many, if not most, economically relevant decision processes are characterized by certain other features such as automatic, fast and effective cognitive processes, which are not under direct volitional control (Bargh and Chartrand, 1999). Second, they are under the influence of unrecognized and finely tuned affective mechanisms, which often play a decisive role in action (Damasio et al., 1996; Davidson et al., 1990; Panksepp, 2004). Third, many of these processes have been shaped by evolution in order to serve social purposes (BCacioppo, 2002). Thus, decision-making and evaluation in economic contexts will be influenced by mechanisms dedicated to social interaction (Braeutigam et al., 2004). Hence, the incorporation of neuroimaging into the decision-making sciences has generated the interest not only of economists but also of marketing scientists. The line of reasoning was that if the neuroscientist could "image" the brain at work during the perception of particular concepts of value for the economists, such as the "perceived value" of choices to be made, then the same could be attempted for concepts useful in marketing. In fact, images of activated brain areas during the appreciation of concepts, such as the "brand" or simply during the observation of advertisements, was attempted and successively largely explored by different brain imaging devices. Todays, there are high hopes that neuroimaging technology could solve some of the problems that marketers faced. In particular, a prominent hope is that neuroimaging will both streamline marketing processes and save money. Another hope is that neuroimaging will reveal information about consumer preferences that is unobtainable through conventional methods. In this context, the most popular brain imaging method adopted in the neuromarketing field is the functional Magnetic Resonance Image (fMRI), a technique that returns a sequence of images of the cerebral activity by means of the measure of the cerebral blood flow. It is very well known that the hemodynamic measurements of the brain activity allow a level of localization of the activated brain structures on the order of few cubic millimeters, being capable to detect activations also in deep brain structures. However, the lack of time resolution, due to the intrinsic nature of the measured signal, makes the fMRI unsuitable to follow the brain dynamics on the base of its sub seconds activity. However, there are other brain imaging techniques that allow to follow on a millisecond base the brain activity during the exposition to relevant marketing stimuli, such as electroencephalography (EEG), although characterized by a spatial resolution of square centimeters. To overcome this problem, high-resolution EEG technology (hrEEG) has been developed to enhance the poor spatial information content of the EEG activity in order to detect the brain activity with a spatial resolution of a square centimeter and the unsurpassed time resolution of milliseconds (Babiloni et al., 2004). In addition, EEG devices are also relatively inexpensive, robust and even wearable by the subject, making such technology of interest for the evaluation of marketing stimuli. Nowadays, researchers are attempting to investigate the signs of the brain activity correlated with an increase of attention, memory and emotional engagement during the observation of commercial advertisements (Vecchiato et al., 2010; Langleben et al., 2009). In fact, indirect variables of emotional processing could be gathered by tracking variations of the activity of specific anatomical structures linked to the emotional processing activity in humans, such as the pre- and frontal cortex (PFC and FC respectively; Davidson, 2004). The PFC region is structurally and functionally heterogeneous but its role in the generation of the emotions is well recognized. Specifically, findings suggest that the left PFC is an important brain area in a widespread circuit that mediates appetitive approach, while the right PFC appears to form a major component of a neural circuit that instantiates defensive withdrawal (Davidson, 2004; Balconi et al., 2009). In addition, it is very well know the role of the frontal areas in cognitive processes such as memory and attention in complex tasks (Werkle-Bergner et al., 2006; Klimesch, 1999). Moreover, by monitoring autonomic activity such as the heart rate (HR) it is possible to assess the emotional state of the subject (Montano et al., 2009).

The way in which such amount of information from neuroscience could be useful to the marketing studies in retail point of sale is suggested by the well-known fact that at least the 70% of the new products launched worldwide (including cars, shoes, clothes etc.) failed within the first six months. This happens simply because people are not saying (or are not able to say) the truth when they are interviewed with respect to the product they have tasted or watched previously. There are a lot of reasons for that behavior, but this disability to proper report the own feelings related to the product is usually a huge obstacle to the correct collection of information from consumers, at the base of solid marketing actions. Thus, the application of neuroscience-based methodologies allows to the researchers to gain information not available otherwise related to the unconscious and spontaneous reaction of the consumers in front of the product stimuli. Of course such kind of information is just a part of the whole story related to the acquisition of consumer's habit in front of the product.

In the present context, the object of our research was to describe methodologies available to analyze the brain activity and the emotional engagement occurring during the "naturalistic" observation of products in the point of sale. Hence, the final goal was to link significant variations of the EEG and autonomic variables with cognitive and emotional reactions to the presented and observed products. In order to do that, different indexes could be employed to summarize the performed measurements, later used in the statistical analysis.

CEREBRAL PROCESSES RELEVANT IN THE RETAIL

Main Cerebral Events that Occur in a Visit of a Shop

The aim of the application of neuroimaging techniques in cognitive neuroscience is to understand how brain functioning mediates cognition and human behavior such as decision-making, reward processing, memory, attention, approach and withdrawal motivation and emotional processing.

The use of advanced neuroimaging techniques allows an increasingly precise identification of neural responses across specific brain regions. The importance of such anatomical localization is due to the potential relevance of certain brain regions for consumer research given the role they play in different cognitive and emotional functions.

The idea is to use the amount of knowledge already developed so far in neuroscience to interpret the cerebral events related to purchase occurring during the visit of a store or a shop. In the following, it will be described some main cerebral processing that have place during a purchase in a store with a description of the possible cerebral regions involved in such processing.

In the successive section it will be described how such processes could be measured in a store.

Decision-Making

A central question within consumer research is about consumer decision-making and the way consumers assess different product alternatives based on perceived benefits and costs. Several regions of the prefrontal cortex (PFC), situated in the frontal lobe of the brain, might be related to this question due to the critical role they play in the underlying processes of human decision-making. Specifically, there is evidence of both the orbitofrontal cortex (OFC) and the ventromedial prefrontal cortex (VMPFC) being involved in the processing of different alternatives and potential outcomes through the assessment of their (perceived) value (Tremblay & Schultz, 1999). In particular, the OFC is associated with the evaluation of trade-offs and the expected capacity of outcomes to satisfy one's needs (Wallis, 2007). It also plays a central role in making choices about appropriate behaviors, especially when one is faced with unpredictable situations (Elliott, Dolan, & Frith, 2000). Finally, the dorsolateral prefrontal cortex (DLPFC) also plays a critical role in decision-making as it is known for being involved in cognitive control over emotions (Rilling, King-Casas, & Sanfey, 2008). In particular, it is involved in the control of impulses for complying with social norms while the ventrolateral prefrontal cortex (VLPFC) could play a role in motivating this social norm compliance by representing the threat of punishment from others (Rilling & Sanfey, 2011). Interestingly, the cognitive effort in the PFC appears to be lower when a sure gain is expected com- pared with risky decisions. Measuring the activity of these regions may therefore provide useful insights about the neural foundations of consumer choices and marketing constructs such as perceived value.

Reward Processing

Another set of brain regions of particular interest for consumer research consists of those part of the reward system of the brain. They indeed respond to subjectively attractive rewards such as food (Berridge, 1996), money (Knutson, Adams, Fong, & Hommer, 2001) and drugs (Wise & Rompré, 1989). In other words, an attractive product design or a favourite brand that is able to act as a rewarding stimulus within consumers' brains may trigger the psychological motivations that influence purchase behaviour. A way of monitoring the processing of rewarding stimuli is to measure the activation of the striatum, a striped mass of white and grey matter located in the basal ganglia inside the forebrain. Although a major function of the striatum is movement planning and control, it also plays a role in the reward system of the brain. Thus, there is evidence of the role of the striatum and its components (putamen, caudate nucleus and nucleus accumbens) in the evaluation of actual rewards with respect to one's expectations (Knutson & Wimmer, 2007) and the influence of social factors on the reward-related activity in this region (Fliessbach et al., 2007). Also part of the reward system is the ventral tegmental area (VTA). It is responsible for transmitting dopamine, a neurotransmitter, to other brain regions, thus enabling the modulation of decision-making and playing a role in goal-seeking behaviors (Fields, Hjelmstad, Margolis, & Nicola, 2007).

Attention and Memory

The amount of information consumers are exposed to is enormous, yet our processing capacity is limited. Each second we are exposed to an estimated 11 million bits of information that reach us through all our senses, yet humans are capable of processing only around 50 bits of that information, letting most of the input go by unnoticed (Wilson, 2002). How consumers represent, attend to, and perceive incoming information may have a profound influence on their behavior. One of the key questions at this stage, discussed next, is what consumers pay attention to (i.e., focus on) once there are exposed to a number of rapidly identified choice alternatives (i.e., brand, communications). Attention is the mechanism responsible for selecting the information that gains preferential status above other available information. Recent review of attention in neuroscience indicates that four conceptual components are fundamental to attention: bottom-up or saliency filters, top-down control, competitive visual selection, and working memory (Plassmann et al., 2012). The brain mechanisms involved in attention and visual processing could also be of interest to consumer research since advertising, logos and product designs all possess an important aesthetic component (e.g. shapes, colours) that is processed by the brain in the form of a visual stimulus. The aforementioned prefrontal cortex is responsible for directing and focusing attention and has been shown to connect with the neurons in charge of processing visual stimuli in the vision centre of the brain, called the occipital lobe (Armstrong, Fitzgerald, &Moore, 2006). Similarly, the study of memory-related mental processes might provide useful findings in relation to variables influencing consumer behaviour such as product experience, brand awareness and advertising recall. Most importantly, the hippocampus, located in the temporal lobe, plays a major role in the formation of different forms of memory and is a central region for memory processing and consolidation (McGaugh, 2000), thus involved in long-term memory. In particular, it is key to the acquisition and recall of declarative memory (Eichenbaum, 2000). Additionally, the amygdala, situated next to the hippocampus to which it is closely related, plays an important role as a modulator of the memory system, particularly in the consolidation of memory (McGaugh, 2000).

Approach and Withdrawal Motivation

Human fMRI studies have shown that activity in the orbitofrontal cortex (OFC), in particular its medial parts, at the time a reward is being enjoyed correlates with subjective reports about the pleasantness or valence of the experience. An interesting open question is which neural systems encode negative experiences. Several studies have found that unpleasantness of taste might the correlated with brain activity in the lateral OFC and left dorsal anterior insula/operculum. O'Doherty and colleagues found that the size of abstract punishments (i.e., losing money) activated lateral parts of the ODC (O'Doherty, Kringelbach, Rolls, Hornak, & Andrews, 20101). One problem in investigating negative experience is to dissociate it from intensity. This problem arises due to the negativity bias of intensity: negative experiences is to dissociate it from intensity. This problem arises due to the negativity bias of intensity: negative experiences are usually also perceived to be more intense and thus are often confounded, in particular for visual stimuli such as facial or object attractiveness. Using a different methodological approach to investigate positive vs. negative emotional experiences, neuromarketing studies are based on the idea that there is a left-right asymmetry of the frontal electroencephalography (EEG) signals (Davidson, Rkman, Saron, Senulis, & Friesen, 1990). These and related studies suggest that relatively greater activity in the left frontal region is associated with either positive emotional experience or the motivational drive to approach an object (Harmon-Jones, 2003). Although there are strong correlations between frontal EEG asymmetry and personality traits, the degree to which the asymmetry changes from one moment to another is questionable. Some studies have applied this approach to measure moment-to-moment fluctuations in emotional responses to advertisements without accounting for autocorrelations in time or multiple statistical comparisons. However, the validity of such approaches is unclear, as hemispheric asymmetry is also an index of working memory load (Habib, Nyberg, & Tulving, 2003). Further research to investigate the neural representation of positive vs. negative experienced values is needed (Plassmann et al., 2012).

According to the neuroscience literature, human behaviour relies on two motivational systems, whether behaviour is motivated by an anticipated desirable outcome or a possible aversive outcome. Specifically, EEG studies have provided important empirical support about the left frontal cortex being involved in a system that motivates approach behaviour (Davidson, 2004). Reversely, the right frontal cortex has been shown to be involved in a system that motivates withdrawal behaviour (Davidson, 2004). Thus, studies suggest that a greater activity in the left frontal region with respect to the right frontal region is associated with approach motivation, which is in turn generally associated with positive emotions, although the latter should be applied with caution as not always the case (Harmon-Jones, 2003). On the contrary, a relatively greater activity in the right frontal region is associated with withdrawal motivation, which is in turn generally associate d with negative emotions (Harmon-Jones, 2003). Consequently, the interest for consumer research is the possibility to monitor the left–right asymmetry of consumers' brains so as to interpret their motivational responses to marketing stimuli and thus potentially infer the attractiveness of products and brands.

Emotional Processing

A central brain region for emotional responses is the amygdala. In particular, it is involved in the processing of negative emotions and unknown stimuli, as well as in aversive responses to inequity (Rilling & Sanfey, 2011). It is also known as a locus of aversive and fear memory (Maren & Quirk, 2004). To a minor extent, it has equally been shown to process positive emotions, usually in relation to rewarding stimuli. Another key emotion-related region is the insula (or insular cortex). It plays a role in the processing of negative experiences such as the perception and expectation of risks, especially when making decisions for which a social or financial risk is expected (Preuschoff, Quartz, & Bossaerts, 2008; Knutson & Bossaerts, 2007). Likewise, the activation of the insula has been associated with anger and disgust in response to unfair economic situations. Additionally, as well as being involved in the evaluation process of stimuli, the aforementioned OFC plays a role in experiencing and anticipating the emotion of regret when outcomes differ from expectations (Coricelli et al., 2005). Finally, it is worth mentioning another area involved in emotional processing, the cingulate cortex, which includes the cingulate gyrus. Studies indeed suggest that the anterior cingulate evaluates emotional and motivational information and integrates it in the decision-making process (Bush, Luu, & Posner, 2000). Moreover, the anterior cingulate has been associated with the experience of an internal conflict between alternative options, its activation possibly being due to a conflict between cognitive and emotional motivations. The role of emotions in decision-ma king has been further explained through neurological and cognitive frameworks such as the somatic marker theory. Overall, the study of these brain mechanisms is likely to be central in consumer neuroscience due to the importance of the emotional component of purchase decisions in traditional consumer research. However, this will need to account for the complexity of the interconnected cerebral networks involved in emotions as there is not a single brain region responsible for emotional processes and no single brain region is activated in relation to one particular type of emotions (Phan, Wager, Taylor, & Liberzon, 2002).

It is worth of noticing as the emotional processing will impact on the variation of some physiological variables, such as the heart rate and the activity of the eccrine glands of the palm of the hands. This last variable is called Galvanic Skin Response (GSR). Variation of the heart rate and of the GSR return information about the "emotion" perceived by the user according to the theory of the affective emotional plane proposed by Barrett and Russel (1999). In such theory the attributes of a perceived "emotion" are related to its arousal level and its valence. These arousal and valence qualities are the independent components of all the emotions perceived by humans. These different qualities of the emotion could be gathered in humans by measuring the variation of the heart rate for the valence of the emotion, and by measuring the activity of the eccrine glands of the hand for the arousal level.

MEASURING THE DECISION-MAKING AND THE EMOTION IN THE RETAIL STORES

So far several cerebral processes relevant in the retail have been described: decision-making, approach and withdrawal motivation, emotion are neural activities that play a role during the retail visit. The aim of this particular section of the chapter is to describe how such cerebral processes could be measured at the state of the art actually. Once such processes could be measured in the store, with the needed time resolution, the intensity of such processes could be put in relation with the specific events occurring within the store. For instance, the measurement of the emotional level before the decision to purchase a product, or before the removal of a product from the basket could return valuable insight about the mechanisms of acceptance or rejection of the goods offered on the shelf within the retail.

Among the different neuroimaging methods available, the measure of the neuroelectrical activity of the brain by using a net of electrodes disposed on the head (electroencephalogram, EEG) presents several advantages when compared to the other available neuroimaging methodologies. In particular, the measurement of the cerebral activity with the use of EEG allows to follows with an appropriate time-resolution the development of the brain processes during the shop visit. The consumer could wear a cap with the electrodes and the cerebral activity will be gathered by a lightweight device (the EEG acquisition device). In such a way the collection of the brain activity could be related in time to the events occurring in the shop to the consumer. In fact, the EEG is able to track the cerebral activity on the base of the second time resolution or less. Such kind of time-relation cannot be possible by using other neuroimaging methodology such for instance the functional Magnetic Resonance Imaging (fMRI), that was unable to obtain an image of the cerebral activity with a time resolution lower than tens of seconds. In addition to the EEG, that return information about the possible occurrence of cerebral events related to the decision-making as well as to the approach/withdrawal activity in the store, there is also the possibility to follow the emotional processes occurring there. Such emotional processes are followed by measuring the alteration of body markers including the heart rate (with the electrocardiogram, ECG) and the activity of the hand sweat glands. These body markers are strictly related to the variation of the emotions perceived by the consumers during the visit of the shop and the relative purchases. Of course, the measuring of the cerebral activity and the related emotions could be performed also by using other neuroimaging devices, such as for instance the functional Magnetic Resonance Imaging (fMRI). However, in this case since the fMRI device cannot be moved (it is an installation of the weight of thousands kilograms) the retail visit could be performed just virtually by the consumer. In such condition the consumer lies supine within the fMRI device while a retail scenario would be projected on his/her pair of glass. Although possible, the measurement of the retail visit with fMRI device is then quite cumbersome to realize and without the needed time resolution to solve the fine temporal details of the decision making related to the product purchase.

The EEG and the physiological signals (heart rate, GSR) could be then measured in normal consumer during a visit of a retail. Information returned by these measurements are related to the time-varying appreciation of the goods reviewed by the consumer within the shop during the visit, as indexed by the approach-withdrawal EEG activity (Davidson, 2000). In fact, with this particular cerebral index it is possible to evaluate the profile of appreciation of the different goods during the visit of a shelf.

In particular, the methodology to perform such data collection are explained technically in the following paragraph.

EEG Recordings and Signal Processing

The cerebral activity was recorded by means of a portable EEG system (BEmicro and Galileo software, EBneuro, Italy). Informed consent was obtained from each subject after explanation of the study, which was approved by the local institutional ethics committee. Electrodes were arranged according to an extension of the 10-10 international system. Recordings were initially extra-cerebrally referred and then converted to an average reference off-line. We collected the EEG activity at a sampling rate = 256 Hz while the impedances kept below 5 kΩ. Each EEG trace was then converted into the Brain Vision format (BrainAmp, Brainproducts GmbH, Germany) in order to perform signal pre-processing such as artefacts detection, filtering and segmentation. The EEG signals have been band pass filtered at 1-45 Hz and depurated of ocular artefacts by employing the Independent Component Analysis (ICA). The EEG data have been re-referenced by computing the Common Average Reference (CAR). Individual Alpha Frequency (IAF) has been calculated for each subject in order to define the alpha as alpha = [IAF-2, IAF+2].

Approach-Withdrawal Index

In order to define an approach-withdrawal index (AW) according to the theory related to the earlier introduced EEG frontal asymmetry theory, we computed such imbalance as difference between the average EEG power of right and left channels. The formula we used is the following:

$$AW = \frac{1}{N_P}\sum_{i\in P} x_{\alpha i}^2(t) - \frac{1}{N_Q}\sum_{i\in Q} y_{\alpha i}^2(t) == AveragePower_{\alpha right,frontal} - AveragePower_{\alpha left,frontal} \quad (1)$$

where $x_{\alpha i}$ and $y_{\alpha i}$ represent the ith EEG channel in the alpha band that have been recorded from the right and left frontal lobes, respectively. In addition, P = {Fp2,AF6,AF4,F4} and Q = {Fp1,AF7,AF3,F5}, N_P and N_Q represent the cardinality of the two sets of channels. In such a way, an increase of AW will be related to an increase of interest and vice versa. The AW signal of each subject has been z-score transformed and then averaged to obtain an average waveform.

Emotional Index

The emotional index is defined by taking into account the GSR and HR signals. As far as the construction of such variable concerns, we refer to affects circumplex (Russel and Bartlett, 1999) where the coordinates of a point in this space are defined by the HR (horizontal axis) and the GSR (vertical axis). As presented in the Introduction section, several studies have highlighted that these two autonomic parameters correlate with valence and arousal, respectively.

In order to have a mono-dimensional variable, we describe the emotional state of a subject by defining the following Emotional Index (EI):

$$EI = 1 - \frac{\beta}{\pi} \quad (2)$$

where

$$\beta = \begin{cases} \frac{3}{2}\pi + \pi - \vartheta, GSR_z \geq 0, HR_z \leq 0 \\ \frac{\pi}{2} - \vartheta, otherwise \end{cases} \quad (3)$$

GSR_Z, HR_Z represent the z-score variables of GSR and HR respectively; ϑ, in radians, is measured as arctang (HRz, GSRz). Therefore, the angle β is defined in order to transform the domain of ϑ from [-π, π] to [0, 2 π] and obtain the EI varying between [-1, 1]. This is why we have two ways to calculate β. According to Equations 2 and 3 and the affect circumplex, negative *($HR_Z < 0$)* and positive *($HR_Z >$*

0) values of the EI are related to negative and positive emotions, respectively, spanning the whole affect circumplex.

However, it remains mandatory to understand which particular good on the shelf is responsible for the variation of the approach-withdrawal activity recorded in the user. To this purpose the use of a particular device able to return the information about the eye-gaze of the consumer within the shop is highly recommendable. Such device is called eye-tracking and it is able to collect the movement of the eye-gaze with a frequency of 30 up to 60 times per second. In this way it is possible to collect "where" the consumer looks within the shop and at the same time which particular cerebral activity is experienced. In addition to the EEG measurements and the eyetracker device, a couple of electrodes on the hand could return in real time also the profile of the variation of the heart rate of the consumer during the retail visit (for the valence of the perceived emotion) as well as the variation of the perceived arousal level. Such variation of the heart rate and arousal are then correlated to the quality of the emotion perceived by the consumer during the observation of the good in the store, with a time resolution of a couple of seconds. It is of interest to underline that the emotional information provided by the heart rate collection are different from those obtained by the approach-withdrawal evaluation, being the same different in perception that exists between the words "emotional" and the word "feeling" in the English vocabulary. In particular, the "emotional" attitude of the consumer in front of a good is indexed by the variation of his/her heart rate and eccrine glands activity (according to a well-established theory in psychology for valence and arousal of an emotion) while the "cerebral" interest of the consumer for the good is instead indexed by the approach-withdrawal EEG value.

The experimental setup needed to perform such multimodal recording of the cerebral and eye activity of a consumer within the visited shop is presented in Figure 1, in a case for a consumer within the retail food store. Different information are available by the acquisition of the cerebral and emotional feedback of the consumer during the decision-making moments related to the choice of a bottle of wine: his appreciation of the particular bottle, as indexed by the approach-withdrawal EEG activity and the appreciation of particular of the bottle's shape, that attracts more the eye-gaze. In addition, the emotion perceived related to the wine is then also measured by the electrodes measuring heart rate and cutaneous

Figure 1. Experimental setup

hand impedance (for arousal). As a consequence, an entire complete set of information about the perceived emotion and interest of the consumer within the store is available with a time-resolution of one second.

It is possible then to follow the cerebral and emotional processes of a consumer within a store during his/her visit. Such processes are indexed by specific variables that could be quantitatively characterized and analyzed. In addition, through a standard normalization procedures (Zar, 1994) the values of such emotional and cerebral indexes could be averaged across different consumers, without be subjected to the large inter-subject variability usually exhibited by the biological indexes. This is obtained by shifting from the original variables collected to their z-score values, obtained by estimating the average and the standard deviation values of such variables in the analyzed sample. By using these normalization procedure and the related z-scores, it is then possible to obtain average cerebral and emotional profile of appreciation of goods during a predetermined visit of the shop for one population. In addition to that, it is possible to describe the perceived cerebral and emotional appreciation of the shop visit for two distinct population groups, such as a group of normal consumers (group A) and another group of consumers that have a fidelity card of the shop (group B). In this case, the possibility to characterize the perceived emotion and cerebral processes of these two groups of consumers within the shop will open new perspectives about the understanding of their behavior beyond the mere description of their purchases.

For instance, Figure 2 represents the possible outcome of an experiment performed by using the recording of cerebral and emotional processes during the visit of a food store in two groups of consumers as described above (A and B). The x-axes of the Figure 2 present the predetermined path that all the participants of the groups A and B performed across the shop, while the y-axes represent the levels of the average cerebral appreciation for the persons belonging to the group A and B across different areas of the store

It could be appreciated that the different population have a similar appreciation profile but in the section in which a group could have different benefit from the other (e.g. section wines and liquor). It could be concluded that population A (normal customers) generated levels of appreciation lower than

Figure 2. Possible outcome

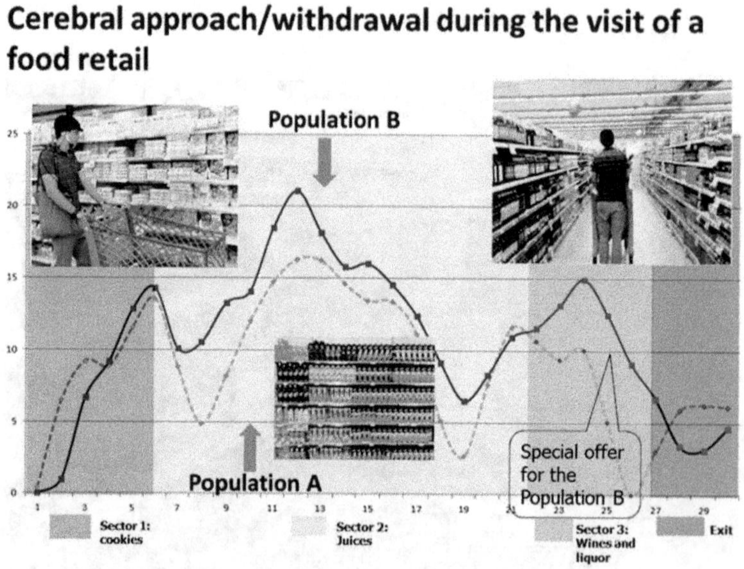

usual customers (group B) during a visit in the shop. Such information could be used by the management of the retail to promote specific actions to increase the level of satisfaction of the customers belongs to the A group.

CONCLUSION AND FUTURE RESEARCH DIRECTIONS

In this chapter we have demonstrated the possibility to collect cerebral and emotional activity during the visit of a point of sale. The technology employed relies on the use of advanced indexes pointing at two particular psychological brain states, the appreciation/withdrawal from a stimulus and the emotion measured with the body markers, through EEG, EKG and GSR. Another important point is that in the present chapter a contribution to the literature of the estimation of the emotion from the collection of EKG and GSR has been described, through the use of the Equations 2-3. Such approach could return valuable information as derived from normal subjects visiting the point of sale.

It became evident that the possibility of measuring emotional and cerebral activity in consumer during their visit of the shop will open new areas of research in order to better understand the cerebral processes underlying the decision of purchase in the point of sale.

While up to some years ago such mobile measurements were cumbersome for the consumer, forced to bring with him/her a very heavy computer on the back of shoulder, today the situation appear much better. Lightweight devices are available for the measurements of the cerebral activity during the visit of shops or even museums and other educational or entertainment sites (such as cinemas, theatre, etc.) as demonstrated in Figure 3. In such Figure the equipment employed for the collection of cerebral and emotional activity during the visit of an art gallery is displayed. The future for the collection of the biometrical reactions of the consumers during the visit of a retail will be of very light (ideally of the size of a modern mobile smartphone) and will return even in real time the engage of the users during the

Figure 3. Lightweight devices

shopping experience. Such amount of data will be of paramount importance to characterize from the "consumers" point of view the most interesting areas of the point of sale from the cerebral and from an emotional perspective.

ACKNOWLEDGMENT

This work was in part supported by a grant of the Italian Minister of Foreign Affairs with a bilateral project between Italy and China "Neuropredictor" and by a project supported by PRIN2012 to Fabio Babiloni and Giovanni Vecchiato.

REFERENCES

Armstrong, K. M., Fitzgerald, J. K., & Moore, T. (2006). Changes in visual receptive fields with microstimulation of frontal cortex. *Neuron*, *50*(5), 791–798. doi:10.1016/j.neuron.2006.05.010 PMID:16731516

Babiloni, F., Mattia, D., Babiloni, C., Astolfi, L., Salinari, S., Basilisco, A., & Cincotti, F. et al. (2004). Multimodal integration of EEG, MEG and fMRI data for the solution of the neuroimage puzzle. *Magnetic Resonance Imaging*, *22*(10), 1471–1476. doi:10.1016/j.mri.2004.10.007 PMID:15707796

Balconi, M., Brambilla, E., & Falbo, L. (2009). BIS/BAS, cortical oscillations and coherence in response to emotional cues. *Brain Research Bulletin*, *80*(3), 151–157. doi:10.1016/j.brainresbull.2009.07.001 PMID:19591907

Berridge, K. C. (1996). Food reward: Brain substrates of wanting and liking. *Neuroscience and Biobehavioral Reviews*, *20*(1), 1–25. doi:10.1016/0149-7634(95)00033-B PMID:8622814

Braeutigam, S., Rose, S. P., Swithenby, S. J., & Ambler, T. (2004). The distributed neuronal systems supporting choice-making in real-life situations: Differences between men and women when choosing groceries detected using magnetoencephalography. *The European Journal of Neuroscience*, *20*(1), 293–302. doi:10.1111/j.1460-9568.2004.03467.x PMID:15245501

Bush, G., Luu, P., & Posner, M. I. (2000). Cognitive and emotional influences in anterior cingulate cortex. *Trends in Cognitive Sciences*, *4*(6), 215–222. doi:10.1016/S1364-6613(00)01483-2 PMID:10827444

Cacioppo, J. T. (2002). Social neuroscience: Understanding the pieces fosters understanding the whole and vice versa. *The American Psychologist*, *57*(11), 819–831. doi:10.1037/0003-066X.57.11.819 PMID:12564179

Chartrand, T. L., & Bargh, J. A. (1999). The chameleon effect: The perception-behavior link and social interaction. *Journal of Personality and Social Psychology*, *76*(6), 893–910. doi:10.1037/0022-3514.76.6.893 PMID:10402679

Coricelli, G., Critchley, H. D., Joffily, M., O'Doherty, J. P., Sirigu, A., & Dolan, R. J. (2005). Regret and its avoidance: A neuroimaging study of choice behavior. *Nature Neuroscience*, *8*(9), 1255–1262. doi:10.1038/nn1514 PMID:16116457

Damasio, A. R., Everitt, B. J., & Bishop, D. (1996). The somatic marker hypothesis and the possible functions of the prefrontal cortex. *Philosophical Transactions of the Royal Society of London. Series B, Biological Sciences*, *351*(1346), 1413–1420. doi:10.1098/rstb.1996.0125 PMID:8941953

Davidson, R. J., Ekman, P., Saron, C. D., Senulis, J. A., & Friesen, W. V. (1990). Approach-withdrawal and cerebral asymmetry: Emotional expression and brain physiology. I. *Journal of Personality and Social Psychology*, *58*(2), 330–341. doi:10.1037/0022-3514.58.2.330 PMID:2319445

Davidson, R. J., Shackman, A. J., & Maxwell, J. S. (2004). Asymmetries in face and brain related to emotion. *Trends in Cognitive Sciences*, *8*(9), 389–391. doi:10.1016/j.tics.2004.07.006 PMID:15350238

Eichenbaum, H. (2000). A cortical-hippocampal system for declarative memory. *Nature Reviews. Neuroscience*, *1*(1), 41–50. doi:10.1038/35036213 PMID:11252767

Elliott, R., Dolan, R. J., & Frith, C. D. (2000). Dissociable functions in the medial and lateral orbitofrontal cortex: Evidence from human neuroimaging studies. *Cereb Cortex*, *10*(3), 308–317. doi:10.1093/cercor/10.3.308 PMID:10731225

Fields, H. L., Hjelmstad, G. O., Margolis, E. B., & Nicola, S. M. (2007). Ventral tegmental area neurons in learned appetitive behavior and positive reinforcement. *Annual Review of Neuroscience*, *30*(1), 289–316. doi:10.1146/annurev.neuro.30.051606.094341 PMID:17376009

Habib, R., Nyberg, L., & Tulving, E. (2003). Hemispheric asymmetries of memory: The HERA model revisited. *Trends in Cognitive Sciences*, *7*(6), 241–245. doi:10.1016/S1364-6613(03)00110-4 PMID:12804689

Harmon-Jones, E. (2003). Early career award. Clarifying the emotive functions of asymmetrical frontal cortical activity. *Psychophysiology*, *40*(6), 838–848. doi:10.1111/1469-8986.00121 PMID:14986837

Klimesch, W. (1999). EEG alpha and theta oscillations reflect cognitive and memory performance: A review and analysis. *Brain Research. Brain Research Reviews*, *29*(2-3), 169–195. doi:10.1016/S0165-0173(98)00056-3 PMID:10209231

Klimesch, W., Doppelmayr, M., & Hanslmayr, S. (2006). Upper alpha ERD and absolute power: Their meaning for memory performance. *Progress in Brain Research*, *159*(1), 151–165. doi:10.1016/S0079-6123(06)59010-7 PMID:17071229

Knutson, B., & Bossaerts, P. (2007). Neural antecedents of financial decisions. *The Journal of Neuroscience*, *27*(31), 8174–8177. doi:10.1523/JNEUROSCI.1564-07.2007 PMID:17670962

Knutson, B., Fong, G. W., Adams, C. M., Varner, J. L., & Hommer, D. (2001). Dissociation of reward anticipation and outcome with event-related fMRI. *Neuroreport*, *12*(17), 3683–3687. doi:10.1097/00001756-200112040-00016 PMID:11726774

Knutson, B., Rick, S., Wimmer, G. E., Prelec, D., & Loewenstein, G. (2007). Neural predictors of purchases. *Neuron*, *53*(1), 147–156. doi:10.1016/j.neuron.2006.11.010 PMID:17196537

Knutson, B., & Wimmer, G. E. (2007). Splitting the difference: How does the brain code reward episodes? *Annals of the New York Academy of Sciences*, *1*(1104), 54–69. doi:10.1196/annals.1390.020 PMID:17416922

Langleben, D.D., Loughead, J.W., Ruparel, K., Hakun, J.G., Busch-Winokur, S., Holloway, M.B., … Lerman, C. (2009). Reduced prefrontal and temporal processing and recall of high "sensation value" ads. *Neuroimage*, *46*(1), 219-25.

Maren, S., & Quirk, G. J. (2004). Neuronal signalling of fear memory. *Nature Reviews. Neuroscience*, *5*(11), 844–852. doi:10.1038/nrn1535 PMID:15496862

McGaugh, J. L. (2000). Memory-A century of consolidation. *Science*, *287*(5451), 248–251. doi:10.1126/science.287.5451.248 PMID:10634773

Montano, N., Porta, A., Cogliati, C., Costantino, G., Tobaldini, E., Casali, K. R., & Iellamo, F. (2009). Heart rate variability explored in the frequency domain: A tool to investigate the link between heart and behavior. *Neuroscience and Biobehavioral Reviews*, *33*(2), 71–80. doi:10.1016/j.neubiorev.2008.07.006 PMID:18706440

O'Doherty, J., Kringelbach, M. L., Rolls, E. T., Hornak, J., & Andrews, C. (2001). Abstract reward and punishment representations in the human orbitofrontal cortex. *Nature Neuroscience*, *4*(1), 95–102. doi:10.1038/82959 PMID:11135651

Panksepp, J., Nocjar, C., Burgdorf, J., Panksepp, J. B., & Huber, R. (2004). The role of emotional systems in addiction: A neuroethological perspective. *Nebraska Symposium on Motivation*, *50*(1), 85–126. PMID:15160639

Phan, K. L., Wager, T., Taylor, S. F., & Liberzon, I. (2002). Functional neuroanatomy of emotion: A meta-analysis of emotion activation studies in PET and fMRI. *NeuroImage*, *16*(2), 331–348. doi:10.1006/nimg.2002.1087 PMID:12030820

Plassmann, H., Zoëga Ramsøy, T., & Milosavljevic, M. (2012). Branding the brain: A critical review and outlook. *Journal of Consumer Psychology*, *5*(1), 85–115.

Preuschoff, K., Quartz, S. R., & Bossaerts, P. (2008). Human insula activation reflects risk prediction errors as well as risk. *The Journal of Neuroscience*, *28*(11), 2745–2752. doi:10.1523/JNEUROSCI.4286-07.2008 PMID:18337404

Rilling, J. K., King-Casas, B., & Sanfey, A. G. (2008). The neurobiology of social decision-making. *Current Opinion in Neurobiology*, *18*(2), 159–165. doi:10.1016/j.conb.2008.06.003 PMID:18639633

Rilling, J. K., & Sanfey, A. G. (2011). The neuroscience of social decision-making. *Annual Review of Psychology*, *62*(1), 23–48. doi:10.1146/annurev.psych.121208.131647 PMID:20822437

Russell, J. A., & Barrett, L. F. (1999). Core affect, prototypical emotional episodes, and other things called emotion: Dissecting the elephant. *Journal of Personality and Social Psychology*, *76*(5), 805–819. doi:10.1037/0022-3514.76.5.805 PMID:10353204

Tremblay, L., & Schultz, W. (1999). Relative reward preference in primate orbitofrontal cortex. *Nature*, *398*(6729), 704–708. doi:10.1038/19525 PMID:10227292

Vecchiato, G., De Vico Fallani, F., Astolfi, L., Toppi, J., Cincotti, F., Mattia, D., & Babiloni, F. et al. (2010). The issue of multiple univariate comparisons in the context of neuroelectric brain mapping: An application in a neuromarketing experiment. *Journal of Neuroscience Methods*, *191*(2), 283–289. doi:10.1016/j.jneumeth.2010.07.009 PMID:20637802

Wallis, J. D. (2007). Orbitofrontal cortex and its contribution to decision-making. *Annual Review of Neuroscience*, *30*(1), 31–56. doi:10.1146/annurev.neuro.30.051606.094334 PMID:17417936

Werkle-Bergner, M., Müller, V., Li, S. C., & Lindenberger, U. (2006). Cortical EEG correlates of successful memory encoding: Implications for lifespan comparisons. *Neuroscience and Biobehavioral Reviews*, *30*(6), 839–854. doi:10.1016/j.neubiorev.2006.06.009 PMID:16904180

Wilson, M. A. (2002). Hippocampal memory formation, plasticity, and the role of sleep. *Neurobiology of Learning and Memory*, *78*(3), 565–569. doi:10.1006/nlme.2002.4098 PMID:12559835

Wise, R. A., & Rompre, P. P. (1989). Brain dopamine and reward. *Annual Review of Psychology*, *40*(1), 191–225. doi:10.1146/annurev.ps.40.020189.001203 PMID:2648975

Zar, J. H. (2000). *Biostatistical analysis*. New York: Prentice Hall.

ADDITIONAL READING

Giovanni Vecchiato. (2013). *Arianna Trettel, Patrizia Cherubino and Fabio Babiloni, Neuroelectrical Brain Imaging Tools for the Study of the Efficacy of TV Advertising Stimuli and their Application to Neuromarketing*. Springer International.

Stephen Genco. (2013). *Andrew Pohlmann, Peter Steidl, Neuromarketing For Dummies*. Wiley.

Thomas Ramsoy. (2014). *Introduction to Neuromarketing & Consumer Neuroscience*. Kindle Edition.

KEY TERMS AND DEFINITIONS

Electroencephalography (EEG): The recording of electrical activity along the scalp. EEG measures voltage fluctuations resulting from ionic current flows within the neurons of the brain.

Emotion: In psychology and philosophy, emotion is a subjective, conscious experience characterized primarily by psychophysiological expressions, biological reactions, and mental states. Emotion is often associated and considered reciprocally influential with mood, temperament, personality, disposition, and motivation.

Eyetracker: A non-invasive device able to follow the human eye-gaze during the exploration of a picture, a shelf or other goods. Mobile eye-tracker are wear on the head as glass while non-portable eyetracker devices are solidal with the computer screen.

Eyetracking: The process of measuring either the point of gaze (where one is looking) or the motion of an eye relative to the head.

Functional Magnetic Resonance Imaging or Functional MRI (fMRI): A functional neuroimaging procedure using Magnetic Resonance Imaging technology that measures brain activity by detecting associated changes in blood flow. This technique relies on the fact that cerebral blood flow and neuronal activation are coupled.

Neuromarketing: A new field of marketing research that studies consumers' sensorimotor, cognitive, and affective response to marketing stimuli.

Neuroscience: The scientific study of the nervous system. Traditionally, neuroscience has been seen as a branch of biology. However, it is currently an interdisciplinary science that collaborates with other fields such as chemistry, computer science, engineering, linguistics, mathematics, medicine and allied disciplines, philosophy, physics, and psychology.

Chapter 22
Rational, Emotional, and Neural Foundations of Economic Preferences

Harish C. Chandan
Argosy University, USA

ABSTRACT

Classical economics assumes human economic decision making is completely rational and dominated by self-interest. Behavior economics emerged to account for the fact that human economic preferences are often influenced by emotional and psychological factors leading to inconsistent, intransitive, and irrational decisions that fail to maximize utility and minimize cost and transcend only self-interest. Both rationality and emotions are seated in the human brain in the prefrontal cortex and limbic system, respectively. The brain imaging methods of neuroscience help in understanding the interplay between economic behavior and neural mechanisms. The human economic decision making behavior involves computational and neurobiological processes and is related to the psychological processes. Classical Economics, Psychology, and Neuroscience converge in Neuroeconomics to better understand and predict human economic decision-making. Neuromarketing is an emerging field that uses neuroscience techniques to understand economic preferences of consumers.

INTRODUCTION

Classical economics and game theory assume that human economic decision-making is motivated *only* by self-interest and utility maximization through rational choice in a context of scarce resources. The assumption is that humans utilize all the information available in the market to decide without being influenced by biases and emotions and then act in a logical and rational manner. They weigh risk and return factors by considering the *likelihood* of an outcome and the *value* of the outcome to maximize benefits while minimizing the costs. However, in real life, individuals do *not* always make rational, utility-maximizing decisions. People do deviate from rationality in a systematic manner and have biases.

DOI: 10.4018/978-1-5225-5478-3.ch022

The biases, rules of thumb, or heuristics and emotions often overpower rational thinking and result in irrational behavior (Kahneman, 2003a, b).

Economic decision-making is often influenced by unconscious biases, implicit assumptions, and errors in judgment. These systematic biases have origins in human psychology. Human economic decisions are often influences by contextual and psychological factors leading to inconsistent, intransitive, and irrational decisions that fail to maximize utility and minimize cost (Summerfield & Tsetos, 2015). Behavior economics emerged to account for the cognitive biases, heuristics, emotions, and other psychological processes which often override reason unconsciously and automatically. Behavior economics and social neuroscience converged in neuroeconomics to better understand and predict human economic behavior realistically (Popescu & Nica, 2014). Neural activity causally determines economic choices and the neural computations necessary for making economic choices are stochastic (Corsani, 2013). Thus, economic decision-making has roots in economics, psychology and social neuroscience.

This chapter reviews the rational, emotional, and neural foundations of economic decisions. First, the classical perspective of rational choice in economic decision-making is presented and bounded rationality in real-life economic decisions is discussed. The special human capacity to represent others' intentions, beliefs and desires (i.e., mentalizing or the theory of mind – TOM) is discussed and the relationship between economic preferences and the big five personality traits are explored. The role of emotions, biases, and rules of thumb or heuristics in economic decision-making is reviewed leading to behavioral economics or behavioral finance. The neural perspective of economic decision-making that involves mentalizing and empathy is presented. This integrative neuroeconomic view allows one to better understand the human economic decision making in a way that combines rational, emotional, psychological, social and neural mechanisms. The promise and ethics of neuromarketing are discussed. Finally, policy implications and suggestions for future research are discussed.

CLASSICAL ECONOMIC DECISION MAKING: RATIONAL CHOICE AND UTILITY MAXIMIZATION

Classical economics studies individual and aggregate human economic decision-making and behavior. The individual decisions that people and businesses make regarding the allocation of resources and prices of goods and services (considering taxes and regulations created by governments) is called microeconomics which focuses on supply and demand and investigates how a specific company could maximize its production so it could lower prices and better compete in its industry. The aggregate human economic decision-making is covered in macroeconomics which studies the behavior of the economy as whole and not just specific companies.

Macroeconomics views economy-wide phenomena such as gross domestic product (GDP) and how it is affected by changes in unemployment, national income, rate of growth, and price levels. For example, macroeconomics would examine how a change in net exports would affect a nation's capital account or how GDP would be affected by unemployment rate. An understanding of both microeconomic and the macroeconomic behavior is required to create policies for social equity and justice. Due to the social nature of humans, individual economic decision making behavior is influenced by aggregate economic decision-making. For example, when an influential investor sells his stock for a particular company, smaller investors tend to do the same even though they did not know the reasons that led the influential investor to sell. This irrational behavior is called "herding" which is covered in behavior economics (Kudryavtsev, et al., 2013).

Economic decision-making can be regarded as a problem-solving activity leading to a solution that is considered satisfactory by the individual. This involves individuals selecting one option from many available options based on cost-benefit analysis. According to the rational choice theory in classical economics, people make decisions by reasoning to maximize benefits while minimizing costs. A rational decision is made by considering the 'likelihood' of a potential outcome and the 'value' of the outcome. The individual chooses the outcome that represents a higher 'value' for the *combined* effect of the 'likelihood' and the 'value' (i.e., the product of 'likelihood' and 'value') (Samuelson, 1983).

However, in real life, consumers take risks based on what appears to be a likely future outcome in an instance of random chance such as in a coin toss. Their logic is based on seeing the same outcome occur several times in a row and assuming a different outcome is due. This logic is not rational since the odds are independent and remain the same for either outcome in every instance. This is known as the "gambler's fallacy" (Ayton & Fischer, 2004).

Classical economics is based on the assumption that humans always make rational decisions to maximize expected utility. However, the rational choice assumption is one of the limitations of classical economics. Some examples of irrational economic behavior include *gambler's fallacy*, *groupthink*, and *herding*. For example, if some influential person sells his stocks at the hint of a financial decline, others decide to do the same out of fear of an economic collapse. People may rationally understand that selling their stock may not make sense for them in the long-term. They sell it anyway because everyone else is doing the same thing. This has its basis in the human emotions and mirror neurons: "monkey see, monkey do". Due to groupthink and herding, people follow whatever is popular at the time, thinking as a group of people instead of as individuals (Moorhead & Montanari, 1986).

In the framework of utility maximization, the assumption is that humans behave rationally and make optimal choices. However, the concept of utility is *subjective* and is a measure of the satisfaction the individual will derive from the consumption of a particular item or service. An individual's subjective perception of utility is difficult to assess. Since utility cannot be measured directly, it has to be inferred from the observed individual choice. These revealed preferences appear in people's willingness to pay for a product or service. The subjective nature of economic preferences means the concept of utility does not necessarily always involve receiving the most monetary or material benefit because the satisfaction received could be purely emotional. For example, while it would be more financially lucrative for an executive to remain at a company rather than to retire early, it would still be considered rational behavior for her to seek an early retirement if she feels the benefits of retired life outweigh the utility from the pay check that she receives (Simon, 1955).

An individual's concept of utility is shaped by preferences such as background risk, time, and social preferences. These preferences, in combination with expectations of future events, perceptions, beliefs, strategic consideration, prices, and constraints shape behavior. People usually face several risks and the interaction between them can affect the willingness to bear any one of them. Social preference captures the idea that an individual's utility does not depend only on his own material payoff, but that it is also shaped by others' behavior and material payoff. Social preferences include altruism and negative and positive reciprocity. Trust describes an individual's beliefs about others' trustworthiness combined with a preference to take social risks (Fehr, Fischbacher, & Kosfeld, 2005). Time preference describes how an individual trades off utility at different point in time (Frederick et al., 2002).

BOUNDED RATIONALITY: ECONOMIC DECISION MAKING IN REAL LIFE

The human mind is finite and is bound by cognitive limits. Humans do not have access to unlimited information to solve problems, nor do they have all the time in the world to ponder them. The theory of bounded rationality states people are not always able to obtain all the information they would need to make the best possible decision. The decision-making process involves information gathering, evaluating alternatives, and deciding on an optimal choice. In an information-rich world, a wealth of information creates a poverty of attention and a need to allocate that attention efficiently among the overabundance of information sources that might consume it.

Contrary to the principles of classical economics, individuals do not seek to maximize their benefit from a particular course of action since they cannot assimilate and digest all the information that would be needed to achieve such a thing. They do not obtain access to all the information required, but even if they could, their minds would be unable to process it properly. Instead of trying to maximize utility, the human mind seeks what is called "satisficing" – a combination of two words: "satisfy" and "suffice". People in many different situations seek something that is satisfactory and "good enough", rather than maximizing benefit (Kahneman, 2003a).

Simon's concept of bounded rationality suggests that the concept of utility or self-interest may be bounded as well. Due to subjective perception of the concept of utility, humans struggle to analyze problems *objectively* when the outcomes directly affect themselves. They view the economic choices through a "frame" of personal experience warped by social or cultural bias. The subjective perception of utility and the concept of bounded rationality result in humans coming up with their own 'rules of thumb', or "heuristics," when making quick decisions. The economic decisions that are made with relatively simple 'rules of thumb' do not make excessive demands upon one's thinking. While it is inherently rational to do so, the rules themselves and the behavior they lead to may not be. Heuristics are inherently irrational. In addition to utility maximization, other factors may drive consumers' economic choices such as altruism or self-control (Kahneman, 2003b).

Two additional examples of irrational economic preferences include mental accounting and retail therapy. The concept of mental accounting illustrates how people behave irrationally by placing greater value on some dollars than others even though all dollars have the same value. They might drive to another store to save $10 on a $20 purchase, but they would not drive to another store to save $10 on a $1,000 purchase Another example of irrational behavior is retail therapy in which consumers shop impulsively on credit to improve their mood. The field of behavioral economics is based on the idea that individuals often make irrational decisions and explores why they do so (Thaler, 1999).

PSYCHOLOGICAL AND EMOTIONAL FACTORS IN DECISION MAKING: BEHAVIORAL ECONOMICS

The classical perspectives on economics and securities assume that everyone takes full account of the available information in the market and behaves rationally. However, behavior economists have identified a large number of situations in which people deviate from the rational choice theory of standard economics and make decisions based on *psychological* elements (Kahneman, 1994; Brahmana et al., 2012).

An example is the rejection of low offers from the other person in the ultimatum game where a pair of partners must split the money in order to keep it. Rationally, accepting the low offer would enrich the person, but he is enraged at the unfairness of the low offer and rejects it even though he will not have the money he would have if he had accepted it. The person rejects the offer even though it is not in his self-interest to get pleasure at the thought of punishing the person who made the unfair offer. Such anomalies that occur in economic life cannot be explained by conventional economic theories. Therefore, the emergence of behavioral economics provides realistic psychological foundations to enhance the explanatory power of economics (Tversky & Kahneman, 1986).

Behavioral economics has been applied to many areas such as macroeconomics, labor economics, finance, and marketing. Behavioral economics integrates social, cognitive, and emotional factors in understanding economic decisions and considers the effect of emotions and psychology on economic decision-making (i.e., how emotions and thoughts affect monetary decisions) (Heukelom, 2007). Behavioral economics combines the rational perspective with psychological perspective that includes the role of emotions.

Emotions play a very significant role in economic decision-making. The use of heuristics (rules of thumb) and influence of framing on economic decisions cannot be explained based on rational choice alone. Emotion or intuition influences human decision-making and actions. The human brain has both rational and emotional centers that come into play for understanding offers, making choices, and taking actions. In terms of decision-making, the human brain can respond automatically fast (System I) or deliberately slow (System II). System I corresponds to unconscious, automatic processing leading to quick, intuitive response to a problem. System II corresponds to deliberate and slow response. It monitors the quality of answers provided by System I, considers the pros and cons, and sometimes can correct or override the decision from System I (Kahneman, 2003a, b).

Pure neoclassical economic theory uses mathematical models and a few basic assumptions like rational choice and maximizing utility. Behavior economics theory allows for variance in consumer logic and behavior, but behavioral economists cannot rely as heavily on mathematical models to predict outcomes. Instead, they collect real world data on past consumer behavior and conduct experiments involving real transactions to gauge how consumers might behave in future situations. The goal in collecting such data is to eliminate unlikely outcomes so that the likely ones come into focus. Although not as exact of a science compared to using mathematical equations, behavioral economists often manage to make startlingly accurate economic predictions. Economists have found that making realistic assumptions about human nature generally leads to a more precise result (Kahneman & Thaler, 1991).

Nevertheless, behavioral economics has succeeded in explaining market anomalies where neoclassical economics could not. For instance, it has been used to examine the roles played by human greed and fear in the 2008 global financial crisis. The promise of windfall profits lead financial companies to create and sell highly complex credit default swaps without fully understanding their risk. When the stock market crashed, fear drove usually adventurous hedge fund investors to withdraw their money from the market, even when they could have bought good stocks at record-low prices. Behavioral economics can explain other phenomena as well, such as why some prices or wages refuse to change with market forces -price stickiness, why stock markets perform worse on Mondays (the "calendar effect"), and why some investors choose to hold onto poorly performing stocks while selling high-performing ones (the "disposition effect") (Singh, 2009).

Rational economics suggests that people's choice of an economic behavior should not change over time. However, people act more impatiently when making decisions about shorter-term options than longer-term ones. For example, people will often pay extra for express delivery of a desirable but non-essential item that is in stock, but not for one that is out-of-stock and, therefore, will not be deliverable until later. When people are offered an option for immediate reward, limbic structures are activated significantly more than when both options in the choice involved future delivery. The selection of the latter option was associated with significantly greater prefrontal cortex activity than the choice of the immediate one. Behavioral economics also deals with more complicated constructs like prospect theory of risky choice, hyperbolic time discounting, level-k models of games, and social preferences corresponding to internal rewards based on what happens to other agents (Volk et al., 2012).

PSYCHOLOGICAL BIASES AND HEURISTICS IN ECONOMIC DECISION MAKING

Globally, two commonly observed current market behaviors include excessive trading and the tendency to disproportionately hold on to losing investments while selling winners. Excessive trading is rooted in human emotion, overconfidence bias, and ease of trading due on-line trading. The tendency to hold on to losers while selling winners is rooted in loss aversion bias. Theoretically, markets are efficient but in practice, the biased crowd behavior of the investors can move the stock price. Human biases are rooted in their beliefs and are mostly unconscious. The human perception of reality is very subjective and unique based on one's past experiences. There are many biases affecting decision-making abilities and causing people to make irrational decisions.

An example of a bias is *anchoring and adjustment* where the decision is unduly influenced by initial information that shapes our view of subsequent information. Another example is *availability bias*, which is the tendency for some items that are more readily available in memory to be judged as more frequently occurring. For example, someone who watches a lot of movies about terrorist attacks may think the frequency of terrorism to be higher than it actually is. Due to *loss-aversion bias*, a loss of $100 hurts more than the joy of gaining $100. Another example is the *recency bias* in which people pay more attention to recent information to make decisions and ignore the more distant information even though it may be more important. These psychological biases lead to market anomalies and investor behavior that is not reconciled with the traditional finance paradigms (Park & Sohn, 2013).

Some of the ideas in behavioral economics include rule of thumb, framing, and market inefficiencies. People generally act on "rules of thumb" as opposed to rational thought. The *heuristics* (rule of thumb) and *framing* in making economic preferences cannot be explained by a purely rational approach. An example of a rule of thumb is "you get what you pay for", which is mostly true. However, sometimes a cheaper product is as good as or even better than an expensive version. It would be rational in this case to buy the cheaper product of equal quality. Most people, however, would buy the more expensive product, thinking it is superior.

Framing represents how a problem is presented and influences people's thoughts on a problem. An example is the higher sales of a product when the discount is presented as 75% versus $3 even though the discounted amount is the same in both cases. The third idea in behavioral economics is market inefficiencies which explain outcomes when something other than the expected happens market efficiency is the idea that price of a stock reflects all the known information available about a stock. No investor knows what will happen to the price before all of the other investors. Market inefficiency is anything that happens to challenge that idea, in a non-rational way (Table 1). For example, by selling overvalued

Table 1. Psychological Biases in Economic Decision Making (Adapted from Kahneman & Tversky, 1974, 1991, 2002; Kudryavtsev et al., 2013)

Bias	Description
Heuristic bias	'Rules of thumb' to make quick decisions.
Loss Aversion Bias	Actions are focused on avoiding loss rather than trying to make gains
Framing bias	This is best avoided by using numeracy with absolute measures of efficacy
Recency bias	Paying more attention on more recent information and either ignore or forget more distant information
Anchoring bias	Decisions are unduly influenced by initial information that shapes our view of subsequent information.
Group think bias	Peer pressure to conform to the opinions held by the group.
Confirmation Bias	Selective search for evidence, People tend to be willing to gather facts that support certain conclusions but disregard other facts that support different conclusions. Individuals who are highly defensive in this manner show significantly greater left prefrontal cortex activity as measured by EEG than do less defensive individuals, Premature termination of search for evidence, People tend to accept the first alternative that looks like it might work.
Choice-supportive bias	People distort their memories of chosen and rejected options to make the chosen options seem more attractive.
Primacy bias	Paying more attention to the first set of information
Repetition bias	A willingness to believe what one has been told most often and by the greatest number of different sources.
Source credibility bias	A tendency to reject a person's statement on the basis of a bias against the person, organization, or group to which the person belongs. People preferentially accept statement by others that they like
Correspondence bias	People tend to assume that outcomes are caused by dispositional factors, for example a person's constitution or personality, even when the actual cause is due to situational factors, for example, luck or coincidence

stocks, and using that money to buy undervalued stocks, if done correctly, investors can make a lot of money this way, even if it does not seem rational (Kahneman & Tversky, 1979).

Research on the biases of risk-aversion and seeking is mainly focused on the choices about risk to oneself. The choices about risks to a group of others include biases of inequality aversion or fairness. Several of the same biases of risky decision-making also apply to the choices about inequality aversion or fairness suggesting that similar heuristics are being employed. This means that decisions about fairness are being made in very similar ways even though there is no personal risk involved. People will often feel differently about identical outcomes because of the presentation of how that outcome is reached. The framing and reference points matter in choices about fairness and inequality (Michaelson, 2015).

BIG FIVE PERSONALITY TRAITS AND ECONOMIC PREFERENCES

Personality traits are the relatively enduring patterns of thoughts, feelings, and behaviors that reflect the tendency of an individual to respond in certain ways under certain circumstances. The big five domains of psychological personality constructs include: openness, conscientiousness, extraversion, agreeableness, and neuroticism. Openness represents individual differences in the tendency to be open to new aesthetic, cultural, and intellectual experiences. Conscientiousness represents the tendency to be organized, responsible, and hard working. Extraversion represents the quality of being more outgoing, gregarious, sociable and openly expressive. Agreeableness represents the tendency to act in a cooperative, unselfish manner. Neuroticism represents a chronic level of emotional instability and proneness to psychological distress (Roberts, 2006).

Both economic preferences and personality traits predict a wide range of life outcomes. If the economic preferences and personality traits are closely linked, they are expected to be substitutes in explaining heterogeneity in behavior. However, if economic preferences and personality traits capture different aspects of behavior, the two measurement systems may have complementary predictive power for important life outcomes. In other words, the economic preferences and personality traits, individually or jointly, may explain life outcomes. Data from incentivized laboratory experiments and representative samples found only low degrees of association between economic preferences and personality.

However, on regressing life outcomes – such as labor market success, health status and life satisfaction, earnings, unemployment, and education – simultaneously on preference and personality measures, it was found that economic preferences and personality are complementary constructs when it comes to explaining heterogeneity in important life outcomes and behavior. However, no clear picture emerges regarding the relationship between measures of personality and economic preferences (Becker et al., 2012).

A core element of economic theory is the assumption of stable preferences. This assumption was tested in public goods games by repeatedly eliciting cooperation preferences in a fixed subject pool over a period of five months. It was found that cooperation preferences are very stable at the aggregate level, and, to a smaller degree, at the individual level, allowing us to predict future behavior fairly accurately. Furthermore, these results provide evidence on the psychological foundations of cooperation preferences. The personality dimension 'Agreeableness' is closely related to both the type and the stability of cooperation preferences. (Volk, Thoni, & Ruigrok, 2012).

SOCIAL FACTORS INFLUENCING ECONOMIC PREFERENCES: THEORY OF MIND

Economics and game theory assume that people are capable of predicting the actions of others. This requires people to be able to understand others' motives and beliefs and view the perspective of others. The capacity to represent others' intentions, beliefs, and desires is referred as mentalizing or the theory of mind (TOM) mentioned earlier and is a special intelligence that can assess not only one's own desires and beliefs, but also those of others. Economic TOM entails a hierarchy of beliefs: "I know my preferences, and I have beliefs about your preferences, beliefs about your beliefs about my preferences". Humans have an ability to view an economic situation from the other person's perspective based on their understanding of the motives and beliefs of others. Human economic preferences are based on both self-interest and regards for others.

People with a well-developed TOM have a better understanding of how other people think, feel, and react and can take this into account when making a decision resulting in a greater sense of fairness. They have a better understanding of how their decisions impact others (Robalino & Robson, 2012). TOM requires one to be able to understand the intentions, beliefs and desires of others, share their feelings, and predict their behavior regarding their economic decisions. The ability to share others' feelings or to empathize has consequences regarding one's own economic preferences. In addition, one's own emotions, biases, framing and heuristics influence the economic preferences and decision-making.

Sometimes the human decision-making process is a psychological one dictated by emotions. Some anomalies like heuristics and framing that occur in economic decision-making cannot be explained by conventional economic theories using rational choice. Empathy is the ability to understand another person's perceptions on a deep level without reference to one's own perceptions—which may be a culminating feature of TOM development. However, economists do not have any direct insight into empathy and its interactions with one's own economic decision-making (Singer & Fehr, 2005).

The human mind is an adaptive classification system in which the external stimuli is ordered and classified in terms of neural connections that are the result of biological inheritance and individual's past experiences unique to the individual. The ongoing classification and coding processes in the human mind create neural connections that allow the individual to continuously update a model of the environment as experienced by the individual and leads to economic preferences, decisions, and behavior.

Hayek's (1952) TOM and his economic theory helped develop modern cognitive and behavioral economics where the effect of cognitive biases on economic behavior is recognized. Entropy TOM is an economic theory of mind in which the value and cost of information processing is integrated into the overall picture of economic decision-making (Chen, 2011). Behavioral economics emerged to account for these anomalies by integrating social, cognitive, and emotional factors in understanding economic decisions (Kahneman, 2003a). Behavioral economics combines the rational perspective with psychological perspective that includes the role of emotions and personality attributes.

Evolutionary psychology provides a means of explaining constructs that deal with mental capacities like TOM and Machiavellian or social brain. Hayek's (1952) TOM shows how human codify information and use dispersed knowledge to make economic choices. TOM deals with a mental capacity that allows one to be able to read others' behaviors, linguistic and non-verbal cues, and analyzes their intentions. To have a social brain means being able to assess, compete with and, where necessary, outwit others. TOM states that humans have the innate ability to intuitively understand one's own and other people's minds or mental states which include thoughts, beliefs, perceptions, knowledge, intentions, desires, and emotions.

Human behavior is influenced by these mental states. This ability to recognize one's own state of mind and those of others is central to human consciousness. The understanding that others have mental states different from one's own makes it possible to infer what others are thinking and to predict their behavior (Loasby, 2004).

Entropy TOM is an economic theory of mind in which the value and cost of information processing is integrated into the overall picture of economic decision-making and includes a theory of judgment which provides a common framework to integrate behavioral and informational theories of investment. The theory of judgment provides a quantitative link between investors' judgment and their trading activities. It offers a simple and unified understanding of patterns in market activities and investor behaviors (Chen, 2011).

The cognitive ability of TOM develops by degrees from infancy through childhood and adolescence into adulthood. It forms a basis for a child's acquisition of language and the development of appropriate social behavior and skills. It includes the ability to understand that the representations generated in one's mind do not necessarily reflect the real world and that other people may hold different representations. Mind is involved in all aspects of daily living and social interactions. It underlies learning and teaching and the ability to follow directions and understand socially based information. Mind skills are essential for working cooperatively. They also underlie the ability to manipulate and deceive others.

A lack of mind skills is considered by many to be a core deficit in autism (Singer & Fehr, 2005). In interactive decision-making, TOM reasoning is not of a low level in all competitive game settings. Individuals display effective TOM reasoning, reflecting realistic assumptions about their opponents in both competitive and relatively simple games (Goodie et al., 2012). Human judgment and decision-making involve both cognitive and emotional. In risk perception and decision-making, one must extend beyond the deliberate, rational evaluation of the various alternatives to see if it feels right. Intuition and affect play an important role in decision-making (Bohm & Brun, 2008).

NEURAL MECHANISMS OF ECONOMIC DECISION MAKING: NEUROECONOMICS

Human economic decision-making is rooted in emotional, rational, and neural mechanisms. Both reason and emotion reside in the brain and their interaction leads to economic preferences (Miller & Cohen, 2001). Economic, goal-directed decision-making behavior has basis in neurobiological and computational processes in the human brain and is causally determined by the neural activity in human brain. The neural computations necessary for making choices are stochastic (Corsani, 2013). Based on neurobiological evidence, human behavior is motivated by seeking gain and avoiding loss. During anticipation of gain versus loss, different neural circuits may generate emotion since the computational and neurobiological processes behind the goal-directed decision-making and behavior are also related to the psychological processes (Cohen, 2005). Neuroscience studies provide insight into human economic preferences and decision-making by linking the neural mechanisms to the cognitive and psychological processes in the human brain. Neuroeconomics seeks to explain human decision-making in an integrative way by combining the neural, emotional, and cognitive mechanisms in the human brain.

Economic decision-making involves the ability to process multiple alternatives to determine a course of action that leads to gain and avoids loss. The perspectives of classical, experimental and behavioral economics, cognitive and social psychology, neurobiology, endocrinology, mathematics, and computer science are combined together in the interdisciplinary field of neuroeconomics which seeks to explain human economic decision-making behavior in an integrative fashion in terms of neural mechanisms, cognition, and emotion. Using a combination of tools from many fields, Neuroeconomics avoids the shortcomings that arise from a single perspective approach (Fehr & Rangel, 2011).

Neuroeconomics uses brain imaging to investigate processes of economic decision-making that involves the computation and the comparison of decision values underlying goal-directed behavior (Popescu & Nica, 2014). The neurons in the orbitofrontal cortex (OFC) encode values of the choices available and the choice made. The individual must quickly estimate the reward value of the stimulus and predict when the reward will occur. The rational or self-interest focused decision-making is related to this prior processing of the expected reward or punishment. Emotions provide quick, automatic ways to react to new situations through rules of thumb and heuristics.

Humans do have the capacity to override the quick emotional response to a new situation and develop a delayed rational response. This capacity resides in the prefrontal cortex which is the most recently evolved part of human brain that supports higher cognitive processes. The limbic system is responsible for emotional reactions. The role of emotion in human behavior is rooted in our evolutionary past and is helpful for survival even today through the quick fight or flight response. Dopamine is the key neurotransmitter in the limbic system for reward processing which allows humans to engage in "approach" type behaviors. Serotonin is important in the "avoidance" type behaviors (Schultz, 2006).

Non-invasive brain stimulation and imaging techniques provide insight into the neural networks involved in the decision-making and generation of choices. The regions of brain activation while making these decisions provide insight into human economic preferences (Glimcher et al., 2008). Understanding the neural implementation of the psychological processes involved in economic decision-making provides insight into the cognitive and motivational processes driving human behavior. Experimentally-induced variations in neural activity in specific regions of people's brain changes people's willingness to make an economic decision. Neurological change influences the mental state, but relating the change to mental state is difficult because the latter is not directly observable (Hunter, 2013).

The human brain is best understood as a confederation of two mechanisms- cognition (reason) in the prefrontal cortex and emotion in the limbic system. The prefrontal cortex has the reasoning abilities and the capacity for cognitive control. It has the ability to guide thought and action in association with abstract goals or intentions, especially when this requires overcoming countervailing habits or reflexes. A human decision can be viewed as a joint decision by a committee of emotion and reason. Cognition has the ability to guide the thought and action in association with abstract goals or intentions especially when it requires overcoming emotional reactions. Cognition and emotion usually act together and guide human behavior.

At times, emotion competes with cognition offering a different evaluation of the same situation and emotion can override reason quickly and unconsciously. Humans do have the capacity to override the quick emotional response to a situation and develop a delayed rational response. This capacity resides in the prefrontal cortex which is the most recently evolved parts of our brain that supports higher cognitive processes whereas the role of emotion is rooted in our evolutionary past and is helpful for survival through the quick fight or flight response.

The parts of the brain that decide approach/avoidance behavior in survival settings are also involved in financial decision making. In brain imaging studies, activation in the nucleus accumbens is a positive predictor of choosing a risky asset. Activation in the Anterior Insula is a positive predictor of choosing the risk-neutral asset. Excessive activation in these areas causes deviations from optimal choice of a risk-neutral asset and correlates with making risk-seeking and risk-averse investments (Kuhnen & Knotson, 2005).

Economic decision-making assumes that people are capable of predicting others' actions through mentalizing and empathy. Mentalizing allows one to predict others' intentions, beliefs, desires and motives and make choices that promote self-interest. Empathy represents the capacity to share the feelings of others. Through mentalizing and empathy, people are able to view the economic transactions from the perspectives of both the self-interest and from other participant's perspective. Neuroscience provides insight into the neural mechanisms underlying mentalizing and empathy. The ability to mentalize develops in humans by age five and is impaired in adults with autism. Using a wide range of stimuli representing the intentions, beliefs, and desires of others, brain imaging studies reveal that the same part of the medial prefrontal lobe called anterior paracingulate cortex is involved when mentalizing about the thoughts, intentions or beliefs of others or when people are attending to their own states. The ability to empathize allows one to make decisions that also consider the interests of others. Due to empathy, humans can share their feelings and emotions in the absence of any direct emotional stimulation to themselves. The neuroscientific model of empathy suggests that observation of another person in a particular emotional state automatically activates a representation of that state in the observer with its associated autonomic and somatic responses (Singer & Fehr, 2005).

The empathic responses can be measured using functional Magnetic Resonance Imaging (fMRI). The automatic empathic response does not require conscious effortful processing but can be inhibited or controlled (Preston & de Waal, 2002). In the automatic empathic response, there is active engagement of some explicit judgments about others' feelings and one represents the goals of others in terms of one's own goals without even being aware of it. There are individual differences in empathic abilities. People with stronger empathic abilities are better predictors of others' motives and actions and are more likely to exhibit altruistic behaviors (Singer, et al., 2004).

Social neuroscience provides insights into the theory of mind or 'mentalizing' the neural mechanisms underlying our capacity to understand others' intentions, beliefs, and desires and capacity to share the feelings of others (i.e., empathy). Empathy and mentalizing are useful to individuals to make self-interested choices because one is able to predict others' actions more accurately. While pursuing self-interest, empathy allows one to be considerate of others and encourages other-regarding behavior. Neuroscientific empathy experiments indicate that the same affective brain circuits are automatically activated when we feel pain and others feel pain (Frith & Frith, 2003).

Using Positron Emission Tomography (PET) and fMRI, it has been shown that the perception and production of motor action has a common code in the brain (Grezes & Decety, 2001). Brain activity studies in choice tasks and belief formation tasks have established that a part of the medial prefrontal lobe called the anterior paracingulate cortex is involved in mentalizing others' beliefs, thoughts, and intentions of others. The same area is involved when people are attending to their own mental states – thoughts, beliefs, and intentions (Singer & Fehr, 2005).

Another finding is that the neural mechanisms underlying our ability to represent others' goals and intentions by the mere observation of their motor actions. The neural basis for imitation has been found to be the "mirror neurons". The discovery of mirror neurons demonstrates that a translation mechanism is present in the brain and is automatically elicited when viewing others' actions. This mirror system might underlie our ability to share others' mental states providing us with automatic simulation of their actions, goals, and intentions (Gallagher & Frith, 2003).

Neuroeconomics aims to discover mechanisms of economic decisions, and express them mathematically to explain and predict observed choice. The economic, psychological, and neural levels of explanation can be linked. The neural evidence is quite strong for a process of learning valuations through prediction error, and a simple model of neural valuation and comparison that corresponds to random utility. Neuroimaging studies of decision-making have generally related neural activity to objective measures (such as reward magnitude, probability or delay), despite choice preferences being subjective. However, economic theories posit that decision-makers behave as though different options have different subjective values.

Using fMRI, it has been shown that the neural activity in several brain regions—particularly the ventral striatum, medial prefrontal cortex and posterior cingulate cortex—tracks the revealed subjective value of delayed monetary rewards. This similarity provides unambiguous evidence that the subjective value of potential rewards is explicitly represented in the human brain. Neuroscientific discoveries can constrain and guide the models of economic preferences and the economic behavior can shape our understanding of the brain (Singer & Fehr, 2005).

In summary, although expected utility (EU) and the concept of rational agents are still being used in mainstream economics there is growing awareness about the role of emotions in economic decision making. Behavioral economics emerged by integrating social, cognitive, and emotional factors in understanding economic decisions. Neuroeconomics adds another layer by using neuroscientific methods in understanding the interplay between economic behavior and neural mechanisms. The economic preferences of time, risk, reciprocity, trust, and altruism have a neural basis. Neural activity has causal effects on economic behavior. The brain areas that generate emotional states also process information about risk, rewards and punishment. This suggests that emotions influence financial decisions. Positive emotional states such as excitement induce people to take risks and to be confident about their ability to evaluate investment options. Negative emotions like anxiety have the opposite effect on risk taking.

Emotions change behavior by changing risk preferences as well as learning (Kuhnen & Knutson, 2011). Activation in the Nucleus Accumbens is a positive predictor of choosing a risky asset. Activation in the Anterior Insula is a positive predictor of choosing the riskless event. Excessive activation in these areas causes deviations from optimal choice of a risk-neutral agent. It correlates with making risk-seeking and risk-averse mistakes (Kuhnen & Knutson, 2005). One can exogenously change one's risk taking behavior by triggering the brain areas that process emotions.

Anticipation of viewing highly rewarding stimuli promotes financial risk taking. This effect is partially mediated by increased Nucleus Accumbens activation. Neuroeconomics has contributed to our understanding of financial investing and consumer behavior (Knutson, et al., 2008).

NEUROMARKETING: ADVERTISING, MARKETING, AND BRAND LOYALTY

Neuromarketing is the application of neuroscience techniques to understand the decision processes of consumers in marketing, advertising, and brand loyalty. It involves scanning a consumers' brain using fMRI and recording their responses to stimuli to draw conclusions about how to sell to customers more effectively (Lindstrom, 2010). The purchasing behavior is determined at both conscious and subconscious levels. The traditional conscious measures such as surveys and focus groups miss the subconscious contribution which is 75% to 95%. To understand the subconscious level contribution towards the consumer purchasing behavior, one has to use neuroscience techniques.

Neuromarketing focuses on the reward system placed at the heart of the midbrain, which is related to release of dopamine. This motivates humans for simple behaviors like seeking food or sex or complex behaviors like choosing a brand, a taste, or an investment. Dopamine induces subjective feelings of pleasure and positive emotions. Cultural objects like cars that reflect a degree of wealth and social status and money activate the reward system. The human brain gets activated differently by gain or loss, both during anticipation and receiving (Touhami et al., 2011).

Just like in medicine, a placebo can be 30% effective; there is a marketing placebo effect. An unconscious expectation about the relationship between price and quality is an example of a marketing placebo effect. A change in price can affect neural representations on experiences pleasantness and effectiveness of a product (Shiv et al., 2005). Increasing the price of the same wine increases subjective reports of flavor pleasantness as well as blood-oxygen-level dependent activity in medio orbitofrontal cortex which is the brain area associated with experienced pleasantness during experiential tasks. This can be measured using the functional brain imaging techniques (Plassmann et al., 2008).

Using brain imaging, the neural circuits that are involved in identifying a brand preference have been identified. The preference of a brand is based on two factors including the intrinsic components of a product and "branding". The "branding" plays a key role in the mechanism of preference which influences buying behavior and brand loyalty (Plassman, Kenning, & Ahler, 2007). Two separate systems are involved in generating brand preference. When the judgments are based solely on sensory information (e.g. taste), the ventromedial prefrontal cortex (VMPFC) is involved. When the judgments are based on sensory and cultural information, hippocampus, dorsal prefrontal cortex and the midbrain are activated. In a taste test where the subjects knew that they were tasting coke, there was significant more activation than that for Pepsi. However, in blind taste tests, both coke and Pepsi tasting led to similar activation (McLure et al., 2004).

There is a debate about the ethics of neuromarketing. The critics of neuromarketing question the its ethics and recommend focusing on the heart rather than involving the brain. In spite of ethical concerns, neuromarketing is here to stay to provide insights into human cognition and emotion. Neuroscience techniques are just instruments to better understand our wants, drives and motivations. We have an ethical responsibility not to use neuromarketing to manipulate consumers. (Flores, Baruca, & Saldivar, 2014).

The most promising application of neuromarketing may be in product development before the product is released in the marketplace (Ariely & Berns, 2010). Table 2 lists the companies that use neuromarkeitng for various purposes ranging from customer satisfaction, ad campaigns and packaging design.

SUMMARY AND IMPLICATIONS FOR SOCIAL POLICIES

The human economic decision-making is an integrative social process involving interaction of reason, emotion, feelings, and the ability to mentalize and empathize in the human mind. The human mind is a complex and *adaptive* classification system since it is modified by both a process of natural selection in forming the organism's set of neural connections and by individual's life experiences. To make sense of the world, human mind classifies and orders the external information inputs in terms of neural connections that are the result of the biological inheritance of the individual and the individual's unique past experiences.

The individual views the world through the lens of his or her past experiences. This information classification process in the human mind is dynamic and contextual. The order attributed to each particular information input depends on the other inputs that arrive with it. The inputs include both rational and psychological aspects. There is no one-to-one relationship between each given stimulus and the sensory order which is unique for each person. The sensory order depends on the external stimulus and the specific neural system involved in the process. Each individual has a unique neural system although there are common features among individuals. The ongoing classification and coding processes create neural connections that allow the individual to continuously update a model of the environment as experienced by the individual, and leads to economic preferences, decisions, and behavior (Ambrosino, 2014).

Table 2. Companies and the Purpose of Neuromarketing Research (adapted from Flores, Baruca and Saldivar, 2014; Babu and Vidyasagar, 2012)

Company	Industry	Neuromarketing Research Purpose
Microsoft	Technology/Software	Consumers' satisfaction and frustration in their interaction with computers
EBay	Online auction	Ad campaigns
Frito-Lay (Pepsi Co)	Food	Commercials, Products and Packaging
Daimler	Automotive	Consumer reaction to car headlight characteristics
Hyundai	Automotive	Consumer reaction when viewing a sports car
Viacom	Media	Consumer reaction to advertising
Neurofocus	Research	Conducted neuromarketing research for Google, Chevron and Walt Disney
Delta Airlines	Airlines	Understanding customer travel experience

Classical economics which is based on rational choice and utility maximization does not adequately explain human economic decisions in real life. Instead of conducting due diligence to gather all the information available to make rational economic decisions, people make decisions based on their emotions, biases, and social influences. In other words, economic decisions are based on rational, emotional, social and neural factors. Behavioral economics emerged to account for the effect of emotions, biases, heuristics, and psychology on economic decision-making. The body's response to the external stimuli begins with emotions and feelings. Human mind begins at the level of feelings. Even with intact reason, if emotions are impaired, humans are not able to make good decisions.

Emotions are central to the life-regulating processes of humans. "We are not thinking machines that feel, we are feeling machines that think". Feelings are mental experiences of body states, which arise as the brain interprets emotions. The self-concept of humans emerges from emotions and feelings (Jason, 2014).

Behavioral economics includes both the rational and emotional aspects of economic decision-making which seeks to unite the basic principles of classical economics with the realities posed by human psychology. Behavioral economics has been applied to many areas including macroeconomics, labor economics, finance, and marketing. Behavioral finance combines the individual economic behavior with group economic behavior that includes social influence. Human economic decisions are influenced by what others do. Stock prices move up and down without any substantial change in the fundamentals of the companies.

People trade in herds and that influences stock prices. Behavioral economics includes the social factors including people's ability to predict others' intentions that influence self-interested economic preferences (Singh, 2009). The root cause of the 2008 financial crisis was a psychological phenomenon, not a fundamental one. Risk-seeking behaviors were evident in loss-dominant markets while excessive optimism and confirmation bias drove the crisis (Shefrin, 2009, 2000; Park & Sohn, 2013; Sadeghnia & Habibniko, 2013).

The interaction of behavioral economics with social neuroscience led to the development of neuroeconomics which seeks to explain human economic decision making in an integrative way by combining insights from psychology, economics, and neuroscience. From a neuroscientific perspective, human decision-making is computational and involves the ability to process multiple alternatives and to follow a course of action. Economic preferences are the result of individuals assigning *values* to the offered choices in their brain (Fehr & Rangel, 2011). The brain imaging methods of neuroscience help in understanding the interplay between economic behavior and neural mechanisms.

Human economic decision-making combines rational and intuitive involving cognition and emotion which are seated in the prefrontal cortex and limbic system respectively in the human brain. Economic decision-making reflects the interaction of automatic and controlled parts of human brain. The automatic system is fast, unconscious, and emotional and requiring little effort while the controlled system is slow, rational, and deliberate and requires effort. People rely on rules of thumb or heuristics to make fast, automatic judgments especially when they are pressed for time, under cognitive load, in a good mood or lacking motivation (Sanfey & Chang, 2008). Understanding neural mechanisms contributes to our understanding of economic decision-making and studying economic behavior has the potential to contribute to understanding of neural mechanisms. Some of the criticisms of neuroeconomics include exaggeration of new findings, misuse of brain imaging data and poor testing of hypotheses (Craver & Alexandrova, 2008; Rubinstein, 2008).

In addition to self-interested rational and emotional economic decisions, humans do make *other-regarding* or altruistic economic decisions. Social neuroscience provides insight into the neural mechanisms that underlie the human capacity to predict others' actions (*mentalize*) and share feelings of others (*empathy*). The ability to mentalize promotes self-interested behavior and the ability to empathize promotes the other-regarding behavior or altruism. Rationality, emotions, and the ability to mentalize and empathize, are all seated in the human brain (Singer & Fehr, 2005).

The understanding of the interaction between emotion and cognition has great relevance to economists, sociologists, and policy makers. The social policies are aimed at protecting the society from emotional responses that may have served us well in our evolutionary past, but can interfere with the smooth functioning of modern society. The development of human capacity for rational thought has given rise to powerful technologies that can threaten human survival if human behavior is guided by the emotions alone (Cohen, 2005).

The findings from behavioral economics and judgment and decision making highlight many behavioral impediments including lack of self-control, emotions, and choice architecture that influence the savings behavior of individuals which impacts financial security in retirement. Policy makers can use this information to incentivize savings in the short-term, reframing the decision-context and shifting reference points to help individuals save more and spend less (Knoll, 2010).

Future research on neural mechanisms for economic decision-making can focus on gender and generation differences in the responses to the same stimuli. Another area for future research would be to understand the differences in the neural mechanisms in economic decision making for the labor force and the owners of the capital to get insight into the income-inequality issues that can generate social unrest. Insights into customer satisfaction, ad campaigns and packaging through Neuromarketing can help the firms and customers alike. Understanding the rational, emotional and neural mechanisms in economic decision-making enables us to predict human behavior at both the micro and macroeconomic levels.

REFERENCES

Ambrosino, A. (2014). A Cognitive Approach to Law and Economics: Hayek's Legacy. *Journal of Economic Issues*, *48*(1), 19–48. doi:10.2753/JEI0021-3624480102

Ariely, D., & Berns, G. S. (2010). Neuromarketing: The Hope and Hype of Neuroimaging in Business. *Nature Reviews. Neuroscience*, *11*(4), 284–292. doi:10.1038/nrn2795 PMID:20197790

Ayton, P., & Fischer, I. (2004). The Hot Hand Fallacy and the Gambler's Fallacy: Two Faces of Subjective Randomness? *Memory & Cognition*, *32*(8), 1369–1378. doi:10.3758/BF03206327 PMID:15900930

Babu, S. S., & Vidyasagar, T. P. (2012). Neuromarketing: Is Campbell in Soup? *The IUP Journal of Marketing Management*, *9*(2), 76–100.

Becker, A., Deckers, T., Dohmen, T., Falk, A., & Kosse, F. (2012). *The Relationship Between Economic Preferences and Psychological Personality Measures*. Retrieved On January 8, 2015, from: http://www.iame.uni-bonn.de/people/thomas-dohmen/iza-dp-6470

Bohm, G., & Brun, W. (2008). Intuition and Affect in Risk Perception and Decision Making. *Judgment and Decision Making*, *3*(1), 1–4.

Borghans, L., Duckworth, A. L., Heckman, J. J., & Ter Weel, B. (2008). The Economics and Psychology of Personality Traits. *The Journal of Human Resources*, *43*(4), 972–1059. doi:10.3368/jhr.43.4.972

Brahmana, R. K., Hooy, C. W., & Ahmad, Z. (2012). Psychological Factors on Irrational Financial Decision Making. *Humanomics*, *28*(4), 236–257. doi:10.1108/08288661211277317

Chen, J. (2011). The Entropy Theory of Mind and Behavioral Finance. *The IUP Journal of Behavioral Finance*, *8*(4), 6–39.

Cohen, J. D. (2005). The Vulcanization of the Human Brain: A Neural Perspective on Interactions between Cognition and Emotion. *The Journal of Economic Perspectives*, *19*(4), 3–24. doi:10.1257/089533005775196750

Corsani, A. (2013). Rent and Subjectivity in Neoliberal Cognitive Capitalism. *Knowledge Cultures*, *1*(4), 67–83.

Craver, C. F., & Alexandrova, A. (2008). No Revolution Necessary: Neural Mechanism For Economics. *Economics and Philosophy*, *24*(03), 381–406. doi:10.1017/S0266267108002034

Fehr, E., Fischbacher, U., & Kodfeld, M. (2005). Neuroeconomic Foundations of Trust and Social Preferences: Initial Evidence. *The American Review*, *95*(2), 346–351.

Fehr, E., & Rangel, A. (2011). Neuroeconomic Foundations of Economic Choice- Recent Advances. *The Journal of Economic Perspectives*, *25*(4), 3–30. doi:10.1257/jep.25.4.3 PMID:21595323

Flores, J., Baruca, A., & Saldivar, R. (2014). Is Neuromarketing Ethical? Consumers Say Yes, Consumers Say No. *Journal of Legal. Ethical and Regulatory Issues*, *17*(2), 77–91.

Frederick, S., Lowenstein, G., & O'Donoghue, T. (2002). Discounting and Time Preference: A Critical Review. *Journal of Economic Literature*, *40*(1), 351–401. doi:10.1257/jel.40.2.351

Frith, U., & Frith, C. D. (2003). Development and Neurophysiology of Mentalizing. *Philosophical Transactions of the Royal Society of London. Series B, Biological Sciences*, *358*(1431), 459–473. doi:10.1098/rstb.2002.1218 PMID:12689373

Gallagher, H. L., & Frith, C. D. (2003). Functional Imaging of Theory of Mind. *Trends in Cognitive Sciences*, *7*(2), 77–83. doi:10.1016/S1364-6613(02)00025-6 PMID:12584026

Glimcher, P. W. Colin, Camerer, E. F., Fehr, E., & Poldrack, R. A. (2008). Introduction: A Brief History of Neuroeconomics. In P. Glimcher, E. Fehr, C. Camerer, & R. A. Poldrack (Eds.), Neuroeconomics: Decision Making and the Brain (pp. 1-12). London: Academic Press.

Goodie, A. S., Doshi, P., & Young, D. L. (2012). Levels of Theory-of-Mind Reasoning in Competitive Games. *Journal of Behavioral Decision Making*, *25*(1), 95–108. doi:10.1002/bdm.717

Grezes, J., & Decety, J. (2001). Functional Anatomy of Execution, Mental Simulation, Observation and Verb Generation of Actions: A Meta Analysis. *Human Brain Mapping*, *12*(1), 1–19. doi:10.1002/1097-0193(200101)12:1<1::AID-HBM10>3.0.CO;2-V PMID:11198101

Hayek, F. A. (1952). *The Sensory Order*. Chicago: Chicago University Press.

Heukelom, F. (2007). *Kahneman and Tversky and the Origin of Behavioral Economics.* Tinbergen Institute Discussion Paper, TI 2007-003/1. Retrieved on May 9, 2013, from: http://www.tinbergen.nl

Hunter, M. (2013). Creativity and Making Connections: The Patchwork of Entrepreneurial Opportunity. *Psychological Issues in Human Resources Management, 1*(2), 7–51.

Jason, P. (2014). Questions and Answers with Antonio Damasio. *MIT's Technology Review, 117*(4), 48–51.

Kahneman, D. (1994). New Challenges to the Rationality Assumption. *Journal of Institutional and Theoretical Economics, 150*(1), 18–44.

Kahneman, D. (2003a). Perspective on Judgment and Choice: Mapping Bounded Rationality. *The American Psychologist, 58*(9), 697–720. doi:10.1037/0003-066X.58.9.697 PMID:14584987

Kahneman, D. (2003b). Maps of Bounded Rationality: Psychology for Behavioral Economics. *The American Economic Review, 93*(5), 1449–1475. doi:10.1257/000282803322655392

Kahneman, D., & Thaler, R. (1991). Economic Analysis and the Psychology of Utility: Applications to Compensation Policy. *The American Economic Review, 81*(2), 341–346.

Kahneman, D., & Tversky, A. (1974). Judgment Under Uncertainty: Heuristics and Biases. *Science, 185*(4157), 1124–1131. doi:10.1126/science.185.4157.1124 PMID:17835457

Kahneman, D., & Tversky, A. (1979). Prospect Theory: An Analysis of Decision Under Risk. *Econometrica, 47*(2), 313–327. doi:10.2307/1914185

Kahneman, D., & Tversky, A. (1991). Loss Aversion in Riskless Choice: A Reference Dependent Model. *The Quarterly Journal of Economics, 106*(4), 1039–1061. doi:10.2307/2937956

Kahneman, D., & Tversky, A. (Eds.). (2002). *Choices, Values and Frames.* Cambridge, MA: Cambridge University Press.

Knoll, M. A. Z. (2010). The Role of Behavioral Economics and Behavioral Decision- Making in Americans' Retirement Savings Decisions. *Social Security Bulletin, 70*(4), 1–23. PMID:21261167

Knutson, B., Wimmer, G. E., Kuhnen, C. M., & Winkielman, P. (2008). Nucleus Accumbens Activation Mediates the Influence of Reward Cues on Financial Risk Taking. *Neuroreport, 19*(5), 509–513. doi:10.1097/WNR.0b013e3282f85c01 PMID:18388729

Kudryavtsev, A., Cohen, G., & Hon-Snir, S. (2013). Rational or Intuitive: Are Behavioral Biases Correlated Across Stock Market Investors? *Contemporary Economics, 7*(2), 31–53. doi:10.5709/ce.1897-9254.81

Kuhnen, C. M., & Knutson, B. (2005). The Neural Basis for Financial Risk Taking. *Neuron, 47*(5), 763–770. doi:10.1016/j.neuron.2005.08.008 PMID:16129404

Kuhnen, C. M., & Knutson, B. (2011). The Influence of Affect on Beliefs, Preferences, and Financial Decisions. *Journal of Financial and Quantitative Analysis, 46*(3), 605–626. doi:10.1017/S0022109011000123

Lindstrom, M. (2010). *Buy Ology: Truth and Lies about Why We Buy.* New York: Broadway Books.

Loasby, B. J. (2004). Hayek's Theory of the Mind. *Evolutionary Psychology and Economic Theory. Advances in Austrian Economics, 7,* 101–134. doi:10.1016/S1529-2134(04)07006-1

McLure, S. M., Li, J., Tomlin, D., Cypert, K. S., Montague, L. M., & Montague, P. R. (2004). Neural Correlates of Behavioral Preference for Culturally Familiar Drinks. *Neuron, 44*(2), 379–387. doi:10.1016/j.neuron.2004.09.019 PMID:15473974

Michaelson, Z. (2015). Biases in Choices About Fairness: Psychology and Economic Inequality. *Judgment and Decision Making, 10*(2), 198–203.

Miller, E. K., & Cohen, J. K. (2001). An Integrative Theory of Prefrontal cortex Function. *Annual Review of Neuroscience, 24*(1), 167–202. doi:10.1146/annurev.neuro.24.1.167 PMID:11283309

Moorhead, G., & Montanari, J. R. (1986). An Empirical Investigation of the Groupthink Phenomenon. *Human Relations, 39*(5), 399–410. doi:10.1177/001872678603900502

Park, H., & Sohn, W. (2013). Behavioral Finance: A Survey of the Literature and Recent Development. *Seoul Journal of Business, 19*(1), 3–42.

Plassman, H., Kenning, P., & Ahlert, A. (2007). Why Companies Should Make Their Customers Happy: The Neural Correlates of Customer Loyalty. *Advances in Consumer Research. Association for Consumer Research (U. S.), 34*, 735–739.

Plassman, H., O'Deherty, J., Shiv, B., & Rangel, A. (2008). Marketing Actions can Modulate Neural Representations of Experienced Pleasantness. *Proceedings of the National Academy of Sciences of the United States of America, 105*(3), 1050–1054. doi:10.1073/pnas.0706929105 PMID:18195362

Popescu, G. H., & Nica, E. (2014). Neuroeconomic Models of Decision Making. *Knowledge Horizons – Economics, 6*(1), 63–66.

Preston, S. D., & de Waal, F. B. M. (2002). Empathy: Its Ultimate and Proximate Bases. *Behavioral and Brain Sciences, 25*(1), 1–72. PMID:12625087

Robalino, N., & Robson, A. (2012). The Economic Approach to theory of mind. *Philosophical Transactions of the Royal Society of London. Series B, Biological Sciences, 367*(1599), 2224–2233. doi:10.1098/rstb.2012.0124 PMID:22734065

Roberts, B. W. (2006). Personality Development and Organizational Behavior. In B. M. Staw (Ed.), *Research on Organizational Behavior* (pp. 1–41). Amsterdam: Elsevier Science/ JAI.

Rubinstein, A. (2008). Comments on Neuroeconomics. *Economics and Philosophy, 24*(03), 485–494. doi:10.1017/S0266267108002101

Sadeghnia, M., & Habibniko, A. (2013). Behavioral Finance and NeuroFinance and Research Conducted in This Area. *IJCRB, 4*(12), 793–801.

Sanfey, A. G., & Chang, L. J. (2008). Multiple Systems in Decision Making. *Annals of the New York Academy of Sciences, 1128*(1), 53–62. doi:10.1196/annals.1399.007 PMID:18469214

Schultz, W. (2006). Behavioral Theories and the neurophysiology of reward. *Annual Review of Psychology, 57*(1), 87–115. doi:10.1146/annurev.psych.56.091103.070229 PMID:16318590

Shefrin, H. (2000). *Beyond Greed and Fear: Understanding the Behavioral Finance and the Psychology of Investing*. Boston, MA: Harvard Business School Press.

Shefrin, H. (2009). How Psychological Pitfalls Generated the Global Financial Crisis. In L. Siegel (Ed.), *Understanding the Global Financial Crisis*. Boston, MA: Harvard Business School Press. doi:10.2139/ssrn.1523931

Shiv, B., Carmon, Z., & Ariely, D. (2005). Placebo Effects of Marketing Actions: Consumers may Get What They Pay For. *JMR, Journal of Marketing Research*, *42*(4), 383–393. doi:10.1509/jmkr.2005.42.4.383

Simon, H. (1955). A Behavioral Model of Rational Choice. *The Quarterly Journal of Economics*, *69*(1), 99–188. doi:10.2307/1884852

Singer, T., & Fehr, E. (2005). Neuroscientific Foundations of Economic Decision Making – the Neuroeconomics of Mind Reading and Empathy. *The American Economic Review*, *95*(2), 34–345. doi:10.1257/000282805774670103

Singer, T., Seymour, B., O'Doherty, J. P., Kaube, H., Dolan, R. J., & Frith, C. D. (2004). Empathy for Pain Involves the Affective but not Sensory Components of Pain. *Science*, *303*(5661), 1157–1162. doi:10.1126/science.1093535 PMID:14976305

Singh, R. (2009). Behavioral Finance – The Basic Foundations. *ASBM Journal of Management*, *II*(1), 89–98.

Summerfield, C., & Tsetsos, K. (2015). Do Humans Make Good Decisions? *Trends in Cognitive Sciences*, *19*(1), 27–34. doi:10.1016/j.tics.2014.11.005 PMID:25488076

Thaler, R. (1999). Mental Accounting Matters. *Journal of Behavioral Decision Making*, *12*(3), 183–206. doi:10.1002/(SICI)1099-0771(199909)12:3<183::AID-BDM318>3.0.CO;2-F

Touhami, Z. O., Benlafkih, L., Jiddane, M., Cherrah, Y., Malki, O. E., & Benomar, A. (2011). Neuromarketing: Where Marketing and Neuroscience Meet. *African Journal of Business Management, 5*(5), 1528-1532.

Tversky, A., & Kahneman, D. (1986). Rational Choice and the Framing of Decisions. *The Journal of Business*, *59*(4), S251–S278. doi:10.1086/296365

Volk, S., Thoni, C., & Ruigrok, W. (2012). Temporal stability and psychological Foundations of cooperation preferences. *Journal of Economic Behavior & Organization*, *81*(2), 664–676. doi:10.1016/j.jebo.2011.10.006

KEY WORDS AND DEFINITIONS

Behavioral Economics: It combines the rational perspective of classical economics with the psychological perspective that includes the role of emotions and biases.

Big Five Personality Traits: They are five domains of psychological personality traits that are the relatively enduring patterns of thoughts, feelings and behaviors that reflect the tendency to respond in certain ways under certain circumstances and affect outcomes.

Bounded Rationality: People have cognitive limits and they are not always able to obtain all the information available they need to make the best possible decisions.

Emotional Response: An automatic response based on fight or flight response.

Heuristics: Rules of thumb people use as a short cut to make a decision in lieu of rational thought.

Neuroeconomics: It uses brain-imaging techniques of neuroscience to understand rational and emotional economic decision-making in humans.

Neuromarketing: Application of neuroscience techniques like brain imaging to understand the decision processes of consumers in marketing, advertising and brand loyalty.

Rational Choice: A choice arrived at using logic and reason alone.

Theory of Mind or Mentalizing: The capacity to represent others' intentions, beliefs and desires.

Utility: A subjective perspective of the benefits of a choice.

This research was previously published in Neuroeconomics and the Decision-Making Process edited by Bryan Christiansen and Ewa Lechman, pages 45-65, copyright year 2016 by Business Science Reference (an imprint of IGI Global).

Chapter 23
Economic Decision Making, Emotion, and Prefrontal Cortex

Salim Lahmiri
ESCA School of Management, Morocco

ABSTRACT

How diverse regions of the brain are coordinated to produce objective-directed decision is the essence of neuroeconomics. Indeed, the latter is a formal framework to describe the involvement of numerous brain regions including frontal, cingulate, parietal cortex, and striatum in economic and financial decision-making process. The purpose of this chapter is to explain the relationship between economic decision making and emotion on one hand, and the relationship between economic decision making and prefrontal cortex on the other hand.

INTRODUCTION

Understanding and explaining human decision-making process is approached differently in economics and neuroscience. For instance, traditional economic theory explains behaviour primarily through theoretical foundations such as utility, preferences, and axioms. On the other side, neuroscience considers the physiological aspects and somatic variables that affect decision-making. In recent years, neuroeconomics has emerged as a multidisciplinary research area that integrates knowledge from neuroscience, psychology, and economics to better understand economic decision-making and to specify more accurate models of choice and decision. Multiple options always characterize decisional situations by multiple options, each carrying potential rewards, risks and related outcome probabilities.

Therefore, real life decision making requires the ability to make decisions effectively. The purpose of the chapter is to explain the relationship between economic decision making and emotion on one hand, and the relationship between economic decision making and prefrontal cortex on the other hand. In particular, based on a brief literature review, we aim to present the role of psychological factors, especially emotions, in economic decision making and presenting cortex regions involved in financial decision making. We essentially focus on physiological and neuroimaging aspects of economic and

DOI: 10.4018/978-1-5225-5478-3.ch023

financial decision making. The chapter is organized as follows. Section 1 focuses on economic decision making and emotion. Section 2 covers economic decision making and the prefrontal cortex. Section 3 provides directions for future works as suggested in the literature. Section 3 presents the Conclusion.

Economic Decision Making and Emotion

Classical and neoclassical economics states that decisions are made based on valuable information and cost-benefit analysis. In particular, the main purpose in making a decision is to maximize the gains or outcome expected value expressed as a linear multiplicative relationship between probability and utility. Basically, classical economic theory assumes that all agents are aggregating and weighing information accurately and consistently. However, in real life problems, individuals have not been found to make decisions following a normative model (Beach & Lipshitz, 1993). In addition, human decision making process is affected by heuristics, biases, and framing effects (Tversky & Kahneman, 1974). More importantly, it involves psychological factors; especially emotions (Bechara et al., 2000; Bechara et al., 2002). Therefore, classical economic theory fails to explain human behaviour and decision-making.

Several psychological biases have been proposed in economic literature to explain economic decision making process including theory of beliefs (Barberis & Thaler, 2003), belief perseverance (Barberis & Thaler, 2003; Rabin, 1998), confirmatory bias (Rabin, 1998), overconfidence (Barberis & Thaler, 2003), optimism (Barberis & Thaler, 2003), representativeness (Barberis & Thaler, 2003; Kahneman, 2003), prototype heuristics (Kahneman, 2003), law of small numbers (Rabin, 1998), conservatism (Barberis & Thaler, 2003), anchoring and adjustment (Barberis & Thaler, 2003; Kahneman, 2003; Rabin, 1998), and availability biases (Barberis & Thaler, 2003; Kahneman, 2003). They are described in Table 1.

However, the most studied factor is emotion. Indeed, According to Somatic Marker Hypothesis (Bechara et al., 2000; Bechara et al., 2002), decision-making is a process guided by emotions. It provides a neuroanatomical and cognitive framework for decision-making and suggests that the process of decision-making depends on neural substrates that regulate homeostasis, emotion, and feeling. In order

Table 1. Biases in financial decision making

Biases	Description
Theory of beliefs	How agents form their expectations in the market
Belief perseverance	People form their own hypotheses and are likely to less attentive to information which contradicts their hypotheses
Confirmatory bias	People are influenced by initial judgements. They tend to use information to affirm their initial hypotheses
Overconfidence	Estimated quantities or probabilities are affected by overconfidence
Optimism	People's unrealistic positive view about themselves
Representativeness	People tend to determine something by the characteristics of the group or class to which it belongs.
Prototype heuristics	Broader view of the representativeness heuristic
Law of small numbers	People exaggerate the behaviour of small samples
Conservatism	Hesitation of many investors to act on new information
Anchoring and adjustment	Values are estimated similar to the initial values of uncertain quantities
Anchoring and adjustment	Probability of an event is estimated based on memories and personal experiences

to investigate decision making process in neuroeconomics, a series of brain imaging modalities are used to record electrical activity and metabolic activity of the human brain including electroencephalograph (EEG), transcranial magnetic stimulation (TMS), functional magnetic resonance imaging (fMRI), and positron emission tomography (PET). These modalities help understanding and explaining neurobiological mechanisms of the decisional process (Pradeep, 2010). Indeed, at anatomical level, neuronal activation is observed in caudate nucleus in relation with satisfaction (de Quervainet al., 2004), and in the ventral striatum related to exhibitions of trust in an investment game (King-Casas et al., 2005). In general, emotional process is associated with activation in ventral tegmental, nucleus accumbens, ventromedial frontal, orbitofrontal, anterior cingulated cortex, and amygdala and insular cortex (Dalgleish, 2004).

The relationship between the cortex regions was also investigated to understand the decision making process. For instance, by examining communication across brain regions to examine motivational decision making, evidence was discovered of intercommunication between associative regions, mainly within the prefrontal cortex, and striatum, insula, cingulated cortex and amygdale (Alexander & Brown, 2011; Andersen & Cui, 2009; Clark, 2010; Kable & Glimcher, 2009; Rangel & Hare, 2010; Wallis & Kennerley, 2010). In addition, these works related subcomponents of decision making processes to specific brain regions (Alexander & Brown, 2011; Andersen & Cui, 2009; Clark, 2010; Kable & Glimcher, 2009; Rangel & Hare, 2010; Wallis & Kennerley, 2010). In particular, dorsal and ventral lateral prefrontal cortices and parietal lobe are implicated in integrative computation of expected value.

However, ventromedial and dorsomedial prefrontal cortices and adjacent orbitofrontal cortex are involved in abstract value representation. Univariate data analysis is the most common technique to establish direct relationships between specific economic variables and regional brain activity. However, they treat each region in isolation rather in dynamical framework where it is assumed to interplay with other regions. Therefore, multivariate techniques are becoming more popular to study the dynamic connectivity between cortical nodes. In this regard, several approaches have been used in the literature; including psychophysiological interaction (PPI) tests (Friston et al., 1997), dynamic causal modeling (Friston et al., 2003; Stephan et al., 2007), and graph-based network analysis (Camara et al., 2009; Zamora-López et al., 2010). These methods are described in Table 2.

Using univariate analysis to study financial decision making tasks, Camara et al., (2008) found that activity in the ventral striatum is closely mirrored by responses across the insula, amygdala, hippocampus, and orbitofrontal cortex during reward processing. Hare et al. (2010) concluded that ventromedial prefrontal cortices activity integrated inputs from the anterior insula and temporal cortex to reflect the amount donated during a study of charitable giving. Park et al. (2011) found that sub-genual cingulate and amygdale are coupled when monetary rewards are linked to physical pain.

Table 2. Methods descriptions in economic decisional task

Methods	Description
Psychophysiological interaction (PPI)	It tests for the changes in the (physiological) influence of a given source region on the activity of a second target region as a result of a specific external (psychological/economic/financial) factor.
Dynamic causal modeling	It is used to gain quantitative estimates of the magnitude and direction of causal interaction between brain regions when decision-maker is subject to external perturbation.
Graph-based network analysis	When decision-maker is peforming a task, its brain areas are parcelled PPI is applied across parcels to identify networks of connections modulated by relevant economic factors.

In a recent study, Minati et al. (2012) used graph-based network analysis and concluded that: (i) expected value is positively correlated with activity within medial prefrontal and occipital cortices; (ii) value-sensitive effective connections are arranged as a unitary characterized by dense connectivity and high shortest-path centrality; (iii) observed effective connectivity effects are highly pertinent to dichotomous gain/loss comparisons. For a summary of cortex regions involved in decision-making making process, we have adapted Table 3 from Gutnik et al. (2006) to show cortex regions along with their functional classifications.

Economic Decision Making and Prefrontal Cortex

The prefrontal cortex is one of the human brain regions that attract attention of researchers in economic and financial decision making given its role in executive functions such as reasoning, planning and decision-making (Spinella et al., 2007). For instance, individuals with neurological insults to prefrontal systems often show poor financial management (Spinella et al., 2007), and healthy individuals consistently show activation of prefrontal systems during financial tasks (Spinella et al., 2007). Table 4 shows the results of neuroimaging in financial decision processing.

In order to examine the relationship between self-rated executive functions and personal income, Spinella et al. (2007) administered a sample of subjects under study the Barratt Impulsiveness Scale (Patton et al., 1995) which is a self-rating scale that measures three factors (Patton et al., 1995): non-planning (orientation towards the present rather than the future), motor impulsivity (or acting without thinking), and attentional impulsivity (lack of concentration). Spinella et al. (2007) found evidence of an inverse relationship between income and total impulsiveness, and that association remains significant after controlling for sex, age and education. In addition, they found that the motor impulsivity was the subscale with the most consistent associations. Therefore, they concluded that self-rating measure of executive dysfunction is inversely correlated with income.

In order to study other aspects of financial behaviour, Spinella et al. (2007) examined the relationship between prefrontal systems and credit card debt. They found that credit card debt is related to the ability to mentally track finances. They concluded that a prefrontal system is involved in executive functions that

Table 3. Decision-making functions of brain areas (adapted from Gutnik et al., 2006)

Area of Brain	Subregion	Function
Prefrontal cortex	Orbitofrontal cortex /ventromedial	Incentive gain, best-guess estimation
	prefrontal cortex	Reversal learning
	Dorsolateral prefrontal cortex	Manipulation of decision-related information and conscious; deliberation during decision-making; working memory functions; Decision-making under uncertainty
	Anterior cingulate cortex	Cognitive conflict processing
Amygdala		Reward recognition; perception of emotion; emotion in general
Basal ganglia	Mesolimbic and mesocortical dopaminergic pathways	Reward and addiction and their decision processing
	Prefrontal serotonergic pathways	Reinforcement-driven learning and decision-making

Table 4. Prefrontal–subcortical systems in response to monetary reward and punishment

Study	Impulse and Region
Delgado et al. (2000)	Rewards activate nucleus accumbens.
Elliott et al. (2000)	Ventral striatum activated during reward but not during punishment.
Knutson et al. (2000)	Monetary rewards and punishments activate the caudate and putamen
O'Doherty et al, (2001)	Monetary rewards activate the orbitofrontal cortex
O'Doherty et al, (2001)	Magnitude of orbitofrontal activation correlates with the magnitude of rewards and punishments.
O'Doherty et al, (2001)	Medial prefrontal cortex also responds to monetary rewards
Knutson et al. (2001)	Activation of the ventromedial prefrontal cortex when reward was contrasted against non-reward outcomes
Gehring & Willoughby (2002)	Increases in cerebral blood flow and electrophysiological activation is observed in the medial prefrontal cortex during both monetary rewards and punishments
Gehring & Willoughby (2002)	After loss, choice is riskier and yields greater loss-related activity in the medial prefrontal cortex.
Elliott et al. (2000)	Subcallosal regions of the medial prefrontal cortex activated during monetary reward
Ernest et al (2002)	Monetary decision making causes predominantly right-sided activation across multiple prefrontal Areas: dorsolateral, orbitofrontal, anterior cingulated.
Elliott et al. (2003)	Striatal and brainstem monoamine structures activated in response to financial reward.

relate to financial behaviour. Since damage in prefrontal cortex affects negatively short and long-term decision making (Spinella et al., 2008), the authors collected information from a sample of individuals about their attitudes toward money and credit cards and a measure of their possible diminished neural functioning. In particular, the purpose was to study the relationship between measures of prefrontal cortical functioning and credit card and money attitudes by using the frontal systems behaviour scale.

The latter was developed by Grace et al. (1999) in order to evaluate behavioural and personality traits associated with the function of the prefrontal cortex. Grace et al. (1999) found that traits and habits associated with psychological impairment due to prefrontal cortical dysfunction are significantly associated with credit card and money attitudes and behavior. In particular, subjects with high scores on the measure of prefrontal cortical dysfunction consider money as a means to impress and influence others and as a symbol of success. In addition, subjects with high scores had hesitancy, doubt and suspicion regarding money transactions. Finally, they consider money as a source of worry and anxiety and were less concerned with producing financial security for the future Grace et al. (1999).

More recently, Cloutier and Gyurovski (2014) investigated ventromedial prefrontal cortex (VMPFC) activity during impression formation of individuals varying on distinct dimensions of social status. For instance, participants were asked to form an impression of the targets, but were not instructed to explicitly evaluate them based on social status. They found evidence of interaction of status dimension and level in ventromedial prefrontal cortex. In particular, results from functional magnetic resonance imaging (fMRI) showed evidence of preferential response to targets with higher compared to lower moral status as previously demonstrated, and also greater response to targets with lower compared to higher financial status.

Evans and Hampson (2014) examined if sex differences on cognitive tasks dependent on the prefrontal cortex. In particular, they examined the idea that the prefrontal cortex is sexually differentiated by comparing men's and women's performance on tasks which are associated to activation in discrete regions of the frontal cortex. Results from three experiments (self-ordered Pointing, Iowa gambling task, probabilistic reversal learning task) showed that functions carried out by the ventromedial prefrontal cortex and orbitofrontal cortex are sexually differentiated in humans. Ricardo et al. (2015) tested if resting state connectivity may predict choice behaviour in monetary exchange game. The experimental results from twenty-nine healthy adults resting state fMRI showed an increased functional connectivity between the salience (bilateral insula/ anterior cingulate) and central executive (dorsolateral prefrontal cortex/ posterior parietal cortex) networks significantly predicted the monetary choice.

In summary, results found in Spinella et al. (2007, 2008) are in accordance with prior studies (Goel, 1997; Ernest et al., 2002) confirming that prefrontal systems play a role in specific aspects of finances including income and credit card use.

FUTURE RESEARCH DIRECTIONS

There are several directions for future research in the context of prefrontal cortex and financial decision making. For instance, Spinella et al. (2008) recommended identifying which financial attitudes and decisions are associated with prefrontal cortex dysfunction and which are not. In addition, Spinella et al. (2008) recommended examining the association between neuropsychological measures of brain dysfunction and decision-making using econometric tests that economist under conditions of risk and uncertainty. Furthermore, Spinella et al. (2008) suggested investigating the association between abnormal electrcoencephalograms and income and other economic variables.

Because of its interdisciplinary character and approaches, neuroeconomics are now gaining attention in marketing and consumer research to explain buying-decisions. Indeed, as a contiguous to neuroeconomics, neuromarketing combines neuroscience, economics and psychology (Lee et al., 2007) to study price perception, processing and storage (Plassmann et al., 2008), present effects such as framing or anchoring (Deppe et al., 2007), examining advertising appeal (Kenning et al., 2007), studying product policy (Stoll et al., 2008) and analyzing of buying-decisions (Grosenick, Greer, & Knutson, 2008).

CONCLUSION

Economic theory is based on the fact that the economic agent is rational, self-interested, unemotional, and it seeks to maximize utility. However, it was observed that agent always deviate from rationality and show some decision biases or anomalies. Consequently, neuroeconomics has emerged as a multi-disciplinary science to analyze economic decision and behaviour from a neurobiological perspective by using advanced neuroscientific methods; including procedures for measuring electromagnetic activity of the brain and those sensitive to changes of cerebral blood flow or metabolism. In this regard, studies have shown that prefrontal cortex is significantly associated with financial decision making process. In addition, the latter is highly affected by emotions.

REFERENCES

Alexander, W. H., & Brown, J. W. (2011). Medial prefrontal cortex as an action-outcome predictor. *Nature Neuroscience, 14*(10), 1338–1344. doi:10.1038/nn.2921 PMID:21926982

Andersen, R. A., & Cui, H. (2009). Intention, action planning, and decision making in parietal–frontal circuits. *Neuron, 63*(5), 568–583. doi:10.1016/j.neuron.2009.08.028 PMID:19755101

Barberis, N., & Thaler, R. (2003). A survey of behavioral finance. In G. M. Constantinides, M. Harris, & R. M. Stulz (Eds.), *Handbook of the Economics of Finance*. Amsterdam: Elsevier.

Beach, L. R., & Lipshitz, R. (1993). Why classical decision theory is an inappropriate standard for evaluating and aiding most human decision making. In G. A. Klein & J. Orasanu (Eds.), *Decision making in action: models and methods*. Westport, TX: Ablex Publishing.

Bechara, A., Damasio, H., & Damasio, A. R. (2000). Emotion, decision-making, and the orbitofrontal cortex. *Cerebral Cortex, 10*(3), 295–307. doi:10.1093/cercor/10.3.295 PMID:10731224

Bechara, A., Tranel, D., & Damasio, A. R. (2002). The *somatic marker hypothesis and decision-making*. In F. Boller & J. Grafman (Eds.), *Handbook of neuropsychology: Frontal lobes, 7(2)*. Amsterdam: Elsevier.

Camara, E., Rodriguez-Fornells, A., & Münte, T. F. (2008). Functional connectivity of reward processing in the brain. *Frontiers in Human Neuroscience, 2*, 19. doi:10.3389/neuro.09.019.2008 PMID:19242558

Camara, E., Rodriguez-Fornells, A., Ye, Z., & Münte, T. F. (2009). Reward networks in the brain as captured by connectivity measures. *Frontiers in Neuroscience, 3*(3), 350–362. doi:10.3389/neuro.01.034.2009 PMID:20198152

Clark, L. (2010). Decision-making during gambling: An integration of cognitive and psychobiological approaches. *Philosophical Transactions of the Royal Society of London. Series B, Biological Sciences, 365*(1538), 319–330. doi:10.1098/rstb.2009.0147 PMID:20026469

Cloutier, J., & Gyurovski, I. (2014). Ventral medial prefrontal cortex and person evaluation: Forming impressions of others varying in financial and moral status. *NeuroImage, 100*, 535–543. doi:10.1016/j.neuroimage.2014.06.024 PMID:24936688

Dalgleish, T. (2004). The emotional brain. *Nature Reviews. Neuroscience, 5*(7), 583–589. doi:10.1038/nrn1432 PMID:15208700

de Quervain, D. J. F. et al.. (2004). The neural basis of altruistic punishment. *Science, 305*(5688), 1254–1258. doi:10.1126/science.1100735 PMID:15333831

Delgado, M. R., Nystrom, L. E., Fissell, C., Noll, D. C., & Fiez, J. A. (2000). Tracking the hemodynamic responses to reward and punishment in the striatum. *Journal of Neurophysiology, 84*, 3072–3077. PMID:11110834

Deppe, M., Schwindt, W., Pieper, A., Kugel, H., Plassmann, H., Kenning, P., & Ringelstein, E. B. et al. (2007). Anterior cingulate reflects susceptibility to framing during attractiveness evaluation. *Neuroreport, 18*, 1119–1123. PMID:17589310

Elliott, R., Friston, K. J., & Dolan, R. J. (2000). Dissociable neural responses in human reward systems. *The Journal of Neuroscience*, *20*, 6159–6165. PMID:10934265

Elliott, R., Newman, J. L., Longe, O. A., & Deakin, J. F. (2003). Differential response patterns in the striatum and orbitofrontal cortex to financial reward in humans: A parametric functional magnetic resonance imaging study. *The Journal of Neuroscience*, *23*, 303–307. PMID:12514228

Ernst, M., Bolla, K., Mouratidis, M., Contoreggi, C., Matochik, J. A., Kurian, V., & London, E. D. et al. (2002). Decision-making in a risk-taking task: A PET study. *Neuropsychopharmacology*, *26*(5), 682–691. doi:10.1016/S0893-133X(01)00414-6 PMID:11927193

Evans, K. L., & Hampson, E. (2015). Sex differences on prefrontally-dependent cognitive tasks. *Brain and Cognition*, *93*, 42–53. doi:10.1016/j.bandc.2014.11.006 PMID:25528435

Friston, K. J., Buechel, C., Fink, G. R., Morris, J., Rolls, E., & Dolan, R. J. (1997). Psychophysiological and modulatory interactions in neuroimaging. *NeuroImage*, *6*(3), 218–229. doi:10.1006/nimg.1997.0291 PMID:9344826

Gehring, W. J., & Willoughby, A. R. (2002). The medial frontal cortex and the rapid processing of monetary gains and losses. *Science*, *295*(5563), 2279–2282. doi:10.1126/science.1066893 PMID:11910116

Goel, V., Grafman, J., Tajik, J., Gana, S., & Danto, D. (1997). A study of the performance of patients with frontal lobe lesions in a financial planning task. *Brain*, *120*(10), 1805–1822. doi:10.1093/brain/120.10.1805 PMID:9365372

Grace, J., Stout, J. C., & Malloy, P. F. (1999). Assessing frontal behavior syndromes with the Frontal Lobe Personality Scale. *Assessment*, *6*(3), 269–284. doi:10.1177/107319119900600307 PMID:10445964

Grosenick, L., Greer, S., & Knutson, B. (2008). Interpretable classifiers for fMRI improve prediction of purchases. *IEEE Transactions on Neural Systems and Rehabilitation Engineering*, *16*(6), 539–548. doi:10.1109/TNSRE.2008.926701 PMID:19144586

Gutnik, L. A., Hakimzada, A. F., Yoskowitz, N. A., & Patel, V. L. (2006). The role of emotion in decision-making: A cognitive neuroeconomic approach towards understanding sexual risk behaviour. *Journal of Biomedical Informatics*, *39*(6), 720–736. doi:10.1016/j.jbi.2006.03.002 PMID:16759915

Hare, T. A., Camerer, C. F., Knoepfle, D. T., & Rangel, A. (2010). Value computations in ventral medial prefrontal cortex during charitable decision making incorporate input from regions involved in social cognition. *The Journal of Neuroscience*, *30*, 583–590.

Kable, J.W, & Glimcher, P.W. (2009). The neurobiology of decision: Consensus and controversy. *Neuron*, *63*, 733–745.

Kahneman, D. (2003). Maps of Bounded Rationality: Psychology for Behavioral Economics. *The American Economic Review*, *93*(5), 1449–1475. doi:10.1257/000282803322655392

Kenning, P., Plassmann, H., & Ahlert, D. (2007). Applications of functional magnetic resonance imaging for market research. *Qualitative Market Research*, *10*(2), 135–152. doi:10.1108/13522750710740817

King-Casas, B. et al.. (2005). Getting to know you: Reputation and trust in a two-person economic exchange. *Science*, *308*(5718), 78–83. doi:10.1126/science.1108062 PMID:15802598

Knutson, B., Adams, C. M., Fong, G. W., & Hommer, D. (2001). Anticipation of increasing monetary reward selectively recruits nucleus accumbens. *The Journal of Neuroscience*, *21*, RC159. PMID:11459880

Knutson, B., Westdorp, A., Kaiser, E., & Hommer, D. (2000). FMRI visualization of brain activity during a monetary incentive delay task. *NeuroImage*, *12*(1), 20–27. doi:10.1006/nimg.2000.0593 PMID:10875899

Lee, N., Broderick, A. J., & Chamberlain, L. (2007). What is 'neuromarketing'? A discussion and agenda for future research. *International Journal of Psychophysiology*, *63*(2), 199–204. doi:10.1016/j.ijpsycho.2006.03.007 PMID:16769143

Minati, L., Grisoli, M., Seth, A. K., & Critchley, H. D. (2012). Decision-making under risk: A graph-based network analysis using functional MRI. *NeuroImage*, *60*(4), 2191–2205. doi:10.1016/j.neuroimage.2012.02.048 PMID:22387471

O'Doherty, J., Kringelbach, M. L., Rolls, E. T., Hornak, J., & Andrews, C. (2001). Abstract reward and punishment representations in the human orbitofrontal cortex. *Nature Neuroscience*, *4*(1), 95–102. doi:10.1038/82959 PMID:11135651

Park, S. Q., Kahnt, T., Rieskamp, J., & Heekeren, H. R. (2011). Neurobiology of value integration: When value impacts valuation. *The Journal of Neuroscience*, *31*(25), 9307–9314. doi:10.1523/JNEUROSCI.4973-10.2011 PMID:21697380

Patton, J. H., Stanford, M. S., & Barratt, E. S. (1995). Factor structure of the Barratt impulsiveness scale. *Journal of Clinical Psychology*, *51*(6), 768–774. doi:10.1002/1097-4679(199511)51:6<768::AID-JCLP2270510607>3.0.CO;2-1 PMID:8778124

Plassmann, H., O'Doherty, J., Shiv, B., & Rangel, A. (2008). Marketing actions can modulate neural representations of experienced pleasantness. *Proceedings of the National Academy of Sciences of the United States of America*, *105*(3), 1050–1054. doi:10.1073/pnas.0706929105 PMID:18195362

Pradeep, A. K. (2010). *The buying brain: Secrets for selling to the subconscious mind*. John Wiley & Sons.

Rabin, M. (1998). Psychology and Economics. *Journal of Economic Literature*, *36*, 11–46.

Rangel, A., & Hare, T. (2010). Neural computations associated with goal-directed choice. *Current Opinion in Neurobiology*, *20*(2), 262–270. doi:10.1016/j.conb.2010.03.001 PMID:20338744

Ricardo, C., Andrew, J. G., Gutman, D. A., & Kilts, C. D. (2015). Organization of intrinsic functional brain connectivity predicts decisions to reciprocate social behavior. *Behavioural Brain Research*. doi:10.1016/j.bbr.2015.07.008

Spinella, M., Yang, B., & Lester, D. (2007). Lester D. Prefrontal systems in financial processing. *Journal of Socio-Economics*, *36*(3), 480–489. doi:10.1016/j.socec.2006.12.008

Spinella, M., Yang, B., & Lester, D. (2008). Prefrontal cortex dysfunction and attitudes toward money: A study in neuroeconomics. *Journal of Socio-Economics*, *37*(5), 1785–1788. doi:10.1016/j.socec.2004.09.061

Stephan, K. E., Harrison, L. M., Kiebel, S. J., David, O., Penny, W. D., & Friston, K. J. (2007). Dynamic causal models of neural system dynamics: Current state and future extensions. *Journal of Biosciences*, *32*(1), 129–144. doi:10.1007/s12038-007-0012-5 PMID:17426386

Stoll, M., Baecke, S., & Kenning, P. (2008). What they see is what they get? An fMRI-study on neural correlates of attractive packaging. *Journal of Consumer Behaviour*, *7*(4-5), 342–359. doi:10.1002/cb.256

Tversky, A., & Kahneman, D. (1974). Judgement under uncertainty: Heuristics and biases. *Science*, *185*(4157), 1124–1131. doi:10.1126/science.185.4157.1124 PMID:17835457

Wallis, J. D., & Kennerley, S. W. (2010). Heterogeneous reward signals in prefrontal cortex. *Current Opinion in Neurobiology*, *20*(2), 191–198. doi:10.1016/j.conb.2010.02.009 PMID:20303739

Zamora-López, G., Zhou, C., & Kurths, J. (2010). Cortical hubs forma module formulti sensory integration on top of the hierarchy of cortical networks. *Frontiers in Neuroinformatics*, *4*, 1. PMID:20428515

ADDITIONAL READING

Anselme, P. (2015). Does reward unpredictability reflect risk? *Behavioural Brain Research*, *280*, 119–127. doi:10.1016/j.bbr.2014.12.003 PMID:25496783

Bereczkei, T., Papp, P., Kincses, P., Bodrogi, B., Perlaki, G., Orsi, G., & Deak, A. (2015). The neural basis of the Machiavellians' decision making in fair and unfair situations. *Brain and Cognition*, *98*, 53–64. doi:10.1016/j.bandc.2015.05.006 PMID:26093237

Canessa, N., Pantaleo, G., Crespi, C., Gorini, A., & Cappa, S. F. (2014). The impact of egocentric vs. Allocentric agency attributions on the neural bases of reasoning about social rules. *Brain Research*, *1581*, 40–50. doi:10.1016/j.brainres.2014.06.001 PMID:24928617

Dixona, M. L., & Christoff, K. (2014). The lateral prefrontal cortex and complex value-based learning and decision making. *Neuroscience and Biobehavioral Reviews*, *45*, 9–18. doi:10.1016/j.neubiorev.2014.04.011 PMID:24792234

Yasuno, F., Matsuoka, K., Kitamura, S., Kiuchi, K., Kosaka, J., Okada, K., & Kishimoto, T. et al. (2014). Decision-making deficit of a patient with axonal damage after traumatic brain injury. *Brain and Cognition*, *84*(1), 63–68. doi:10.1016/j.bandc.2013.11.005 PMID:24316983

KEY TERMS AND DEFINITIONS

Economic Decision Making Process: Identifying and selecting appropriate actions or sequence of actions to achieve an ideal solution.

Economic/Financial Reward: Outcome or benefit of an economic/financial decision.

Financial Gambles: A tool to investigate how motivational decisions are reached in the presence of risk.

Functional Magnetic Resonance Imaging: It is a technique for measuring brain activity by detecting the changes in blood oxygenation and flow that occur in response to a specific neural activity.

Magnetic Resonance Imaging: It is a medical imaging technique used in radiology to detect structural anatomy of the body. In particular, it makes pictures of organs and structures inside the body.

Neuroeconomics: A discipline used to apply findings from neuroscience to the study of economic behaviour.

Prefrontal Cortex: The most anterior portions of the frontal lobe.

Risk: It is a known probability distribution of possible outcomes.

Uncertainty: It refers to an unknown probability distribution of possible outcomes.

This research was previously published in Neuroeconomics and the Decision-Making Process edited by Bryan Christiansen and Ewa Lechman, pages 122-131, copyright year 2016 by Business Science Reference (an imprint of IGI Global).

Chapter 24
Neuroscience Applications in Financial Markets:
A Practitioner's Perspective

Alessia Falsarone
PineBridge Investments, USA

ABSTRACT

The 2007-2008 Global Financial Crisis provided strong evidence of social dynamics in global financial markets which challenge basic rationality of independent decision makers when choices invoke short term trade-offs as opposed to sustainable long term socio-economic value creation and its ethical implications. Individual market participants are drawn into a Darwinian review of today's decision making environment in which they are called to make choices under systemic uncertainty. Sophisticated visualization tools from fMRI to PET scans allow for brain functions and decision making behaviors to be analyzed in an evolutionary context. From the theory of bounded rationality to its modern critiques by behavioralists, this chapter investigates how social learning may be at its best right at this moment, when ambiguity and known unknowns are extensive and expected utility frameworks are being replaced by local optima findings and "good enough" solutions.

INTRODUCTION

As an emerging field of research, *neurofinance* brings together socioeconomic theories and novel experimental findings from the cognitive behavioral sciences. Through the study of individual biases, social learning, cognitively engrained behaviors, and mere intuition, decision makers today are urged to overcome ambiguity and uncertainty by establishing cooperative networks to learn from, while allowing the simplest, evidence-based heuristics to guide them toward satisfactory outcomes. Social capital theory and mindfulness studies introduce a powerful means to an end: "good enough" outcomes at the individual decision level are indeed most likely to contribute to the rise of cooperative networks in today's fast-paced, fully interconnected system represented by global capital markets.

DOI: 10.4018/978-1-5225-5478-3.ch024

Background

In the context of this chapter, the term *neurofinance* encompasses all financial applications of neuroscience disciplines, ranging from market microstructure to the behavior of the decision makers that commit to transactions through capital markets. Among practitioners, neurofinance has emerged as an independent field of study in the aftermath of the Global Financial Crisis (GFC) of 2007–2008. The ethical aspects of short-term seeking behaviors faced unparalleled scrutiny as a result of the crisis. In particular, the GFC brought about the need to reassess the essence of decision making and what makes decisions value accretive or not and to whom. Since then, the focus of market participants (e.g., investors, regulators, and the broader investment community) has shifted from the evaluation of short-term outcomes to the assessment of long-term consequences on economic growth, with the ultimate goal of benefiting society at large. As a result, the market participant him/herself goes from being thought of as a pure utility-driven economic agent to a human decision maker. Personality traits, heuristic tendencies, and inner biases take center stage. The commonly accepted assumption of fully rational economic agents comes to question at the same time as new frontiers in academic research blend empirical findings with existing intuition from a wide variety of disciplines (e.g., biology, psychology, chemistry, ecology, and sociology, to name just a few).

For over a century, the founding principle of traditional finance, the *Efficient Market Hypothesis (EMH)*, had postulated the maximization of expected utility measures as a key driver behind individual decision making. New empirical processes, which directly challenge the EMH, continue to be evaluated and tested. Challenging economic theories are nothing new to industry practitioners. Businesses that adopted the computational principles of mathematical finance, as well as its complexities, throughout the 1970s and 1980s found themselves challenged by the myth of full rationality assumptions when behavioral finance studies brought about the principle of bounded rationality (BR) in the 1990s. Coined by Stanford University professor Herbert A. Simon, BR encompasses all the theoretical work and experimentations that deal with "constraints on the information processing capacity of the actor" (Simon, 1972, p. 162).

Today, neurofinance aims to address BR by bringing the economic agent and his/her learning models (or inhibitors thereof) to the forefront of an astounding multidisciplinary field of research. Thanks to advances in computational ability, as well as the theory of decision making coupled with sociology findings, neurofinance promises to have wide implications for industry practitioners. The future of its direct applications has just begun.

MAIN FOCUS OF THE CHAPTER

Issues, Controversies, Problems

Homo Economicus and the Environment

Some of the immediate applications of neurofinance studies deal with the concept of *value creation* in capital markets and how to improve the impact of investment decisions by learning more about the physiological functions that support decision-making behaviors. Specifically, the latest empirical findings that support the theory of the brain and its functions have successfully investigated the process of selective information gathering, which each homo economicus (the economic agent) undergoes to make decisions.

Figure 1. Structure of neuroscience reasoning
Source: Chapter author, 2015

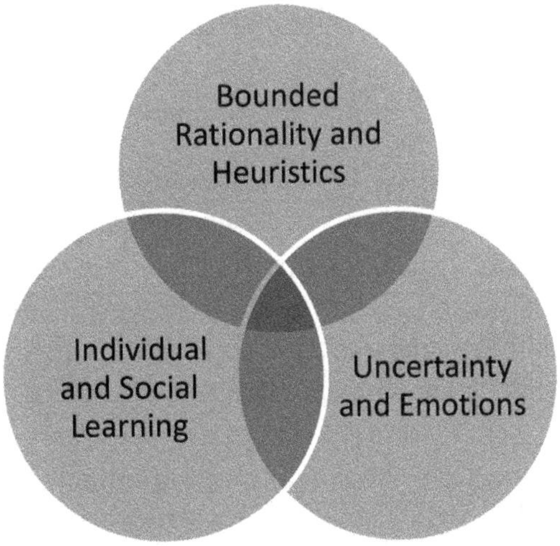

BR considerations were introduced in the early 1990s, so as to arrive at feasible solutions for otherwise highly computationally complex decision problems. By recognizing the cognitive limitations that affect human behaviors when making decisions, BR does not question the ability of economic agents to arrive at locally optimal answers (or to agree on the existence of at least one local optimum). Local optima are indeed acceptable and constitute the best decision within a possible solution set.

When emotions are brought into the picture, they are frequently associated with triggers of human irrationality and introduce a negative connotation into the optimal analysis of a selection of choices. "Emotions can be useful or valuable", cites Tseng (2006, p. 10). Yet, as pointed out by Lo (2002), the impact of emotions in the face of unforeseen risks may not reflect the cognitive assessment of risk carried out by the homo economicus.

According to Rubinstein (2001), the collective irrationalities of the investment community do not directly flow into prices. By identifying environmental factors that improve efficient information sharing and processing, neurofinance research has contributed practical insights on how to improve decision-making behaviors at both individual and organizational levels.

Many of the structural challenges faced by finance practitioners are now being recreated in research laboratories with the goal of assessing how individuals attempt to resolve these challenges and learn from them. By capturing and analyzing changes in the cerebral blood flow at times when subjects are involved in cognitive tasks, advanced visualization techniques, such as functional magnetic resonance imaging (fMRI), and their specialized imaging outputs have led researchers to identify apparent deviations from rationality and how they might be linked to environmental considerations. In this context, the work of Prof. Payzan-LeNestour, from the University of New South Wales in Australia, is of particular importance for practitioners. Prof. Payzan-LeNestour's research focuses on identifying the kind of information that is processed by the brain most effectively, and under which exogenous conditions this information is retained as learning. Portfolio managers and traders have to process information in rapidly changing environments with the added complexity of the majority of financial transactions occurring in an over-the-counter transaction model, where environmental considerations brought about the structural design of the market play a major role in assessing risk factors during decision making. Behavioralists (i.e., proponents of the BR principle) acknowledge an element of chance by recognizing that economic agents and their utility-maximizing minds are subject to what Herbert Simon called *cognitive limitations*.

Adaptive Social Capital and Value Creation

The social dynamics of capital markets have been widely researched. Evolutionary finance and its Darwinian successors point to the importance of "social interaction" and "social capital", which market participants build through their trading networks over the course of their careers. The social capital accumulated by professional traders and the depth of the underlying relationships they establish seem to take priority over individual skills and knowledge. It appears useful to think of this in the context of an inner network value theory, in which factors such as trust and reciprocity ultimately affect the outcome of a trading decision.

In less than a decade, brain imaging technologies such as positron emission tomography (PET) have successfully contributed to a wide variety of controlled experiments on brain activities. The more recent introduction of evolutionary elements among European academics has shifted the focus from individual decisions to system decisions, as exemplified by an aggregate of investment decisions (e.g., investment strategies) and how their day to day implementation results in a dynamic system. Evolutionary finance in particular uses genetic programming to calibrate the evolution of investment strategies via simulation techniques. So far, the target of its analyses has been the investment management community – namely, institutional money managers. Over the most recent decade, MIT's Prof. Andrew Lo, among others, has introduced elements of sociobiology and the theory of evolution with adaptive characteristics to behavioral studies. Going forward, by rethinking BR theory, and on the back of the introduction of modeling in continuous time, instances that would have been labeled market "abnormalities" under the proponents of EMH are likely to be representative of what "normal" market conditions look like in today's highly interconnected world.

What is likely to characterize value creation in this new norm of social networks and adaptive behaviors?

The 1980s witnessed the emergence of theories of short term overreaction, which drives long-term profits or gains in investments. One of the most interesting findings revealed how recently acquired information becomes more relevant during decision making, and it is weighted higher than prior beliefs by the human mind. In the late 1990s, Prof. Terrance Odean, from the University of California at Berkeley, conceptualized this in the *Theory of Existing Beliefs*, according to which strongly acquired information results in the individual becoming overly confident. From a Behavioral Finance workshop hosted at the J.F. Kennedy School of Harvard University in the winter of 2007, Prof. Odean called overconfidence a trap in the mind of the decision maker. The perceived scarcity of information, and its private versus public status, contributes to building said degree of *overconfidence*. The state of emotions in a decision maker helps steer his/her ability to select which piece of information to process first. Industry best practices have long taken such findings to heart and included so-called "best execution" tests in the day-to-day surveillance of capital market participants. As an example, traders that are involved in the over-the-counter marketplace (e.g., bond traders) have to provide evidence of fair value to calibrate the price accepted in a bilateral transaction vis-a-vis (offer) prices that have been rejected. In this instance, value creation is preserved in monetary terms if net wealth after the closing of a transaction is the highest among a set of offer prices placed by different agents for the same asset.

Back in the early 1970s, when Prof. Kahneman and Prof. Tversky pioneered the study of Prospect Theory, the definition of value was thought of in terms of net wealth (i.e., the aggregate of gains and losses). As a consequence, it has become intuitively acceptable to think of risk-seeking behaviors as most likely after having realized a gain; this is in contrast to risk-averting behaviors, which would follow losses. In his publication, *Irrational Exuberance* (2000), Prof. Shiller points to "*context dependent*" factors leading through the bull market of 1982–2000. This was followed by the passage of the Sarbanes-Oxley Act (Corporate and Auditing Accountability and Responsibility Act) in 2002, which served as the most tangible attempt to correct biased behaviors of market participants at the time of an extreme market. While evidence of professional traders behaving inherently as "intuitive Bayesians" is still in question, the over-the-counter nature of trading a variety of financial instruments shifts the focus to the social context in which the primary task is to establish measures of fair value, resulting in the maximization of bilateral wealth.

The Role of Uncertainty in Learning

The speed of learning is traditionally thought of in terms of a social agent's ability to dissect uncertainty by assigning a weight and expected value to its outcome. At the cognitive level, individuals are able to do that by themselves. According to Prospect Theory, newer information cues hold greater weight in decision making than older cues, as evidenced by a decision maker's willingness to adapt to an ever changing environment. In such context, risk is loosely defined as a form of expected uncertainty ("known unknown"). Pre-existing beliefs and long established convictions would therefore contribute to a second order of perceived uncertainty in the eye of the beholder.

Intuitively, readily available learning seems to occur when the social agent is faced with *unexpected* uncertainty. Recent findings suggest that the brain is able to categorize "uncertainty" into at least three distinct breeds:

- **Pure Risk:** Expected uncertainty (or known unknown),
- **Estimation Uncertainty:** The imprecision of the learner's existing beliefs (ambiguity, or unknown known), and
- **Unexpected Uncertainty:** Irreducible, systemic volatility (or unknown unknown).

Recent empirical research has focused on the investigation of the so-called "unexpected uncertainty" and how it is related to individual learning. A belief-updating mechanism inherent in the hardwiring of the human mind has emerged as the main provider of individual learning stimuli.

Overall, the perception of risk plays a key factor in determining whether unexpected uncertainty is present or whether the social agent is confusing that with system-wide "volatility". The reason behind this intuition is that distinct areas of the brain are involved in these two instances. In addition, both visual and auditory attention plays a role in determining learning rates. Prof. Payzan-LeNestour and others point out how unexpected uncertainty results in (experiential) learning when no explicit representation of reward is readily available. Specifically, the decision maker is capable of adjusting his/her rate of learning as a function of increased unexpected uncertainty when sequential trials are presented. Unexpected uncertainty is characterized as exhibiting sudden changes (jumps) in outcome probabilities (volatility), while estimation (or model) uncertainty is closer to the concept of "ambiguity" – i.e., payoff probabilities are known and need to be estimated. It is noteworthy to mention that researchers

Figure 2. Uncertainty matrix
Source: Chapter author, 2015

Risk (irreducible)	**Ambiguity**
• Assessment of risk • Cost-benefit trade-off ➡ Cooperative learning recurs to expert consensus	• Interactive modeling • Multi-criteria mapping ➡ Individual learning active; yields individual input in group deliberation
Uncertainty	**Ignorance**
• Interval analysis • Scenario methods ➡ Cooperative learning allows for evaluative judgement	• Monitoring and surveillance • No feedback mechanism ➡ Individual learning disabled; social learning may allow for adaptability and flexibility

report a widespread aversion to ambiguity on behalf of socioeconomic agents, whereby restraint from ambiguity leads to a pattern of inconsistent (yet unpredictable) choices. Ambiguity somehow disenables "exploration" to acquire more knowledge and it hinders real learning. This same ambiguity is defined as "stochastic volatility" in mathematical finance terms and its computationally intensive realizations are highly researched by practitioners.

Learning Models: Ambiguity, Emotions, and Memory

Unexpected uncertainty and ambiguity are ultimately defined as those properties of the environment that need to be studied in an effort to facilitate design and adoption of the most effective learning method. Experimental researchers focus on two prevailing methods:

- Bayesian learning, in which the learning rate varies. This method allows for slow forgetting, including the ability to discard prior convictions and long-held beliefs in favor of newer ones.
- Reinforcement learning, in which the learning rate remains constant.

Recent evidence suggests that human behavior toward decision making is Bayesian in nature. In the real world, reward/loss outcomes are indeed random and, as such, subject to sudden, discrete changes. Ambiguity is thus believed to be a learning enabler, as social agents will adjust their learning rate to the new aspects of the outcome-generating process they are experiencing, given an element of structural uncertainty about the environment to begin with.

Neural imaging studies report that humans have the ability to distinguish between risk, ambiguity, and unexpected uncertainty. In essence, the human brain is geared toward distinguishing between perceived risk, structural uncertainty, and expected volatility in presence of a net wealth factor. Nevertheless, the primary takeaway of experimental evidence is that any model of learning needs to successfully distinguish between a situation of structural (unexpected) uncertainty, as opposed a situation of plain ambiguity (stochastic volatility) for the learning to be constructive and lasting. The human brain cannot, by itself,

Figure 3. Emotions <> memory <> environment
Source: Chapter author, 2015

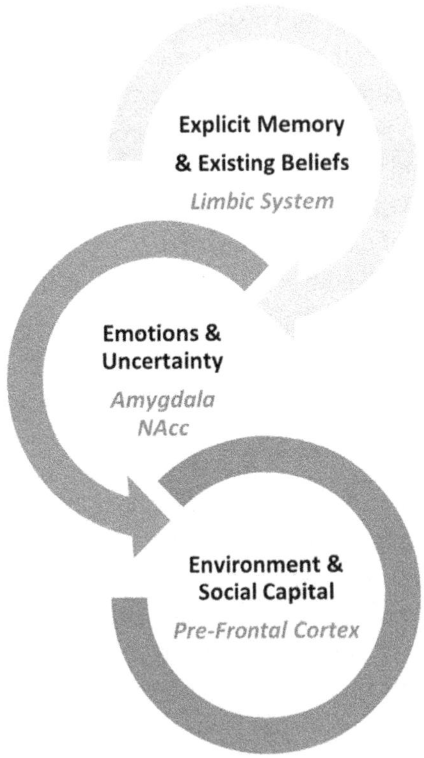

consistently single out structural uncertainties and it may ultimately confuse systemic risks with actual net losses (i.e., realizations of probabilistic outcomes of uncertainty).

Historically, brain studies have neglected the analysis of emotions when considering neural connections and, consequently, emotions tend to be ignored or associated with irrational behaviors. Although regarded as less objective, emotions play the same role as "memories" or "perceptions" in affecting the rate of learning in human agents. This is mainly due to the brain and body been treated as separate entities, rather than as part of a system. Under the recent evolutionary perspective, cognitive studies of brain functions look at the brain as the integral part of a system, which emotions are also part of.

Hence, without accounting for emotion and the experience of emotion, we leave out an understanding of the bioregulation of the human organism and its interrelation with the external environment (i.e., society and culture). One such example of systemic interaction is that which occurs through the *limbic system*, which has been found to be involved in the processing of emotions and, more broadly, feelings. It involves other inner structures within the human brain–body dynamic by virtue of neurons and blood streams, which act as main connectors. In this context, behaviors are thought of as explicit expressions of an emotional state (which are then captured by the external environment), while feelings materialize as experiences of an emotional state, and that impact future behaviors (such as the thinking and planning functions of the brain, which is also capable of selective involvement in the external environment). Emotions play a role in the memory process, defined as "a property of living systems that are bent on survival", by Damasio (1998, p. 84). They indeed feature a thinking component.

There exists an important link between exposure to information/external environment, which favors opportunities to practice memory building (i.e., learning) versus the speed and depth of retention. In this context, empirical research has predominantly focused on the occurrence of failures of human memory. "Memory" itself is defined as a system that tries to dissect and catalog enormous amounts of data, while making sense of it. Anderson and Schooler (1991) introduce models of memory and the environment, and they unveil how recent exposure to a particular instance plays a non-trivial role in the pattern of memory.

One question that comes to mind is whether human memory makes "optimal" use of past information. In other words, does the probability of a memory vary more or less in line with the pattern of past use?

Recent studies support the use of a power function to represent the delay in retention and, as such, a strongly decaying memory function (i.e., P = AT^-b, where P is the so-called performance measure of choice, A is an estimation parameter, and T time). By taking the logarithms of both sides, the equation becomes:

$$\text{Log } P = \text{Log } A - b \text{ Log } T$$

The above relationship seems to suggest that there is a slow delay in memory function over time. Therefore, retention is likely to satisfy power function properties. As practitioners may guess, testing whether practice functions follow a power or an exponential form reveals that the former continues to be a better fit; in this case, the rate of exposure (b) represents the learning rate.

We are also reminded that the so called spacing effect is quite an important property that defines how memory works as a system. When enough time has elapsed between one occurrence of a memory and the next, the context changes from being in nature a learning environment. That being said, when there are long lags between memories, it is most likely that the memories are associated with different contexts, indicating that memory can, anecdotally, adapt to the environment over time.

Memory is capable of retaining only those instances most likely to turn "useful". In mathematical terms, one would argue that the brain reasons via a set of statistical inferences to build its own reservoir of memories under *a need probability structure*. The brain function discards memories with very low-need probability:

$$C > pG,$$

where C is the cost of retaining a memory, p is the need probability, and G is the gain/reward associated with the successful retention (and use) of memories. Some of the key findings that result from the intersection of a memory as a system and the externalities brought about by the prevailing environment can be summarized as follows:

- State of memory and, as such, learning, is linked to the environment, most likely in a "causal" fashion.
- Human memory follows power laws and reflects the properties of the surrounding environment.
- Spacing and retention effects are intertwined; faster retention takes place for more frequent occurrences.
- The human mind is capable of cleaning up the clutter by itself!

Learning Models for Practitioners

Classic decision-making models are based on the assessment of the balance between risks and rewards in the form of utility functions which, for analytical ease, are typically designed as single choice. Symmonds, Bossaerts, and Dolan (2010) introduce the real-life idea of an operating environment with sequential choice, where human agents update their decision utilities by re-evaluating risk and reward assumptions through a belief of an updating mechanism as new information is acquired. The mind continuously tracks and re-evaluates choices in a probabilistic manner, as learning is affected by a perception of gain versus loss and first versus second best. Analyses of neural responses reveal that the statistical concepts of "mean" and "variance" are present in brain activities related to decision making. This idea holds true in sequential choice experiments, where economic agents/decision makers develop assumptions based on their personal risk/reward expectations. It also introduces *dynamic inconsistency* in individual decision making in the sense that future choices seem to become dependent upon the order in which choices are presented to the decision maker.

Do decision makers evaluate the distribution of potential outcomes when in planning mode, and do they assume "future" choices will be "optimal"?

The analysis of neural activity supports a degree of bias for above-average outcomes, although at low chance (extreme positives) and with aversion toward below-average outcomes (even if an outcome is associated with low probability events). Therefore, increasing or decreasing risk pattern association determines choice outcome. The mind is capable of analyzing representations of expected value, variance, and skewness in a context-dependent fashion, where strategic outcomes do not necessarily follow an "optimal strategy". Rather, they follow the strongest signal generated by an *emotional* representation of risk, rather than the realization of a specific risk/reward utility.

Prior research by Beherens and Woolrich (2007) had already established that the human mind is able to de-clutter itself through neural activity. The open question is how. Generally speaking, theories of learning are centered on how human agents make connections with past actions and their outcomes. The result is that decision making relies on the outcome of similar decisions that were made in the past (recent) history.

Bayesians promote the idea that the belief-updating mechanism will be such that the learning rate is scaled by the degree of uncertainty of the ultimate reward (function) at the time of action. For this reason, more recent experience yields a stronger predictive power. At the neural level, the part of the brain responsible for housing such mechanism is the anterior cingulate cortex. From the decision maker's standpoint, changes in the external environment are more relevant than obtaining better intuition of the reward probabilities (or, of the behavior of the reward function associated with a certain action). Such changes are reflected in different areas of the human brain, which ultimately drive the decision-making process.

Volatility in external conditions drives effective learning rates, and thus the updating of beliefs by the neural system, independently of the task at hand. Ultimately, as a learning rate pattern is established, the optimal decision is reached in a continuous fashion. This pattern is a better representation of the latest environmental changes, rather than only a representation of past decisions and their outcomes (as achieved via memory). Unexpected uncertainty and its representations are correlated with levels of "norepinephrine" in the brain. A focus on current volatility factors plays a role in successfully reaching an optimal strategy and executing it effectively.

Figure 4. The neural systems enablers of learning
Source: Chapter author, 2015

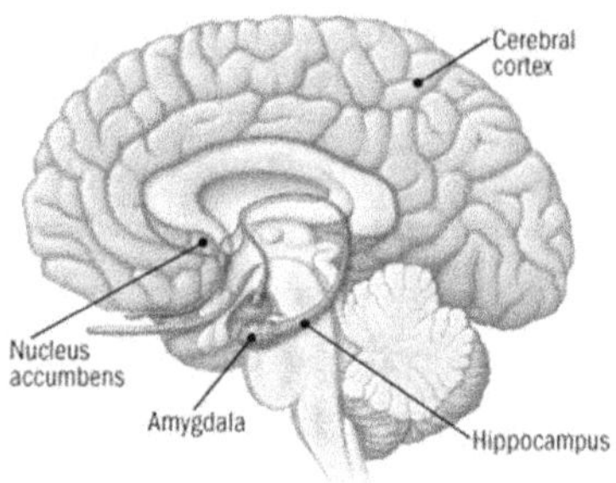

NEURAL BASIS OF COGNITIVE ACTIVITY

- **Amygdala (social behavior/ emotion and reward learning):** processing of unpredictability; fear of betrayal; inequity aversions; reward learning
- **Cerebral cortex (action)**: memory-controlled learning put to work
- **Hippocampus (learning and memory)**: experience storing in spatial context; conscious memory of experience
- **Nucleus accumbens (balance)**: balance between cognition, emotion and action

In their experimental work on financial risk-taking behaviors, Kuhnen and Knutson (2005) bring forward ample evidence from the world of investments, which points to systematic deviations from rational decision making i.e., entailing the maximization of "financial utility" of the economic agent. A question that naturally emerges is whether further analysis of neural activity prior to choice selection is likely to embed such deviations from rationality. In fact, empirical findings seem to suggest that "inconsistencies" from "optimal" choices stem from either risk-seeking or risk-averting behaviors (which ultimate translate into human error). Are there preconditions that steer the brain away from what rationally makes the most sense and leads it to settle for suboptimal outcomes? Researchers show that neural activity, which occurs prior to making the ultimate choice, engages distinct circuits in the brain, one connected with risk-taking (i.e., the expectation of gain/reward, housed in nucleus accumbens or NAcc) and another connected with risk-aversion behaviors (i.e., the expectation of losses, housed in the anterior insulate cortex or AIC).

Applied Heuristics: From Inferences to Learning

Gigerenzer et al. (2009) define heuristics as an ecological field of study, as its accuracy depends on the environment and relies primarily on an inductive process. The practice of heuristics advocates the opposite of resource-intensive algorithms, yet it involves a resourceful/adaptive side of the brain.

Recent research suggests that drawing learning ex post from the experience of one's most successful peers may affect the ability of economic agents to perform learning tasks individually and, as a result, it decreases the retention rate. There is evidence that learners that imitate based on the successful outcome of peers' decisions perform, on average, worse than both individual learners and non-selective imitators (i.e., imitators that copy the group's selection, regardless of whether the outcome is desirable or not).

When individual, experiential learning (trial and error based) is coupled with group dynamics, a peer exchange can lead to individual learning through a variety of mechanisms:

- **Imitation:** Imitate-the-best rule is presented. Its correlation with sample size and length of observed decision-making patterns may result in the most recent observations being most telling because successful decision makers may change their "cue" order as time goes by. The performance of imitators deteriorates as group size increases.
- **Plurality:** The option that receives the most votes is the preferred choice. There is a maximum number of group members above which new additions to the number of decision makers do not add or subtract from the method's efficiency and effectiveness (in terms of time/quality of the ultimate outcome).

- **Averaging:** This entails multidisciplinary analysis of all empirical evidence surrounding the task at hand. This learning method is the most successful, as group size increases, although it exhibits decreasing marginal returns.

The *Less is More* approach is one example of widely used heuristics, where processing is significantly reduced (time and burden) and (human) decision making is based on a set of "adaptive" answers. The Less is More heuristic emerged in 2008, the same time at which a systematic science of heuristics emerged (this was same time that the GFC occurred) and was brought into deeper consideration. It represents the human cognitive reaction to dealing with uncertainty and ambiguity. In those instances, the human agent resorts to its own "adaptive toolbox".

During the 1970s, Prof. Simon coined the verb "to satisfice", which contrasted the "maximize" principle; the former involves the process of finding a good enough solution. In an environment of choice that is made under sequential options, this is the process by which the decision maker sets an "aspirational" level of what "good enough" means and then terminates the search (finds a solution). A "satisficing" mind is reassured to always find a solution. The human mind is cognitively constrained by computationally intensive optimization exercises. The 1970s also mark the rise of the "negative" connotation of heuristics. According to some researchers, behaviors steered by "heuristical" knowledge ignore the basic laws of probability and accepted principles of rationality. Computational techniques based on heuristics brought about evidence of second-best approximations, rendering "fast and frugal" decisions as the best alternative to purely rational cognitive processes.

An important discovery of the current times is that the accuracy–effort trade off (i.e., the idea that less information gathering and processing leads to less accurate outcomes) does not always hold. In fact, there are situations in which more information and computation lead to worse inferences and, therefore, less accuracy.

"Forgetting" is analytically proven to improve the degree of accuracy in drawing inferences that lead to decision making. Recent analytical and empirical findings support effects such as less-is-more, they stress the need to incorporate heuristics in the decision making process, and they help one understand when/where heuristics improve decision making. The when/where answer is thus in the environment.

The concept of heuristics is often associated with bias because it is based on induction. "A biased mind that operates with simple heuristics can thus not only be more efficient in the sense of less effort but also more accurate that a mind that bets on avoiding bias." (Giginzer et al., 2008, p. 125)

To understand the importance of "bias" on the predictive power of an algorithm, let us break up the prediction error into three components:

$$\text{Total Error} = (\text{Bias})^2 + \text{Variance} + \text{Noise}$$

where:

- Bias represents the difference of the underlying (population) function versus the observed mean function of a (randomly drawn) sample.
- Variance represents the sum of the square deviations of the population function versus individual samples.
- Noise represents a catch-all bucket.

Heuristics is a proponent of Bias versus Variance for decision making. The importance of the Variance term will come back in the context of the Less is More heuristic.

The human agent's adaptive toolbox relies on memory selecting the heuristic that yields the desired outcome fastest and with recourse to minimal information processing, thus introducing an "ecological" rationality as a novel field of applied studies. Gigerenzer et al. have provided evidence that individuals utilize heuristics in an adaptive way.

Why can heuristics deliver an acceptable solution – i.e., when heuristics are accurate? Does that not imply an accuracy trade off? Answer: determining a solution is context bound and environment dependent. Under conditions of uncertainty and within small samples, heuristics yield "satisficing" results without compromising on accuracy, and they are delivered with low variance characteristics.

In adaptive strategy selection based on "simple" heuristics, "frugal" heuristics employ only a fraction of all available information, and, inherently, they are thought of as "fast". Decision making, from probabilistic mental models to fast and frugal ("psychological inference"), opens the door to "adaptive" decision making, which deals with "human" rather than "statistical" inference works.

One example of this type of heuristic is the Take the Best heuristic, which points to a decision-making process that is natural to the human mind (i.e., non-compensatory, fast, and frugal).

As the cost of information gathering increases, the use of the Take the Best heuristic is prevalent. This is not a result of the nominal price of information gathering and processing being high; rather, this is the result of when the cost becomes higher than the "perceived" utility of the information to the decision maker. In this case, local "satisficing" leads to global optimization (e.g., lower long-term cost).

Social Networks and Team Decision Making

The role of empathy in enhancing organizational effectiveness, ultimately serving as a key feature of leadership, emerged in 1998, when Harvard Business School professor Daniel Goleman introduced the concept of emotional intelligence, which continues to be at the center of social neuroscience investigations. Most recently, the idea that people rely on instance knowledge (e.g., people in a person's social network) has taken precedence over sole leadership. As an example, communicating a shared past on the basis of what is remembered by a group of economic agents impacts individual decisions and their effectiveness. Popularity can also influence impression formation in a different way, through its effect on sampling. In some cases, people may feel compelled to try out popular alternatives, even if they do not believe these options are of the highest quality.

The advance of social media in the financial field and its role in real-time information sharing has a direct impact on the way collaborative networks enable learning and value creation. Measures of so-called "network engagement" have been found to be a meaningful indicator of collaborative traits in driving decisions most effectively. Peer groups established through social media have led the way to engage collaboration through the immediacy of impact, which is embedded in newly established social channels. The emergence of "wisdom of the crowd" implications on all aspects of behavioral sciences, as applied to decision making, is undeniable. There is, indeed, recurring evidence of herd-like behaviors with the negative connotations associated with these instances.

Bentley et al. (2014) propose an interesting analysis on the subject in a principal component framework by mapping the degree of learning (individual versus social) versus the transparency of outcomes in terms of a risks and payoff function. Their bi-dimensional representation of collective knowledge gathering defines four behavioral patterns.

Figure 5. Individual learning in a social context
Source: Bentley et al., 2014

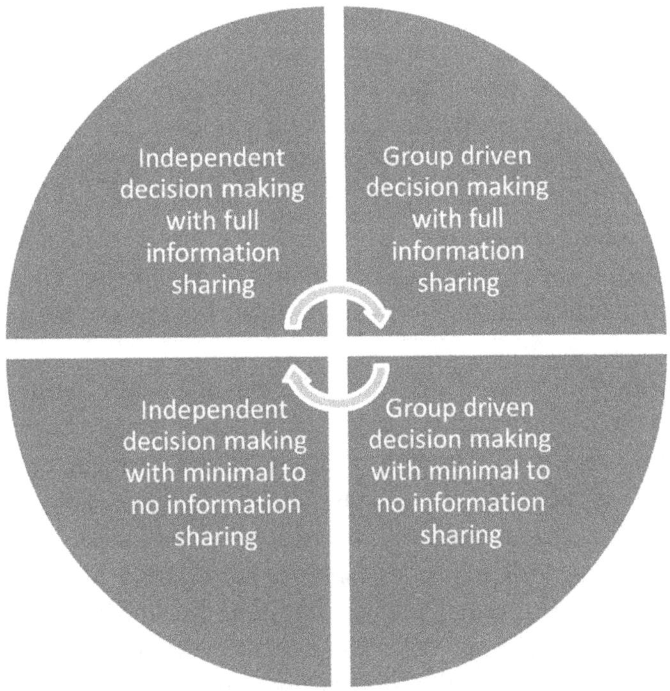

The quadrants need to be read as part of a dynamic mechanism, which evolves in time and depends on the intensity of their popularity. In fact, it is worthwhile noting that big data and real-time dissemination and access to information have contributed to grow the decision universe exponentially, including the set of possible outcomes of a decision. As such, there is a cognitive outcome that corresponds to individuals developing some sort of trust in the ultimate decisions of more experienced "choosers" and a higher likelihood of following others' behaviors.

A description of each quadrant follows:

- **Individual Learner Under Transparent Payoff:** Individual actions are driven by benefit/cost maximization behavior. The theme of bounded rationality prevails as payoff is reward driven (full transparency leads to reward-seeking optimization). Which dynamics originate in this state? "Maximal" behavior until new information gathering or externalities destabilize the systemic equilibrium, thereby pushing the economic agent to reassess local optima given a "newer" information set.
- **Social Learner Under Transparent Payoff:** A copying mechanism prevails, as transparency of the outcome/payoff fades away. "Success" or "perceived fitness" drives behavioral dynamics. Adaptive behaviors are more common in this quadrant, although they are not necessarily conducive to finding local optima (an example of communal fishing and exploration returning better yield away from the wisdom of crowds). Such behavioral dynamics arte highly dependent on the population (peer group) depth and size, and on how such characteristics affect the efficiency of value creation. The ultimate threat of this type of environment is reflected in the limited impact

and potential for small-scale, adaptive societies where positive social influence exists (those where the Law of Large Numbers and the Central Limit Theorem break down).

- **Social Learner Under Fuzzy Payoff:** Imitation mechanism based only on popularity/full-blown herding mode, which may reflect a random choice ("I will have what she is having" mindset). Characterized by boom and bust cycles (climate change and inertia surrounding policy action are represented here, as are other public debates such as those on genetically modified organisms [GMOs], where gaps in one group's knowledge versus the other result in a polarized divide).
- **Individual Learner Under Fuzzy Payoff:** An independent guess state that is widespread in the consumer products advertising industry, where given the sheer size of products and feedback marketed, the probability of any particular choice becomes similar to that of a (repeated) lottery to the point of individual choices seemingly being uncorrelated to one another.

One question that emerges is whether big data is facilitating a faster transition to the social learner under the Fuzzy Payoff model. Information sharing may seem to be transitioning in that direction, yet actual "learner"' networks may not follow right away. Social learning may become better informed to begin with. As the number of successful innovations increases, multiple choices become value accretive and, as such, the "need" for transparency of payoffs decreases. Big data and its innovations may ultimately rediscover the value of the most basic Bayesian analysis of human behavior: one's experience. Individuals learn by going through their experience and benchmarking the outcome to that of their closer peers.

An important aspect of mapping human behavior toward decision making is ideally to gain knowledge on what triggers dynamics among quadrants – such as, for example, how human agents move from social learners to individual learners as more knowledge spreads and less uncertainty surrounds the ultimate outcome.

In addition, the more self-aware the collective behavior, the more active and fast are the retention and dissemination of information, as well as the shifts in demand for what matters, thereby affecting the behavior of new learners. Ultimately, the crowd guides itself as it proactively generates new knowledge.

Social Decisions

In the words of Castells (2000), we live in a network society, defined by a set of horizontal relationships. A network society is most likely characterized by a multitude of values. As a result, it is more difficult to arrive to a commonly shared, group-driven vision. Koppenjan and Klijn (2004) notice how today's highly interconnected and inherently transparent network society, where a multitude of interests are represented, has funneled many more uncertainties in the process leading to decision making by a group. This is the direct consequence of introducing stakeholders with potentially diverging or asymmetrical goals in mind. Through network theory, such a framework can be reproduced and analyzed, and it can further specifically study how structural uncertainties should be solved by practitioners in the form of a "network strategy", thereby introducing a network perspective. How does one manage uncertainty in the presence of networks? What social infrastructure is capable of building a set of network rules that are most likely to facilitate the building of trust within the network? Traditional learning methods that deal with uncertainty (in the form of ambiguity or, more broadly, of systemic risk) may most likely fail to adapt to the structural intricacies of an established network.

In this context, decision making becomes an issue of interaction among those actors who have their interest at stake with the ultimate goal of generating a joint action plan. The underlying assumption is that (economic) actors do not select strategies (action plans) at random. Therefore, cooperation is presented as a learning process.

Klijn and Edelenbos (2007) investigate the topic of trust in complex decision-making networks. Review the value of trust as a coordination tool in a modern network-oriented society – i.e., how to influence processes and outcomes through trust. In an effort to derive empirical evidence on the inner workings of "trust" and its ability to be a driver of group decision making under uncertainty, "trust" is defined very narrowly by its key features:

- **Vulnerability:** A lack of trust brings vulnerability.
- **Risk:** Risk taking is possible within one's comfort zone ("calibrated") when trust is established.
- **Expectation:** Trust lowers ambiguity and brings about transparency surrounding one's respective behaviors and actions.

Qualitative intuition stresses the fact that trust enables lower cost investments (both in terms of manpower and financial costs) in cooperation, since relationships based on trust appear to be associated with lower transaction costs (they are thus associated with organizing a network and steering decision making within its members).

Scharpf (1997, p. 89) says that "Being able to trust is an advantage – but exploiting trust may even be more advantageous". Trust introduces a risk when network cooperation is based on anticipated profits from the relationship (as measured by increased net wealth). It enhances information exchange and value creation, yet it may lead to an exploitation of it. Ultimately, this seems to suggest that a trust-based network would require infrastructure monitoring, which decreases the occurrence of opportunistic behaviors by removing their incentives.

Rilling and Sanfey (2011) contribute to the topic of decisions carried out in the context of social interaction. The authors point out how the vast majority of decisions are made in a social context, and how the outcomes are somehow related to the behaviors/choices of other agents (those that are part of the same network or affiliated networks). Similarly to individual decisions, social decisions involve trade-offs, in which our cognitive biases and social norms come into play.

- Short-term outcomes versus long-term rewards
- Emotion versus reason
- Self-interest versus altruism

Rilling and Sanfey's research reiterates how social learning (defined as learning from others) is the foundation of human behavior and holds more influence over individual behaviors than individual learning itself. Information that has been acquired through a network mechanism appears to be more lasting than individual, experience-based learning. The area of the brain that appears to be receptive of the social stimuli is again the ACC, where information processing occurs.

In the context of value creation, the most crucial decision the social agent will ever make appears to be who to learn from.

SOLUTIONS AND RECOMMENDATIONS

From their early days, neurofinance studies are contributing valuable insights to how the structure and organization of the decision-making process itself affects final outcomes and their likelihood of reoccurrence. Technological advances in real-time data infrastructure are likely to give individual market participants the false intuition that *optimal* decisions ought to be unbiased. Clearly, the empirical evidence brought about by the literature review presented in this chapter seems to suggest that there could be a comeback to Prof. Herbert Simon's BR theory as a conscious decision of the individual in his/her role as a decision maker. Yet, in today's highly interconnected environment of social learning and with a renewed focus on benefiting society at large, when dealing with systemic risk and ambiguity, the social agent wins over the economically rational negotiator by virtue of simple heuristics.

Can the human brain be trained for sound decision making? The concepts of "prospective memory" (i.e., the act of remembering to perform a planned action or intention at some future point in time) and "experience-based" as opposed to "description-based" decisions (e.g., outcomes and probabilities are given upfront) have the most relevance in understanding the microstructure of capital markets. The introduction of feedback-enhancing mental models would likely bridge the gap between confidence and competence. With reference to the work of Prof. John Sterman from MIT:

Generating reliable evidence through scientific methods requires the ability to conduct controlled experiments, discriminate among rival hypotheses, and replicate results. But the more complex the phenomenon, the more difficult are these tasks . . . . Decisions are the result of applying a (decision) rule or policy to information about the world as we perceive it. (Sterman, 2006, pp. 505-509)

Incorporating feedback from experience and from others at a faster rate than the underlying changes in a society and the environment is at the core of value creation through collective learning. The concept of the "fair value" of financial instruments will have to adapt as well to reflect the "perceived value" of that instrument in the context of a marketplace, which allows for the interests of the buyer and the interests of the seller to be represented under a collective value creation effort.

FUTURE RESEARCH DIRECTIONS

The future of learning among market participants is likely to focus on new experimentation. Future research will most likely attempt to draw conclusions on how information is selectively processed and, ultimately, retained by the brain in the context of full transparency of information. Advances in cognitive psychology affecting people's judgment will allow for a more accurate snapshot of human behavior under different degrees of stress.

The need for applications of learning models that incorporate heuristics-based decision-making processes and the emergence of big data phenomena are likely to yield tremendous contributions to determine how humans make decisions based on an "acceptable" amount of information. One caveat to consider is that researchers themselves are biased designers of laboratory studies. As a result, their observation of empirical evidence is somewhat tainted by their own educational background and field of expertise, as well as by the time constraints under which these studies are being conducted.

Practitioners will likely applaud future research initiatives that answer the following questions: Do market participants need to operate in an environment of lower (perceived) volatility? Is it up to market participants (human economic agents) or regulatory bodies to determine the acceptable level (and measure) of volatility under which value creation occurs and society at large benefits?

CONCLUSION

There are ample applications of priority heuristics in capital markets, specifically in trading and high performance-generating activities. "Satisficying" solutions (local optima) are within reach by means of establishing a transparent framework that decision makers can trust. Therefore, by leveraging social capital, a local maximum is most likely guaranteed.

REFERENCES

Anderson, J. R., & Schooler, L. J. (1991). Reflections of the environment in memory. *Psychological Science*, *2*(6), 396–408. doi:10.1111/j.1467-9280.1991.tb00174.x

Beherens, T. E. J., Woolrich, M. W., Walton, M. E., & Rushworth, M. F. S. (2007). Learning the value of information in an uncertain world. *Nature Neuroscience*, *10*(9), 1214–1221. doi:10.1038/nn1954 PMID:17676057

Bentley, R. A., O'Brien, M. J., & Brock, W. A. (2014). Mapping collective behavior in the big data-era. *Behavioral and Brain Sciences*, *37*(01), 63–76. doi:10.1017/S0140525X13000289 PMID:24572217

Bentley, R. A., O'Brien, M. J., & Brock, W. A. (2014). Mapping collective behavior in big-data era. *Behavioral and Brain Sciences*, *37*(01), 63–119. doi:10.1017/S0140525X13000289 PMID:24572217

Brandstaetter, E., & Gigerenzer, G. (2008). Risky choice with heuristics: Reply to Birnbaum (2008), in Johnson, Schulte-Mecklenbeck, and Willemsen (2008), and Rieger and Wang (2008). *Psychological Review*, *115*(1), 281–290. doi:10.1037/0033-295X.115.1.281 PMID:18211205

Brandstaetter, E., Gigerenzer, G., Hertwig, R. (2006). The priority heuristic: making choices without trade-offs. *Psychological Review*, *113*(2), 409-432.

Broeder, A. (2003). Decision making with the "adaptive toolbox": Influence of environmental structure, intelligence, and working memory load. *Journal of Experimental Psychology. Learning, Memory, and Cognition*, *29*(4), 611–625. doi:10.1037/0278-7393.29.4.611 PMID:12924862

Bruguier, A. J., Quartz, S.R., Bossaerts, P. (2010). Exploring the nature of "trader intuition". *The Journal of Finance*, *45*(5).

Castells, M. (2000). *The rise of the network society: economy, society and culture*. Cambridge, MA: Blackwell Publishers.

Damasio, A. (1998). Emotion in the perspective of the integrated nervous system. *Brain Research. Brain Research Reviews*, *26*(2-3), 83–86. doi:10.1016/S0165-0173(97)00064-7 PMID:9651488

Dodge, K. A. (1991, July-August). Emotions and social information processing. Harvard Business Review. doi:10.1017/CBO9780511663963.009

Eustigneev, I. V., Hens, T., & Schenk-Hoppe', K. R. (2008). *Evolutionary finance*. Swiss Finance Institute Research Paper Series No. 08-14.

Falsarone, A. (2006). *Fundamental reality vs. market perception: modeling US equities*. Stanford University, Financial Mathematics Seminar.

Fan, J. D., & Suchow, J. W. (2014). The crowd is self-aware. *Behavioral and Brain Sciences*, *37*(1), 1. doi:10.1017/S0140525X13001714 PMID:24572223

Fenton-O'Creevy, M., Nicholson, N., Soane, E., & Willman, P. (2004). *Traders: risks, decisions, and management in financial markets*. Oxford, UK: Oxford University Press. doi:10.1093/acprof:oso/9780199269488.001.0001

Gigerenzer, G. (2001). The adaptive toolbox: Toward a Darwinian rationality. *Nebraska Symposium on Motivation*, *47*, 113–143. PMID:11759346

Gigerenzer, G., & Brighton, H. (2009). Homo Heuristicus: Why biased minds make better inferences. Max Planck Institute for Human Development. *Topics in Cognitive Science*, *1*(1), 107–143. doi:10.1111/j.1756-8765.2008.01006.x PMID:25164802

Henrich, J., & McElreath, R. (2003). The evolution of cultural evolution. *Evolutionary Anthropology*, *12*(3), 123–135. doi:10.1002/evan.10110

Hens, T. (2002). *Evolutionary portfolio theory*. Merrill Lynch Asset Allocation Almanac, Special Report #4.

Hertwig, R. (in press). *Decisions from experience* (G. Keren & G. Wu, Eds.). Oxford, UK: Blackwell. doi:10.1002/9781118468333.ch8

Hertwig, R., & Frey, R. (in press). How representations of knowledge shape our actions. In P. Meusburger, B. Werlen, & L. Suarsana (Eds.), *Knowledge and action: Knowledge and space* (Vol. 9). Dordrecht, The Netherlands: Springer.

Jha, P., Krompinger, J., & Baime, M. (2007). Mindfulness training modifies subsystems of attention. *Cognitive, Affective & Behavioral Neuroscience*, *7*(2), 109–119. doi:10.3758/CABN.7.2.109 PMID:17672382

Klijn, E.H., & Edelenbos, J. (2007). *Trust in complex decision making networks*. Academic Press.

Koppenjan, J., & Klijn, E. H. (2004). *Managing uncertainties in networks. A network approach to problem solving and decision making*. Routledge/ Taylor & Francis.

Kuhnen, C. M., & Knutson, B. (2005). The neural basis of financial risk taking. *Neuron*, *47*(5), 763–770. doi:10.1016/j.neuron.2005.08.008 PMID:16129404

Lejarraga, T., Lejarraga, J., & Gonzalez, C. (2014). Decisions from experience: How groups and individuals adapt to change. *Memory & Cognition*, *42*(8), 1384–1397. doi:10.3758/s13421-014-0445-7 PMID:25102774

Lo, A. W. (2005). Reconciling efficient markets with behavioral finance: The adaptive markets hypothesis. *The Journal of Investment Consulting*, *7*, 21–44.

Lo, A. W., & Repin, D. V. (2002). The psychophysiology of real-time financial risk processing. *Journal of Cognitive Neuroscience*, *14*(3), 323–339. doi:10.1162/089892902317361877 PMID:11970795

Pachur, T. (2012). *Modeling bounded rationality*. Basel, Switzerland: Habilitationsschrift.

Pachur, T., Hertwig, R., & Rieskamp, J. (2013). The mind as an intuitive pollster: Frugal search in social spaces. In *Simple heuristics in a social world* (pp. 261–291). New York, NY: Oxford University Press.

Pachur, T., & Kellen, D. (2013). *Modeling gain-loss asymmetries in risky choice: The critical role of probability weighting*. Conference Paper (Artikel, die in Tagungsbänden erschienen sind), Universitaet Basel

Pachur, T., & Olsson, H. (2012). Type of learning task impacts performance and strategy selection in decision making. *Cognitive Psychology*, *65*(2), 207–240. doi:10.1016/j.cogpsych.2012.03.003 PMID:22575684

Payzan-LeNestour, E., & Bossaerts, P. (2011). Risk, unexpected uncertainty, and estimation uncertainty: Bayesian learning in unstable settings. *PLoS Computational Biology*, *7*(1), e1001048. doi:10.1371/journal.pcbi.1001048 PMID:21283774

Payzan-LeNestour, E., Dunne, S., Bossaerts, P., & O'Doherty, J. (2013). The neural representation of unexpected uncertainty during value-based decision making. *Neuron*, *79*(1), 191–201. doi:10.1016/j.neuron.2013.04.037 PMID:23849203

Pfeffer, J., & Sutton, R. I. (2006). Evidence-based management. *Harvard Business Review*, *84*(1), 62–74. PMID:16447370

Pleskac, T. J. (in press). Learning models in decision making. In G. Keren & G. Wu (Eds.), *The Wiley-Blackwell handbook of judgment and decision making*. Chichester, UK: Wiley-Blackwell. doi:10.1002/9781118468333.ch22

Rilling, J. K., & Sanfy, A. G. (2011). The neuroscience of social decision-making. *Annual Review of Psychology*, *62*(1), 23–48. doi:10.1146/annurev.psych.121208.131647 PMID:20822437

Rubinstein, M. (2001). Rational markets: yes or no? The affirmative case. *Financial Analysts Journal*, *57*(3).

Sako, M. (1998). Does trust improve business performance? In C. Lane & R. Backmann (Eds.), *Trust Within and Between Organizations*. Oxford University Press.

Shiller, R. (2000). *Irrational exuberance*. Princeton, NJ: Princeton University Press.

Siegel, D. (1999). *The developing mind: how relationships and the brain interact to shape who we are*. New York: Guilford Press.

Siegel, D. (2007). *The mindful brain: reflection and attunement in the cultivation of well-being*. W. W. Norton & Company.

Simon, H. A. (1972). Theories of bounded rationality. In Decision and Organization. C.B. McGuire and R. Radner Publishers.

Symmonds, M., Bossaerts, P, Dolan, R. J. (2010). A behavioral and neural evaluation of prospective decision making under risk. *J Neurosci., 27.*

Tomasello, M. (1999). *The cultural origins of human cognition.* Cambridge, MA: Harvard Univ. Press.

Tseng, K.C. (2006). Behavioral finance, bounded rationality, neuro-finance, and traditional finance. *Investment Management and Financial Innovations, 3*(4).

Van den Bos, W., Golka, P. J. M., Effelsberg, D., & McClure, S. M. (2013). Pyrrhic victories: The need for social status drives costly competitive behavior. *Frontiers in Neuroscience, 7,* 189. doi:10.3389/fnins.2013.00189 PMID:24167468

Woike, J. K., Bonardi, J. P., & Garcia-Retamero, R. (2013). When groups should not imitate their most successful members. *Proceedings of the 35th Annual Conference of the Cognitive Science Society* (pp. 3795-3800). Austin, TX: Cognitive Science Society

Yu, A. J. (2012). Change is in the eye of the beholder. Nature Neuroscience, 15(7).

ADDITIONAL READING

Breitner, H. C., Aharon, I., Kahneman, D., Dale, A., & Shinzgal, P. (2001). Functional imagining of neural responses to expectancy and experience of monetary gains and losses. *Neuron, 30*(2), 619–639. doi:10.1016/S0896-6273(01)00303-8 PMID:11395019

Camerer, C., Loewenstein, G. F., & Prelec, D. (2005). Neuroeconomics: How neuroscience can inform economics. *Journal of Economic Literature, 43*(1), 9–64. doi:10.1257/0022051053737843

DiMauro, V., Bulmer, D. (2010). Social media's role in decision making by business professionals (Survey Results), *Society for New Communications Research*, ERTips and JDEtips. January

Grüne-Yanoff, T., & Hertwig, R. (in press). Nudge versus boost: How coherent are policy and theory? *Minds and Machines.*

Kahneman, D. (2013).Thinking fast and slow, Farrar, Straus and Giroux / New York Knauff, M., Pauen, M., Sebanz, N., Wachsmuth, I. (Eds.), Cooperative minds: Social interaction and group dynamics. Proceedings of the 35th Annual Conference of the Cognitive Science Society (pp. 3205-3210). Austin, TX: Cognitive Science Society.

Lee, D. (2005). Neuroeconomics: Making risky choices in the brain. *Nature Neuroscience, 8*(9), 1129–1130. doi:10.1038/nn0905-1129 PMID:16127442

Liebermann, M. D. (2007). Social cognitive neuroscience: A review of core processes. *Annual Review of Psychology, 58*(1), 259–289. doi:10.1146/annurev.psych.58.110405.085654 PMID:17002553

Thaler, R., & Sunstein, C. R. (2008). *Nudge.* Penguin Group.

KEY TERMS AND DEFINITIONS

Bayes' Rule/Bayesian Learning: $P(A|B) = P(B|A)P(A)/P(B)$. Using this idea of conditional probability to express what we want Bayes' Theorem to discover, we say that $P(A|B)$, the probability that A is true given that B is true, is the posterior probability of A. The idea is that $P(A|B)$ represents the probability assigned to A after taking into account the new piece of evidence, B. The idea of Bayesian learning is to compute the posterior probability of the target features of a new sample conditioned on its input features and all of the training samples.

Nucleus Accumbens (NAcc): A region in the brain that is part of the basal forebrain, specifically of the ventral striatum. There is a NAcc in each hemisphere of the brain. The primary function associated with the NAcc is its role in the reward circuit of the brain.

Over the Counter: In applied capital markets science, this refers to a marketplace where buyers and sellers privately negotiate terms of individual transactions, as opposed to having them intermediated through an organized exchange. The result is multiple prices for the same asset, as opposed to one price, in exchanges. Fixed income securities (e.g., bonds) are transacted in over-the-counter markets. Equities are transacted through an exchange mechanism.

Prefrontal Cortex: The anterior part of the frontal lobe of the brain. This area is involved in regulating complex cognitive, emotional, and behavioral functions.

This research was previously published in Applying Neuroscience to Business Practice edited by Manuel Alonso Dos Santos, pages 115-141, copyright year 2017 by Business Science Reference (an imprint of IGI Global).

Index

D

E

F

G

H

I

J

L

M

N

O

P

Q

R

S

T

U

V

www.ingramcontent.com/pod-product-compliance
Lightning Source LLC
LaVergne TN
LVHW060351300926
849162LV00006B/1246